Geomathematics: Theoretical Foundations, Applications and Future Developments

Quantitative Geology and Geostatistics

VOLUME 18

For further volumes:
http://www.springer.com/series/6466

Frits Agterberg

Geomathematics: Theoretical Foundations, Applications and Future Developments

Springer

Frits Agterberg
Ministry of Natural Resources Canada,
  Geological Survey of Canada (NRCan-GSC)
Ottawa, ON, Canada

ISSN 0924-1973
ISBN 978-3-319-06873-2   ISBN 978-3-319-06874-9 (eBook)
DOI 10.1007/978-3-319-06874-9
Springer Cham Heidelberg New York Dordrecht London

Library of Congress Control Number: 2014944795

© Springer International Publishing Switzerland 2014
This work is subject to copyright. All rights are reserved by the Publisher, whether the whole or part of the material is concerned, specifically the rights of translation, reprinting, reuse of illustrations, recitation, broadcasting, reproduction on microfilms or in any other physical way, and transmission or information storage and retrieval, electronic adaptation, computer software, or by similar or dissimilar methodology now known or hereafter developed. Exempted from this legal reservation are brief excerpts in connection with reviews or scholarly analysis or material supplied specifically for the purpose of being entered and executed on a computer system, for exclusive use by the purchaser of the work. Duplication of this publication or parts thereof is permitted only under the provisions of the Copyright Law of the Publisher's location, in its current version, and permission for use must always be obtained from Springer. Permissions for use may be obtained through RightsLink at the Copyright Clearance Center. Violations are liable to prosecution under the respective Copyright Law.
The use of general descriptive names, registered names, trademarks, service marks, etc. in this publication does not imply, even in the absence of a specific statement, that such names are exempt from the relevant protective laws and regulations and therefore free for general use.
While the advice and information in this book are believed to be true and accurate at the date of publication, neither the authors nor the editors nor the publisher can accept any legal responsibility for any errors or omissions that may be made. The publisher makes no warranty, express or implied, with respect to the material contained herein.

Printed on acid-free paper

Springer is part of Springer Science+Business Media (www.springer.com)

To Codien

Foreword

Our world has entered the "Big Data" era, in which new data are produced with unprecedented speed and in large quantities. The new knowledge obtained through digital analysis and the novel methods of data mining are greatly benefitting the process of human decision-making. The arrival of this new era is gradually changing people's life, work and thinking. In the field of science, "Big Data" has led to the emergence of "Data Science" which will affect, to some extent, all fields of science, *by performing scientific research by using digital data* and *by using scientific methods to study digital data*.

Earth Science and geological work are data-intensive. It is no exaggeration to say that different kinds of geological data have to be obtained and aggregated by different means, if we want to solve geological problems and use the results. No geological problem can be solved correctly and effectively without the in-depth study of a variety of geological inputs. Geological data are characterized by the four "deeps": they are deep in the Earth, deep under the sea, deep in outer space and deep in time. Consequently, it is not easy to obtain useful geological data in practice, and data collection can be expensive. Much attention has to be paid to costs and benefits.

Earth Scientists should always do their best to define "populations" and ensure that truly representative samples are collected from these target populations. Because they differ from place to place, geological samples can never fully comprise the entire population of study; there is no "overall data completeness" or "comprehensive data" in geological science and practice. Because of this, other methods of approach have to be developed in order to make random sampling results fit the target populations as closely as possible, so that information loss because of restricted sampling access is minimized.

Ultimately, the purpose of Earth Science is to promote the progress and development of human society: it explores the course, procedures, and products of the Earth's evolution, in order to make use of the advantages of and guard against the disadvantages that the Earth's evolution brings to us. Geological data have significant characteristics reflecting time of origin, space and genesis. They can manifest different outcomes of genetic processes depending on space and time conditions.

Therefore, when faced with geological data, one should not only know the "what?" but also the "why?" and "how?" about what the data truly mean and how they should be used. One should not only establish "correlations" but also "causality" and "spatiotemporal relations". In this, geology differs from some other areas in the Big Data era where the focus is on the "what?" and correlations without the "why?" and causality.

All of the preceding considerations aim to illustrate the complexity and particularities of geological data as well as their usefulness and importance. Fully comprehensive geological data collection, their effective treatment, rational analysis and translation into "digital" knowledge, all depend on guidance provided by powerful theory and applications of efficient methods. Mathematical geosciences precisely provide the powerful tools needed to realize these aims. In the early days, monographs in mathematical geoscience were published under the name of "statistical analysis of geological data" or by using similar titles. However, in the information age and Big Data era, we prefer the word "geomathematical" for increased emphasis and clarification.

The book in front of us is a new masterpiece by Canadian scholar F. Agterberg: *Geomathematics: Theoretical Foundations, Applications and Future Developments*. It follows *Geomathematics: Mathematical Background and Geo-science Applications*, another masterpiece of his published in 1974. There are 12 chapters in this new book and it contains a large number of case-history studies to illustrate the application of mathematical methods in researching basic geological problems in stratigraphy, tectonics, igneous petrology and geochronology, in order to solve practical problems such as the description and characterization of various types of ore deposits and the prediction and evaluation of mineral resources.

Honoring the complexity and characteristic features of geological data, this book illustrates the application of geomathematical methods in detail. Although it also discusses the processing of traditionally structured data, this book focusses more on the processing of unstructured data such as that arise in the analysis and processing of various geological map and image data. Besides introduction of theory and methods of classical probability calculus and statistics, this book extensively covers non-linear process theory as applied to the geosciences. The book fully describes the latest developments and achievements of mathematical geosciences over the past decade. In his last chapter, the author suggests a number of topics for further development of the mathematical geosciences showing his deep interest in nonlinear process simulation and applications in future geological research. The solution of these new problems will undoubtedly raise the research level in geology and open up new avenues of approach in the mathematical geosciences.

Professor Agterberg's ideas, theories, methods and practical applications of geomathematics have significant impact on the development of mathematical geosciences worldwide. During his tenure as President of the International Association for Mathematical Geosciences (IAMG), he paid much attention to and promoted the development of mathematical geosciences in all countries. Especially, as an old friend of the Chinese people, F. Agterberg has offered valuable concern and help in the development of the mathematical geosciences in China. It is my great honour to write the foreword for this book, and I would like to take this opportunity to extend my warm congratulations to F. Agterberg for the publication of this book. I also thank him sincerely for his personal contributions to the development of the mathematical geosciences in China.

China University of Geosciences Pengda Zhao
Academician of the Chinese Academy of Sciences
8th March 2014

Preface

This book is intended for mathematical geoscientists including graduate students, professionals and teachers. It differs from my earlier book: *Geomathematics: Mathematical Background and Geoscience Applications* (published by Elsevier in 1974) in that mathematical background has been restricted to the bare essentials. Emphasis is on applications of geomathematical methods in case history studies. The previous book was translated into the Russian and Chinese and, in total, over 10,000 copies were sold. Daniel Merriam had urged me repeatedly to write another book with case history studies only. I finally got around to following up on his suggestion. The underlying mathematics can be found in original publications, on the internet, or in other geomathematical books of which there now exist many, especially in the field of geostatistics. Readers of this book should have basic knowledge of calculus and elementary statistics.

Some techniques described in this book have become more widely accepted than others. These include probabilistic mineral resource estimation, trend surface analysis, ranking and scaling of biostratigraphic events, construction of numerical timescales, Weights-of-Evidence modelling and the logistic model. Other topics treated in this book are consideration of edge effects in 2-D map studies, time series analysis of sedimentary data related to ice sheet retreat, Alpine tectonic reactivation along pre-existing Hercynian schistosity planes in the basement of the Italian Dolomites, use of the grouped jackknife method for bias reduction, downward mineral potential extrapolation, and use of permanent, volume-independent, frequency distributions. During the past 20 years, primarily in collaboration with Qiuming Cheng, his colleagues and students at the China University of Geosciences in Wuhan and Beijing and at York University, Toronto, I have worked on applications of multifractals to study the spatial distribution of metals in rocks and orebodies. This topic also will be discussed in the last three chapters. The aim of this book is to present the preceding variety of topics within a coherent framework.

Originally, I studied geology and geophysics at Utrecht University in the Netherlands. This education provided me with a good mathematical and geoscience background. However, in those days there were no digital computers; neither was mathematical statistics taught specifically for geoscience applications. As a new graduate student in 1957, I happened to read an introductory textbook on probability and statistics by Hans Freudenthal, professor in the mathematics department of Utrecht University. His outline of probability calculus and statistical inference fascinated me, so later that year I began to apply statistics, initially on attitudes of schistosity planes and minor fold axes in Paleozoic rocks during my first 3 months fieldwork in the basement of the Italian Dolomites. These attitudes differ randomly from outcrop to outcrop, but regional patterns could be established by averaging.

In 1961, I became a postdoctorate fellow at the University of Wisconsin in Madison where my main assignment was to statistically analyze paleocurrent measurements from Atokan and Desmoinesian rocks in the Ouachita Mountains of Oklahoma. My final report was sent out for review to William Krumbein who provided encouragement, recommended publication and later invited me to visit him at Northwestern University. This was the time that chemical analysis of rock samples was becoming more widespread and statistical analysis became more needed than before. Also, digital computers were becoming available for numerical analysis. Altogether, there was an exponential increase in geoscientific research. The great period of quantification in the earth sciences had commenced. Because of my statistical interests, I became "petrological statistician" in my first job at the Ministry of Natural Resources Canada, Geological Survey of Canada (NRCan-GSC) in Ottawa in 1962.

I was asked to create a Geomathematics Section in 1971. Its original staff included Andrea Fabbri, Chang-Jo Chung and Rao Divi. Later I have enjoyed much fruitful collaboration with GSC colleagues, especially with Graeme Bonham-Carter and Felix Gradstein. I have always maintained close links with universities. In 1968, I commenced teaching a course on statistics in geology at the University of Ottawa over a 25-year period. As Adjunct Professor, I have supervised graduate students during the 1980s and 1990s including Andrea Fabbri, Eric Grunsky, Mark D'Iorio, Danny Wright and Qiuming Cheng. Additionally, I have lectured, often jointly with colleagues, in more than 50 short courses worldwide. I am grateful for the many friendships formed and the invaluable help received during the past 55 years. In 2008 I felt honoured by the publication of the Springer Festschrift entitled *Progress in Geomathematics* edited by Graeme Bonham-Carter and Qiuming Cheng with 30 papers written by 58 colleagues and friends.

I have had the good fortune to personally know and work with most of the early mathematical geologists who were the pioneers in our field including William Krumbein, John Griffiths, Georges Matheron, Andrew Vistelius, Walther Schwarzacher, Daniel Merriam, Hans de Wijs, Váslav Němec, Timothy Whitten, Jean Serra, Felix Chayes and Zhao Pengda. Three times I made the pilgrimage to Fontainebleau, cradle of geostatistics. At the same time I was profoundly influenced by prominent mathematical statisticians including John Tukey who on three occasions went to the length of publishing helpful further inputs on my statistical projects in Ottawa. Special mention also should go to Geoffrey Watson, who became a good friend in 1968 and who stimulated me to improve my mathematical skills during the 1970s.

I had the privilege of knowing Danie Krige as a friend and esteemed colleague for more than 50 years. As a graduate student at the University of Utrecht I had read

Preface

Krige's MSc thesis on microfilm in the library preparing for an economic geology seminar on the skew frequency distribution of mining assays. Danie wrote me a letter with comments on the resulting paper. He visited me in Ottawa on his way to the 3rd APCOM meeting held at Stanford University in 1963. APCOM is the acronym of *Applications of Computers and Operations Research in the Mineral Industries*. Danie persuaded NRCan-GSC management that I should attend the 4th APCOM hosted by the Colorado School of Mines in 1964. Originally, APCOM meetings provided an important forum for mathematical geologists. Many of us attended several APCOMs in the early days.

John Griffiths invited me to give two papers (instead of one) at the APCOM he was organizing at Penn State University in 1966. Later, like Dan Merriam, he systematically read through every chapter of my earlier geomathematics book before it went to the printer. Richard Reyment invited me to serve on the founding committee of the International Association for Mathematical Geology (IAMG). After the inaugural IAMG meeting in Prague a year later, Geof Watson, Andrew Vistelius, Jean Serra and I went to Amsterdam to participate in a meeting of the International Statistical Institute which, with the International Union of Geological Sciences, continues to be one of the parents of the IAMG now called International Association for Mathematical Geosciences.

In 1976 Dan Merriam asked me to become Leader of the International Geological Correlation Program's Quantitative Stratigraphy Project. Felix Gradstein and I had become interested in quantitative stratigraphy and later in the numerical geological timescale. We organized 1-week short courses in eight different countries under the auspices of IGCP and IUGS. This resulted in fruitful collaborations with the other lecturers including Walther Schwarzacher, Ian Lerche, Jim Brower and Jan van Hinte.

During preparation of the script for this book, I have received help from NRCan-GSC management; especially Cathryn Bjerkelund was most encouraging. Several diagrams were newly drafted by Kim Nguyen. Graeme Bonham-Carter and Eric Grunsky are thanked for being my critical readers of the script. I am grateful for support from my colleagues Eric de Kemp and Ernst Schetselaar in the Geoscience Integration and Analysis Section in which I am an Emeritus Scientist. I am most grateful to my wife Codien, who always has supported my scientific endeavours, which were kept a high priority in our household, often at the expense of other activities. Thanks are due to our four sons, their spouses and the seven grandchildren for their tolerance and support. Very special mention goes to son Marko who has provided invaluable help in keeping our computer lab at home going for many years and in the painstaking preparation of colour diagrams for this book. Petra van Steenbergen of Springer SMM NL has been most encouraging and provided valuable guidance during the preparation of the script. Hermine Vloemans and Laura van Zon at the Springer Office in Dordrecht have been helpful, and so have S. Madhuriba and the staff of the Springer Production Department. Thanks are also due to copyright owners who graciously allowed me to reproduce figures and material from tables that had been published before.

Ottawa, Canada   Frits Agterberg

Contents

1	**Complexity of the Geological Framework and Use of Mathematics**	1
	1.1 Use of Mathematics in Geology	2
	1.2 Geological Data, Concepts and Maps	3
	1.2.1 Map-Making	6
	1.2.2 Geological Cross-Sections	8
	1.2.3 Scientific Method in the Geosciences	9
	1.2.4 Quality of Predictions	11
	1.3 Use of Curves	12
	1.3.1 Trend-Lines	12
	1.3.2 Elementary Differential Calculus	14
	1.3.3 Graphical Curve-Fitting	15
	1.4 Use of Surfaces	16
	1.4.1 Automated 3-Dimensional Map-Making: Central Baffin Example	18
	1.4.2 Folds and Faults	18
	1.5 Image Analysis	22
	1.5.1 Geometrical Covariance, Intercept and Rose Diagram	23
	1.5.2 Minkowski Operations: Bathurst Acidic Volcanics Example	26
	1.5.3 Boundaries and Edge Effects	31
	References	38
2	**Probability and Statistics**	41
	2.1 History of Statistics	42
	2.1.1 Emergence of Mathematical Statistics	43
	2.1.2 Spatial Statistics	44
	2.2 Probability Calculus and Discrete Frequency Distributions	48
	2.2.1 Conditional Probability and Bayes' Theorem	48
	2.2.2 Probability Generating Functions	49

		2.2.3	Binomial and Poisson Distributions	50
		2.2.4	Other Discrete Frequency Distributions	53
	2.3	Continuous Frequency Distributions and Statistical Inference		56
		2.3.1	Central-Limit Theorem	56
		2.3.2	Significance Tests and 95 %-Confidence Intervals	58
		2.3.3	Sum of Two Random Variables	60
	2.4	Applications of Statistical Analysis		61
		2.4.1	Statistical Inference: Grenville Potassium/Argon Ages Example	61
		2.4.2	Q-Q Plots: Normal Distribution Example	64
	2.5	Sampling		66
		2.5.1	Pulacayo Mine Example	66
		2.5.2	Virginia Mine Example	70
	References			70
3	**Maximum Likelihood, Lognormality and Compound Distributions**			**73**
	3.1	Applications of Maximum Likelihood to the Geologic Timescale		74
		3.1.1	Weighting Function Defined for Inconsistent Dates Only Model	75
		3.1.2	Log-Likelihood and Weighting Functions	76
		3.1.3	Caerfai-St David's Boundary Example	77
		3.1.4	The Chronogram Interpreted as an Inverted Log-Likelihood Function	79
		3.1.5	Computer Simulation Experiments	81
		3.1.6	Mesozoic Timescale Example	87
	3.2	Lognormality and Mixtures of Frequency Distributions		88
		3.2.1	Estimation of Lognormal Parameters	89
		3.2.2	Muskox Layered Intrusion Example	91
		3.2.3	Three-Parameter Lognormal Distribution	95
		3.2.4	Graphical Method for Reconstructing the Generating Process	96
	3.3	Compound Random Variables		99
		3.3.1	Compound Frequency Distributions and Their Moments	99
		3.3.2	Exploration Strategy Example	100
	References			102
4	**Correlation, Method of Least Squares, Linear Regression and the General Linear Model**			**105**
	4.1	Correlation and Functional Relationship		106
	4.2	Linear Regression		109
		4.2.1	Degree of Fit and 95 % – Confidence Belts	109
		4.2.2	Mineral Resource Estimation Example	112

		4.2.3	Elementary Statistics of the Mosaic Model	115
	4.3	General Model of Least Squares	116
		4.3.1	Abitibi Copper Deposits Example	118
		4.3.2	Forward Selection and Stepwise Regression Applied to Abitibi Copper	123
	4.4	Abitibi Copper Hindsight Study	126
		4.4.1	Incorporation of Recent Discoveries	127
		4.4.2	Comparison of Weight Frequency Distributions for Copper Metal and Ore	131
		4.4.3	Final Remarks on Application of the General Linear Model to Abitibi Copper	134
	References	137
5	**Prediction of Occurrence of Discrete Events**	139
	5.1	Weights-of-Evidence Modeling	140
		5.1.1	Basic Concepts and Artificial Example	141
		5.1.2	Meguma Terrane Gold Deposits Example	144
		5.1.3	Flowing Wells in the Greater Toronto Area	150
		5.1.4	Variance of the Contrast and Incorporation of Missing Data	152
	5.2	Weighted Logistic Regression	161
		5.2.1	Meguma Terrane Gold Deposits Example	163
		5.2.2	Comparison of Logistic Model with General Linear Model	164
		5.2.3	Gowganda Area Gold Occurrences Example	168
		5.2.4	Results of the Gowganda Experiments	174
		5.2.5	Training Cells and Control Areas	176
	5.3	Modified Weights-of-Evidence	179
		5.3.1	East Pacific Rise Seafloor Example	180
	References	185
6	**Autocorrelation and Geostatistics**	189
	6.1	Time Series Analysis	190
		6.1.1	Spectral Analysis: Glacial Lake Barlow-Ojibway Example	191
		6.1.2	Trend Elimination and Cross-Spectral Analysis	198
		6.1.3	Stochastic Modeling	202
	6.2	Spatial Series Analysis	204
		6.2.1	Finite or Infinite Variance?	205
		6.2.2	Correlograms and Variograms: Pulacayo Mine Example	207
		6.2.3	Other Applications to Ore Deposits	212
		6.2.4	Geometrical Probability Modeling	215
		6.2.5	Extension Variance	216
		6.2.6	Short-Distance Nugget Effect Modeling	220
		6.2.7	Spectral Analysis: Pulacayo Mine Example	223
		6.2.8	KTB Copper Example	225

	6.3	Autocorrelation of Discrete Data	228
		6.3.1 KTB Geophysical Data Example	229
	References		230

7 2-D and 3-D Trend Analysis ... 235
	7.1	2-D and 3-D Polynomial Trend Analysis	236
		7.1.1 Top of Arbuckle Formation Example	237
		7.1.2 Mount Albert Peridotite Example	239
		7.1.3 Whalesback Copper Mine Example	243
	7.2	Kriging and Polynomial Trend Surfaces	250
		7.2.1 Top of Arbuckle Formation Example	252
		7.2.2 Matinenda Formation Example	256
		7.2.3 Sulphur in Coal: Lingan Mine Example	259
	7.3	Logistic Trend Surface Analysis of Discrete Data	265
	7.4	Harmonic Trend Surface Analysis	266
		7.4.1 Virginia Gold Mine Example	267
		7.4.2 Whalesback Copper Deposit Exploration Example	268
		7.4.3 East-Central Ontario Copper and Gold Occurrence Example	273
	References		274

8 Statistical Analysis of Directional Features ... 277
	8.1	Directed and Undirected Lines	278
		8.1.1 Doubling the Angle	278
		8.1.2 Bjorne Formation Paleodelta Example	279
		8.1.3 Directed and Undirected Unit Vectors	283
	8.2	Unit Vector Fields	284
		8.2.1 San Stefano Quartzphyllites Example	285
		8.2.2 Arnisdale Gneiss Example	289
		8.2.3 TRANSALP Profile Example	290
		8.2.4 Pustertal Tectonites Example	295
		8.2.5 Tectonic Interpretation of Unit Vector Fields Fitted to Quartzphyllites in the Basement of the Italian Dolomites	297
		8.2.6 Summary of Late Alpine Tectonics South of Periadriatic Lineament	300
		8.2.7 Defereggen Schlinge Example	302
	References		302

9 Quantitative Stratigraphy, Splining and Geologic Time Scales ... 305
	9.1	Ranking and Scaling	306
		9.1.1 Methods of Quantitative Stratigraphy	306
		9.1.2 Artificial Example of Ranking	310
		9.1.3 Scaling	314
		9.1.4 Californian Eocene Nannofossils Example	317
	9.2	Spline-Fitting	323
		9.2.1 Smoothing Splines	324

		9.2.2	Irregularly Spaced Data Points	327
		9.2.3	Tojeira Sections Correlation Example	329
	9.3	Large-Scale Applications of Ranking and Scaling		332
		9.3.1	Sample Size Considerations	334
		9.3.2	Cenozoic Microfossils Example	334
	9.4	Automated Stratigraphic Correlation		339
		9.4.1	NW Atlantic Margin and Grand Banks Foraminifera Examples	339
		9.4.2	Central Texas Cambrian Riley Formation Example	346
		9.4.3	Cretaceous Greenland-Norway Seaway Microfossils Example	349
	9.5	Construction of Geologic Time Scales		352
		9.5.1	Timescale History	354
		9.5.2	Differences Between GTS2012 and GTS2004	357
		9.5.3	Splining in GTS2012	358
		9.5.4	Treatment of Outliers	360
		9.5.5	Early Geomathematical Procedures	361
		9.5.6	Re-proportioning the Relative Geologic Time Scale	362
	References			363
10	**Fractals**			369
	10.1	Fractal Dimension Estimation		370
		10.1.1	Earth's Topography and Rock Unit Thickness Data	372
		10.1.2	Chemical Element Concentration Values: Mitchell-Sulphurets Example	375
		10.1.3	Total Metal Content of Mineral Deposits: Abitibi Lode Gold Deposit Example	381
	10.2	2-D Distribution Patterns of Mineral Deposits		383
		10.2.1	Cluster Density Determination of Gold Deposits in the Kirkland Lake Area on the Canadian Shield	386
		10.2.2	Cluster Density Determination of Gold Deposits in the Larger Abitibi Area	389
		10.2.3	Worldwide Permissive Tract Examples	391
	10.3	Geochemical Anomalies Versus Background		393
		10.3.1	Concentration-Area (C-A) Method	394
		10.3.2	Iskut River Area Stream Sediments Example	394
	10.4	Cascade Models		397
		10.4.1	The Model of de Wijs	399
		10.4.2	The Model of Turcotte	401
		10.4.3	Computer Simulation Experiments	402
	References			406
11	**Multifractals and Local Singularity Analysis**			413
	11.1	Self-Similarity		414
		11.1.1	Witwatersrand Goldfields Example	416
		11.1.2	Worldwide Uranium Resources	417

	11.2	The Multifractal Spectrum	419
		11.2.1 Method of Moments	420
		11.2.2 Histogram Method	425
	11.3	Multifractal Spatial Correlation	427
		11.3.1 Pulacayo Mine Example	429
	11.4	Multifractal Patterns of Line Segments and Points	432
		11.4.1 Lac du Bonnet Batholith Fractures Example	433
		11.4.2 Iskut River Map Gold Occurrences	439
	11.5	Local Singularity Analysis	442
		11.5.1 Gejiu Mineral District Example	445
		11.5.2 Zhejiang Province Pb-Zn Example	448
	11.6	Chen Algorithm	449
		11.6.1 Pulacayo Mine Example	453
		11.6.2 KTB Copper Example	459
	References		463
12	**Selected Topics for Further Research**		467
	12.1	Bias and Grouped Jackknife	468
		12.1.1 Abitibi Volcanogenic Massive Sulphides Example	470
	12.2	Compositional Data Analysis	473
		12.2.1 Star Kimberlite Example	475
	12.3	Non-linear Process Modeling	480
		12.3.1 The Lorentz Attractor	481
	12.4	Three-Parameter Model of de Wijs	483
		12.4.1 Effective Number of Iterations	484
		12.4.2 Au and As in South Saskatchewan till Example	485
	12.5	Other Modifications of the Model of de Wijs	490
		12.5.1 Random Cut Model	491
		12.5.2 Accelerated Dispersion Model	497
	12.6	Trends, Multifractals and White Noise	499
		12.6.1 Computer Simulation Experiment	500
	12.7	Universal Multifractals	504
		12.7.1 Pulacayo Mine Example	506
	12.8	Cell Composition Modeling	509
		12.8.1 Permanent Frequency Distributions	510
		12.8.2 The Probnormal Distribution	512
		12.8.3 Bathurst Area Acidic Volcanics Example	514
		12.8.4 Abitibi Acidic Volcanics Example	522
		12.8.5 Asymmetrical Bivariate Binomial Distribution	526
	References		529
Author Index			533
Subject Index			543

Chapter 1
Complexity of the Geological Framework and Use of Mathematics

Abstract The geosciences continually benefit from the use of mathematics. Probability calculus and statistical inference have contributed significantly to the solution of many geological problems. Statistical considerations form part of almost every application of mathematics in the geosciences. In the Earth sciences, the object of study usually is an aggregate of many smaller objects, which can be studied individually, but often only the properties of the aggregate are of interest. Geomathematics, in its broadest sense, includes all applications of mathematics to solve problems in studies of the Earth's crust.

An important objective of geology is to construct three-dimensional maps of the upper part of the Earth's crust with hypothetical delineation of various rock units including ore and hydrocarbon deposits. Earth imaging and three-dimensional map construction rely heavily on geophysical methods and drilling but use of geological concepts is essential in this endeavor requiring a good understanding of the underlying physical and chemical processes. This involves detailed knowledge of the ages of rock units and processes with use of the fossil record. Classical statistics is important because it helps to quantify the large uncertainties geoscientists normally have to cope with. Quantitative geometry and image analysis also provide essential tools for 3-D map making. In assessing quality of geometrical reconstructions based on limited data, it should be kept in mind that goodness of fit of models with respect to the limited information that is available is not the only criterion to be used; predictive potency is more important and, frequently, this aspect can only be checked by application of a model developed for one area in other areas that are geographically distinct. This chapter deals with the nature of facts and concepts in the geosciences and how these can be quantified. Later the emphasis is shifted to the use of geometry in 2-D and 3-D for map making and analysis. Various geometrical operations on maps are possible only because of the widespread availability of Geographic Information Systems (GIS's) and newly developed 3-D geoscience software.

Keywords Geomathematics • 3-D map-making • Geological cross-sections • Geometry of folds • Faults • Minkowski operations • Edge effects • Central Baffin Island supracrustal rocks • Bathurst acidic volcanics • Leduc reef complex

1.1 Use of Mathematics in Geology

As pointed out previously in Agterberg (1974), geomathematics, in its broadest sense, includes all applications of mathematics in the studies of the Earth's crust. In this book, the emphasis is on use of mathematics to solve 3-D prediction problems with applications to actual examples in selected case history studies. Of course, mathematics is widely used in geophysics, traditionally by formulating problems in terms of differential equations for deterministic processes but during the past 20 years also by means of non-linear process modeling. However, geologists also need a variety of geomathematical techniques, especially in the following fields of activity:

1. *Data acquisition and processing.* Systematic recording, ordering and comparison of data, and methods for graphical display of results.
2. *Data analysis.* Identification of trends, clusters and simple or complex correlations for which geological explanations are needed.
3. *Sampling.* Design of statistical procedures for data acquisition.
4. *Hypothesis testing.* Verification of concepts or models of processes believed to explain the origin and provenance of specific phenomena (includes computer simulation models).
5. *Quantitative prediction in applied geology.* Provision of solutions to specific problems such as estimating probabilities of occurrence of specific types of mineral deposits, volcanic eruptions, earthquakes and landslides.

The advent of digital computers catalyzed a quantitative revolution in geology, because it became possible to apply mathematical and statistical models to large volumes of data (Merriam 1981, 2004). When mathematics is used in geology, parameters must be defined in a manner sufficiently rigorous to permit nontrivial derivations. The initial hurdle is to choose parameters that are substantially meaningful. During model design it is important to keep in mind Chamberlin's (1899) warning: "The fascinating impressiveness of rigorous mathematical analysis, with its atmosphere of precision and elegance, should not blind us to the defects of the premises that condition the whole process. There is, perhaps, no beguilement more insidious and dangerous than an elaborate and elegant mathematical process built upon unfortified premises."

The geosciences continually benefit from the use of mathematics. An important objective of geology is to construct three-dimensional maps of the upper part of the Earth's crust with hypothetical delineation of various rock units including ore and hydrocarbon deposits. Uncertainties in predictive geology are very large but need to be quantified nevertheless. Use of geological concepts is essential in this endeavor involving a good understanding of the underlying physical and chemical processes with detailed knowledge of the ages of rock units and processes. As discussed in Agterberg (2013), geologists have had a long history of interaction with mathematical statisticians. These collaborations continue to be fruitful. During his long and distinguished career, William Krumbein regularly consulted with John Tukey. Krumbein also wrote one of the first geomathematical textbooks together with the mathematical statistician Franklin Graybill (Krumbein and Graybill 1965). These authors distinguished between three types of models in geology: (1) scale-models; (2) conceptual models; and (3) mathematical models. Traditionally, geologists have been concerned mainly with scale-models and conceptual models.

John Griffiths (1967), who was one of the other pioneers of geomathematics, based much of his work on advanced statistical sampling techniques, especially as they had been developed by Ronald Fisher. In the Soviet Union, Andrew Vistelius (1967) worked closely with Andrey Kolmogorov, already when he was preparing his PhD thesis in the 1940s. Felix Chayes (1956, 1971) developed modal analysis in petrography and took to heart Karl Pearson's admonishment regarding spurious correlations that could result from closed-number systems. This later led John Aitchison (1986) to develop compositional data analysis, which continues to be an important research topic (Egozcue et al. 2003).

In turn, geoscientists have inspired statisticians to pursue new research directions. Georges Matheron (1962, 1965) introduced the idea of regionalized random variables and this has helped to found the important field of spatial statistics (Cressie 1991). Ronald Fisher (1953) became interested in statistics of directional features when a geophysics student at Cambridge University asked him for help in dealing with greatly dispersed paleomagnetic measurements. It led to the establishment of the cone of confidence for unit vectors. Geoffrey Watson (1966) was instrumental in developing statistical significance tests for directional features, which are similar to those that were in existence for ordinary data analysis. Geoscientists often work with very large data sets to be subjected to exploratory data analysis with use of jackknife and bootstrap techniques for uncertainty estimation. John Tukey (1962, 1970, 1977) pioneered these approaches and advised mathematical geologists to use them. GIS (Bonham-Carter 1994) and its 3-D extensions, for which credit is owed to engineers and computer scientists, are widely used in geological map modeling and image analysis. Promising new developments include projection pursuit (Friedman and Tukey 1974; Xiao and Chen 2012), boosting (Freund and Shapire 1997), radial basis function theory (Buhmann 2003) and bi-dimensional empirical mode decomposition (Huang et al. 2010).

1.2 Geological Data, Concepts and Maps

The first geological map was published in 1815. This feat has been well documented by Winchester (2001) in his book: *The Map that Changed the World – William Smith and the Birth of Modern Geology*. According to the concept of stratigraphy, strata were deposited successively in the course of geological time and they are often uniquely characterized by fossils such as ammonites that are strikingly different from age to age. It is remarkable that this fact remained virtually unknown until approximately 1,800. There are several other examples of geological concepts that became only gradually accepted more widely, although they were proposed much earlier in one form or another by individual scientists. The best-known example is plate tectonics: Alfred Wegener (1966) had demonstrated the concept of continental drift fairly convincingly as early as 1912 but this idea only became acceptable in the early 1960s. One reason that this theory initially was rejected by most geoscientists was lack of a plausible mechanism for the movement of continents.

Along similar lines, Staub (1928) had argued that the principal force that controlled mountain building in the Alps was crustal shortening between Africa and Eurasia. However, other interpretations including the gravitational concept of van Bemmelen (1960) continued to provide plausible explanations before the theory of plate tectonics became well established. Figure 1.1 (from Agterberg 1961, Fig. 107) shows tectonic sketch maps of two areas in the eastern Alps. The Strigno area in northern Italy shown at the top of Fig. 1.1 is 1600× smaller in area than the region for the eastern Alps at the bottom. The tectonic structure of these two regions is similar. Both contain overthrust sheets with older rocks including crystalline basement overlying much younger rocks. It is now well known that the main structure of the map at the bottom was created during the Oligocene (33.9–23.0 Ma) when the African plate moved northward over the Eurasian plate. On the other hand, the main structure of the relatively small Strigno area was created much later during the late Miocene (Tortonian, 11.6–7.3 Ma) when the Eurasian plate moved southward overriding the Adria microplate that originally was part of the African plate. Incidentally, Fig. 1.1 also provides an example of the concept of similarity of geological patterns at different scales. As another example of self-similarity of geological patterns, Fig. 11.1 shows multifractal gold mineralization patterns that are strikingly similar in two areas that differ in size by a factor of 400. Self-similarity or scale-independence will be discussed in later chapters in the context of fractals and multifractals.

Another geological idea that was conceived early on, but initially rejected as a figment of the imagination, is what later became known as the Milankovitch theory. Croll (1875) had suggested that the Pleistocene ice ages were caused by variations in the distance between the Earth and the Sun. Milankovitch commenced working on astronomical control of climate in 1913 (Schwarzacher 1993). His detailed calculations of orbital variations were published nearly 30 years later (Milankovitch 1941) showing quantitatively that amount of solar radiation drastically changes our climate. This theory was immediately rejected by climatologists because the changes in solar radiation due to orbital variations are miniscule, and by geologists as well because, stratigraphically, their correlation with ice ages apparently was not very good. However, in the mid-1950, new methods helped to establish the Milankovitch theory beyond any doubt. Subsequently, it resulted in the establishment of two new disciplines: cyclostratigraphy

1.2 Geological Data, Concepts and Maps

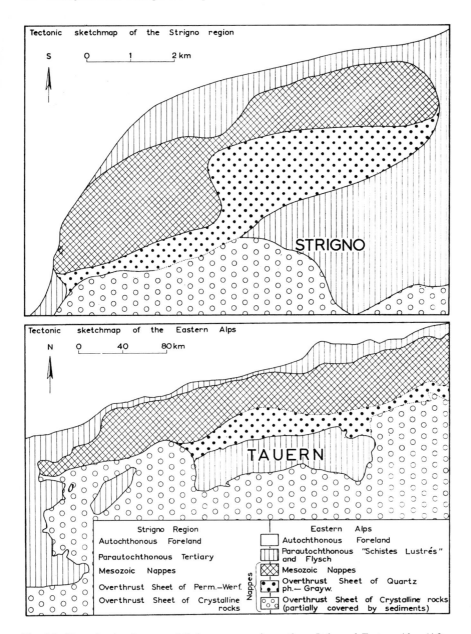

Fig. 1.1 Tectonic sketch maps of Strigno area on in northern Italy and Eastern Alps (After Agterberg 1961). Although area sizes differ by a factor of 1,600, the patterns showing overthrust sheets are similar. Originally gravity sliding was assumed to be the major cause of these patterns. However, after development of the theory of plate tectonics in the 1960s, it has become apparent that plate collision was the primary driving force of Alpine orogeny. The Strigno overthrust sheets are due to Late Miocene southward movement of the Eurasian plate over the Adria micro-plate, and the eastern Alpine nappes were formed during the earlier (primarily Oligocene) northward movement of the African plate across the Eurasian plate. For more discussion, see Chap. 8 (Source: Agterberg 1961, Fig. 107)

and astrochronology (Hinnov and Hilgen 2012). The latest time scales of the Neogene (23.0–2.59 Ma) and Paleogene (66.0–23.0 Ma) periods are entirely based on astronomical calibrations and "floating" astrochronologies are available for extended time intervals (multiple millions of years) extending through stages in the geologic timescale belonging to the Triassic, Jurassic and Cretaceous periods (Gradstein et al. 2012).

1.2.1 Map-Making

A field geologist is concerned with collecting numerous observations from those places where rocks are exposed at the surface. Observation is often hampered by poor exposure. In most areas, 90 % or more of the bedrock surface is covered by unconsolidated overburden restricting observation to available exposures; for example, along rivers (*cf.* Agterberg and Robinson 1972). The existence of these exposures may be a function of the rock properties. In formerly glaciated areas, for example, the only rock that can be seen may be hard knobs of pegmatite or granite, whereas the softer rocks may never be exposed. Of course, drilling can help and geophysical exploration techniques provide additional information, but bore-holes are expensive and geophysics provides only partial, indirect information on rock composition, facies, age and other properties of interest. It can be argued that to-day most outcrops of bedrock in the world have been visited by competent geologists. Geological maps at different scales are available for most countries. One of the major accomplishments of stratigraphy is not only that the compositions but also the ages of rocks nearly everywhere at the Earth's surface now are fairly well known. However, as Harrison (1963) has pointed out, although the outcrops in an area remain more or less the same during the immediate past, the geological map constructed from them can change significantly over time when new geological concepts become available. A striking example is shown in Fig. 1.2. Over a 30-year period, the outcrops in this study area repeatedly visited by geologists had remained nearly unchanged. Discrepancies in the map patterns reflect changes in the state of geological knowledge at different points in time.

Many geologists regard mapping as a creative art. From scattered bits of evidence, one must piece together a picture at a reduced scale, which covers at least most of the surface of bedrock in an area. Usually, this cannot be done without a good understanding of the underlying geological processes that may have been operative at different geological times. A large amount of interpretation is involved. Many situations can only be evaluated by experts. Although most geologists agree that it is desirable to make a rigorous distinction between facts and interpretation, this is hardly possible in practice, partly because during compilation results for larger regions must be represented at scales of 1: 25,000 or 1: 250,000, or less. Numerous observable features cannot be represented adequately in these scale models. Consequently, there is often significant discrepancy of opinions among

1.2 Geological Data, Concepts and Maps

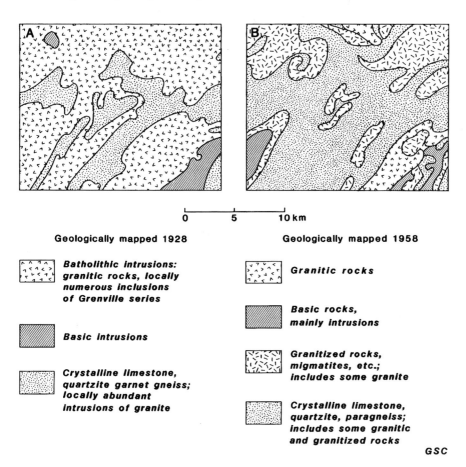

Fig. 1.2 Two geological maps for the same area on the Canadian Shield (After Harrison 1963). Between 1928 and 1958 there was development of conceptual ideas about the nature of metamorphic processes. This, in turn, resulted in geologists deriving different maps based on the same observations

geologists that can be bewildering to scientists in other disciplines and to others including decision-makers in government and industry.

Van Bemmelen (1961) has pointed out that the shortcomings of classical methods of geological observation constrain the quantification of geology. Much is left to the 'feeling' and experience of the individual geologist. The results of this work, presented in the form of maps, sections and narratives with hypothetical reconstructions of the geological evolution of a region, do not have the same exactitude as the records and accounts of geophysical and geochemical surveys which are more readily computerized even though the results may be equally accurate in an interpretive sense. Geophysical and geochemical variables are determined by the characteristics of the bedrock geology which, in any given area, is likely to be non-uniform because of the presence of different rock types, often with abrupt changes at the contacts between them. The heterogeneous nature of the geological framework will be reflected or masked in these other variables.

Geologists, geophysicists and geochemists produce different types of maps for the same region. Geophysical measurements are mainly indirect. They include gravity and seismic data, variations in the Earth's magnetic field, and electrical conductivity. They generally are averages of specific properties of rocks over large volumes with intensities of penetration decreasing with distance and depth. Remotely sensed data are very precise and can be subjected to a variety of filtering methods. However, they are restricted to the Earth's surface. Geochemists mainly work with element concentration values determined from chips of rocks *in situ*, but also from samples of water, mud, soil or organic material.

1.2.2 Geological Cross-Sections

The facts observed at the surface of the Earth must be correlated with one another; for example, according to a stratigraphic column. Continually, trends must be established and projected into the unknown. This is because rocks are 3-D media that can only be observed in 2-D. The geologist can look at a rock formation but not inside it. Sound geological concepts are a basic requirement for making 3-D projections.

During the past two centuries, after William Smith, geologists have acquired a remarkably good capability of imagining 3-D configurations by conceptual methods. This skill was not obtained easily. In Fig. 1.3, according to Nieuwenkamp (1968), an example is shown of a typical geological extrapolation problem with results strongly depending on initially erroneous theoretical considerations. In the Kinnehulle area in Sweden, the tops of the hills consist of basaltic rocks; sedimentary rocks are exposed on the slopes; granites and gneisses in the valleys. The first two projections into depth for this area were made by a prominent geologist (von Buch 1842). It can be assumed that today most geologists would quickly arrive at the third 3-D reconstruction (Fig. 1.3c) by Westergård et al. (1943). At the time of von Buch it was not yet common knowledge that basaltic flows can form extensive flows within sedimentary sequences. The projections in Figs. 1.3a, b reflect A.G. Werner's pan-sedimentary theory of "Neptunism" according to which all rocks on Earth were deposited in a primeval ocean. Nieuwenkamp (1968) has demonstrated that this theory was related to philosophical concepts of F.W.J. Schelling and G.W.F. Hegel. When Werner's view was criticized by other early geologists who assumed processes of change during geological time, Hegel publicly supported Werner by comparing the structure of the Earth with that of a house. One looks at the complete house only and it is trivial that the basement was constructed first and the roof last. Initially, Werner's conceptual model provided an appropriate and temporarily adequate classification system, although von Buch, who was a follower of Werner, rapidly ran into problems during his attempts to apply Neptunism to explain occurrences of rock formations in different parts of Europe. For the Kinnehulle area (Fig. 1.3) von Buch assumed that the primeval

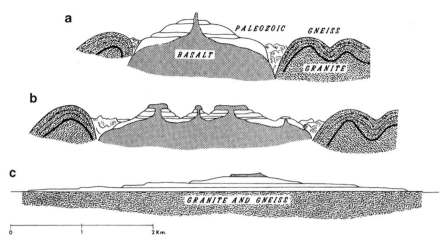

Fig. 1.3 Schematic sections originally compiled by Nieuwenkamp (1968) showing that a good conceptional understanding of time-dependent geological processes is required for downward extrapolation of geological features observed at the surface of the Earth. Sections **a** and **b** are modified after von Buch (1842) illustrating his genetic interpretation that was based on combining Abraham Werner's theory of Neptunism with von Buch's own firsthand knowledge of volcanoes including Mount Etna in Sicily with a basaltic magma chamber. Section **c** is after Westergård et al. (1943) (Source: Agterberg 1974, Fig. 1)

granite became active later changing sediments into gneisses, while the primeval basalt became a source for hypothetical volcanoes.

1.2.3 Scientific Method in the Geosciences

According to van Bemmelen (1972, p. 8), the computerization and adoption of mathematics in geoscience presents a wide-ranging and challenging field of future developments but is fraught with organizational, educational and technical difficulties. The usage of mathematics has progressed further in geophysics and geochemistry than in geology because of differences in the nature of the data and other sampling methods. The relative weights to be assigned to data collected from geological observations at the surface of the Earth are often unknown. A fact can be important and applicable at other places; frequently at great distances from the original observation point, or it may be unique and unimportant (*cf.* Agterberg 1979a). Most geological facts are not unique but greatly influenced by other observations and by deductions from specific concepts on geological processes. Basic measurements, such as chemical analysis of a rock sample or determination of the strike and dip of a structural plane, might represent local exceptions and their importance can be evaluated only against a background of regional data. Satisfactory statistical averaging may not be possible due to lack of exposure.

Van Bemmelen (1961) pointed out that geologists can be distinguished as two types: one type considers geology as a creative art; the other regards it as an exact science. Some Earth scientists aim for classification of their objects of study wishing to force the rigid disciplines of well-established schemes upon them (Wright 1958); others are more receptive of new concepts and less rigorous in schematizing the objects. Ostwald (1910) distinguished between "classicists" and "romanticists". Wegmann (1958) called Werner a classicist contrary to the romanticist J. Hutton, who postulated "Plutonism" instead of "Neptunism". Geology has known an exceptionally large amount of controversy and polemics closely related to the personalities and experiences of the opponents. For example, Neptunism commenced in areas where sedimentary rocks were lying on top of granites and gneisses, whereas Plutonism originated with Hutton in Scotland where tectonics with granite intrusions is more apparent.

The question of whether geology should become more quantitative has been considered continually in the past by geologists as well as other scientists. Fisher (1954) suggested that geology with Lyell (1833) was evolving as a more quantitative science, but opposition to this development quickly grew to the extent that Lyell was forced to omit his elaborate tables and statistical arguments to divide the Tertiary into stages from later editions of his *Principles of Geology*. Most geologists agree that Chamberlin's (1897) scientific method of "multiple working hypotheses" is ideal for geology. A multiple approach is needed because of the great complexity of geological processes. The sheer diversity of observations combined with irregularity of rock exposures entailed by this method make it difficult to apply. However, it is to be preferred to the practice of synthesizing from relatively few broadly relevant observations in order to develop a preferred hypothesis. Griffiths (1962) has stressed the analogy between Chamberlin's method and Fisher's (1960) description of statistical analysis. By using different statistical models and formal inference, it is possible to test different hypotheses for solving the same problem provided that the geological facts can be expressed adequately in numerical form.

The basic principle of producing useful geoinformation from observations is to capture original data in such a form that they become quantitative and can be used for a variety of purposes including map-making and 3-D geomodeling. The step from data recording to production of useful geoinformation has to keep up with the continuous stream of technological innovations. The term "ontology" sometimes is used for a domain model that provides a vocabulary about key geological concepts, their interrelationships and the theoretical processes and principles that are relevant within the geoscientific subdiscipline under which the basic data are collected.

The arrangement of observations into patterns of relationships is a mental process of "induction". It involves postulating hypotheses or theory that is in agreement with the facts. Working hypotheses initially based on intuition gain in functional validity if they lead to "deductions" that can be verified. Logical deductions from a theory provide predictions for facts that have not yet been observed. The validity of a hypothesis is tested on the basis of new facts that had not been considered when the hypothesis was postulated. The result is a recurrent cycle of inductions and deductions. Van Bemmelen (1961) referred to this approach as the prognosis-diagnosis method of research.

The Logic of Scientific Discovery is the title of an influential book by Popper (2002) of which the German Edition was published in 1934 and the first English Edition in 1959. Popper (2002, p. 3) explains that a scientist, whether theorist or experimenter, puts forward statements or systems of statements and tests them step by step. In the field of the empirical sciences, which would include geology, the scientist constructs hypotheses, or systems of theories, and tests them against experience by observation and experiment. Popper opposed the widely held view (e.g., by F.W.J. Schelling and G.W.F. Hegel; and also by I. Kant and F. Bacon) that the empirical sciences can be characterized by the fact that they use inductive methods only. Instead of this Popper advanced the theory of "the deductive method of testing" in that a hypothesis can only be empirically tested and only after it has been postulated. Demarcation by falsification is an essential element in Popper's approach. However, we have to keep in mind the role played by probability theory in which falsification does not necessarily result in rejection of a test hypothesis. Instead of this, the end conclusion may be a relative statement such as "there is more than a 95 % probability that the hypothesis is true". Because of the great uncertainties and possibility of multiple explanations, geoscientists should proceed in accordance with the axioms of the theory of probability as they were advanced, for example, by Kolmogorov (1931) as will be discussed in more detail in the next chapter. It is a remarkable feature of geology that a theory that turns out to be true as proven by irrefutable evidence often was already assumed to be true much earlier by some geologists when conclusive evidence did not yet exist. Examples of this form of anticipation previously discussed in this chapter were the origin of ice ages (Kroll), plate tectonics (Staub), and continental drift (Wegener).

1.2.4 Quality of Predictions

The question of how good is your prediction is continuously asked in the geosciences, particularly in economic geology. For example, in the oil industry it is known from experience that most holes, especially the wildcats, will remain dry. Nevertheless, the geologist is asked to provide an opinion as to whether it is worthwhile drilling a hole at a particular site. This problem has been discussed in detail by Grayson (1960) and de Finetti (1972). The geologist does not have any say in the final decision of whether or not to drill. This is the responsibility of the decision-maker who will reach a conclusion after considering all different pieces of information available of which the geologist's report is just one piece. The geologist cannot state categorically that oil is present or absent. Neither can she restrict herself to a mere listing of reliable facts. A conclusion about the probable outcome of the drilling is precisely what the geologist is called upon to provide. Grayson (1960) found that most geologic reports contain probabilistic answers that are disguised in vague adjectives ("fairly good prospect", "favorable", "permissive", "it's difficult to say", etc.). He has proposed methods for translating a geologist's opinion into subjective probabilities. Grayson's (1960) book deals primarily with drilling decisions by oil and gas operators. A sign hanging in the office of one of the operators interviewed by him states: "Holes that are going to be dry shouldn't be drilled", although some years previously this particular operator had drilled 30 consecutive dry holes. De Finetti (1972) used this paradox to urge the geologist to express his predictions in a probabilistic manner rather than translating them into the inadequate logic of absolute certainty.

The human mind allows the formulation of hypotheses which are flexible to the extent that they may immediately incorporate all new facts before the hypotheses could be properly tested. On the other hand, the advantage of using the logic of mathematics is that it is indisputable and, when random variables are used, it is possible to check the deductions against reality. P.A.M. Dirac (in Marlow 1978) has advocated the use of mathematics in physics as follows: "One should keep the need for a sound mathematical basis dominating one's search for a new theory. Any physical or philosophical ideas that one has must be adjusted to fit the mathematics. Not the other way around. Too many physicists are inclined to start from preconceived physical ideas and then try to develop them and find a mathematical scheme that incorporates them. Such a line of attack is unlikely to lead to success". This is good advice for all scientists.

1.3 Use of Curves

1.3.1 Trend-Lines

It can be argued that all geological processes are deterministic. However, it is usually not possible to use purely deterministic expressions in the mathematical equations used for representation because of uncertainties or unknown causes. In many geological situations, the spatial variability of measurable features can be divided into a regional systematic component (loosely called "trend" or "drift") and more local, unpredictable fluctuations (residuals from the trend). The trends may have been generated by broad-scale deterministic processes. For example, average grain size of sand particles increases towards a beach. However, larger and smaller particles may coexist everywhere in the sampled area in different proportions.

Systematic variations called "trends" are, in part determined by the density of sampling points. If many more measurements are performed locally, a residual from a more regional trend or drift can become the trend for a more local survey area. It is appropriate to use deterministic functions for "trends" when it is kept in mind that they do not necessarily describe the result of deterministic processes. We have a choice of using empirical functions such as polynomials or functions corresponding to curves or surfaces that are theoretical predictions for geological phenomena. For example, Vistelius and Janovskaya (1967) pointed out that the

1.3 Use of Curves

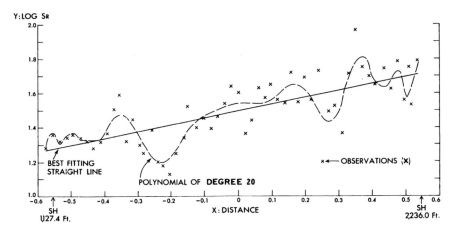

Fig. 1.4 Plot of 38 strontium determinations along a drill-hole through dunite layer of Muskox Layered Intrusion, Northwest Territories, Canada. Two continuous polynomial curves were fitted. Although the polynomial of degree 20 has lesser residual variance, it is not to be preferred because of autocorrelation of the residuals (Source: Agterberg 1974, Fig. 3)

partial differential equations for diffusion processes that caused chemical element concentration variations in some rocks, usually have exponential functions as a solution. For trend fitting they suggested the use of exponential functions with polynomials placed in the exponent. In that situation, one may not know the exact shape of the trends to be fitted to the data but by using a specific class of functions a satisfactory fit is obtained more readily.

Any continuous function for a geometrical shape can be expanded into an infinite series (e.g., polynomial or Fourier series) and a restricted number of terms of the infinite series can provide an adequate approximation for all possible geometrical configurations. Theoretically, even shapes with discontinuities (e.g., breaks related to faults or contacts) can be represented by truncated series. In practice, however, this approach may not be feasible because of the large number of terms and data points that would be required. In general, the fitting of trends should be restricted to geological entities with features that are subject to gradational change without sudden breaks.

Suppose that a curve is fitted to n data points in a diagram. If the curve-fitting is performed by using the method of least squares, every data point x_i ($i = 1, 2, \ldots, n$) deviates from the best-fitting trend-line t_i by a residual r_i ($= x_i - t_i$) and the sum of squared residuals $\Sigma\, r_i^2$ is not necessarily as small as possible. This qualification is illustrated by means of the following example (Fig. 1.4). The original data represent concentration of the element strontium (in parts per million) determined for equally spaced rock samples along a drillhole through a dunite-serpentinite layer of the Muskox ultramafic intrusion, District of Mackenzie (original data from Findlay and Smith 1965).

The logarithm of Sr was plotted against distance and polynomials of degrees 1 and 20 were fitted to the logarithmically transformed data by the method of least squares. The "trend" is rather well approximated by a straight line in Fig. 1.4. The logarithm of Sr content decreases gradationally in the upward direction of the

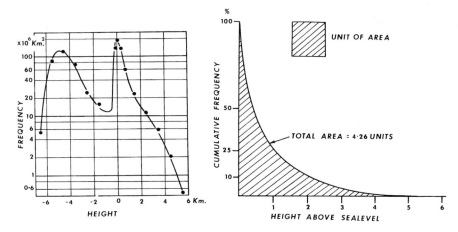

Fig. 1.5 First derivative of hypsometric curve for Earth's surface (After Scheidegger 1963); and hypsometric curve for continents only (Source: Agterberg 1974, Fig. 11)

borehole. This implies that untransformed Sr content decreases exponentially in this direction. The data points deviate from the trend line. Their residuals are irregular but not randomly distributed along the best-fitting straight line. In order to improve the closeness of fit, a polynomial of a higher degree could be fitted. It would be shown by statistical inference (*cf.* variance analysis, Chap. 2) that the fit of the polynomial of degree 20 is "significantly" better. It is based on solving an equation with 21 unknown coefficients whereas only two coefficients were needed to obtain the straight line. It is unlikely that the fluctuations in the second curve are meaningful. Polynomials of degree 10 or 30 show different patterns and, for lack of data, one could not say which fit is best.

1.3.2 Elementary Differential Calculus

The purpose of this section and the next one is to show by means of simple examples how elementary methods of calculus using differentiation and integration can be applied for the analysis of various contour maps and cross-sections. Of course, these methods have been implemented in various special-purpose software systems but a good understanding of what is being done in software applications remains important.

Geophysicists have been interested in studying the surface of the Earth by using hypsometric curves. A hypsometric curve for an area is a plot of the percentage of area above a certain height level against height. The diagram on the right side of Fig. 1.5 is the hypsometric curve for continents. The first derivative of the hypsometric curve for the Earth's surface is shown on the left side of Fig. 1.5. These curves, which were fitted by hand, to observed heights, and their absolute or relative frequencies can be used to determine quantities such as average height by

1.3 Use of Curves

elementary methods of differentiation and integration. For example, average height satisfies the equation Ave $(h) = \int h \cdot dA'$ where A' represents relative area that can be expressed as a percentage. Average height on continents can be estimated rapidly as follows: Firstly the area under the curve on the left of Fig. 1.5 is determined graphically as 4.26 times as large as the unit of area (e.g., by counting squares on graph paper). This value must be divided by 5 to account for the vertical exaggeration. The result is 0.852 km for average height on continents, which, after rounding off, duplicates the value of 850 m reported by Heiskanen and Vening Meinesz (1958). This approximate equality may be fortuitous because the error of the value estimated graphically from Fig. 1.5 probably exceeds 2 m.

A problem analogous to the one solved in the previous example consists of calculating the average value over a given area for a variable of which the contour map is given. For example, in the contour map of Fig. 7.28, which will be discussed later, the contours are for percent copper. A problem that can be solved in that application is to determine the average percentage copper or copper bounded by the 0.5 contour on the map. By constructing a hypsometric curve with height replaced by percentage copper, it was estimated that the average grade for the larger area is approximately 0.99 % copper.

1.3.3 Graphical Curve-Fitting

The following example illustrates graphical integration in an application to an isoclinal fold in cross-Section. A number of attitudes of strata suggesting existence of an anticline were observed along a line across the topographic surface The observation points are labelled 0, 1, 2, ..., n ($n=4$ in Fig. 1.6). The objective is to reconstruct a complete pattern for this fold in vertical cross-section.

It will be assumed that the isoclines (lines of equal dip) in the profile are parallel to a line (Y-axis) through the origin (O) that is set at the first observation point (Fig. 1.6). The projections of points 0. 1, 2, ..., n on the corresponding X-axis through O are called x_i ($i = 1, 2, ..., n$). Let α_i be the angles of dip with respect to the X-axis. Then the curve for a function $f(x) = \tan \alpha$ can be constructed by using the $(n+1)$ known values of α_i. The points P_i in Fig. 1.6 have ordinates equal to $\tan \alpha_i$. The function $f(x)$ plots as a smooth curve passing through the points P_i. Suppose integration of $f(x)$ gives the function $y = \int f(x)\,dx = F(x) + C$ where C is an arbitrary constant. In Fig. 1.6 use is made of the method of graphical integration. The curve $f(x)$ is replaced by a staircase function $f^*(x)$ with the property:

$$\int_{x_i}^{x_{i+1}} f(x)dx = \int_{x_i}^{x_{i+1}} f^*(x)dx$$

where x_i and x_{i+1} are the abscissae of two consecutive points P_i and P_{i+1} ($i = 1, 2, ..., n$-1). The function $f^*(x)$ is readily integrated yielding $F^*(x) + C$. The constant C is specified by letting $F^*(x)$ pass through O. Note that $F^*(x)$ consists of a succession

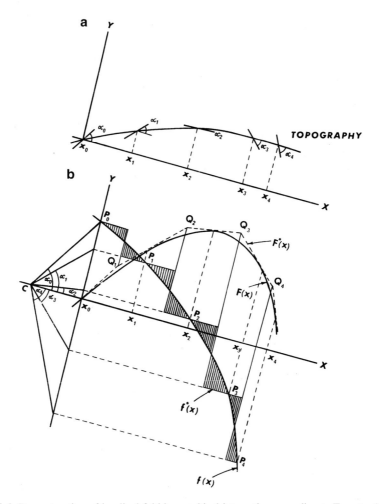

Fig. 1.6 Reconstruction of isoclinal fold by graphical integration according to Trooster (1950). (**a**). Five dip angles along topographic surface are given: (*XY*)-coordinate system is constructed with Y parallel to isoclines. (**b**). Curve $F(x)$ has the known dips at intersection points with isoclines through observation points (Source: Agterberg 1974, Fig. 10)

of straight line segments between points Q_i and Q_{i+1} and with dips equal to α_i. $F(x)$ is a smooth curve that coincides with $F^*(x)$ at points with abscissae x_i.

1.4 Use of Surfaces

As recently pointed out by Hillier et al. (2013), interpreting and modeling geometries of complex geological structures from strike/dip measurements using manually-drafted structural form lines is labor intensive, irreproducible and

1.4 Use of Surfaces

inherently limited to two dimensions. These authors have presented the structural field interpolation (SFI) algorithm to overcome these limitations by constructing 3-D structural form lines from the vector components of strike/dip measurements. It is beyond the scope of this book to explain SFI in detail but an example of its application will be given in the next section to illustrate how geological structures can be modeled in 3-D. A few introductory remarks on representation of lines and planes in 3-D space are as follows.

Any point $P = (x, y, z)$ in a 3-D Cartesian coordinate system with origin O determines a vector OP. Suppose that α, β, and γ are the angles of this vector with X-, Y-, and Z- axes. The cosines of these angles $\lambda = \cos \alpha$, $\mu = \cos \beta$ and $\nu = \cos \gamma$ are the direction cosines of OP. They satisfy the relation $\lambda^2 + \mu^2 + \nu^2 = 1$. The equation $\lambda x + \mu y + \nu z = |OP|$ represents a plane in 3-D. The line OP is the normal of this plane. OP is a unit vector if its length $|OP| = 1$. In structural geology, a plane is characterized by its strike and dip. If the north direction points in the negative X-direction, the strike δ of the plane satisfies $\tan \delta = \mu/\lambda$ and its dip angle is equal to γ.

Statistics of directional features will be discussed in more detail in Chap. 8. However, for a better understanding of the SFI example in the next section, it is pointed out here that one method of estimating the mean direction of n unit vectors in 3-D is to maximize $\Sigma \cos^2 \theta_i$ ($i = 1, 2, \ldots, n$) where θ_i represents the angle between the i-th observed unit vector and the mean to be estimated. This method was first applied by Scheidegger (1964) in connection with the analysis of fault-plane solutions of earthquakes and by Loudon (1964) for orientation data in structural geology. It can be shown that the resulting average unit vector has direction cosines equal to those of the first (dominant) eigenvector of the following matrix:

$$M = \begin{bmatrix} \Sigma \lambda_i^2 & \Sigma \lambda_i \mu_i & \Sigma \lambda_i \nu_i \\ \Sigma \mu_i & \Sigma \mu_i^2 & \Sigma \mu_i \nu_i \\ \Sigma \nu_i \lambda_i & \Sigma \nu_i \mu_i & \Sigma \nu_i^2 \end{bmatrix}$$

A useful interpolation method in 2-D or 3-D is inverse distance weighting. It means that the value of an attribute of a rock such as a chemical concentration value or the strike and dip of a plane at an arbitrary point are estimated from the known values in their surroundings by weighting every value according to the inverse of a power of its distance to the arbitrary point. In SFI this method is called IDW (inverse distance weighted) interpolation. Usually the weights are raised to a power called IDW exponent before they are applied. This exponent is often set equal to 2. The SFI algorithm can employ an anisotropic inverse distance weighting scheme derived from eigen analysis of the poles to strike/dip measurements within a neighborhood of user defined dimension and shape (ellipsoidal to spherical). When the matrix M is used, all its nine elements are multiplied by the same weight that is different for every strike and dip depending on the distance from the arbitrary point and on direction of the connecting line if an anisotropic weighting scheme is used.

The SFI algorithm generates 3-D structural form lines using the anisotropic inverse weighted (IDW) interpolation approach. These structural form lines follow the orientation of planar structural elements, such as bedding and foliation. The structural form lines are iteratively propagated from point to point. At each point the vector field is interpolated from vector components derived from the structural measurements while keeping the continuity of the structural form line intact. Further information on the SFI algorithm can be found in Hillier et al. (2013).

1.4.1 Automated 3-Dimensional Map-Making: Central Baffin Example

Hillier et al. (2013) have applied the SFI algorithm to a 15,000 km^2 study area in the Central Baffin Region, Nunavut, Canada that contains 1,774 structural measurements taken at the surface from supracrustal rocks (Fig. 1.7). The region is marked by near-cylindrical, tight to open East-west shallow fold plunges. A vector field modeling bedding from the region was calculated using all these measurements simultaneously (Fig. 1.8). Structural data and form lines representing the vector field within a 1 km buffer zone were projected perpendicularly onto each section. The resulting vector field demonstrates the capability of SFI to capture the regional folding trends while at the same time detecting the local variability of the data. At locations where the data are relatively dense and highly variable the SFI tool makes it easier to carry out structural interpretation. Structural trends, fold patterns and various scales of anisotropy are clearly visualized, and if needed more detailed models could be calculated from local sub-sets of the data. Additionally, SFI can visualize relationships with crustal features, in this region supporting the compatibility of supracrustal fold patterns within turbiditic units in the West with patterns of basement culminations in the East. Broad regional scale doubly plunging folds are similar in both parts of the region, indicative of a dome and basin pattern, reflected in culminations along section EE'; however, superimposed on this is the higher frequency tighter folding as expected in layered lower grade supracrustals that are structurally above these culminations.

1.4.2 Folds and Faults

As pointed out by Mallet (2004), interpolation of properties of rocks in the subsurface is a recurrent problem in geology. In sedimentary geology, the geometry of the layers is generally known with a precision which is superior to that of the rock properties such as composition. The geometry of layers normally is affected by folding as well as faulting that took place after the time of deposition, whereas the distribution of the rock properties had largely been determined at the time of deposition.

1.4 Use of Surfaces

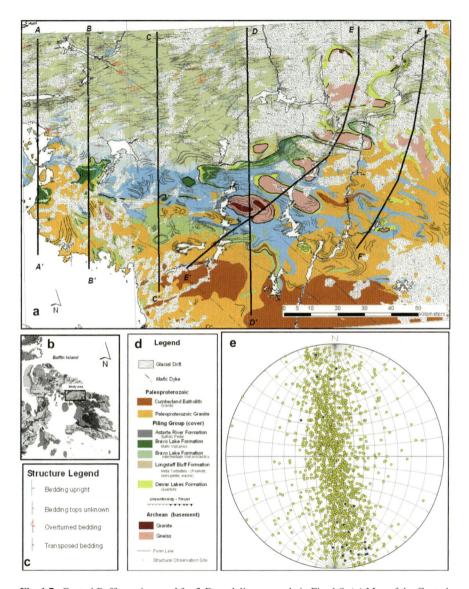

Fig. 1.7 Central Baffin region used for 3-D modeling example in Fig. 1.8. (**a**) Map of the Central Baffin region contains 1,774 structural measurements (tablets) taken at surface from supracrustal rocks (St-Onge et al. 2005). The region is approximately 150 km East-west by 100 km North-south (Note scale for context of 3-D images). (**b**) Reference map. (**c**) Structural legend for supracrustal bedding observations. (**d**) Stratigraphic legend accompanying map (**a**). (**e**) All poles for 1,774 supracrustal dip measurements were used in Structural Field Interpolation (Note shallow near East-west rotation axis for these bedding poles). *Yellow* poles are upright, *blue* poles are overturned (Source: Hillier et al. 2013, Fig. 8)

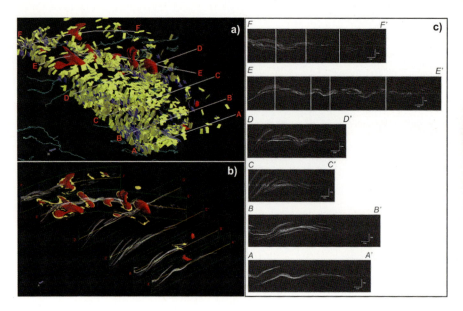

Fig. 1.8 3-D Structural Field Interpolation (SFI) modeling example from Hillier et al. (2013). View of data set and SFI cross sections. *Yellow* and *blue* sides of tablets indicate stratigraphic tops and bottoms. Semi-transparent *red* domes and *yellow* fringe surface represent previously interpreted Archean basement granitic gneiss domes with unconformably overlying Paleoproterozoic Dewar Lakes Formation quartzites depicted in *yellow* around margins of dunes (de Kemp and St-Onge 2007). (**a**) Perspective view of dip data (tablets), location of cross-sections, hydrography and previously interpreted basement domes. (**b**) Perspective view of cross-sections with dip measurements and SFI calculation results on all Sections. (**c**) Vertical sections shown on regional map (Fig. 1.7) and 3-D views Figs. 1.8a, c, with data and STI models projected from 1 km normal to section. Structural form lines of SH are in white. Apparent dip measurements depicted as oriented lines, input parameters used for the calculation were: IDW exponent = 2, type of neighborhood = ellipsoidal oriented north-south, number of nearest neighbors used = 25, formline step length = 100 m, total formline length = 30 km (Source: Hillier et al. 2013, Fig. 9)

As a strategy it is therefore wise to first model the geometry of the layers and then to "simplify the geological equation" by removing the influence of that geometry. One can go about this by defining, mathematically, a new space where all horizons are horizontal planes and where faults, if any, have been eliminated. If the layers are folded, one can use a curvilinear coordinate system (u, v, t) with the (u, v) axes parallel to the layering and the t-axis orthogonal to the layers. In geomodeling, Mallet (2002) introduced the "geological space" (G-space) with such a curvilinear coordinate system. Examples in physical geology where choosing a curvilinear coordinate system is appropriate include the following: (1) Equations of flow through porous media in reservoir engineering become greatly simplified with a curvilinear coordinate system (u, v, t) with (u, v) defining the iso-pressure surfaces and the t-axis aligned with the streamlines; and (2) propagation of seismic front waves in the subsurface also is simplified if one chooses a 3-D curvilinear coordinate system (u, v, t) where (u, v) matches the seismic front with t corresponding to the ray paths (see, e.g., Mallet 2002).

1.4 Use of Surfaces

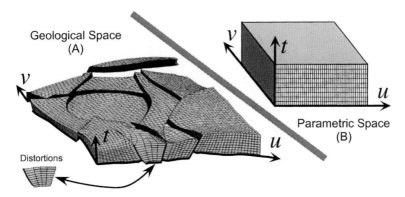

Fig. 1.9 Example of decomposition of a reservoir into a set of hexagonal adjacent 3-D cells adapted to the specific needs of the flow simulator. The edges of these cells cross neither the horizons nor the faults and are aligned to constitute a 3-D curvilinear coordinate system (u, v, t). There are distortions of the lengths of horizontal edges between the top and bottom cells of the reservoir (**a**) Geological space is transformed into (**b**) Parametric space (Source: Mallet 2004, Fig. 1)

As illustrated in Fig. 1.9a, the first generation of flow simulators used in reservoir engineering is based on a decomposition of the subsurface into a set of adjoining hexahedral cells whose edges nowhere cross the horizons and the faults. These cells can be aligned to generate a stratigraphic grid whose edges induce a curvilinear coordinate system (u, v, t) with t oriented in the vertical direction and (u, v) parallel to the bedding (Mallet 2002). As a result, each point in the subsurface has an image in the (u, v, t) parametric domain and images of the nodes of the stratigraphic grid in this parametric space constitute a rectangular grid where bedding is horizontal and not faulted as illustrated in Fig. 1.9b. The curvilinear coordinates account for the shapes of the horizons which themselves controlled geological continuity. However, this original geomodeling approach had two drawbacks: (1) Distortions of horizontal distances could not be avoided when faults are oblique relative to the horizons; as shown in Fig. 1.9a: a pair of faults with V shape in the vertical direction can generate significant distortions of horizontal cell sizes increasing from top to bottom of a reservoir; and (2) the new generation of flow simulators uses unstructured grids based on decomposition of the subsurface into polyhedral cells that cannot be used to compute curvilinear distances.

Mallet (2004) later developed a mathematical "geo-chronological" (GeoChron) model in which the original G-space is replaced by a $\overline{G}$-space with the property that distortions and difficulties related to the curvilinear coordinate (u, v, t) system downward from the bedding planes do not occur (Fig. 1.10). Mallet's (2004) GeoChron model in the $\overline{G}$-space improves upon the earlier G-based approach by eliminating its inherent drawbacks. In $\overline{G}$-space the rectangular (Cartesian) coordinate (x, y, z) system can be used allowing, for example, standard geostatistical estimation of rock properties. In G-space the (x, y, z) system only applies in the immediate vicinity of the folded and unfaulted horizons but in $\overline{G}$-space any curvature-related distortions do not exist. Mallet (2004) uses the expression

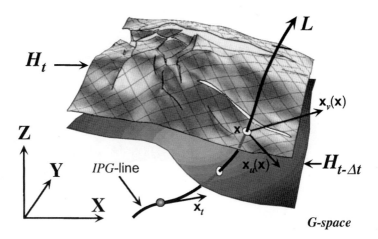

Fig. 1.10 Intrinsic constraints are controlled by the geometry of the horizons and IPG-lines in the G-space (Note that the u-lines and v-lines are contained in H_t and are tangent to x_u and x_v) (Source: Mallet 2004, Fig. 7)

"Iso-Paleo-Geographic" (IPG) line for a t-line represented in G-space (Fig. 1.10). The surface H_t in Fig. 1.10 was horizontal in G-space but has become curved in $\overline{G}$-space. The (u, v) coordinates have become curved u-lines and v-lines in $\overline{G}$-space that are tangent to the components x_u and x_v of the 3-D location vector x, respectively. The x_t component is tangent to the IPG-line also called L in Fig. 1.10.

Mallet (2002, 2004) and colleagues (see e.g., Caumon 2010) have developed powerful methods for the analysis of sedimentary rocks that are both folded and faulted. Most of these techniques have been incorporated in the Gocad software package (http://www.gocad.org) that is used worldwide, especially by oil companies. Other methods for modeling sedimentary systems are discussed in Harff et al. (1999).

1.5 Image Analysis

The theory of textural analysis which deals with the size, shape, orientation and spatial distribution of objects was advanced significantly by Matheron (1975). Serra (1976) implemented these methods on a texture analyzer with hexagonal logic for the study patterns in the plane. Sagar (2013) has published a book on mathematical morphology with many practical applications to systematically analyze the great variety of features observed at the surface of the Earth. This includes use of digital elevation models (DEMs) and digital bathymetric maps (DBMs). The study of shapes and sizes of objects and their interrelationships has become paramount in Geoinformation Science (GISci) that is a new flourishing field of scientific endeavor (*cf.* Wu et al. 2005).

1.5 Image Analysis

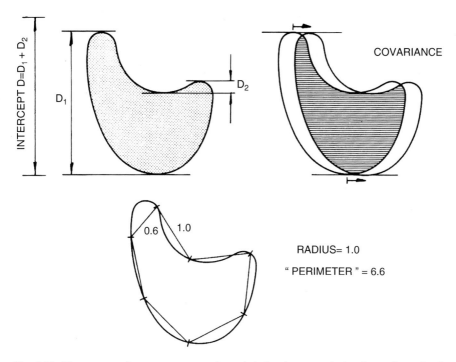

Fig. 1.11 Three types of measurements performed during image analysis of two-dimensional objects (Source: Agterberg 1980, Fig. 2)

1.5.1 Geometrical Covariance, Intercept and Rose Diagram

Two basic concepts of image analysis illustrated in Fig. 1.11 are the intercept D_α for direction α and the geometrical autocovariance (or covariance) $K_\alpha(x)$ for direction α and displacement x. In practice, the intercept of an image which consists of black and white picture points is measured by counting the number of times a black picture point is adjacent to a white picture point in a given direction. The geometrical covariance is obtained by shifting the pattern with respect to itself and measuring the area of overlap after each displacement. The area of any original or derived pattern is determined by counting the number of picture points it contains. For zero displacement, the covariance simply measures the area A which is independent of direction, or $K_\alpha(0) = A$. The change in area which arises from an infinitely small displacement dx in direction α yield the intercept because

$$D_\alpha = \left[\frac{dK_\alpha(x)}{dx}\right]_{x=0}$$

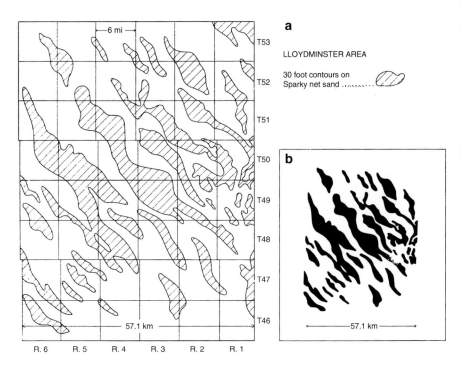

Fig. 1.12 (a). Contour map of Sparky sandstone thickness after Burnett and Adams (1977). (b). Pattern used as input for Quantimet 720 (After Agterberg 1980, Fig. 3)

In practice, the intercept in direction α can be determined by subtracting from A the covariance for a single picture point in direction α.

A practical example is as follows: Fig. 1.12a was taken from a study by Burnett and Adams (1977) concerned with the Lloydminster Sparky Pool, Alberta. The Sparky sand accumulations in this area are northwest-southeast trending bodies with maximum thickness of 9–16 m. Most commonly the Sparky unit consists of an upper and lower sand separated by a shale bed 1–3 m thick. The 30 feet (9.14 m) contour shown in Fig. 1.12a does not include the thickness of this intermediate shale bed. Comparison of the pattern to present-day features in the North Sea and elsewhere suggests that the contours delineate tidal-current ridges formed parallel to the tidal current. Oil production in the Lloydminster area of east-central Alberta and adjacent Saskatchewan is from sands in the Mannville Group of Early Cretaceous (Albian) age. The middle Mannville Sparky sandstone is the main oil producing horizon. Originally, high viscosity coupled with fine-grained unconsolidated nature of the sand kept primary production relatively low.

The black and white pattern of Fig. 1.12b was extracted from Fig. 1.12a and used as input for the Quantimet 720 textural analyzer in order to determine the intercept, a smoothed version of which is shown in Fig. 1.13a. In Agterberg (1979b) an algorithm is presented to estimate the frequency values of rose diagrams for

1.5 Image Analysis

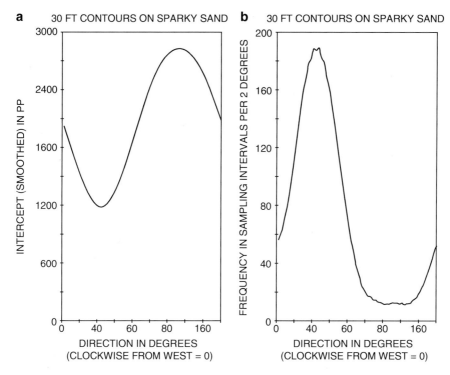

Fig. 1.13 Smoothed intercept and corresponding rose diagram for Fig. 1.12b (Source: Agterberg 1980, Fig. 5)

boundaries of map features. The input for this FORTRAN program consists of intercept measurements which are smoothed for any direction α by replacing them by values on successive parabolas each fitted by least squares to all intercepts contained in the sector $\alpha + \beta$ where β is a search angle set equal to 30° to derive Fig. 1.13a. From each smoothed value D_α and the corresponding second derivative of D_α with respect to α (D_α'') which is also obtained from the fitted parabola, it is possible to compute a frequency value (ΔS_α) of the rose diagram by using A-M. Legendre's formula $\Delta S_\alpha = (D_\alpha + D_\alpha'') \cdot \Delta \alpha$ (cf. Agterberg 1979b) where $\Delta \alpha$ is a small angle set equal to 2° for the example shown in Fig. 1.13b.

A result similar to Fig. 1.13b could have been obtained by the more laborious method of approximating the contours by many very short straight-line segments and representing these straight-line segments in a rose diagram using the method commonly applied for the treatment of vectorial data and lineaments.

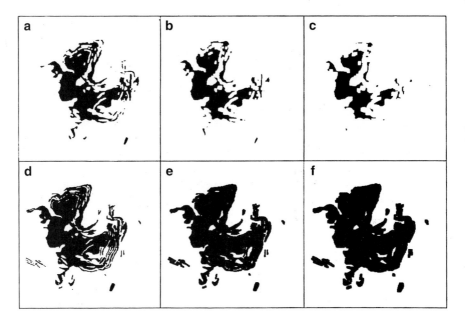

Fig. 1.14 Erosions and dilatations of pattern of acidic volcanics in Bathurst area, New Brunswick. Original pattern A is shown in Fig. 1.14d. Writing B for the operator set of eight-neighbor square logic, these patterns are: (**a**) $A \ominus B$; (**b**) $A \ominus 2B$; (**c**) $A \ominus 3B$; (**d**) A; (**e**) $A \oplus B$; (**f**) $A \oplus 2B$. Dimensions of frame are 84 km × 84 km. North direction points upward (Source: Agterberg and Fabbri 1978, Fig. 1)

1.5.2 Minkowski Operations: Bathurst Acidic Volcanics Example

In mathematical morphology (Serra 1976; Watson 1975; Sagar 2013) various other operations can be performed on black and white patterns, in addition to those described in the previous section. An example which involves pattern erosions and dilatations is as follows: The geology of the Bathurst area in New Brunswick has been described by Skinner (1974). This area contains volcanogenic massive sulphide deposits that are related genetically to the occurrence of acidic volcanics of the Tetagouche Group of Middle-Late Ordovician age. These acidic volcanics were coded from 2-mi geological maps (scale approximately 1:125,000) for an experimental data base described in Fabbri et al. (1975). This pattern was also quantified on a Flying Spot Scanner at the National Research Council of Canada in Ottawa as a set of 18,843 black pixels on a square grid with in total 324 × 320 binary (black or white) picture points spaced 259 m apart in the north-south and east-west directions (Agterberg and Fabbri 1978). The resulting binary image is shown in Fig. 1.14d.

According to Skinner (1974), the stratigraphy and structure of the Tetagouche Group had not been determined and, originally, mapping was based on lithological units without stratigraphic significance. These units were characterized by

1.5 Image Analysis

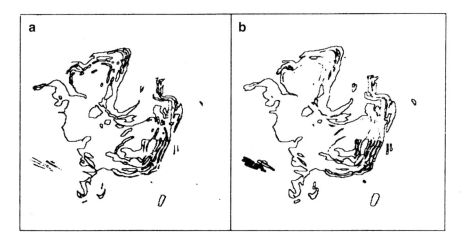

Fig. 1.15 (a) Black picture points removed from original pattern after first erosion; (b) and those added to it after first dilatation (Source: Agterberg and Fabbri 1978, Fig. 2)

(1) sedimentary rocks, (2) metabasalt, and (3) rhyolitic rock. Only the rhyolitic rock is shown in Fig. 1.14d. It comprises rhyolite tuff, augen schist, rhyolite crystal tuff and quartz-sericite schist, rhyolite, minor phyllite and granophyre. Skinner (1974, p. 15) said: "A conspicuous feature of the map of the Tetagouche Group in the Bathurst-Newcastle area is the C-shaped area of rhyolitic rock surrounded by metabasalt and sedimentary rock. The rhyolitic core has been referred to by some geologists as a basin structure, and by others, as a dome. The writer believes the C-shape is the result of two periods of folding. Apparently the Tetagouche Group was folded into northwesterly trending recumbent folds overturned toward the southwest during the late Ordovician Taconic Orogeny, then refolded about northeasterly trending axis during the Devonian Acadian Orogeny. If this is so, the rhyolitic core (map-unit 3) is the youngest part of the group and the surrounding sedimentary rock (map-unit 1) is the oldest." The acidic volcanics underlie an area of about 1,259 km^2 but probably covered an area several times larger when originally deposited. Skinner (1974, p. 28) suggested an ignimbritic (pyroclastic flow) origin for most of these rocks.

Every picture point or pixel on a square grid or raster can be accessed individually. The eight pixels around any black pixel belonging to the binary image of Fig. 1.14d are either white or black. Suppose that they are changed into black pixels if they are white. This operation is termed dilatation by eight neighbor square logic. The result is a new pattern with 23,976 pixels (see Fig. 1.14e). The difference between Fig. 1.14d and e consists of 23,976–18,843 = 5,133 pixels shown separately in Fig. 1.15b. A second dilatation gives the pattern of Fig. 1.14f. The reverse process which consists of replacing black pixels that surround white pixels by white ones is called "erosion". Three successive erosions of the original pattern (Fig. 1.14d) result in Fig. 1.14c, 1.14b and 1.14a, respectively. The black pixels lost during the first erosion (from Fig. 1.14d to 1.14c) are shown separately in Fig. 1.15a.

In order to continue discussion of these operations it is convenient to further adopt terminology as developed by Serra (1976) and Watson (1975). Suppose that the original

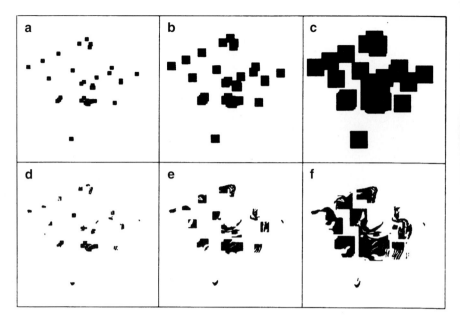

Fig. 1.16 Dilatations of set C for 40 deposit points and intersection of resulting sets with original pattern: (**a**) $C \oplus 4B$; (**b**) $C \oplus 9B$; (**c**) $C \oplus 19B$; (**d**) $A \cap (C \oplus 4B)$; (**e**) $A \cap (C \oplus 9B)$; (**f**) $A \cap (C \oplus 19B)$ (Source: Agterberg and Fabbri 1978, Fig. 3)

pattern of Fig. 1.14d is set A with measure mes A. This measure is the area and can be expressed either as mes $A = 18,843$ pixels or mes $A = 1,264.01$ km^2, because one pixel represents an area of 259 m $\times$ 259 m. Let B be the operator set of the eight-neighbor square logic. B has an origin which is located in the center of the square described by the eight neighboring points. The patterns of Fig. 1.14e, f can be represented as the Minkowski sums $A \oplus B$ and $A \oplus 2B$, respectively. A new set nB can be defined by induction with $nB = [(n-1)B] \oplus B$ for $n = 2,3, \ldots$. It is seen readily that operating on A with the set nB is identical to applying the successive operations $A \oplus nB = [A \oplus (n-1)B] \oplus B$ for $n = 2,3, \ldots$. By using the concept of Minkowski subtraction, the patterns of Fig. 1.14a–c can be written as $A \ominus B$, $A \ominus 2B$; and $A \ominus 3B$, respectively.

If the superscript c denotes complement of a set with respect to the universal set T which consists of all pixels in use, then the patterns of Fig. 1.15a, b are $A \cap (A \oplus B)^c$ and $A \oplus B \cap A^c$, respectively. A set C was formed by assigning each of the 40 massive sulphide deposits in the area to the pixel closest to it on the grid with 259 m-spacing used for the binary images of Figs. 1.14 and 1.15. C consists of 40 black pixels which can be subjected to successive dilatations by use of B. The sets $C \oplus 4B$, $C \oplus 9B$ and $C \oplus 19B$ are shown in Fig. 1.16a–c, respectively. Because each pixel is representative for a cell of 259 m on a side, the length of a cell generated by n dilatations is equal to $(2n + 1) \times 259$ m. Hence the cells obtained by 4, 9 and 19 dilatations of a single pixel are 2.33, 4.92, and 10.10 km, respectively. The latter two cell sides can be used to approximate (5 km $\times$ 5 km) cells and (10 km $\times$ 10 km) cells, respectively. The patterns of Fig. 1.16a–c can be intersected with that of Fig. 1.14d. The resulting sets are shown in Fig. 1.16d–f, respectively.

1.5 Image Analysis

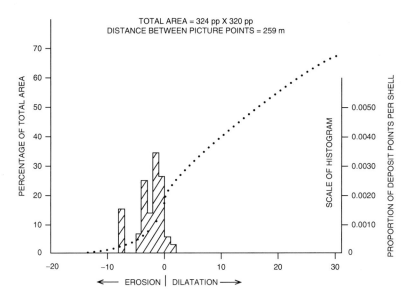

Fig. 1.17 Percentage of total area occupied by patterns obtained after successive erosions and dilatations. Histogram indicates probability that deposit point coincides with arbitrary pixel in shell added after single dilatation or removed after single erosion (Source: Agterberg and Fabbri 1978, Fig. 4)

The erosion of Fig. 1.14d can be continued onward until not a single black point remains. Likewise, the dilatations can be continued from Fig. 1.14e, f onward until most or all of the study area (set T) consists of black pixels. The relative areas of these erosions and dilatations are shown in Fig. 1.17. For dilatations each relative area can be interpreted as the probability $P(nB)$ with

$$P(nb) = \frac{\text{mes } A \oplus nB}{\text{mes } T}, \quad n = 1, 2, \ldots$$

that a random cell with side $(2n+1) \times 259$ m contains one or more black pixels belonging to the original pattern (Fig. 1.14d). The probability $Q(nB)$ that a cell with size mes $C \oplus nB$ contains no acidic volcanics is equal to $Q(nB) = 1 - P(nB)$. Likewise it is possible to measure the probability $P_d(nB)$ that a cell with side $(2n+1) \times 259$ m is a deposit cell containing one or more deposits (see Fig. 1.18) because

$$P_d(nb) = \frac{\text{mes } C \oplus nB}{\text{mes } T}, \quad n = 1, 2, \ldots$$

Another practical result is as follows. A correlation between sets A and C can be carried out by determining how many deposit points are contained in the separate shells added to, or subtracted from the original pattern (Fig. 1.14d) by dilatation or erosion. The original pattern itself contains 36 deposit points or mes $A \cap C = 36$ pixels. The pattern of Fig. 1.15a consists of 5,360 pixels and contains

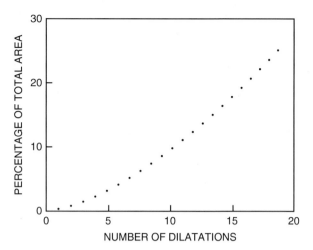

Fig. 1.18 Probability that random square cell measuring $(2n+1) \times 259$ m on a side contains one or more deposits (n denotes number of dilatations of set C) (Source: Agterberg and Fabbri 1978, Fig. 5)

14 deposit points. Hence the probability that an arbitrary pixel in this shell is a deposit point is equal to $14/5{,}360 = 0.00261$. This probability is one of the proportions for separate shells shown in the histogram of Fig. 1.17. The pattern of Fig. 1.14b consists of 9,990 pixels and measures mes $(A \ominus 2B) \cap C = 10$ pixels. This indicates that $36-10 = 26$ of the 40 deposits (or 65 %) occur in the zone identified as acidic volcanic rocks on the geological map and within $(2\sqrt{2} \times 259$ m $=) 733$ m from a contact between acidic volcanics and other rocks on this map. This zone may be favorable relatively for the occurrence of volcanogenic massive sulphide deposits. The probability that a random point in the zone is a deposit point amounts to $26/(18{,}843-9{,}990) = 0.00294$. This is about eight times greater than the probability ($= 0.00039$) that a random point in the entire study area is a deposit point. On the other hand, it is only about 1.5 times greater than the probability ($= 0.00191$) that an arbitrary black pixel of the original pattern (Fig. 1.14d) is a deposit point.

As mentioned before, 36 of the 40 deposit points (or 90 %) coincide with the acidic volcanics of Fig. 1.14d. A generalized form of this ratio for a cell with side $(2n+1) \times 259$ m is:

$$M_{d1} = \frac{\text{mes } A \cap (C \oplus nB)}{\text{mes } C \oplus nB}$$

For the patterns shown in Fig. 1.16, M_{d1} amounts to 0.646 (Fig. 1.16d), 0.537 (Fig. 1.16e), and 0.456 (Fig. 1.16f). It represents a weighted average proportion of acidic volcanics per cell for cells centered about the deposits. The preceding probabilities and ratios follow directly from Minkowski operations on sets. Other problems cannot be solved by measurement only but need a combination of measurement and statistical modeling. An example of such a problem is the determination of the frequency distribution of the random variable X with

1.5 Image Analysis

$$X = \frac{\text{mes } A \cap (R \oplus nB)}{\text{mes } R \oplus nB}$$

where R is a set consisting of a random point in the study area (T). Problems of this type will be solved later in this book (Chap. 14).

The Minkowski operations discussed in this section have been implemented in various Geographic Information Systems (GIS's) and are useful in practical applications. It should, however, be kept in mind that these operations are basically linear. For example, they cannot be used for nonlinear dilatations or erosions as might be desirable if the curve representing the contact between two rock types is the intersection between a curved surface and the topographic surface and one would wish to account for strike/dip of the contact. In such situations, more flexible methods are required along the lines of those discussed in Sect. 1.4.

1.5.3 Boundaries and Edge Effects

Geoscience projects generally are conducted in a study area with a shape that is either rectangular or curved. Various statistical techniques applied to variables observed at points within the study area are subject to edge effects in the vicinity of the boundaries of the study area. Such edge effects arise when observations are used for extrapolations into their immediate neighborhoods. In a simpler situation this kind of problem occurs in 1D as well; for example, in time series analysis edge effects can occur at the beginning and end of a series of observations. In 2-D and 3-D applications, edge effects generally present a more serious problem because 'relatively' many more data points occur near study area boundaries and a relatively simple 1D method such as reflection of a series around its end points in order to obtain extra observations with locations outside the range of observation cannot be used.

In GIS applications, pixels used for representation of attributes are situated on a regular grid. Curved lines on maps also can be represented in vector mode meaning that they are approximated by sequences of densely spaced points on the curves so that these are approximated precisely by strings of very short straight-line segments. In 2-D, edge effects are easier to avoid if the study area is rectangular in shape as it is frequently in remote sensing, geophysical and regional geochemical applications because the data then are averages in which abrupt changes such as those related to contacts between different rock types have been masked out. If map data are averaged in such applications, it is often possible to keep the unit areas for which values are averaged within the boundaries of the study area. In geological map applications, however, the boundaries of study areas often are curvilinear in shape. Examples are the "plays" often used in the oil industry and delineations of permissive areas for occurrence of different types of mineral deposits in economic geology.

Ripley's (1976) estimator $K(r)$ can be used as an example of how undesirable edge effects can be avoided in a map-based statistical application. Statistical theory of spatial point processes is used mainly for the study of patterns of points in the plane (see, e.g., Diggle 1983). The first-order property of a point process is its intensity (λ) which is independent of location for stationary processes. It is estimated by dividing the total number (n) of points within the study area (A) by its total area, or Ave $(\lambda) = n/|A|$. The second-order property of an isotropic, stationary spatial point process can be described by the function $K(r)$ which is proportional to the expected number of points within distance r of an arbitrary point. If the points are distributed randomly according to a Poisson process, the expected number of points within distance r of an arbitrary point is equal to $\pi\lambda r^2$. This model provides a convenient benchmark for complete randomness tests. Clustering of points occurs if there are more points in the vicinity than predicted by the Poisson model; anti-clustering or "regularity" arises if there are fewer points than expected in the vicinity of an arbitrary point. For graphical representation, the function $L(r) = [K(r)/\pi]^{0.5}$ may be used. The Poisson model simply gives $L(r) = r$.

Estimators of $K(r)$ should account for edge effects related to the shape of the study area. There are two reasons why edge effects are important: (1) probability of occurrence of a point cannot be measured directly at a point but only indirectly with respect to neighborhoods that are more strongly affected by boundaries than single points, and (2) the proportion of study area located within distance r from the boundary tends to be large even for small r. It may be possible to work with a guard area inside the boundary of the study area in order to obtain an approximately unbiased estimator but, in general, too much information is lost by doing this.

Ripley's estimator $K(r)$ is applicable to regions that have relatively simple geometrical shapes (e.g., rectangles or circles) or for regions with irregular shapes bounded by polygons. Digitizing the boundary of an irregularly shaped study region in vector mode generally results in a polygon with many sides which are so short that the boundary cannot be distinguished from a smooth curve when it is replotted on the map. An irregular boundary can be compared with a coastline with fjords and peninsulas. An algorithm for obtaining Ripley's estimator originally was developed by Rowlingson and Diggle (1991, 1993). The algorithm of Agterberg (1994) also can be used when there are islands, more than a single enclosed study area, or even lakes within islands. The underlying geometrical rationale is as follows.

Suppose that the second-order properties of an isotropic, stationary point process are characterized by the function $K(r) = \lambda^{-1} E$ [number of further events within distance r of an arbitrary event] where E denotes mathematical expectation. It follows that the expected number of ordered pairs of events within distance r from each other is $\lambda^2 |A| K(r)$ if the first event of each ordered pair falls in area A. Suppose that r_{ij} denotes distance between events i and j in A, and that $I_r(r_{ij})$ is an indicator function assuming the value 1 if $r_{ij} < r$; 0 otherwise, then the observed number of these ordered pairs is $\Sigma_i \Sigma_j I_r (r_{ij})$, $i \neq j$ where the double sum denotes summation over both i and j. This summation excludes pairs of events for which the second event is outside of A. Let the weight w_{ij} represent the proportion of the circumference of the circle around event i with radius r_{ij} that lies within A. Then, as

1.5 Image Analysis

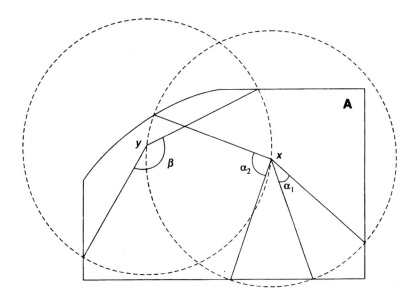

Fig. 1.19 Ripley's (1976) estimator illustrated for region A with events at points x and y. Radius of both circles is equal to $\|x - y\|$; $w_{xy} = (\alpha_1 + \alpha_2)/2\pi$; $w_{yx} = \beta/2\pi$ (Source: Agterberg 1994, Fig. 1)

originally pointed out by Ripley (1976), w_{ij} represents the conditional probability that an event is observed when it is distance r_{ij} away from the i-th event. In general, w_{ij} is not equal to w_{ji} as illustrated in Fig. 1.19. If λ is replaced by the observed intensity $n/|A|$ where n is total number of events in A, then Ripley's estimator for $K(r)$ is obtained with

$$\hat{K}(r) = n^{-2}|A|\sum_{i \neq j} w_{ij}^{-1} I_i(r_{ij})$$

This expression is only valid if r is sufficiently small because for large r the weights may become unbounded. The condition that r should not exceed the radius of the smallest circle about an arbitrary point x that does not intersect the circumference of A. Stoyan et al. (1987, p. 125) discuss the problem of bias for large r and give a method by which it can be avoided. Diggle (1983) argues that the restriction on Ripley's original estimator does not present a serious problem in practice because the dimensions of the region A are generally larger than the distances for which $K(r)$ is of interest. For example, when A is the unit square, r should not exceed $2^{-0.5}$. For larger distances between events, the sampling fluctuations will increase significantly.

As pointed out before, in many types of geological applications, the region A does not have a simple shape. Events (e.g., oil wells or mineral deposits) may only occur within an environment type bounded in 2-D by discontinuities such as intrusive contacts, faults, unconformities or facies changes. An algorithm for polygonal A was programmed originally in the SPLANCS package of Rowlinson

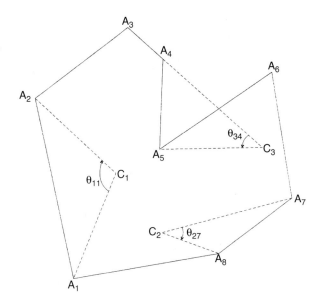

Fig. 1.20 Artificial example of polygon bounding region with events at points C_i ($i = 1, 2, 3$) (Source: Agterberg 1994, Fig. 3)

and Diggle (1991, 1993). In it the study region is approximated by a concave or self-intersecting polygon with m sides. Each circle through two points (cf. Fig. 11.19) intersects this polygon in $2k \leq 2m$ sides. Once these points of intersection have been determined, it is relatively easy to calculate the value of w_{ij} for each of the $n \cdot (n-1)$ circles. This result also can be obtained by successively determining for each side whether it has 0, 1 or 2 points of intersection with one of the circles.

An artificial example with relatively few events is given in Fig. 1.20 for clarification. It is assumed that the n events occur at points labeled C_i ($i = 1, 2, \ldots, n$). The distance between two events satisfies $r_{ij} = |C_i C_j| = |C_j C_i|$. The vertices of the polygon are ordered moving in the clockwise direction. Each point C_i can be related to each side $A_k A_{k+1}$ by a triangle. The angle opposite the side will be written as $\theta_{ik} = \angle A_k C_i A_{k+1}$. This angle is defined to be positive if the entire triangle or a portion of it bounded by the side belongs to the study area; negative otherwise. The sign of the angle will be written as $s(\theta_{ik}) = \theta_{ik}/|\theta_{ik}|$. Two positive angles ($\theta_{11}, \theta_{27}$) and one negative angle (θ_{24}) are shown in Fig. 1.20 for example. In practice, the angle $\theta_{ik} = \angle A_k C_i A_{k+1}$ can be determined as the difference between two angles measured in the clockwise direction from a given direction. Note that $\theta_{ik} < \pi$ ($i = 1, 2, \ldots, n; k = 1, 2, \ldots, m$) because events occur inside (and not on) the boundary, and $\Sigma_k \theta_{ik} = 2\pi$ ($i = 1, 2, \ldots, n$) because negative angles, if they occur, are cancelled out by a surplus of positive angles.

It is convenient to rewrite A_{k+1} as B_k for $k = 1, 2, \ldots, m-1$ setting $B_m = A_1$. Each triangle then has three sides that can be written as $a_{ik} = |C_i B_k|$, $b_{ik} = |C_i A_k|$, and as $c_{ik} = |A_i B_k|$. For individual triangles, the double subscripts can be deleted without creating confusion because each triangle is determined fully by its three corner

1.5 Image Analysis

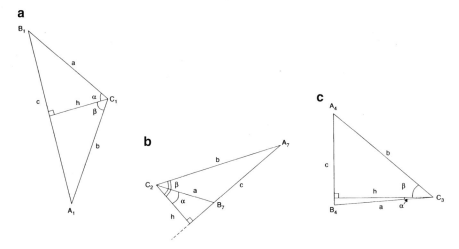

Fig. 1.21 Illustration of definitions of sides, height and top angles for three triangles taken from artificial example of Fig. 1.20. Event points at (**a**) C_1, (**b**) C_2 and (**c**) C_3, respectively (Source: Agterberg 1994, Fig. 5)

Table 1.1 Formulae for ω_{ijk} if $h_{ik} < r_{ij} \leq \max(a,b)$ with further conditions as given in row and column headings

	$|\theta_{ik}| = \alpha_{ik} + \beta_{ik}$	$|\theta_{ik}| < \alpha_{ik} + \beta_{ik}$
$a_{ik} < r_{ik} \leq b_{ik}$	$\omega_{ijk} = (\gamma_{ijk} + \alpha_{ik})s_{ik}$	$\omega_{ijk} = (\gamma_{ijk} - \alpha_{ik})s_{ik}$
$b_{ik} < r_{ik} \leq a_{ik}$	$\omega_{ijk} = (\gamma_{ijk} + \beta_{ik})s_{ik}$	$\omega_{ijk} = (\gamma_{ijk} - \beta_{ik})s_{ik}$
$h_{ij} < r_{ij} \leq \min(a,b)$	$\omega_{ijk} = 2\gamma_{ijk}s_{ik}$	$\omega_{ijk} = 0$

Latter distinguish between triangles with acute angles only (*cf.* Fig. 1.20a, c) and triangles with acute angle along their base (*cf.* Fig. 1.20b); s_{ik} is sign of θ_{ik} (Source: Agterberg 1994, Table 1)

points (see Fig. 1.21). According to elementary trigonometry, the shortest distance (h_{ik}) between C_i and he line through A_k and B_k satisfies:

$$h_{ik} = \frac{2}{a_{ik}b_{ik}}\sqrt{s_{ik}(s_{ik} - a_{ik})(s_{ik} - b_{ik})(s_{ik} - c_{ik})}; s_{ik} = \frac{a_{ik} + b_{ik} + c_{ik}}{2}$$

The following acute angles can be defined for each triangle:

$$\alpha_{ik} = \cos^{-1}\frac{h_{ik}}{a_{ik}}, \beta_{ik} = \cos^{-1}\frac{h_{ik}}{b_{ik}}, \gamma_{ik} = \cos^{-1}\frac{h_{ik}}{r_{ik}}$$

For a given triangle (with C_i, A_k and B_k) only those circles around C_i with radius r_{ij} that intersect the side of the polygon at a point between A_k and B_k are used for the calculation of the weight w_{ij}. Let ω_{ijk} represent the portion of θ_{ik} contributing to the reduction of w_{ij} $1-\Sigma_k \omega_{ijk}/2\pi$ (see Fig. 1.22 for examples). A set of rules is given in Table 1.1. The relative importance of these rules decreases when the sides of the

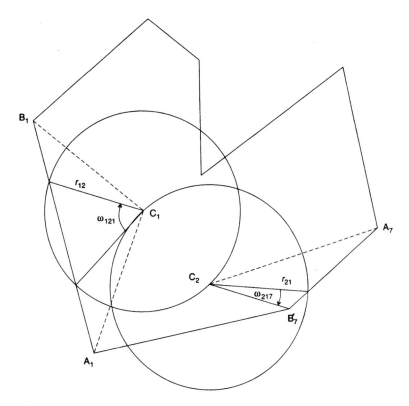

Fig. 1.22 Examples of ω_{ijk}: $\omega_{121} = 2\gamma_{121}$, and $\omega_{127} = \gamma_{127} - \alpha_{27}$ (*cf*. Fig. 1.20 and Table 1.1) (Source: Agterberg 1994, Fig. 6)

polygon become short. In that situation, the solution is almost entirely determined by the relation

$$r_{ij} \leq h_{ik} \rightarrow \omega_{ijk} = 0; r_{ij} > \omega_{ijk} = \theta_{ik}$$

An example of application is given in Figs. 1.23 and 1.24. It is concerned with the Leduc Reef Complex-Windfall Play in Alberta (*cf*. Reinson et al. 1993; Kaufman and Lee 1992). This play has 249 wildcats including 52 gas discovery wells. The play boundary was digitized in vector mode yielding a data set with coordinates for 249 vertices (Agterberg 1994). The area enclosed by the polygon is 12,861 km^2 (15,616 km^2 if the smaller polygon in the center is included in the play). Figure 1.24 shows the function $L_2(r)$ which is for the 52 gas discovery wells only. This function deviates significantly from a straight line through the origin that dips 45° representing complete random distribution of the gas wells within the play area. Obviously they are strongly clustered. This is brought out by the 95° confidence belt for the Poisson model that would probably contain the entire $L_2(r)$ curve if the spatial random distribution model would be satisfied.

1.5 Image Analysis

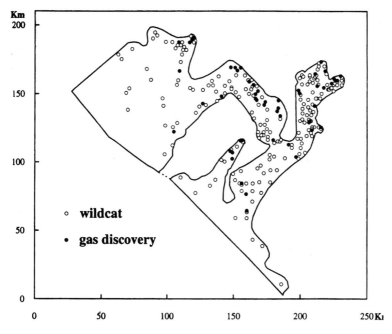

Fig. 1.23 Wildcat locations in Leduc Reef Complex (*cf.* Reinson et al. 1993; Kaufman and Lee 1992). In total, there are 52 gas occurrences among 249 wildcat wells. Boundary of play area of 12,861 km^2 was approximated by polygon with 264 vertices (Source: Agterberg 1994, Fig. 11)

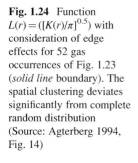

Fig. 1.24 Function $L(r) = ([K(r)/\pi]^{0.5})$ with consideration of edge effects for 52 gas occurrences of Fig. 1.23 (*solid line* boundary). The spatial clustering deviates significantly from complete random distribution (Source: Agterberg 1994, Fig. 14)

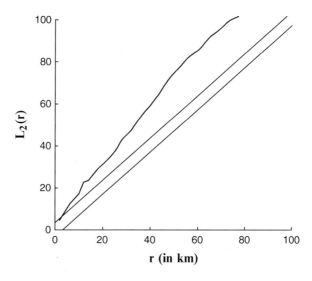

References

Agterberg FP (1961) Tectonics of the crystalline basement of the dolomites in North Italy. Kemink, Utrecht
Agterberg FP (1974) Geomathematics. Elsevier, Amsterdam
Agterberg FP (1979a) Statistics applied to facts and concepts in geoscience. Geol Mijnb 58:201–208
Agterberg F (1979b) Algorithm to estimate the frequency values of rose diagrams for boundaries of map features. Comput Geosci 5:215–230
Agterberg FP (1980) Mineral resource estimation and statistical exploration. In: Miall AD (ed) Facts and principles of world oil occurrence. Can Soc Petr Geol Mem 6, Calgary, pp 301–318
Agterberg FP (1994) FORTRAN program for the analysis of point patterns with correction for edge effects. Comput Geosci 20:229–245
Agterberg FP (2013) At the interface between mathematical geoscience and mathematical statistics. In: Pardo-Igúzquiza E, Guardiola-Albert C, Heredia J, Moreno-Merino L, Durán JJ, Vargas-Guzmán JA (eds) Mathematics of planet earth. Springer, Heidelberg, pp 19–22
Agterberg FP, Fabbri AG (1978) Spatial correlation of stratigraphic units quantified from geological maps. Comput Geosci 4:284–294
Agterberg FP, Robinson SC (1972) Mathematical problems in geology. In: Proceedings of 38th session of the International Statistical Institute, Bulletin 38, pp 567–596
Aitchison J (1986) The statistical analysis of compositional data. Chapman & Hall, London
Bonham-Carter GF (1994) Geographic information systems for geoscientists: modelling with GIS. Pergamon, Oxford
Buhmann JH (2003) Radial basis functions: theory and implementations. Cambridge University Press, Cambridge
Burnett AI, Adams KC (1977) A geological, engineering and economic study of a portion of the Lloydminster Sparky Pool, Lloydminster, Alberta. Bull Can Pet Geol 25(2):341–366
Caumon G (2010) Towards stochastic time-varying geological modeling. Math Geom 42(5):555–569
Chamberlin TC (1897) The method of multiple working hypotheses. J Geol 5:837–848
Chamberlin TC (1899) Lord Kelvin's address on the age of the Earth as an abode fitted for life. Science 9:889–901; 10:11–18
Chayes F (1956) Petrographic modal analysis. Wiley, New York
Chayes F (1971) Ratio correlation. University of Chicago Press, Chicago
Cressie NAC (1991) Statistics for spatial data. Wiley, New York
Croll J (1875) Climate and time in their geological relations. Appleton, New York
De Finetti B (1972) Probability, induction and statistics. Wiley, New York
De Kemp E, St-Onge MR (2007) 3D model basement-cover contact, Central Baffin Island, Nunavut. In: St-Onge MR, Ford A, Henderson I (eds) Digital geoscience atlas of Baffin Island (South of 70°N and East of 80°W), Nunavut; Geological Survey of Canada, Open File 5116
Diggle PJ (1983) Statistical analysis of spatial point patterns. Academic, London
Egozcue JJ, Pawlowsky-Glahn V, Mateu-Figueras G, Barceló-Vidal C (2003) Isometric logratio transformations for compositional data analysis. Math Geol 35(3):279–300
Fabbri AG, Divi SR, Wong AS (1975) A data base for mineral potential estimation in the Appalachian Region of Canada. Geological Survey of Canada Paper 75-1, pp 121–132
Findlay DC, Smith CH (1965) The Muskox drilling project. Geological Survey of Canada Paper 64-44, p 170
Fisher RA (1953) Dispersion on a sphere. Proc R Soc Lond Ser A 217:295–305
Fisher RA (1954) Expansion of statistics. Am Sci 42:275–282
Fisher RA (1960) The design of experiments. Oliver and Boyd, Edinburgh
Freund Y, Shapire RE (1997) A decision-theoretical generalization of on-line learning and an application to boosting. J Comput Syst Sci 55(1):119–139

References

Friedman JH, Tukey JW (1974) A projection pursuit algorithm for exploratory data analysis. IEEE Trans Comput C 23(9):881–890

Gradstein FM, Ogg JG, Schmitz M, Ogg G (eds) (2012) The geologic time scale 2012, 2-volume set. Elsevier, Amsterdam

Grayson CI Jr (1960) Decisions under uncertainty (drilling decisions by oil and gas operators). Harvard Business School, Boston

Griffiths JC (1962) Statistical methods in sedimentary petrography. In: Milner HB (ed) Sedimentary petrography, 4th edn. MacMillan, New York, pp 565–672

Griffiths JC (1967) Scientific method in the analysis of sediments. McGraw-Hill, New York

Harff J, Lemke W, Stattegger K (eds) (1999) Computerised modeling of sedimentary systems. Springer, Heidelberg

Harrison JM (1963) Nature and significance of geological maps. In: Albitton CC Jr (ed) The fabric of geology. Addison-Wesley, Cambridge, MA

Heiskanen WA, Vening Meinesz FA (1958) The earth and its gravity field. McGraw-Hill, New York

Hillier M, de Kemp E, Schetselaar E (2013) 3D form line construction by structural field interpolation (SFI) of geologic strike and dip observations. J Struct Geol 51:167–179

Hinnov IA, Hilgen FJ (2012) Cyclostratigraphy and astrochronology. In: Gradstein FM, Ogg JG, Schmitz M, Ogg G (eds) The geologic time scale 2012. Elsevier, Amsterdam, pp 63–113

Huang J, Zhao B, Chen Y, Zhao PO (2010) Bidimensional empirical mode decomposition (BEMD) for extraction of gravity anomalies associated with gold mineralization in the Tongshi gold field, western Shandong uplifted block, eastern China. Comput Geosci 36:987–995

Kaufman GM, Lee J (1992) Are wildcat well outcomes dependent or independent? Nonrenew Resour Res 1(3):201–233

Kolmogorov A (1931) Ueber die analytischen Methoden in der Wahrscheinlichkeitsrechnung. Math Ann 104:415–458

Krumbein WC, Graybill FA (1965) An introduction to statistical models in geology. McGraw-Hill, New York

Loudon TV (1964) Computer analysis of orientation data in structural geology. Off Naval Res Geogr Branch Rep 13, pp 1–130

Lyell C (1833) Principles of geology. Murray, London

Mallet JL (2002) Geomodeling. Oxford University Press, New York

Mallet JL (2004) Space-time mathematical framework for sedimentary geology. Math Geol 36(1):1–32

Marlow AR (ed) (1978) Mathematical foundations of quantum theory. Academic, New York

Matheron G (1962) Traité de géostatistique appliquée, Mém BRGM 14. Éditions Technip, Paris

Matheron G (1965) Les variables régionalisées et leur estimation. Masson, Paris

Matheron G (1975) Random sets and integral geometry. Wiley, New York

Merriam DF (1981) Roots of quantitative geology, in down-to-earth statistics: solutions looking for geological problems. Syracuse Univ Geol Con 8:1–15

Merriam DF (2004) The quantification of geology: from abacus to Pentium. Earth-Sci Rev 67(1–2):55–89

Milankovitch M (1941) Kanon der Erdbestrahlung und eine Auswendung auf das Eiszeitproblem, vol 32, Royal Serbian Academy Special Publications. Mihaila Ćurčića, Belgrade

Nieuwenkamp W (1968) Natuurfilosofie en de geologie van Leopold von Buch. K Ned Akad Wet Proc Ser B 71(4):262–278

Ostwald (1910) Grosse Männer. Akad Verlagsgesellschaft, Leipzig

Popper K (2002) The logic of scientific discovery. Routledge, New York

Reinson GE, Lee PJ, Warters W, Osadetz KG, Bell LI, Price PR, Trollope, F, Campbell, RI, Barlclay JE (1993) Devonian gas resources of western Canada sedimentological basin – play definition and resource assessment. Geological Survey of Canada, Bulletin 452, pp 1–127

Ripley BD (1976) The second-order analysis of stationary point processes. J Appl Probab 13:255–266

Rowlinson BS, Diggle PJ (1991) Estimating the K-function for a univariate point process on an arbitrary polygon. Tech Report MA91/58, Department of Mathematics, Lancaster University

Rowlinson BS, Diggle PJ (1993) SPLANCS spatial point pattern analysis code in S-plus. Comput Geosci 19(5):627–955

Sagar BSD (2013) Mathematical morphology in geomorphology and GISci. CRC/Chapman Hall, Boca Raton

Scheidegger AE (1963) Principles of geodynamics. Academic, New York

Scheidegger AE (1964) The tectonic stress and tectonic motion direction in Europe and western Asia as calculated from earthquake fault plane solutions. Seismol Soc Am Bull 54:1519–1528

Schwarzacher W (1993) Cyclostratigraphy and the Milankovitch theory. Elsevier, Amsterdam

Serra J (1976) Lectures on image analysis by mathematical morphology. Centre Morphologie Math, Ec Supérieure Mines Paris Rept N-475

Skinner R (1974) Geology of Tetagouche Lakes, Bathurst, and Nepisiguit Falls, New Brunswick. Geological Survey of Canada Memoir 371

St-Onge MR, Scott DJ, Corrigan D, Wodika N (2005) Geological maps and descriptive notes and legend, Central Baffin Island area, Nunavut. Geological Survey of Canada Maps 2077A-2082A, scale 1:100,000

Staub R (1928) Bewegungsmechanismen der Erde. Borntraeger, Stuttgart

Stoyan D, Kendall WS, Mecke J (1987) Stochastic geometry and its applications. Wiley-Interscience, Chichester

Trooster G (1950) Fundamentele beschouwing van profielconstructies. K Ned Akad Wet Proc Ser B 53:913–918

Tukey JW (1962) The future of data analysis. Ann Math Soc 33:1–67

Tukey JW (1970) Some further inputs. In: Merriam DF (ed) Geostatistics. Plenum, New York

Tukey JW (1977) Exploratory data analysis. Addison-Wesley, Reading

Van Bemmelen RW (1960) Zur Mechanik der ostalpinen Deckenbildung. Geol Rundsch 50:474–499

Van Bemmelen RW (1961) The scientific nature of geology. J Geol 453–465

Van Bemmelen RW (1972) Geodynamic models. Elsevier, Amsterdam

Von Buch L (1842) Ueber Granit und Gneiss, vorzüglich in Hinsicht der äusseren Form, mit welcher diese Gebirgsarten auf Erdflache erscheinen. Abh K Akad Berlin IV/2 (18840):717–738

Vistelius AB (1967) Studies in mathematical geology. Consultants Bureau, New York

Vistelius AB, Janovskaya TB (1967) The programming of geological and geochemical problems for all-purpose electronic computers. In: Vistelius AB (ed) Studies in mathematical geology. Consultants Bureau, New York, pp 29–45

Watson GS (1966) The statistics of orientation data. J Geol 74:786–797

Watson GS (1975) Texture analysis. Geol Soc Am Mem 142:367–391

Wegener A (1966) The origin of continents and oceans, 4th edn. Dover, New York

Wegmann E (1958) Das Erbe Werner's und Hutton's. Geologie 7:531–559

Westergård AH, Johansson S, Sundius N (1943) Beskrivning till Kartbladet Lidköping. Sver Geol Unders Ser Aa 182

Winchester S (2001) The map that changed the world. Viking/Penguin Books, London

Wright CW (1958) Order and disorder in nature. Geol Ass Can Proc 69:77–82

Wu C, Liu G, Tian Y, Mao XP, Zhang XL (2005) Discussion on geological information science. Geol Sci Tech Inf 24(3):1–8

Xiao F, Chen J (2012) Fractal projection pursuit classification model applied to geochemical survey data. Comput Geosci 45:75–81

Chapter 2
Probability and Statistics

Abstract From a historical perspective, the theory of statistics was developed relatively recently. Calculating the mean from a number of measurements was a procedure first practiced in 1581 but even this simple method, which is understood by most people to-day, remained highly controversial until the end of the eighteenth century. Traditionally, the upper part of the Earth's crust has been viewed as a complex three-dimensional mosaic of numerous rock units with different compositions and ages. However, as emphasized in this chapter, many geological features display random characteristics that can be modeled by adapting methods of mathematical statistics. The idea that random samples can be taken from statistical populations for the estimation of parameters remains paramount. The main parameters of geoscience data to be estimated are their mean and variance. Frequency distribution analysis is applicable to many different types of geological data. Discrete distributions include the binomial and Poisson, but also the geometric and negative binomial distributions. The normal and lognormal distributions are most important in modeling continuous data although the Pareto distribution is becoming increasingly important because of its close connection to fractal modeling. Many methods of statistical inference including Student's t-test, analysis of variance and the chi-squared test for goodness of fit are based on the normal distribution. These statistical methods remain important in geology if used in an exploratory manner because the random variables considered often are not independent and identically distributed (iid), which is a requirement for statistical problem-solving as practiced in most other fields of science. Especially, numbers of degrees of freedom commonly used in statistical tests are strongly affected by spatial autocorrelation due to the continuous nature of most geological variables.

Keywords Mathematical statistics • Probability calculus • Frequency distributions • Statistical inference • Significance tests • Q-Q plots • Geostatistics • Sampling • Grenville Province age determinations • Pulacayo zinc values • Witwatersrand gold assays

2.1 History of Statistics

Ian Hacking (2006) has pointed out that the concept of "probability", which is a cornerstone of classical statistics, emerged in the middle of the seventeenth century, gradually assuming its dual objective and subjective meanings (*cf.* Agterberg 2013). In its objective sense, probability is related to stable experimental frequencies. Subjective frequencies aim to quantify degrees of belief. Advocates of these two types of probabilities have often disagreed in the past. For example, Ronald Fisher strongly opposed Bayesians who used Thomas Bayes' rule to update initially subjective probabilities in an objective manner. His colleague, the geophysicist Harold Jeffreys, had introduced inductive logic that was later refined by others such as Bruno de Finetti. To-day, primarily deductive and subjective reasoning both continue to be practiced. Jef Caers (2011) argues that "any modeling of uncertainty is only relevant if made dependent on the particular decision question or practical application for which such modeling is called for." Possibility theory as developed by Didier Dubois and colleagues takes a new type of axiomatic approach in uncertainty theory (Dubois and Prade 2000). In many Bayesian approaches, the starting point continues to be based on the concept of equipossibility as originally used by Pierre, Marquis de Laplace.

A simple example of equipossibility followed by deductive reasoning based on traditional axioms of probability theory is Weights-of-Evidence (WofE) modeling to estimate probabilities of occurrences of discrete events such as mineral deposits in a study area. The initial hypothesis of equipossibility is that the probability that a mineral deposit occurs underneath a small unit area on a map is the same everywhere within the study area. It gives the prior probability that only depends on size of unit area. Using Bayes' rule, this prior probability is updated by using as evidence various features of the unit area that differ from place to place in the study area. The final WofE product is a map of posterior probabilities for occurrences of mineral deposits.

The geologist Georgius Agricola (in 1556) developed methods of reading signs on the surface of the Earth such as occurrences of faults or anticlines as indicators for the occurrence of mineralization. Agricola assumed that "sentences" on the Earth's surface tell us what minerals are down below. Nevertheless, as Hacking (2006) points out, Agricola had no idea that there could have been a process of mineralization that took place millions of years ago. Such concepts had not yet been developed. It can be argued that the signs identified by Agricola and several others in the sixteenth century including Paracelsus constituted some early form of what later became known as "probabilities". Paracelsus was active in the field of medicine listing medicines for various illnesses. He developed an early similarity theory that would not meet standards to be developed later. For example, Paracelsus wrote: "Do not the leaves of the thistle prickle like needles. Thanks to this sign, the art of magic discovered that there is no better sign against internal prickling" (Jacobi 1951). The signs of Agricola and Paracelsus were precursors of probabilities. They both knew that the signs they had identified as indicators were not foolproof either leading to certain discovery of new ore or curing disease with complete success. In WofE various theories of process-modeling are taken into account when map layers are selected for improving the posterior probabilities.

2.1 History of Statistics

It is well known that Blaise Pascal around 1650 was adept in solving problems related to the rolling of one or more dice. Hacking (2006) credits Christiaan Huygens with introducing statistical inference in the first probability textbook published in 1657 (also see Kendall 1970, p. 29). Statistical reasoning, however, became only slowly accepted by scientists and much later by the public. This is evident from the history of the arithmetic mean. Some early astronomical calculations that show resemblance to the process of estimating the sample mean are reviewed by Plackett (1970). The first average on record was taken by William Borough in 1581 for a set of compass readings (Eisenhart 1963). The procedure of averaging numbers was regarded with suspicion for a long period of time. Thomas Simpson (1755) advocated the approach in a paper entitled: "On the advantage of taking the mean of a number of observations in practical astronomy", stating: "It is well-known that the method practiced by astronomers to diminish the errors arising from the imperfections of instrument and of the organs of sense by taking the mean of several observations has not so generally been received but that some persons of note have publicly maintained that one single observation, taken with due care, was as much to be relied on, as the mean of a great number."

Originally, the normal distribution was derived from the binomial distribution by Abraham de Moivre in 1718. It became more widely known after its use by Friedrich Gauss (in 1809) and the subsequent derivation of the central-limit theorem, which helped to popularize the idea that many different random errors combine to produce errors that are normally distributed. The normal distribution became another corner stone of mathematical statistics with the development of Student's t-test, analysis of variance and the chi-square test for goodness of fit. During the first half of the twentieth century, many methods of mathematical statistics were developed for statistical populations of independent (uncorrelated) and identically distributed (iid) objects from which random samples can be drawn to estimate parameters such as the mean, variance and covariance. The theory of random sampling became well-established together with rules for determining the exact numbers of degrees of freedom to be used in statistical inference. Generalization to multivariate analysis followed naturally. Krumbein and Graybill (1965) introduced the "general linear model" as a basic tool of mathematical geology.

2.1.1 Emergence of Mathematical Statistics

Karl Pearson (1857–1936) greatly helped to establish the theory of mathematical statistics and to make it more widely known. Many people to-day are familiar with the correlation coefficient and the chi-square test for goodness of fit, which are two of the tools invented by Pearson. R.A. Fisher (1890–1962) was a better mathematician than Pearson (*cf.* Stigler 2008). His earliest accomplishments included finding the mathematical formula for the frequency distribution of the correlation coefficient, and correct usage of degrees of freedom in statistical significance tests including the chi-square test (Fisher Box 1978). Fisher (1960) developed statistical design of experiments using significance tests including the F-tests in analysis of variance. Most of this pioneering work was performed while he was at the Rothamsted Experimental Station in the U.K. (1919–1933).

Fisher's methods entered most statistical textbooks during his lifetime and continue to be taught widely. Degrees of freedom were a subject of disagreement during the 1920s, mainly between Pearson and Fisher. This debate was decidedly won by Fisher. For example, he established that in a chi-square test for goodness of fit the number of degrees of freedom should be decreased by one for every statistical parameter estimated. The mathematical proof of this simple rule is not at all simple. Fisher illustrated the validity of his new result in a 1926 *coup de grace* administered on the basis of 12,000 (2 × 2) contingency tables obtained under random sampling conditions by E.S. Pearson, son of Karl Pearson. Using these data, Fisher calculated that, on average, the chi-square for a 2 × 2 table contingency table has only one degree of freedom instead of the three previously assumed by the Pearsons (*cf.* Fisher Box 1978). These disputes in the 1920s illustrate that the formulae derived by Fisher are not at all that easy to understand. However, many textbooks on applications of statistics in science and engineering are easy to read because they do not contain the formulae underlying the significance tests but only instructions on how to test hypotheses by means of statistical tables such as those for the t-, chi-square and F- distributions. The idea of teaching simple rules only is that practitioners should not be sidetracked by the underlying mathematics. There also exist easy-to-read geostatistical books such as those written by Isobel Clark (1970) and Isaaks and Srivastava (1989) but circulation and acceptance of these 3-D based statistical ideas has been more limited.

It should be kept in mind that most of Fisher's techniques are applicable only if the observations are independent. This requirement was well known to Fisher and other mathematical statisticians including Kolmogorov (1931) who established the axioms of probability calculus. Another consideration, which is easier to understand, is that the assumption of normality (Gaussian frequency distribution curve) for the numbers treated in significance tests has to be approximately satisfied. This is because the tables for Student's t-test, the chi-square test for goodness of fit, analysis of variance and several other well-known significance tests are based on random variables that have frequency distributions derived from the normal distribution.

Rothhamsted continues to be an important research center for statistical research and applications. The methods of Georges Matheron are now used in agricultural research in addition to Fisher's methods; for example, Richard Webster (2001), BAB Rothhamsted Research, published a widely read book on "Geostatistics for Environmental Scientists". This post-Fisher development can be regarded as an extension of traditional statistical theory based on random variables satisfying the axioms set out by Kolmogorov (1931).

2.1.2 Spatial Statistics

Danie Krige (1951) in South Africa first advocated the use of regression analysis to extrapolate from known gold assays to estimate mining block averages

(Krige 1951). This technique can be regarded as a first application of "kriging", which is a translation of the term "Krigeage" originally coined by Georges Matheron (1962) who remarked that use of this word was sanctioned by the [French] Commissariat à l'Energie Atomique to honor work by Krige on the bias affecting estimation of mining block grades from sampling in their surroundings, and on the correction coefficients that should be applied to avoid this bias. Later, Matheron (1967) urged the English-speaking community to adopt the term "kriging" which now is used worldwide.

Krige's original paper was translated into the French and republished in 1955 in a special issue of *Annales des Mines* on the use of mathematical statistics in economic geology. It is followed by a paper by Matheron (1955) who emphasized "permanence" of lognormality in that gold assays from smaller and larger blocks all have lognormal frequency distributions with variances decreasing with increasing block size. Matheron discusses "Krige's formula" for the propagation of variances of logarithmically transformed mining assays, which states that the variance for small blocks within a large block is equal to the variance for the small blocks within intermediate-size blocks plus the variance of the intermediate-size blocks within the large block. This empirical formula could not be reconciled with early theory of mathematical statistics but it constitutes a characteristic feature in the spatial model of orebodies developed in the late 1940s by the Dutch mining engineer Hans de Wijs (1951) whose approach helped Matheron to formulate the idea of "regionalized random variable". Rather than using autocorrelation coefficients as were generally employed in time series analysis under the assumption of existence of a mean and finite variance, Matheron (1962) introduced the variogram as a basic tool for structural analysis of spatial continuity of chemical element concentration values in rocks and orebodies. This is because the variogram allows for the possibility of infinitely large variance as would result from the de Wijsian model for indefinitely increasing distances between sampling points. Aspects of this model were adopted by Krige (1978) in his monograph *Lognormal-de Wijsian Geostatistics for Ore Evaluation* summarizing earlier studies including his successful application to characterize self-similar gold and uranium distribution patterns in the Klerksdorp goldfield in South Africa (also see Sect. 11.1).

Initially, Georges Matheron performed his geostatistical work in the 1950s with mining applications. In the late 1960s his approach caught the attention of mathematical statisticians including Geof Watson and John Tukey. Noel Cressie (1991), a former PhD student of Watson wrote the textbook *Statistics for Spatial Data* casting Matheron's approach into a mathematical statistical context. At the Biennial Session of the International Statistical Institute held in Seoul, 2001, Georges Matheron, John Tukey and Lucien Le Cam jointly were honoured posthumously as great mathematical statisticians from the second half of the twentieth Century. It illustrates that the idea of "regionalized random variable" has become a corner stone of mathematical statistics.

What is kriging variance and why do degrees of freedom not play an important role in spatial statistics? Formulas for kriging variances can be found in all geostatistical textbooks. In general, they are larger than variances that would be estimated by making the simple assumption that the values used for kriging are stochastically independent. This is because element concentration values and realizations from other spatial variables generally are autocorrelated in that values for samples that are close together resemble one another more closely than values that are from farther apart. Random variables with autocorrelation properties have a history of being studied in time series analysis (Bloomfield 2000) of which the statistical theory had become well established in the 1920s and 1930s. Filtering in statistical theory of communication (Lee 1960) can be regarded as a form of kriging. There is also a close connection between splines (Eubank 1988) and kriging. Kriging differs from time series analysis in that the observation points are located in three-dimensional space instead of along a line. Nevertheless, autocorrelation commonly is studied using variograms or correlograms for sampling points with regular spacing along lines. Although Matheron (1965) was most prominent in developing spatial statistics (or "geostatistics" as he preferred to call it), others such as Matérn (1981) in forestry and Gandin (1965) in meteorology, to some extent independently, had advanced the idea of "regionalized random variables" in the 1950s and 1960s as well.

Cressie (1991) reasons as follows to explain why spatial autocorrelation must be considered: Suppose $Z(1),\ldots, Z(n)$ are independent and identically distributed (i.i.d.) observations drawn from a Gaussian distribution with unknown mean μ and known variance σ^2, then the minimum-variance unbiased estimator of μ is equal to the sum of the $Z(i)$ values ($i = 1,\ldots, n$) divided by n, or $M = \{\Sigma\ Z(i)\}/n$. The estimator M is Gaussian with mean μ and variance σ^2/n. It can be used to construct a two-sided 95 % confidence interval for μ, which is $\{M \pm (1.96 \bullet \sigma)/n^{1/2}\}$. To-day many people are familiar with two-sided 95 % confidence intervals on sample means that have been estimated by means of this method. In practice, σ^2 also is unknown and is estimated by taking the sum of squares of the differences between the $Z(i)$ and their average, and dividing this sum by $(n-1)$. However, this standard statistical approach loses its validity when data are not independent but positively correlated. Normally, the extent of positive correlation decreases with distance between locations of points in 2-D or 3-D at which two $Z(i)$ values were measured. Suppose that this distance between observation points is kept the same and all pairs of values $Z(i)$ and $Z(i-1)$ are positively correlated with correlation coefficient $\rho > 0$. Then the variance of M is larger than σ^2/n. This estimator must be multiplied by a factor c that can be estimated although some assumption on the nature of the theoretical autocorrelation function is required. Consequently, the two-sided 95 % confidence interval for μ, is $\{M \pm (c^{1/2} \cdot 1.96 \cdot \sigma)/n^{1/2}\}$ and this is wider than $\{M \pm (1.96 \cdot \sigma)/n^{1/2}\}$. This topic will be discussed in more detail in Sect. 7.1.

It is useful to define $n' = n/c$, which is less than n, as the "equivalent number of independent observations". To provide an illustrative example: suppose chemical element concentration values for $n = 20$ successive drill-core samples (equally spaced along a straight line) are positively correlated with $\rho = 0.5$ and that the space series is equivalent to a time series with the first-order Markov property. Then, $c = 2.95$ and $n' = 6.78$. It means that the 20 values are equivalent to about 7 independent observations and that a 95 % confidence interval neglecting the positive spatial correlation would be 0.58 times too narrow. Obviously, in this situation it would be misleading to set number of degrees of freedom equal to 19. The concept of degrees of freedom then has lost its meaning entirely unless one would base it on the 7 (instead of 20) equivalent "independent" values.

2.1 History of Statistics

What happens when statistical techniques that assume independence of observations are used in situations that there is spatial autocorrelation? In general, the results will be less precise than those obtained by geostatistical methods, variances will be significantly underestimated, and 95 % confidence intervals will be too narrow. Whether or not these shortcomings are serious depends on the characteristics of the subject of application. It is possible that simple averaging provides good results especially in situations where there is the possibility of unforeseen events that cannot be observed immediately. Experienced mining geologists and engineers will know that results obtained by means of exploratory boreholes or channel sampling have limited value depending on circumstances. Geostatistics in mining is most valuable if it is possible and necessary to process large amounts of data that have similar properties such as the hundreds of thousands of gold and uranium determinations in the Witwatersrand goldfields in South Africa. Although the theory of geostatistics primarily was commenced in the field of mining, it should be appreciated that to-day there are many other applications of Matheron's original approach. There now exist numerous applications in environmental sciences, agriculture, meteorology, oceanography, physical geography and other fields. Geostatistical textbooks include Deutsch (2002), Goovaerts (1997) and Olea (1999). Recent new developments in the theory of geostatistics include multiple-point geostatistical simulation based on genetic algorithms (Peredo and Ortiz 2012), sequential simulation with iterative methods (Arroyo et al. 2012) and extensions of the parametric inference of spatial covariances by maximum likelihood (Dowd and Pardo-Igúzquiza 2012).

Box 2.1: Basic Elements of Classical Statistics

The r-th moment of a random variable X is $\mu'_r = E(X^r) = \int_{-\infty}^{\infty} x^r f(x) dx$ where E denotes mathematical expectation. The mean of X satisfies $\mu = \mu'_1 = E(X) = \int_{-\infty}^{\infty} x f(x) dx$. Moments about the mean are defined as: $\mu_r = E[(X-\mu)^r] = \int_{-\infty}^{\infty} (x-\mu)^r f(x) dx$. If c is a constant, $E(X+c) = EX + c$ and $E(cX) = cEX$. When X and Y are two random variables with two-dimensional frequency distribution $f(x,y)$, $E(X+Y) = \int_{-\infty}^{\infty} \int_{-\infty}^{\infty} (x+y) f(x,y) dx dy = EX + EY$. If $\overline{X}$ is a sample mean, $E(\overline{X}) = \mu$. If X is a binary variable with probabilities $P(X=1) = a$ and $P(X=0) = 1-a$ where a is a constant, then $E(X) = P(X=1) = a$. Consequently, probabilities can be treated as expected values. The variance of X is $\sigma^2(X) = \mu_2 = E[(X-\mu)^2] = \int_{-\infty}^{\infty} (x-\mu)^2 f(x) dx = \mu'_2 - \mu^2$. Its properties include $\sigma^2(X+c) = \sigma^2(X)$ and $\sigma^2(cX) = c^2 \sigma^2(X)$. If X and Y are independent, $E(XY) = (EX)(EY)$ and $\sigma^2(X+Y) = \sigma^2(X) + \sigma^2(Y)$. If $\overline{X}$ is the mean of a sample of size n, $\sigma^2(\overline{X}) = \sigma^2(X)/n$. It also follows that $\sigma^2(X-\overline{X}) = \frac{n}{n-1}\sigma^2(X)$. The factor $n/(n-1)$ is known as Bessel's correction. Switching to conventional sample notation, it follows that, $s^2(x) = \dfrac{\sum_{i=1}^{n}(x_i - \overline{x})^2}{n-1}$.

2.2 Probability Calculus and Discrete Frequency Distributions

Kolmogorov (1931) generally is credited with establishing the axioms of mathematical statistics. His formal definitions that involve Borel sets are beyond the scope of this book. However, the rules presented in this section are in agreement with Kolmogorov's original axioms.

Geoscientists should have a basic understanding of probabilities and how to calculate them. The following example provides an illustration of rules of multiplication and addition of probabilities. Suppose that wildcats in a sedimentary basin have a success ratio of $p = 0.2$. One then can answer questions like: What is the probability that 0, 1 or 2 wildcats will strike oil if two new wells are drilled? Writing these probabilities as $p(0)$, $p(1)$ and $p(2)$, the answers are $p(0) = (1-p)^2 = 0.64$, $p(1) = 2p(1-p) = 0.32$ and $p(2) = p^2 = 0.04$, respectively. The sum of these three probabilities is 1. The probability that one or two wildcats will strike oil is $p(1) + p(2) = 0.36$.

2.2.1 Conditional Probability and Bayes' Theorem

A slightly more difficult problem and its solution are as follows: Suppose that $p(D|B)$ represents the conditional probability that event D occurs given event B (e.g., a mineral deposit D occurs in a small unit cell underlain by rock type B on a geological map). This conditional probability obeys three basic rules (cf. Lindley 1987, p. 18):

1. Convexity: $0 \leq p(D|B) \leq 1$; D occurs with certainty if B logically implies D; then, $p(D|B) = 1$, and $p(D^c|B) = 0$ where D^c represents the complement of D;
2. Addition: $p(B \cup C|D) = p(B|D) + p(C|D) - p(B \cap C|D)$; and
3. Multiplication: $p(B \cap C|D) = p(B|D) \cdot p(C|B \cap D)$.

These three basic rules lead to many other rules. For example, replacement of B by $B \cap D$ in the multiplication rule gives: $p(B \cap C|D) = p(B|D) \cdot p(C|B \cap D)$. Likewise, it is readily derived that: $p(B \cap C \cap D) = p(B|D) \cdot p(C|B \cap D) \cdot p(D)$. This leads to Bayes' theorem in odds form:

$$\frac{p(D|B \cap C)}{p(D^c|B \cap C)} = \frac{p(B|C \cap D)}{p(B|C \cap D^c)} \cdot \frac{p(D|C)}{p(D^c|C)}$$

or

$$O(D|B \cap C) = \exp(W_{B \cap C}) \cdot O(D|C)$$

2.2 Probability Calculus and Discrete Frequency Distributions

where $O = p/(1-p)$ are the odds corresponding to $p = O/(1+O)$, and $W_{B \cap C}$ is the "weight of evidence" for occurrence of the event D given B and C. Suppose that the probability p refers to occurrence of a mineral deposit D under a small area on the map (circular or square unit area). Suppose further that B represents a binary indicator pattern, and that C is the study area within which D and B have been determined. Under the assumption of equipossibility (or equiprobability), the prior probability is equal to total number of deposits in the study area divided by total number of unit cells in the study area. Theoretically, C is selected from an infinitely large universe (parent population) with constant probabilities for the relationship between D and B. In practical applications, only one study area is selected per problem and C can be deleted from the preceding equation. Then Bayes' theorem can be written in the form:

$$\ln O(D|B) = W_B^+ + \ln O(D); \quad \ln O(D|B^c) = W_B^- + \ln O(D)$$

for presence or absence of B, respectively. If the area of the unit cell underlain by B is small in comparison with the total area underlain by B, the odds O are approximately equal to the probability p. The weights satisfy:

$$W_B^+ = \ln \frac{p(B \cap D)}{p(B \cap D^c)}; \quad W_B^- = \frac{p(B^c \cap D)}{p(B^c \cap D^c)}$$

As an example of this type of application of Bayes' theorem, suppose that a study area C, which is a million times as large as the unit cell, contains ten deposits; 20 % of C is underlain by rock type B, which contains eight deposits. The prior probability $p(D)$ then is equal to 0.000 01; the posterior probability for a unit cell on B is $p(D|B) = 0.000\ 04$, and the posterior probability for a unit cell not on B is $p(D|B^c) = 2.5 \times 10^{-6}$. The weights of evidence are $W_B^+ = 0.982$ and $W_B^- = -1.056$, respectively. In this example, the two posterior probabilities can be calculated without use of Bayes' theorem. However, the weights themselves provide useful information as will be seen in Chap. 5.

2.2.2 Probability Generating Functions

A random variable X is either discrete or continuous. Some geological frequency distributions are best modeled as compound frequency distributions that require the use of more advanced methods of mathematical statistics including use of probability generating functions. Our treatment of this subject is kept brief for reasons of space. The reader is referred to textbooks of mathematical statistics (e.g., Feller 1968) for further explanations.

> **Box 2.2: Probability Generating Function and Moments**
>
> If X is a discrete random variable with probability distribution $P(X=k)=p_k$ ($k=0, 1, 2, \ldots$), then its probability generating function is:
> $g(s) = p_0 + p_1 x + p_2 x^2 + \ldots$ with $p_k = \frac{1}{k!}\left[\frac{d^k}{ds^k}g(s)\right] = \frac{1}{k!}g^{(k)}(0)$
> The r-th moment of X satisfies: $\mu'_r = \sum_{k=0}^{\infty} p_k = g^{(r)}(1)$. For the mean and variance (cf. Feller 1968, p. 360): $E(X) = g'(1); \sigma^2(X) = g''(1) + g'(1) - [g'(1)]^2$. Suppose that X and Y are two independent discrete random variables with probability distributions $P(X=k)=p_k$; $P(Y=k)=q_k$ ($k=0, 1, 2, \ldots$), then the probability distribution of their sum $Z=X+Y$ satisfies: $g_z(s) = g_x(s) \cdot g_y(s)$.

2.2.3 Binomial and Poisson Distributions

Suppose the outcome of an experiment, like observing presence or absence of a rock type at a point, is either Yes or No. Presence or absence can be denoted as 1 or 0. The experiment is called a Bernoulli trial. If X is a Bernoulli variable with $P(X=0) = 1-p$, $P(X=1) = p$, its generating function is: $g(s) = (1-p) + ps$. The binomial distribution results from n successive Bernoulli trials. Repeated application of the Bernoulli distribution's generating function gives the generating function:

$$g(s) = [(1-p) + ps]^n$$

It follows that the mean and variance of a binomial variable are $E(X) = np$ and $\sigma^2 = np(1-p)$.

For n Bernoulli trials, there are $(n+1)$ possible outcomes: n, (n – 1), …, 2, 1, 0. The $(n+1)$ probabilities for these outcomes are given by the successive terms of the series obtained by expanding the expression $(p+q)^n$. The probability that the outcome is exactly k Yesses in n trials satisfies:

$$P\left(\sum_{i=1}^{n} X_i = k\right) = \binom{n}{k} p^k q^{n-k} \text{ where } \binom{n}{k} = \frac{n!}{k!(n-k)!}$$

This is the binomial frequency distribution. Petrographic modal analysis (Chayes 1956) provides an example of application of the binomial frequency distribution. Suppose that 12 % by volume of a rock consists of a given mineral A. The method of point-counting applied to a thin section under the microscope consists of how many times a mineral is observed to occur at points that together form a regular square grid. Suppose that 100 points are counted. From $p = 0.12$ and $n = 100$ it follows that $q = 0.88$, and for mean and standard deviation: $\mu = np = 12$ and $\sigma = (npq)^{0.5} = 3.25$.

If the experiment of counting 100 points would be repeated many times, the number of times (K) that the mineral A is counted would describe a binomial distribution with mean 12 and standard deviation 3.25. According to the central-limit theorem to be discussed in Sect. 2.3.1, a binomial distribution approaches a normal distribution when n increases. Hence it already can be said for a single experiment that $P(\mu - 1.96 \ \sigma < K < \mu + 1.96 \ \sigma) = 0.95$ or $P(5.6 < K < 18.4) = 95\%$. The resulting value of K is between 5.6 and 18.4 with a probability of 95 %. Precision can be increased by counting more points. For example, if $n = 1{,}000$, then $\mu = 120$ and $\sigma = 10.3$, and $P(99.8 < K < 140.2) = 95\%$. This result also can be written as $k = 120 \pm 20.2$. In order to compare the experiment of counting 1,000 points to that for 100 points, the result must be divided by 10, giving $k' = 12 \pm 2.02$. This number represents the estimate of volume percent for the mineral A. It is $3.25/(10.3/10) = 3.2$ times as precise as the first estimate that was based on 100 points only. This increase in precision is in agreement with the basic statistical result that if $\overline{X}$ is the mean of a sample of size n, $\sigma^2(\overline{X}) = \sigma^2(X)/n$. Since ten times as many points were counted, the result has become $10^{0.5} = 3.2$ times more precise.

Chayes (1956) discusses petrographic modal analysis in more detail. A second example of usefulness of the binomial distribution model is taken from quantitative stratigraphy. In general, most biostratigraphic correlation is based on biozonations derived from range charts using observed oldest and youngest occurrences of microfossil taxa. For example, in exploratory drilling for hydrocarbon deposits in a sedimentary basin, a sequence of borehole samples along a well drilled in the stratigraphically downward direction is systematically checked for first occurrences of new species. When the samples are cuttings taken at a regular interval, there is the possibility that younger material drops down the well so that highest or last occurrences cannot be observed in that situation. The probability of detecting a species in a single sample depends primarily on its abundance. As a measure, relative abundance (to be written as p) of a species in a population of microfossils is commonly used. Together with sample size (n), p specifies the binomial distribution that k microfossils of the taxon will be observed in a single sample. If p is very small, the binomial probability can be approximated by the probability of the Poisson distribution:

$$P(K = k) = \frac{e^{-\lambda}\lambda^k}{k!} \quad (k = 0, 1, \ldots, n)$$

which is determined by its single parameter λ with $E(K) = \sigma^2(K) = \lambda$. The Poisson distribution can be derived from the binomial distribution by keeping $\lambda = np$ constant and letting n tend to infinity while p tends to zero. Like many other frequency distribution models, the approach can be extended to more than a single variable.

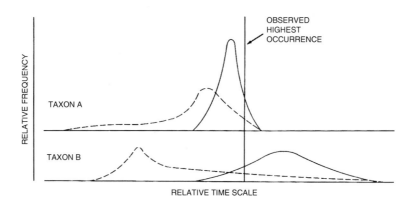

Fig. 2.1 Schematic diagram representing frequency distributions for relative abundance (*broken lines*) and location of observed highest occurrences (*solid lines*) for two fossil taxa. *Vertical line* illustrates that observed highest occurrences of two taxa can be coincident or "coeval" even when the frequency distributions of the taxa are different (Source: Agterberg 1990, Fig. 2.12)

Figure 2.1 shows a hypothetical relationship between relative abundance, observed highest occurrence and relative time for two taxa in a stratigraphic section. This example illustrates that the abundance of a taxon may have changed through time. The range of the frequency curve of the observed highest occurrence is narrower than the range of the abundance curve although both frequency curves end at the same value along the relative time axis. If a systematic sampling procedure is carried out such as obtaining core samples (instead of cuttings) at a regular interval (e.g., 30 ft or 10 m) along a well in exploratory drilling, the highest occurrences of two taxa with overlapping frequency curves can be observed to be coincident. The fact that two taxa have observed last occurrences in the same sample does not necessarily mean that they disappeared from the sedimentary basin at the same time. Rare taxa such as taxon B in Fig. 2.1 are likely to have wider ranges for their highest occurrences. Problems related to well-sampling will be discussed in more detail in Chap. 9.

Figure 2.2 (after Dennison and Hay 1967) shows probability of failure to detect a given species for different values of p as a function of sample size (n). For example, in a sample of $n = 200$ microfossils, a species with $p = 1\%$ has a probability of about 15 % of not being detected. This implies that the chances one or more individuals belonging to the species will be found are good. Unless its relative abundance is small, the first or last occurrence of a species in a sequence of samples then can be established relatively quickly and precisely.

It is noted that the two scales in Fig. 2.2 are logarithmic and that the lines are approximately straight unless p is relatively large. This is because the equation for zero probability of the Poisson distribution, which provides a good approximation to the binomial when p is small, plots as a straight line on double logarithmic graph paper. If 10 is the base of the logarithms, the equation of each straight line in Fig. 2.2 is simply $\log_{10} n = \log_{10} \lambda - \log_{10} p$ with $p = P(K=0) = \exp(-\lambda)$.

2.2 Probability Calculus and Discrete Frequency Distributions

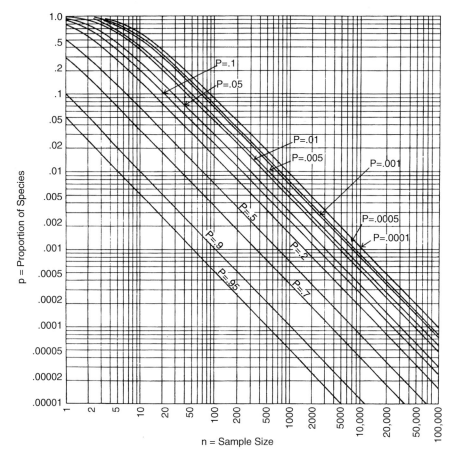

Fig. 2.2 Size of random sample (n) needed to detect a species occurring with proportional abundance (p) with probability of failure to detect its presence fixed at P (After Dennison and Hay 1967) (Source: Agterberg 1990, Fig. 3.1)

2.2.4 Other Discrete Frequency Distributions

Suppose that a succession of Bernoulli trials is started by randomly selecting black or white cells from a black-and-white mosaic (e.g., an array based on pixels). One may ask the question of how many times the trial must be repeated before $X = \Sigma_m X_i = r$ where r is a fixed positive integer number and m represents the number of trials it takes to obtain r. It is convenient to write $m = k + r$ where k is another integer number. The probability of a one (for a black cell) at the m-th trial is p. This probability must be multiplied by the probability that there were exactly k zeros (for white cells) during the preceding m-1 ($=k+r-1$) experiments.

Table 2.1 Fits of Poisson and negative binomial distributions to the spatial distribution of oil deposits for 5 × 5 miles grid areas, Alberta

Deposits	Observed frequency	Poisson frequency	Negative binomial
0	8,586	8,508.5	8,584.3
1	176	303.0	176.8
2	35	5.4	39.1
3	13	0.1	11.3
4	6	0.0	3.6
5	1	0.0	1.2
6 or more	0	0.0	0.7

After Uhler and Bradley (1970, Table 1)
Source: Agterberg (1974, Table XVIII)

The latter probability is binomial. Multiplied by p this gives the negative binomial distribution with:

$$p_k = \binom{r+k-1}{k} p^{k-1} q^k$$

with generating function:

$$g(s) = \left(\frac{p}{1-qs}\right)^r$$

It follows that $E(X) = \mu = rq/p$ and $\sigma^2(X) = rq/p^2$. For the negative binomial: $\sigma^2(X) = \sigma^2 > \mu$. On the other hand, for the ordinary (positive) binomial: $\sigma^2 < \mu$. Comparison of the sample mean and variance of a set of discrete geological data often provides a guideline as to which one of these two distribution (positive or negative binomial) should be fitted (*cf.* Ondrick and Griffiths 1969).

An example of a situation in which the negative binomial provides a better approximation than the Poisson distribution is shown in Table 2.1. It is based on a study by Uhler and Bradley (1970) who divided the Alberta sedimentary basin into 8,811 cells, each measuring 5 miles (8 km) on a side., and counted the number of oil deposits per cell. The results in Table 2.1 show that the negative binomial provides the better fit. Obviously, the Poisson distribution that has only one parameter (λ) is not sufficiently flexible in this application.

If $r=1$, the negative binomial distribution reduces to the geometric distribution that can be illustrated as follows by an application to lithological components in the Oficina Formation, eastern Venezuela. Krumbein and Dacey (1969) found that the thicknesses of these components can be described by geometric distributions. Their data consisted of a series of letters A, B, C, and D, originally obtained by Scherer (1968) who coded the rock types as (A) sandstone, (B) shale, (C) siltstone, and (D) lignite, in a well core at 2-ft (61 cm) intervals. Although the thicknesses of these lithologies are continuous random variables, the sampling scheme reduced them to discrete random variables.

2.2 Probability Calculus and Discrete Frequency Distributions

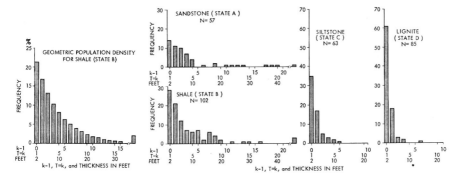

Fig. 2.3 Theoretical frequency distribution for shale and sandstone distributions for lithological components in Oficina Formation, eastern Venezuela (After Krumbein and Dacey 1969). All distributions are geometric by close approximation (Source: Agterberg 1974, Fig. 35)

For each rock type, the frequencies for sequences of successive letters can be determined and plotted in a diagram. For example, one counts how many times the sequences A, AA, AAA, ... occur in the total series. The results are shown in Fig. 2.3. Krumbein and Dacey (1969) fitted geometric distributions to these data and tested the degree of fit by means of chi-square tests for goodness of fit. The theoretical distribution for shale is shown on the left-hand side of Fig. 2.3. The good fit of the geometric distributions indicates that the following stochastic process was controlling the sedimentation.

Suppose that for a lithology, say sandstone, $p(A)$ denotes the probability that at a distance of 2 ft., another lithology (not sandstone) will occur. Obviously, $q(A) = 1 - p(A)$ then represents the probability that the same rock type (sandstone) will occur. If for every lithology p (and q) remained constant during deposition of the entire series, then the probability that a sequence for any lithology is k letters long, satisfies the negative binomial form. Consequently, $r = 1$ because each sequence is terminated at the point where it is replaced by another lithology. The result is a geometric frequency distribution for each lithology. It is possible to divide the probability p for each lithology into three parts, one for each other lithology. All probabilities can be arranged into the following transition matrix:

$$\begin{bmatrix} q(A) & p(AB) & p(AC) & p(AD) \\ p(BA) & q(B) & p(BC) & p(BD) \\ p(CA) & p(CB) & q(C) & p(CD) \\ p(DA) & p(DB) & p(DC) & q(C) \end{bmatrix} = \begin{bmatrix} 0.787 & 0.071 & 0.075 & 0.067 \\ 0.048 & 0.788 & 0.061 & 0.103 \\ 0.105 & 0.316 & 0.430 & 0.149 \\ 0.182 & 0.388 & 0.132 & 0.298 \end{bmatrix}$$

This matrix is the transition matrix of a Markov chain of the first order as demonstrated by Krumbein and Dacey (1969). It illustrates the close connection between geometric frequency distributions and first order Markov chains. Doveton (2008) discusses how Markov mean first-passage time statistics can be obtained for sedimentary successions.

Another discrete random variable to be used later in this book (Chap. 9) has the logarithmic series distribution with:

$$p_k = \frac{\alpha \vartheta^k}{k} \quad (k = 1, 2, \ldots; 0 < \vartheta < 1).$$

Its moment generation function is: $g(s) = [\log_e(1 - \vartheta e^k)]/[\log_e(1 - \vartheta)]$. The first two moments about the origin are: $E(X) = \alpha\vartheta(1 - \vartheta)^{-1}$ and $\mu_2 = \alpha\vartheta(1 - \vartheta)^{-2}$. Therefore, $\sigma^2(X) = \alpha\vartheta(1 - \alpha\vartheta)(1 - \vartheta)^{-2}$. This model is often used in the biosciences for spatial distribution of species.

2.3 Continuous Frequency Distributions and Statistical Inference

Continuous random variables can assume any value on the real line. Traditionally, an important role is played by the normal or Gaussian distribution. It is frequently observed in practice. Theoretically, it is the end product of the central-limit theorem. The normal distribution underlies many methods of statistical inference such as z-test, Student's t-test, chi-square test and analysis of variance.

> **Box 2.3: Moment Generating Function and Characteristic Function**
>
> If X is a continuous random variable, its moment generating function satisfies: $m(u) = E(e^{uX}) = \int_{-\infty}^{\infty} e^{ux} f(x) dx$. Consequently, $\mu'_r = \int_{-\infty}^{\infty} x^r f(x) dx = m^r(0)$; and $E(X) = m'(0); \sigma^2(X) = m''(0) - [m'(0)]^2$. For continuous random variables, characteristic functions $g(u)$ have a wider field of application than moment generating functions. They satisfy the following inverse relationship:
> $g(u) = E(e^{iuX}) = \int_{-\infty}^{\infty} e^{iux} f(x) dx;\ f(x) = \frac{1}{2\pi}\int_{-\infty}^{\infty} e^{iux} g(x) dx$. The moments satisfy: $\mu'_r = E(X^r) = i^{-r} g^{(r)}(0)$. The Pareto distribution with frequency density $f(x) = \frac{ak^a}{x^{a+1}}$ ($k > 0, a > 0; x \geq k$) provides an example. Provided that r is less than a, the r-th moment about zero is: $\mu'_r = \frac{ak^r}{a-r}$. Consequently, $E(X) = \frac{ak}{a-1}$ and $\sigma^2(X) = \frac{a^2}{(a-1)^2(a-2)}$. The Pareto will be used in Chaps. 3 and 11.

2.3.1 Central-Limit Theorem

The central-limit theorem evolved from the de Moivre-Laplace theorem which stated that in the limit (sample size $n \to \infty$) a positive binomial distribution becomes a normal (Gaussian) distribution (*cf.* Bickel and Doksum 2001, p. 470).

2.3 Continuous Frequency Distributions and Statistical Inference

In its later, more general form the theorem states roughly that the sum of many independent random variables will be approximately normally distributed if each summand has high probability of being small (Billingsley 1986, p. 399). An example of how averaging of data with different kinds of frequency distributions is shown in Fig. 2.1. Because of the central-limit theorem, the averages of n measurements produce new random variables that become normally distributed when n increases. Because in Nature and also in the social sciences so many random variables are approximately normally distributed, and the central-limit theorem seems to provide a plausible explanation, the normal distribution became a corner stone of mathematical statistics.

Box 2.4: The Normal or Gaussian Distribution

If X is a normally distributed random variable with expected value equal to μ and variance σ^2, its frequency density $f(x)$ satisfies: $f(x) = \frac{1}{\sigma\sqrt{2\pi}} \exp\left\{-\frac{1}{2}\left(\frac{x-\mu}{\sigma}\right)^2\right\}$ and its cumulative frequency distribution is: $F(x) = \frac{1}{\sigma\sqrt{2\pi}} \int_{-\infty}^{x} \exp\left\{-\frac{1}{2}\left(\frac{x-\mu}{\sigma}\right)^2\right\} dx$. The corresponding equations for the random variable $Z = (X-\mu)/\sigma$, which is the normal distribution in standard form, are: $\varphi(z) = \frac{1}{\sqrt{2\pi}} e^{-\frac{1}{2}z^2}$ and $\Phi(z) = \frac{1}{\sqrt{2\pi}} \int_{-\infty}^{z} e^{-\frac{1}{2}z^2} dz$.

Because of symmetry of the Gaussian density function with respect to $z=0$: $\Phi(z) = 1 - \Phi(-z)$. The graph of $\Phi(z)$ is S-shaped. The fractiles Z_P of the standard normal random variable satisfy $\Phi(Z_P) = P$. Tables of these fractiles are widely available. P is called the probit of Z_P. It is noted that, originally, the term "probit" was coined by Bliss (1935) for a fractile augmented by 5 in order to avoid the use of negative numbers. To-day, however, the term "probit transformation" is widely used for $Z_P = \Phi^{-1}(P)$. Later in this book (Chap. 9) it will be used to transform observed frequencies into their corresponding Z-values.

The sum of a number (f) of Z^2 values is distributed as $\chi^2(f)$ representing the chi-square distribution for f degrees of freedom, which is a particular form of the gamma distribution with probability density:

$$f(x) = \frac{(x-\gamma)^{\alpha-1} \exp\left[-\frac{x-\gamma}{\beta}\right]}{\beta^\alpha \Gamma(\alpha)} \quad (\alpha > 0, \beta > 0, x > \gamma)$$

Later in this book, the gamma distribution will be used to model biostratigraphic events (Sect. 9.3) and amounts of rock types contained in grid cells superimposed on geological maps (Sect. 12.8). The main use of the gamma distribution, however, is that it becomes $\chi^2(f)$ if $\alpha = f/2$, $\beta = 2$ and $\gamma = 0$.

Box 2.5: Frequency Distributions Derived from the Normal

Frequency distributions derived from the normal distribution include the χ^2-, F- and t-distributions. They are defined for f, f_1 and f_2 "degrees of freedom" as:

$$\chi^2(f) = \sum_{i=1}^{f} Z_i^2; \quad F(f_1,f_2) = \frac{\chi_1^2(f_1)/f_1}{\chi_2^2(f_2)/f_2}; \quad t(f) = Z/\sqrt{\chi^2(f)/f}$$

"Degrees of freedom" is a useful concept of classical statistics. It represents the number of values in the calculation of a statistic that are free to vary independently. The following example illustrates this concept. In general, the sample variance satisfies $s^2 = \frac{\sigma^2 \chi^2(f)}{f}$. If the population mean μ is known, a sample of n data has $f = n$ degrees of freedom. However, if μ is not known and estimated by the sample mean $\bar{x} = \sum_{i=1}^{n} x_i$, Bessel's correction (cf. Box 2.1) must be applied and $s^2 = \frac{\sigma^2 \chi^2(n-1)}{n-1}$ so that $f = n - 1$ (Fig. 2.4).

2.3.2 Significance Tests and 95 %-Confidence Intervals

Fractiles $\Phi(Z_P)$ that are widely used in significance tests and to define 95 %-confidence intervals are $\Phi(1.645) = 0.95$ and $\Phi(1.96) = 0.975$. They correspond to setting the level of significance $\alpha = 0.05$ in one-tailed and two-tailed significance tests, respectively. The so-called z-test of significance provides a prototype for all other tests of significance. It works as follows: Suppose that n numbers are drawn from a normal population with mean μ and variance σ^2, then the sample mean $\bar{X}$ is normally distributed about μ with variance σ^2/n. Consequently,

$$Z = \frac{\bar{X} - \mu}{\sigma/\sqrt{n}}$$

is normally distributed about 0 with standard deviation one. Suppose that there is outside information suggesting that $\mu = \mu_0$. Then the test- or null-hypothesis H_0 that the sample mean $\bar{X}$ originates from a normal distribution with mean μ_0 and standard deviation $\sigma/\sqrt{n}$ is rejected if $|Z| > 1.96$ after μ is replaced by μ_0. The value $|Z_P| = 1.96$ represents the significance limit for the test. If $|Z| < 1.96$, the hypothesis $\mu = \mu_0$ is accepted. This test is two-tailed because the absolute value of $\bar{X} - \mu$ is being considered. If there is additional outside information that would allow taking either $\mu > \mu_0$ or $\mu < \mu_0$ as the test hypothesis, a one-tailed test with significance limit $Z_P = 1.645$ should be used.

2.3 Continuous Frequency Distributions and Statistical Inference

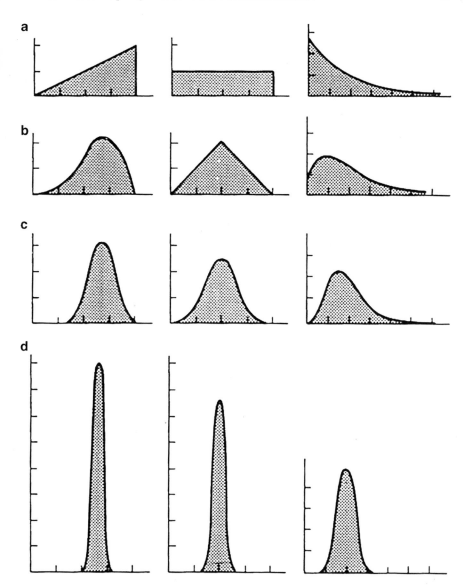

Fig. 2.4 Examples of the effect of averaging illustrate the central limit theorem of mathematical statistics. No matter what shape the frequency distribution of the original observations (**a**), taking the average of two (**b**), four (**c**) or 25 (**d**) observations not only decreases the variance but brings the curve closer to the normal (or Gaussian) limit (After Lapin 1982; Davis 1986) (Source: Agterberg 1990, Fig. 2.18)

2.3.3 Sum of Two Random Variables

Other frequency distribution models include the uniform or rectangular distribution. It can be defined as:

$$f(x) = \frac{1}{b-a} \text{ if } a < x < b; \quad f(x) = 0 \text{ otherwise}$$

This distribution is shown graphically in Fig. 2.2 for $a = 0$ and $b = 5$. Suppose that the normal distribution is written in standard form $\varphi(x)$. It is shown in Fig. 2.2 as well. A more general problem of mathematical statistics that has been discussed by Wilks (1962) and Gnedenko (1963) and at a more advanced level by Loève (1963) is: What is the frequency distribution of the sum of two random variables?

> **Box 2.6: Sum of Uniform and Normal Random Variables**
> Suppose that X_1 and X_2 are two random variables and that their sum is written as $Y = X_1 + X_2$ with frequency distribution function $f(y)$. The joint probability $P(X_1 = x_1 \text{ and } X_2 = x_2)$ can be represented as $f(x_1, x_2)$ for infinitesimal area $dx_1 dx_2$, and: $P(Y < y) = F(y) = \iint_{x_1+x_2<y} f(x_1, x_2) \, dx_1 dx_2$. Integrating the joint probability of X_1 and X_2 over the area where $x_1 + x_2 < y$ provides the total probability that Y is less than $y = x_1 + x_2$. It follows that:
> $F(y) = \int_{-\infty}^{y-x_1} \left[\int_{-\infty}^{y-x_1} f(x_1, x_2) dx_2 \right] dx_1$. After some manipulation, it can be derived that $f(y) = \int_{-\infty}^{\infty} f_2(y-x) f_1(x) dx$. If $h(x)$ represents the sum of the uniform distribution with $f(x)$ as defined at the beginning of this section and the standard normal $\varphi(x)$, it follows that
> $h(x) = \frac{1}{b-a} \int_a^b \varphi(x-y) dy = \frac{1}{b-a} [\Phi(x-a) - \Phi(x-b)]$.

For the example of Fig. 2.5, $h(x) = 0.2 \cdot [\Phi(x) - \Phi(x-5)]$. Single values of $h(x)$ were calculated and connected by a smooth curve in this figure. For example, if $x = 2$, $h(x) = 0.2 \cdot [\Phi(2) - \Phi(3)] = 0.2 \cdot [0.9773 - 0.0014] = 0.1952$. The three curves shown in Fig. 2.2 each have total area under the curve equal to one. The curve for $h(x)$ resembles a Gaussian curve so that it would be difficult to distinguish it from a Gaussian curve on the basis of a sample of n values unless n is large. In the example, the base of the uniform distribution $f(x)$ is five times the standard deviation of the normal curve $\varphi(x)$. In fact, this standard deviation was used as unit of distance along the X-axis. The shape of $h(x)$ will change if the ratio (base of f/standard deviation of φ) is changed. If this ratio is decreased, $h(x)$ approaches the normal form. On the other hand, if it is increased, $h(x)$ begins to develop a flat top. Its shape then approaches that of $f(x)$. More mathematical details

2.4 Applications of Statistical Analysis

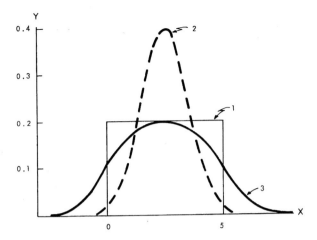

Fig. 2.5 Frequency curve (*3*) for the sum of a rectangular distribution (*1*) and a Gaussian distribution (*2*). This kind of frequency distribution results from random sampling of a space- or time-variable subject to a linear trend changing from 0 to 5 in this artificial example. Residuals from the trend have zero mean and unity variance. The three curves have areas equal to one (Source: Agterberg 1974, Fig. 23)

on derivation of the equation for the curve $g(x)$ in Fig. 2.5 are given in Agterberg (1974, pp. 195–198). In the next chapter the model will be used for maximum likelihood estimation of the age of stage boundaries in the Geologic Timescale. It also will be used as a partial explanation of the Vistelius lognormality model (*cf.* Sect. 3.2).

2.4 Applications of Statistical Analysis

As shown by Schuenemeyer and Drew (2011), many problems of mathematical geoscience are best solved by using the methods of classical statistics. The aim of the next example is to apply statistical inference (F-, t- and χ^2- tests) to a set of age data to be followed by a graphical (Q-Q) test applied to solve the same problem.

2.4.1 *Statistical Inference: Grenville Potassium/Argon Ages Example*

The normal distribution can be used as a starting point for various types of significance tests: notably Student's t-test, the chi-square test and the F-test in analysis of variance. Usage of these tests is illustrated on the basis of age data that were available in 1968 for the Grenville Province on the Canadian Shield

Table 2.2 Potassium/argon age determinations, Grenville Province, Canadian Shield

Biotite		Muscovite	
Age class (m.y)	No. ages	Age class (m.y)	No. ages
770	2		
830	1		
850	6		
870	5	870	3
890	6		
910	7	910	1
930	12	930	3
950	11		
970	5	970	2
990	2	990	4
1,010	3	1,010	2
1,030	3		
1,050	4	1,050	1
1,070	1	1,070	1
1,090	2		
1,110	1		
1,130	1		
1,170	1		
1,190	2		
1,210	1	1,210	1
	$n_1 = 76$		$n_2 = 18$

Source: Agterberg (1974, Table XIII)

(Table 2.2). They are mainly for micas in granitic rocks. A distinction was made between biotite and muscovite ages. The measurements were grouped into classes which are 20 million years wide. Midpoints of these classes are given in the table. For example, there are six ages of 890 Ma meaning that six measured dates were between 880 and 900 Ma. This example originally was given in Agterberg (1974).

The average biotite age for 76 single dates is 954 Ma, and that for muscovite is 976 Ma. The null hypothesis that the biotite and muscovite population means are equal, can be tested by Student's t-test. Strictly speaking this test is based on the assumption that the variance of the population for biotite ages is equal to that for muscovite ages. A sample variance is estimated as $s^2 = [1/(n-1)]\Sigma(x-\bar{x})^2$ (cf. Box 2.1). The sum of squared deviations from the sample mean is divided by $n-1$ instead of by n to account for the variance of the deviation between $\bar{x}$ and its corresponding population mean (μ). The hypothesis of equality of variance will be tested first by using an F-test. Both t- and F-test are based on the assumption that the underlying populations are normal. The normality assumption can be tested separately by using a chi-square test.

Using the subscript 1 for biotite and 2 for muscovite, the sample variances are $s_1^2 = 8,337$ and $s_2^2 = 6,861$. The F-test for comparing two variances consists of calculating the ratio $F(75, 17) = s_1^2/s_2^2 = 1.22$. It is customary to divide the larger variance by the smaller one, so that the F-ratio is greater than one. If the two population variances are equal to one another, the estimated F-ratio would be close

2.4 Applications of Statistical Analysis

Table 2.3 Chi-square test for normality, biotite ages, Grenville Province

Class i	Limits	f_o	z_i	$\Phi(z_i)$	$\Phi(z_i) - \Phi_{(Z_i-1)}$	f_t	$\Delta = |f_o - f_t|$	Δ^2	Δ^2/f_t
1	<860	9	−1.04	0.149	0.149	11.3	2.3	5.29	0.47
2	860–900	11	−0.60	0.274	0.125	9.5	1.5	2.25	0.24
3	900–940	19	−0.16	0.436	0.162	12.3	6.7	44.89	3.65
4	940–980	16	0.28	0.610	0.174	13.2	2.8	7.84	0.59
5	980–1020	5	0.72	0.764	0.154	11.7	6.7	44.89	3.84
6	1020–1060	7	1.16	0.877	0.113	8.6	1.6	2.56	0.30
7	>1060	9	∞	1.000	0.123	9.3	0.3	0.09	0.01
									Sum = 9.1

Source: Agterberg (1974, Table XIV)

to one. For level of significance $\alpha = 0.05$, we should compare it $F_{0.975} = 2.36$ representing the 97.5 % fractile of the cumulative F-distribution in statistical tables. The 97.5 % fractile is used because this is a two-tailed significance test. Since F (75, 17) $< F_{0.975}$ the null hypothesis for equality of variance can be accepted. The two sample variances of Table 2.2 can be combined with one another yielding the overall variance $s^2 = 8{,}064$.

Next, the t-test can be used to test the hypothesis that the two population means are equal. It consists of calculating the quantity:

$$\hat{t}(f_1 + f_2) = \frac{|\bar{x}_1 - \bar{x}_2|}{s(x)\sqrt{\frac{1}{n_1} + \frac{1}{n_2}}}$$

Because the t-test also is two-tailed, this statistic (=0.89) for level of significance $\alpha = 0.05$ can be compared with $t_{0.975}$ (92) $= 1.98$. Because it is smaller than this significance limit, the hypothesis of equality of means can be accepted. The two means in Table 2.2 can be combined with one another yielding the overall mean age of 958.5 Ma.

Strictly speaking, the preceding significance tests only can be applied if the age dates are normally distributed. The chi-square test for goodness of fit can be used to test for normality. Its usage is illustrated in Table 2.3. First class limits are set and the number of observations per class is counted. This gives the observed frequencies (f_o) to be compared with theoretical frequencies (f_e) that are computed by using the normal frequency distribution model. Normal fractiles are determined for all class limits and successive differences between them are multiplied by total number of observations (=76). Next the following statistic is obtained:

$$\hat{\chi}^2 \approx \sum \frac{(f_0 - f_t)^2}{f_t}$$

Two important points must be considered here: (1) The approximation is only valid if all theoretical frequencies are at least 5; care must be taken that all classes are sufficiently wide to allow for this (Cramér 1947); and (2) the sum of the observed frequencies is equal to n. It can be shown (Cramér 1947) that for this reason one degree of freedom must be subtracted from the number of classes m. More degrees of freedom are lost if, in order to obtain the theoretical frequencies (f_e), use is made of parameters estimated from the observations; in general, the number of degrees of freedom is to be reduced further by the number of parameters that were estimated. Consequently, the chi-square test of normality has $(m-3)$ degrees of freedom. For the example of Table 2.3 with $\hat{\chi}^2 = 9.1$, the number of degrees of freedom is 4. From statistical tables, it can be found for $\alpha = 0.05$ that $\chi^2_{0.95}(4) = 9.49$. Hence the normality hypothesis can be accepted. However, it should be kept in mind that $\chi^2_{0.941}(4) = 9.1$. This means that a normal distribution would yield a $\hat{\chi}^2$ equal to or larger than 9.1 in only 5.9 % of events if this particular experiment were to be repeated a large number of times for the same theoretical distribution.

The preceding chi-square test for goodness of fit is well-known. It was originally proposed by Karl Pearson and refined by Ronald Fisher who exactly determined the number of degrees of freedom to be used. A similar test that is at least as good as the chi-square test is the G^2-test (see, e.g., Bishop et al. 1975). Finally, the Kolmogorov-Smirnov test should be mentioned. It consists of determining the largest (positive or negative) difference between theoretical and observed frequencies. In the two-tailed Kolmogorov-Smirnov test, the absolute value of the largest difference should not exceed $1.36/n^{0.5}$ with a probability of 95 % provided that the number of observations exceeds 40. The corresponding confidence for the one-tailed test is $1.22/n^{0.5}$.

2.4.2 Q-Q Plots: Normal Distribution Example

Normality can also be tested graphically by means of a so-called *Q-Q* plot for comparing observed quantiles with theoretical quantiles. When the theoretical frequency distribution is normal, this is the same as using normal probability paper. In Fig. 2.6, the scale along the vertical axis is linear but the horizontal scale has been changed in such a manner that the S-shaped curve for any theoretical cumulative normal distribution plots as a straight line. A normal distribution always becomes a straight line on normal probability paper. Figure 2.6 shows three types of plot for the 76 biotite ages listed in Table 2.3: (1) original data (points); (2) theoretical normal curve (straight line); (3) a 95 % confidence belt on the theoretical normal curve. These three plots have been constructed as follows:

Firstly, cumulative frequencies were determined for the classes of ages shown in Table 2.2. These were converted into cumulative frequency percentage values. If upper class limits are used, it is not possible to plot the value for the 1,200–1,220 Ma class because the last class has cumulative frequency of 100 % that is not part of the probability scale. One may omit plotting this last value but a slight

2.4 Applications of Statistical Analysis

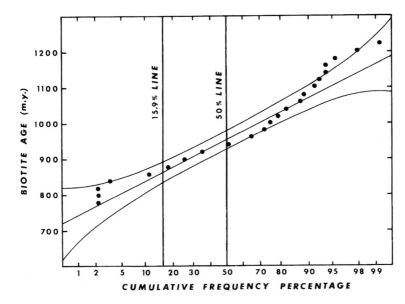

Fig. 2.6 Graphical normality test for 76 biotite ages from Grenville Province, Canadian Shield (After Agterberg 1974). Observed cumulative frequencies (*solid dots*) are not all contained within 95-% confidence belt on the theoretical normal curve (plotting as *straight line* on this Q-Q plot), indicating some departure from normality that is strongest in the upper tail (Source: Agterberg 1974, Fig. 24)

refinement consists of plotting the data against so-called plotting percentages by using the equation:

$$\text{Plotting percentage} = \frac{3 \times \text{cumulative frequency} - 1}{(3n+1)/100}$$

where n represents number of data (Tukey 1962; Koch and Link 1971). In most practical applications, the preceding refinement is not used and cumulative frequencies are used for plotting percentages. This is because the improvement gained by the refinement generally is very small.

Secondly, a straight line was plotted for the normal biotite distribution. This line passes through two points with (1) the mean (=954.5 Ma) with cumulative frequency of 50 %, and (2) the point with abscissa equal to the mean (=954.5 Ma) minus one standard deviation (=91.3 Ma), and value of 15.9 % along the probability scale.

Thirdly, an approximate 95 % confidence belt was constructed as previously used by Hald (1952) and Vistelius (1960). If z_t represents the standardized value for a point on the straight line for the theoretical normal distribution and z an observed value that may range at random about the line, then the 95 % confidence interval satisfies:

$$z_t \pm 1.96\sqrt{\frac{\Phi(z_t) \cdot \Phi(-z_t)}{n \cdot \varphi^2(z_t)}}$$

If the data come from a normal distribution, a 95 % confidence belt, on the average, contains about 95 % of the plotted observations. The pattern in Fig. 2.2 suggests several departures from normality notably in the upper tail. It also can be seen that the absolute value of the largest difference between observed and expected frequencies is 0.14. This is slightly less than 0.16 representing the 95 % confidence limit according to the two-tailed Kolmogorov-Smirnov test.

2.5 Sampling

Difficulties of sampling rocks at the surface of the Earth and performing measurements on them were discussed in Chap. 1. Geophysical measurements generally are indirect and apply to large volumes of rocks. The determination of rock composition data is more direct but often there is a problem on how to decide on what larger rock unit is represented by the piece of rock subjected to chemical analysis. Also, at the microscopic scale most rocks are heterogeneous and volume of rock sample matters. The purpose of the examples given in this section is to illustrate some of the problems often encountered when rock samples are collected for the purpose of chemical analysis. Normally rocks are crushed before their chemical composition is determined on a sample taken from the resulting powder. This aspect of rock sampling has been studied in detail by geochemists, chemists and mineral engineers (see, e.g., Gy 2004). Drill-core samples can be used to produce good results. In this Section, two examples will be given of a technique called "channel sampling" used in the mining industry for sampling ore *in situ* in the past and at present.

2.5.1 Pulacayo Mine Example

De Wijs (1951) used a series of 118 zinc concentration values (Table 2.4) from samples taken at a regular (2 m) interval along a horizontal drift in the Pulacayo Mine in Bolivia (Fig. 2.7). This series has been used extensively for later study by many authors including Matheron (1962), Agterberg (1974, 2012), Chen et al. (2007), Lovejoy and Schertzer (2007) and Cheng (2014). Average zinc value for Table 2.4 is 15.61 %. This example will again be used later in this book. Geological background on the Pulacayo orebody is provided here and consideration paid to the question of how representative this example is of ore

2.5 Sampling

Table 2.4 Classical example of 118 zinc concentration values (in per cent) from Pulacayo Mine, Bolivia (original data according to de Wijs 1951)

17.7, 17.8, 9.5, 5.2, 4.1, 19.2, 12.4, 15.8, 20.8, 24.1, 14.7, 21.6, 12.8, 11.9, 35.4, 12.3, 14.9, 19.6, 10.6,
15.1, 15.6, 9.3, 8.1, 13.5, 30.2, 29.1, 7.4, 12.3, 13.6, 9.5, 13.1, 27.4, 8.8, 11.4, 6.4, 27.4, 8.8, 11.4, 6 .4,
11.0, 11.4, 14.1, 20.9, 10.8, 15.3, 24.0, 12.3, 7.8, 9.9, 20.7, 25.0, 19.1, 13.1, 27.4, 15.2, 12.2, 10.1, 12.3,
16.7, 18.6, 6.0, 10.6, 11.3, 4.7, 10.9, 6.0, 7.2, 5.6, 8.9, 5.8, 8.9, 6.7, 7.2, 9.7, 10.8, 17.9 10.9, 13.7, 22.3,
10.2, 5.1, 13.9, 9.0, 10.6, 13.8, 6.5, 6.5, 10.6, 10.6, 23.0, 21.8, 32.8, 30.2, 30.8, 33.7, 26.5, 39.3, 24.5,
24.9, 23.2, 16.0, 20.9, 10.3, 22.6, 16.2, 22.9, 36.9, 23.5, 18.5, 16.4, 17.9, 18.5, 13.6, 7.9, 31.9, 14.1, 7.1,
3.9, 3.7, 22.5, 27.6, 17.3

Successive assays are for channel samples that were taken 2 m apart on the 446-m level of this mine

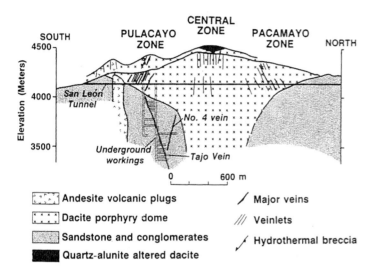

Fig. 2.7 Simplified cross-section of Pulacayo dome with steeply dipping Tajo vein (After Pinto-Vásquez 1993). Mining level depths were measured downward from San Léon Tunnel (Source: Agterberg 2012, Fig. 2)

deposit and rock sampling in general. The geological setting of the Pulacayo Mine and genesis of the sphalerite-quartz ore deposit are briefly described in a scientific communication by Pinto-Vásquez (1993).

The zinc values used by de Wijs (1951) are for channel samples cut at 2-m intervals across the steeply dipping Tajo vein along a horizontal drift on the 446-m level. This level depth was measured downward from elevation of the San Léon Tunnel (Fig. 2.7). The 2.7 km long Tajo vein was discovered in 1883 and mined until 1956. According to Ahlfield (1954), this "silver mine" had the largest annual zinc and second largest annual silver production in Bolivia. On average, the Tajo vein was 1.10 m thick with ore containing 14 % zinc and 0.1 % silver. Relative sphalerite (zinc sulphide) content increased downward in the orebody. According to Turneaure (1971) the age of the Tajo vein was Neogene, probably as young as Pliocene. Figure 2.8 shows ore minerals in Pulacayo massive sulphide at

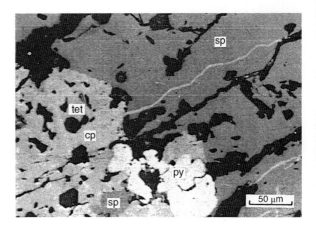

Fig. 2.8 Micrograph of massive sulphide ore in Pulacayo Mine (After Villaipando and Ueno 1987). Ore minerals are *sp* sphalerite, *tet* tetrahedrite, *cp* chalcopyrite, *py* pyrite (Source: Agterberg 2012, Fig. 3)

microscopic scale. The silver was in the form of fine grains associated with tetrahedrite. In a conference report by Villaipando and Ueno (1987) it can be seen that zinc content of sphalerite varied between 65.62 % and 66.03 %. This implies that maximum possible zinc content would be 66 % and this is above the largest value of 39.3 % zinc in Table 2.4. However, because the sampled material consisted not only of massive sulphide but also out of mineralized wall rock, the largest possible value is probably considerably less than 66 %. This upper limit must have constrained maximum possible zinc enrichment.

On the 446-m level, average thickness of massive vein filling averaged only 0.50 m in width but wall rocks on both sides contained disseminated sphalerite, partly occurring in subparallel stringers. The channel samples were cut over a standard width of 1.30 m corresponding to expected mining width. Consequently, each assay value represents average weight percentage zinc for a rod-shaped channel sample of 1.30 m cut perpendicular to the vein (Fig. 2.9). Figure 2.10 is a smoothed version of the 118 values of Table 2.5. The signal-plus-noise method used for this smoothing was described in detail in Agterberg (1974). It assumes that each zinc value is the sum of a "signal" component for continuous change along the series plus a random white-noise component. Together these two components were assumed to produce an autocorrelation function of the type $\rho_h = c \cdot \exp(-ah)$ were h represents distance along the drift as will be discussed in more detail in Chap. 6. Filtering out the noise component produces the signal shown in Fig. 2.10. Various other statistical methods such as simple moving averaging, kriging or inverse distance weighting could be used to produce similar smoothed patterns. The method used to estimate the signal values of Fig. 2.10 will be explained in Sect. 6.1.1.

The fact that the average zinc content (=15.61 %) on the 446-m level differs from the 14 % overall average for the Tajo vein supports Ahlfield's observation that average zinc content increases downwards. Obviously, there existed large-scale zinc-composition "trends" in this ore deposit. In order to capture some of these trends, Agterberg (1961) fitted a sine function to the first 65 values in Table 2.4 but his best-fitting amplitude of 2.77 % is not statistically significant.

2.5 Sampling

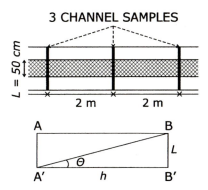

Fig. 2.9 Schematic representation of channel sampling in Pulacayo Mine. Successive channel samples along horizontal mining drift on 446-m level were 1.3-m long and 2-m apart. The Tajo vein, which is 0.5-m wide on average, consists of massive sulphide (hatched pattern) but wall rock on both sides of the vein contained disseminated sulphide and stringers of sulphide ore. Anticipated stoping width was 1.3-m but effective channel width was (L) was set equal to width of vein (=0.5 m). Lag distance (h) is 2 m or multiple of 2 m (Source: Agterberg 2012, Fig. 4)

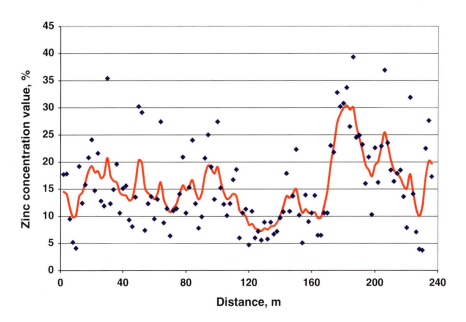

Fig. 2.10 Pulacayo Mine zinc concentration values for 118 2-m channel samples along horizontal drift samples (see Table 2.4 for original data from de Wijs 1951). Sampling interval is 2 m. "Signal" retained after filtering out "Noise" (Source: Agterberg 2012, Fig. 1)

Table 2.5 Example of 61 gold assays (inch-dwt values) taken at 5-ft intervals along a drive in the Virginia Mine

414, 362, 155, 621, 1034, 827, 1034, 1603, 621, 672, 569, 1189, 362, 2741, 776, 6671, 465, 155, 259, 310, 465, 362, 414, 414, 465, 103, 1344, 2637, 155, 569, 155, 52, 52, 155, 2172, 414, 1138, 310, 982, 724, 259, 465, 569, 310, 543, 440, 621, 414, 207, 3051, 2586, 776, 6050, 1086, 362, 517, 776, 621, 827, 724, 362, 284, 259

1 in.-dwt = 3.95 cm-g; digitized from Krige and Ueckermann (1963, Fig. 2)

2.5.2 Virginia Mine Example

Another set of chemical analyses for channel samples (digitized from Krige and Ueckermann 1963, Fig. 2) is shown in Table 2.5. These are 61 gold assays taken at 5-ft intervals along a drive in the Virginia Mine, Witwatersrand gold field, South Africa. Each number is the product of a gold concentration value and thickness of the vein (or gold "reef") reported as an inch-pennyweight value (1 in.-dwt = 3.95 cm-g). Because they were digitized from a graph these values are approximations only. A striking difference between the values in Tables 2.4 and 2.5 is that the gold measures are much more erratic than the zinc values. A measure often used to express degree of variability in a data set is the coefficient of variation (CV) which is the standard deviation divided by the mean. For Table 2.4, CV = 0.51; for Table 2.5, it is 1.33. Nevertheless, one can also see in Table 2.5 that neighboring values in the series tend to be similar. Such autocorrelation is common in chemical element concentration values in rocks, especially for metals. Witwatersrand gold assays also will be used in other examples later in this book.

The spatial autocorrelation of the series shown in Tables 2.4 and 2.5 is similar to that shown in many time series. However, it should be kept in mind that geochemical space series are linear samples taken from 3-D rock bodies that usually display similar or other autocorrelation in other places and directions as well. This fact led Matheron (1962, 1965) to formalize the concept of "regionalized random variable" with properties different from the independent and identically distributed (iid) random variables commonly used in other applications of statistics such as populations of plants and in socio-economic surveys.

References

Agterberg FP (1961) The skew frequency curve of some ore minerals. Geol Mijnb 40:149–162
Agterberg FP (1974) Geomathematics. Elsevier, Amsterdam
Agterberg FP (1990) Automated stratigraphic correlation. Elsevier, Amsterdam
Agterberg FP (2012) Sampling and analysis of element concentration distribution in rock units and orebodies. Nonlinear Process Geophys 19:23–44
Agterberg FP (2013) At the interface between mathematical geoscience and classical statistics. In: Pardo-Igúzquiza E, Guardiola-Albert C, Heredia J, Moreno-Merino L, Durán JJ, Vargas-Guzmán JA (eds) Mathematics of planet earth. Springer, Heidelberg, pp 19–22

References

Ahlfield F (1954) Los Yacimientos minerales de Bolivia. Imp Industrial, Bilbao
Arroyo D, Emery X, Peláez M (2012) Sequential simulation with iterative methods. In: Abrahamsen P, Hauge R, Kolbjørnsen O (eds) Geostatistics Oslo 2012. Springer, Dordrecht, pp 3–14
Bickel PJ, Doksum K (2001) Mathematical statistics, vol 1, 2nd edn. Prentice-Hall, Upper Saddle River
Billingsley P (1986) Probability and measure, 2nd edn. Wiley, New York
Bishop YMM, Fienberg SE, Holland PW (1975) Discrete multivariate analysis. MIT Press, Cambridge, MA
Bliss CI (1935) The calculation of the dosage mortality curve. Ann Appl Biol 22:134–167
Bloomfield P (2000) Fourier analysis of time series. Wiley, New York
Caers J (2011) Modeling uncertainty in the earth sciences. Wiley-Blackwell, Chichester
Chayes F (1956) Petrographic modal analysis. Wiley, New York
Chen Z, Cheng Q, Chen J, Xie S (2007) A novel iterative approach for mapping local singularities from geochemical data. Nonlinear Process Geophys 14:317–324
Cheng Q (2014) Generalized binomial multiplicative cascade processes and asymmetrical multifractal distributions. Nonlinear Process Geophys 21:472–482
Clark I (1970) Practical geostatistics. Elsevier, Amsterdam
Cramér H (1947) Mathematical methods of statistics. Princeton University Press, Princeton
Cressie NAC (1991) Statistics for spatial data. Wiley, New York
Davis JC (1986) Statistics and data analysis in geology, 2nd edn. Wiley, New York
De Wijs HJ (1951) Statistics of ore distribution. Geol Mijnb 30:365–375
Dennison JM, Hay WW (1967) Estimating the needed sampling area for subaquatic ecologic studies. J Paleontol 41:706–708
Deutsch CV (2002) Geostatistical reservoir modeling. Oxford University Press, New York
Doveton JH (2008) Application of Markov mean first-passage time statistics to sedimentary successions: a Pennsylvanian case-study from the Illinois Basin. In: Bonham-Carter G, Cheng Q (eds) Progress in geomathematics. Springer, Heidelberg, pp 411–419
Dowd PA, Pardo-Igúzquiza E (2012) Estimation of the parametric inference of spatial covariances by maximum likelihood. In: Abrahamsen P, Hauge R, Kolbjørnsen O (eds) Geostatistics Oslo 2012. Springer, Dordrecht, pp 129–141
Dubois D, Prade H (2000) Possibility theory, probability theory and multiple-valued logics: a clarification. Ann Math Artif Intell 32:35–66
Eisenhart MA (1963) The background and evolution of the method of least squares. In: Proceedings of the 34th Session International Statistics Institute, Ottawa (preprint)
Eubank RL (1988) Spline smoothing and nonparametric regression. Dekker, New York
Feller W (1968) An introduction to probability theory and its applications. Wiley, New York
Fisher RA (1960) The design of experiments. Oliver and Boyd, Edinburgh
Fisher Box J (1978) R.A. Fisher: the life of a scientist. Wiley, New York
Gandin LS (1965) Objective analysis of meteorological fields. US Department of Commerce Clearinghouse, Springfield
Gnedenko BV (1963) The theory of probability. Chelsea, New York
Goovaerts P (1997) Geostatistics for natural resources evaluation. Oxford University Press, New York
Gy P (2004) 50 Years of Pierre Gy's "theory of sampling – a tribute". Chemometr Intell Lab 74:61–70
Hacking I (2006) The emergence of probability, 2nd edn. Cambridge University Press, Cambridge
Hald A (1952) Statistical theory with engineering applications. Wiley, New York
Isaaks EH, Srivastava RM (1989) An introduction to applied geostatistics. Oxford University Press, New York
Jacobi J (ed) (1951) Paracelsus, selected writings. Guterman, London
Kendall MG (1970) The beginnings of a probability calculus. In: Pearson ES, Kendall MG (eds) Studies in the history of statistics and probability. Hafner, Darien, pp 19–34
Koch GS Jr, Link RF (1971) Statistical analysis of geological data. Wiley, New York
Kolmogorov A (1931) Ueber die analytischen Methoden in der Wahrscheinlichkeitsrechnung. Math Ann 104:415–458

Krige DG (1951) A statistical approach to some basic valuation problems on the Witwatersrand. J S Afric Inst Mining Metallurgy 52:119–139

Krige DG (1978) Lognormal-de Wijsian geostatistics for ore evaluation, South African Institute of Mining and Metallurgy monograph series 1. South African Institute of Mining and Metallurgy, Johannesburg

Krige DG, Ueckermann HJ (1963) Value contours and improved regression techniques for ore reserve valuations. J S Afr Inst Min Metall 63:429–452

Krumbein WC, Dacey MF (1969) Markov chains and embedded Markov chains in geology. J Int Assoc Math Geol 1:79–96

Krumbein WC, Graybill FA (1965) An introduction to statistical models in geology. McGraw-Hill, New York

Lapin LI (1982) Statistics for modern business decisions, 3rd edn. Harcourt, Brace and Jovanovich, New York

Lee YW (1960) Statistical theory of communication. Wiley, New York

Lindley DV (1987) The probability approach to the treatment of uncertainty in artificial intelligence and expert systems. Stat Sci 2(1):17–24

Loève M (1963) Probability theory, 3rd edn. Van Nostrand, Princeton

Lovejoy S, Schertzer D (2007) Scaling and multifractal fields in the solid earth and topography. Nonlinear Process Geophys 14:465–502

Matérn B (1981) Spatial variation, 2nd edn. Springer, Berlin

Matheron G (1955) Application des methods statistiques à l'évaluation des gisements. Ann Min XII:50–74

Matheron G (1962) Traité de géostatistique appliquée, Mém BRGM 14. Éditions Technip, Paris

Matheron G (1965) Les variables régionalisées et leur estimation. Masson, Paris

Matheron G (1967) Kriging or polynomial interpolation procedures? Trans Can Inst Min Metall 70:240–242

Olea RA (1999) Geostatistics for engineers and earth scientists. Kluwer, Norell

Ondrick CW, Griffiths JC (1969) FORTRAN IV computer program for fitting observed count data to discrete distribution models of binomial, Poisson and negative binomial. Kansas Geological Survey Computer Contributions 35

Peredo O, Ortiz M (2012) Multiple-point geostatistical simulation based on genetic algorithms implemented in a shared-memory supercomputer. In: Abrahamsen P, Hauge R, Kolbjørnsen O (eds) Geostatistics Oslo 2012. Springer, Dordrecht, pp 103–114

Pinto-Vásquez J (1993) Volcanic dome associated precious and base metal epithermal mineralization at Pulacayo, Bolivia. Econ Geol 88:697–700

Plackett RI (1970) The principle of the arithmetic mean. In: Pearson ES, Kendall MG (eds) Studies in the history of statistics and probability. Hafner, Darien, pp 121–126

Scherer W (1968) Applications of Markov chains to cyclical sedimentation in the Oficina Formation, eastern Venezuela. Unpublished thesis, Northwestern University, Evanston

Schuenemeyer J, Drew L (2011) Statistics for earth and environmental scientists. Wiley, New York

Simpson T (1755) On the advantage of taking the mean of a number of observations in practical astronomy. Philos Trans R Soc Lond 46:82 (also in vol 10, 1809 ed. Baldwin, London, p 579)

Stigler SM (2008) Karl Pearson's theoretical errors and the advances they inspired. Stat Sci 23:261–271

Tukey JW (1962) The future of data analysis. Ann Math Soc 33:1–67

Turneaure FS (1971) The Bolivian tin-silver province. Econ Geol 66:215–255

Uhler RS, Bradley PG (1970) A stochastic model for petroleum exploration over large regions. J Am Stat Assoc 65:623–630

Villaipando B, Ueno H (1987) El yacimiento argentifero de Pulacayo, Bolivia. Mem Col Inst Geol Econ. University Mayor San Andres, La Paz, pp 203–235

Vistelius AB (1960) The skew frequency distribution and the fundamental law of the geochemical processes. J Geol 68:1–22

Webster R (2001) Geostatistics for environmental scientists. Wiley, New York

Wilks SS (1962) Mathematical statistics. Wiley, New York

Chapter 3
Maximum Likelihood, Lognormality and Compound Distributions

Abstract Estimators obtained by the classical method of moments, although they are unbiased, can have the undesirable property of being relatively imprecise, and more precise estimators can be obtained from the same data by using other methods of estimation. One such method is based on the principle of maximum likelihood that is particularly useful in applications to relatively small samples from positively skewed frequency distributions including the lognormal also to be discussed in this chapter. The maximum likelihood method was invented by Fisher (1922) who developed it as an alternative to Bayesian statistics to which he was strongly opposed. The principle of maximum likelihood is as follows. Suppose that the frequency distribution of a random variable X contains several parameters $\theta_1, \theta_2, \ldots$ that are to be estimated. The probability that a value x_i will fall within a narrow interval Δx is approximately $f(x_i) \Delta x$. The probability that n values will fall where they do is $\Delta x^n \cdot \Pi_n f(x_i)$. As long as Δx is sufficiently narrow, the choice of it does not matter and one can consider the product $\Pi_n f(x_i)$ only. This is the likelihood function. For any given values of the parameters $\theta_1, \theta_2, \ldots$, the likelihood function will assume a specific value. If this value is large it can be assumed that choice of parameters is a good one. The likelihood method consists of maximizing $\Pi_n f(x_i, \theta_j)$ or its logarithm $\Sigma \log f(x_i, \theta_j)$ with respect to the θ_j. The resulting maximum likelihood estimators always have the minimum variance property. However, they can be biased contrary to estimators based on the method of moments. The maximum likelihood method with its rapid convergence to normality for increased sample size is helpful in various geoscience applications including the analysis of age determinations in numerical timescale construction. Other illustrative examples include applications to chemical element concentration values from small sets of observations taken from surface rocks or drill-cores, compositions of larger channel samples from orebodies, thickness measurements of layers in sedimentary rocks, and occurrences of fossils in stratigraphic sections.

Keywords Maximum likelihood • Geologic timescale • Chronogram • Lognormal distribution • Maximum entropy • Generating process • Compound random variable • Exploration strategy • Muskox ultramafic layered intrusion • Witwatersrand gold assays • Canadian Shield mineral deposits

3.1 Applications of Maximum Likelihood to the Geologic Timescale

Several methods have been developed to estimate the ages of boundaries between stratigraphic subdivisions in the Geologic Timescale. Cox and Dalrymple (1967) originally developed an approach for estimating the age of Cenozoic chron boundaries from inconsistent K-Ar age determinations of basaltic rocks. Harland et al. (1982, 1990) adopted this method in their calculations of stage boundary ages for the first two international geologic timescales (GTS82 and GTS90). The basic principle of this approach is as follows: assuming a hypothetical trial age for an observed chronostratigraphic boundary, rock samples stratigraphically above this boundary should be younger, and those below it should be older. An inconsistent date is either an older date for a rock known to be younger, or a younger date for a specimen known to be older. The difference between each inconsistent date and the trial age can be standardized by dividing it by the standard deviation of the inconsistent date. Thus, relatively imprecise dates receive less weight than more precise dates. The underlying assumptions are that: (1) the rock samples are uniformly distributed along the time axis, and (2) the error of each date satisfies a "normal" (Gaussian) error distribution with standard deviation equal to that of the age determination method used. Standardized differences between inconsistent dates and trial age can be squared and the sum of squares (written as E^2) can be determined for inconsistent dates corresponding to the same trial age. Chronograms constructed by Harland et al. (1982) were U-shaped plots of E^2 against different trial ages spaced at narrow time intervals. The optimum choice of age was selected at the trial age where E^2 is a minimum.

Using the maximum likelihood method, Agterberg (1988) made the following improvement to this method. In addition to inconsistent dates, there generally are many more consistent dates for any trial age selected for determination of the age of a chronostratigraphic boundary. The maximum likelihood method can be used to combine consistent with inconsistent dates, resulting in an improved estimate of the age of the chronostratigraphic boundary under consideration. Each standardized difference with respect to a trial age can be interpreted as the fractile of the Gaussian distribution in standard form, and transformed into its corresponding probability. Summation of the logarithmically transformed probabilities then yields the log-likelihood value of the trial date. In this type of calculation, inconsistent dates receive more weight than consistent dates. Consequently, the improvement resulting from using consistent dates, in addition to the inconsistent dates, generally is relatively minor. Only when there are relatively few dates, possibly combined

3.1 Applications of Maximum Likelihood to the Geologic Timescale   75

with a total lack of inconsistent dates, does the use of consistent dates yield significantly better results. The log-likelihood function is beehive-shaped. This topic will be discussed in more detail in the next sections.

3.1.1 Weighting Function Defined for Inconsistent Dates Only Model

Cox and Dalrymple (1967) developed their statistical approach for estimating the age of boundaries between polarity chronozones in the Cenozoic (Brunhes, Matuayana, Gauss and Gilbert chronozones). A slightly modified version of their method was used in Harland et al. (1982) for estimating the ages of boundaries between the stages of the Phanerozoic geologic timescale as follows. Suppose that t_e represents an assumed trial or "estimator" age for the boundary between two stages. Then the n measured ages t in the vicinity of this boundary can be classified as t_y (younger) or t_o (older than the assumed stage boundary). Each age determination t_{yi} or t_{oi} has its own standard deviation s_i. If these standard deviations are relatively large, a number (n_a) of the age determinations is inconsistent with respect to the estimator t_e. Only the n_a inconsistent ages t_{ai} with $t_{oi} < t_e$ and $t_{yi} > t_e$ were used for estimation by Cox and Dalrymple (1967) and Harland et al. (1982). These inconsistent ages may be indicated by letting i go from 1 to n_a. In Harland et al. (1982) the quantity E^2 with

$$E^2 = \sum_{i=1}^{n_a} (t_{ai} - t_e)^2 / s_t^2$$

is plotted against t_e in the chronogram for a specific stage boundary. Such a plot usually has a parabolic form, and the value of t_e for which E^2 is a minimum can be used as the estimated age of the stage boundary.

The preceding approach also can be formulated as follows: Suppose that a stage with upper stage boundary t_1 and lower boundary t_2 is sampled at random. This yields a population of ages $t_1 < t < t_2$ with uniform frequency density function $h(t)$. Suppose further that every age determination is subject to an error that is normally distributed with unit variance. In general, the frequency density function $f(t)$ of measurements with errors that satisfy the density function φ for standard normal distribution satisfies:

$$f(t) = \int_{-\infty}^{\infty} \varphi(t-x) h(t) dx = \frac{1}{t_2 - t_1} \int_{t_1}^{t_2} \varphi(t-x) dx = \frac{1}{t_2 - t_1} [\Phi(t-t_1)] - [\Phi(t-t_1)]$$

For this derivation, the unit of t can be set equal to the standard deviation of the errors. Alternatively, the duration of the stage can be kept constant whereas the standard deviation (σ) of the measurements is changed. Suppose that $t_2 - t_1 = 1$, then

$$f(t) = \Phi\left(\frac{t-t_1}{\sigma}\right) - \Phi\left(\frac{t-t_2}{\sigma}\right)$$

Graphical representations of $f(t)$ for different values of σ were given by Cox and Dalrymple (1967, Fig. 7).

3.1.2 Log-Likelihood and Weighting Functions

Suppose now that the true age τ of a single stage boundary is to be estimated from a sequence of estimator ages t_e by using n measurements of variable precision on rock samples that are known to be either younger or older than the age of this boundary. This problem can be solved if a weighting function $f(x)$ is defined. The boundary is assumed to occur at the point where $x=0$. For the lower boundary (base) of a stage, $\Phi[(t-t_1)/\sigma]$ can be set equal to one yielding the weighting function $f(x>t_e)=1-\Phi(x)$. Alternatively, this weighting function can be derived directly. If all possible ages stratigraphically above the stage boundary have equal chance of being represented, then the probability that their measured age assumes a specific value is proportional to the integral of the Gaussian density function for the errors. In terms of the definitions given, any inconsistent age t_y greater than t_e has $x>0$ whereas consistent ages with $t_y < t_e$ have $x<0$. It is assumed that standardization of an age t_{yi} or t_{oi} can be achieved by dividing either $(t_{yi}-t_e)$ or $(t_{oi}-t_e)$ by its standard error s_i yielding $x_t=(t_{yi}-t_e)/s_i$ or $x_t=(t_{oi}-t_e)/s_i$.

Suppose that x_i is a realization of a random variable X. The weighting function $f(x)$ then can be used to define the probability $P_i = P(X_i - x_i) = f(x_i) \cdot \Delta x$ that x will lie within a narrow interval Δx about x_i. The method of maximum likelihood for a sample of n values x_i consists of finding the value of t_e for which the product of the probabilities P_i is a maximum. Because Δx can be set equal to an arbitrarily small constant, this maximum occurs when the likelihood function

$$L = L(x|t_e) = \prod_{i=1}^{n} f(x_i)$$

is a maximum. Taking the logarithm at both sides of this equation, the model becomes as graphically illustrated in Fig. 3.1a:

$$\log L(x|t_e) = \sum_{i=1}^{n} \log[1 - \Phi(x_i)]$$

If the log-likelihood function is written as y and its first and second derivatives with respect to t_e as y' and y'', respectively; then the maximum likelihood estimator of the true age τ occurs at the point where $y'=0$ and its variance is $-1/y''$ (cf. Kendall and Stuart 1961, p. 43). The log-likelihood function becomes parabolic in

3.1 Applications of Maximum Likelihood to the Geologic Timescale 77

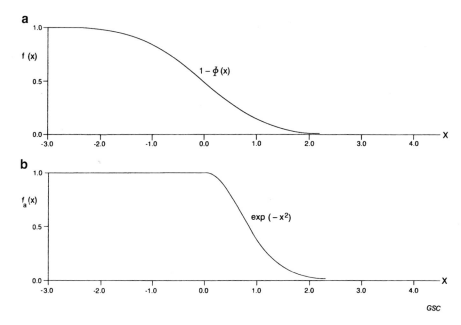

Fig. 3.1 Two different weighting functions on basis of which the likelihood function can be estimated. (**a**). The function $f(x)$ follows from the assumption that every age determination is the sum of random variables for (1) uniform distribution of (unknown) true ages, and (2) Gaussian distribution for measurements. (**b**). The function $f_a(x)$ is for inconsistent ages only with log-likelihood function $-E^2$ (Source: Agterberg 1990, Fig. 3.12)

shape when n is increased. If the equation of the parabola is written as $y = a + b \cdot t_e + c \cdot t_e^2$, the maximum likelihood estimate $MLE(\tau)$ becomes $-b/2c$ with variance $-1/2c$. It will be shown on the basis of a practical example in the next section that a chronogram using E^2 represents the maximum likelihood solution for a different kind of filter with equation:

$$f_a(x|t_e) = \exp\left[-(x - t_e)^2\right]$$

where $x > t_e$ because n_a inconsistent ages are used only. This weighting function is shown in Fig. 3.1b. If the corresponding likelihood function is written as L_a, it follows that $E^2 = -\log_e L_a$.

3.1.3 Caerfai-St David's Boundary Example

For example, the quantity E^2 is plotted in the vertical direction of Fig. 3.2 for the Caerfai-St David's boundary example taken from Harland et al. (1982, Fig. 3.7i). The data on which this chronogram is based are shown along the top. Values of E^2 were calculated at intervals of 4 Ma and a parabola was fitted to the resulting values

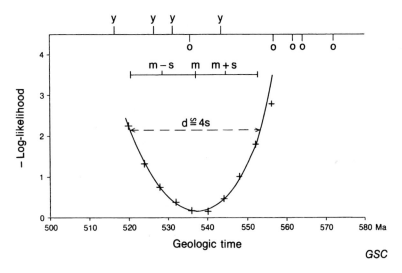

Fig. 3.2 Chronogram for Caerfai St. David's boundary example with parabola fitted by method of least squares. $E^2 = -\log$-likelihood function is plotted in vertical direction creating a basket. Dates belonging to stages, which are older and younger than boundary, are indicated by o and y, respectively. Standard deviation follows from d representing width of parabola for E^2 equal to its minimum value augmented by 2 (Source: Agterberg 1990, Fig. 3.13)

by using the method of least squares. If the log-likelihood function is parabolic with $E^2 = a + b \cdot t_e + c \cdot t_e^2$, it follows that the maximum likelihood estimator is normally distributed with mean $m = b/2c$ and variance $s^2 = 1/2c$. It will be shown in next paragraph that, graphically, s can be determined by taking one fourth of the width of the parabola at the point where E^2 exceeds its minimum value by 2 (see Fig. 3.2). This method differs from the procedure followed by Harland et al. (1982) who defined the error of their estimate by taking one-half the age range for which E^2 does not exceed its minimum value by more than 1. This yields a standard deviation that is $\sqrt{2}$ times larger than the one resulting from L_a.

According to the theory of mathematical statistics (Kendall and Stuart 1961, pp. 43–44), the likelihood function is asymptotically normal, or:

$$e^y = \frac{1}{\sigma\sqrt{2\pi}} \exp\left(-\frac{t^2}{2\sigma^2}\right)$$

Here $e^y = L(x|t_e)$ and $t = t_e - \tau$; σ represents the standard deviation of this asymptotically normal curve that is centered about $t = 0$. Taking the logarithm at both sides gives the parabola $y = \max - t^2/2\sigma^2$ where max represents the maximum value of the log-likelihood function. Setting $y = \max - 2$ then gives $t = 2\sigma$. Consequently, the width of the parabola at two units of y below its maximum value is equal to 4σ. The parabola shown in Fig. 3.2 (and subsequent illustrations) is assumed to provide an approximation of the true log-likelihood function. The standard deviation obtained from the nest-fitting parabola is written as s. In Fig. 3.2 the Y-axis has

3.1 Applications of Maximum Likelihood to the Geologic Timescale

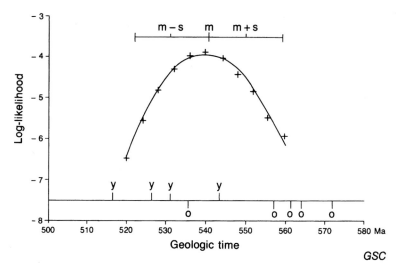

Fig. 3.3 Caerfai St. David's boundary example. Age estimated by maximum likelihood method using L. Standard deviation (s) and width of 95 % confidence interval are approximated closely by results shown in Fig. 3.2 (Source: Agterberg 1990, Fig. 3.14)

been inverted so that $-y = E^2$ points upwards in order to facilitate comparison with chronograms in Harland et al. (1982).

Figure 3.3 shows estimates based on L. The resulting best-fitting parabola is almost equal to the one in Fig. 3.2 that was based on L_a instead of L. Consequently, the estimated ages of the Caerfai-St David's boundary and their standard deviations obtained from L_a and L also are similar. This conclusion will be corroborated by a more detailed comparison of the weighting functions for L and L_a at the end of this section, and by computer simulation experiments to be described in the next section. However, it will be shown that L_a does not provide a good approximation when inconsistent data are missing. When n is small, L also produces better results than L_a because a parabolic chronogram is more readily obtained when the consistent ages are used together with the inconsistent ages as will be illustrated by the following example. An age estimate based on Harland et al.'s (1982, Fig. 3.4h, p. 54) chronogram for the Norian-Rhaetian boundary is 213 Ma. The corresponding standard error reported by Harland et al. (1982) is 9 Ma. The L-based maximum likelihood method using the same data set of only six dates gives an estimated age of 215.5 Ma with standard error of 4.2 Ma.

3.1.4 The Chronogram Interpreted as an Inverted Log-Likelihood Function

The approach taken in this section differs slightly from the one originally taken by Cox and Dalrymple (1967). The basic assumptions that the dates are uniformly distributed through time and subject to measurement errors are made in both methods of approach. Cox and Dalrymple (1967, see their Fig. 4 on p. 2608) demonstrated that, under these conditions, the inconsistent dates for younger rocks have probability of occurrence P_{ly} with:

$$P_{ly}(t) = \tfrac{1}{2}\text{erfc}\left[\frac{t-\tau}{\sigma_m\sqrt{2}}\right]$$

where erfc denotes complementary error function and τ is true age of the chronostratigraphic boundary (=boundary between geomagnetic polarity epochs in Cox and Dalrymple's original paper). The standard deviation for measurement errors is written as σ_m. Setting $\tau=0$ and using the relationship $\tfrac{1}{2}\text{erfc}\,(z/\sqrt{2})=1-\Phi(z)$ it follows that:

$$P_{ly}(t) = 1 - \Phi\left(\frac{t}{\sigma_m}\right) = f\left(\frac{t}{\sigma_m}\right)$$

If t/σ_m is replaced by x, the weighting function shown in Fig. 3.1 is obtained. Consequently, this weighting function can be interpreted as the probability that an inconsistent age t_a is measured for the younger rocks. Likewise, $P_{lo}(t) = f(-t/\sigma_m)$ can be interpreted as the probability that an inconsistent age t_a can be defined for older rocks.

Cox and Dalrymple (1967) next introduced the trial boundary age t_e and defined a measure of dispersion of all inconsistent dates t_a satisfying:

$$D^2(t_a - t_e) = \int_{-\infty}^{\infty} (t - t_e)^2 P_I(t) dt$$

where $P_I(t) = P_{ly}(t)$ if $t \geq 0$; and $P_I = P_{lo}(t)$ if $t < 0$. For $t_e = \tau$, this quantity is a minimum (see Cox and Dalrymple 1967, Fig. 5 on p. 2608). A normalized version of E^2 can be directly compared to the theoretical error for $D^2(t_a-t_e)$ when the number of inconsistent dates is large. This normalization consists of dividing E^2 by average number of dates per unit time interval. It is noted that $P_I(t)$ does not represent a probability density function because it can be shown that

$$\int_{-\infty}^{\infty} P_I(t) dt = \sqrt{2/\pi} = 0.798 < 1 \text{ if } t_e = \tau$$

In this section, E^2 is not interpreted as a quantity that is approximately proportional to $D^2(t_a-t_e)$. Instead of this, it is regarded as the inverse of a log-likelihood function with Gaussian weighting function. For very large samples, good estimates can be obtained using the inconsistent dates only. For small samples, however, significantly better results are obtained by using the consistent dates together with the inconsistent dates by replacing the Gaussian weighting function by $f(x)$.

All Gaussian weighting functions provide the same mean age for a chronostratigraphic boundary when the maximum likelihood method is used. However, the

3.1 Applications of Maximum Likelihood to the Geologic Timescale

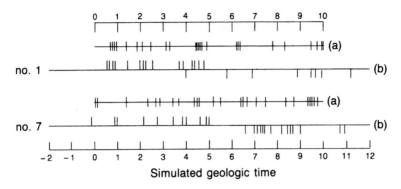

Fig. 3.4 Two examples of runs (No. 1 and No. 7) in computer simulation experiment. True dates (a) were generated first, classified and increased (or decreased) by random amount. Younger and older ages are shown above and below scale (b), respectively (Source: Agterberg 1990, Fig. 3.15)

standard deviation of this mean depends on the choice of the constant p in exp $(-px^2)$. For example, $p = 1.0$ for $f_a(x)$ in Fig. 3.1b. Assuming that $f(x)$ of Fig. 3.1a represents the correct weighting function, one can ask for which value of p the Gaussian function exp $(-px^2)$ provides the best approximation to $f(x)$ with $x \geq 1$. Let u represent the difference between the two curves, so that $\log_e \{1 - \Phi(x)\} = -px^2 + u$. Minimizing Σu^2 for $x_i = 0.1 \cdot k$ ($k = 1, 2, \ldots, 20$) by the method of least squares gives $p_{opt} = 1.13$. Because of the large difference between the two curves near the origin, p_{opt} increases when fewer values x_i are used. It decreases when more values are used. Letting k run to 23 and 24, respectively, yields p_{opt} values equal to 1.0064 and 0.9740, respectively. These results confirm the conclusion reached previously that a Gaussian weighting function with $p = 1.0$ provides an excellent approximation to $f(x)$.

3.1.5 Computer Simulation Experiments

Computer simulation in geoscience has had a long history of useful applications (Harbaugh and Bonham-Carter 1970). Computer simulation experiments were performed by Agterberg (1988) in order to attempt to answer the following questions: (a) does the theory of the preceding sections remain valid even when the number of available dates is very small; (b) how do estimates obtained by the method of fitting a parabola to the log-likelihood function compare to estimates obtained by the method of scoring which is commonly used by statisticians in maximum likelihood applications (see, e.g., Rao 1973); and (c) how do results derived from the chronograms in Harland et al. (1982) compare to those obtained by the maximum likelihood method.

Figure 3.4 and Table 3.1 illustrate the type of computer simulation experiment performed. Twenty-five random numbers were generated on the interval [0, 10].

Table 3.1 Run 1 in computer simulation experiment

τ	t		(x = t−4.6)	z	P	log$_e$P
4.587	A	4.380	−0.220	0.220	0.5871	−0.5325
7.800	B	8.048	3.448	3.448		
2.124	A	2.193	−2.407	2.407	0.9920	−0.0081
0.668	A	2.239	−2.361	2.361	0.9909	−0.0092
6.225	B	5.802	1.202	1.202	0.8853	−0.1218
9.990	B	9.945	5.345	5.345		
4.896	A	4.574	−0.026	0.026	0.5102	−0.6730
4.606	A*	6.487	1.887	−1.887	0.0296	−3.5211
0.796	A	0.553	−4.047	4.047		
1.855	A	2.526	−2.074	2.074	0.9810	−0.0192
6.292	B	6.923	2.323	2.323	0.9899	−0.0101
3.280	A	1.998	−2.602	2.602	0.9954	−0.0046
2.422	A	1.435	−3.165	3.165		
1.397	A	0.912	−3.688	3.688		
4.538	A	4.365	−0.235	0.235	0.5928	−0.5230
0.830	A	0.803	−3.797	3.797		
6.194	B*	4.033	−0.567	−0.567	0.2854	−1.2540
4.545	A	3.930	−0.670	0.670	0.7490	−0.2890
4.774	A*	4.814	0.214	−0.214	0.4154	−0.8786
0.905	A	0.713	−3.887	3.887		
9.763	B	11.197				
8.285	B	8.902	4.302	4.302		
3.131	A	3.676	−0.924	0.924	0.8224	−0.1955
9.987	B	9.435	4.835	4.835		
9.442	B	9.620	5.020	5.020		
					Total =	−8.0397

True dates t were classified as A (younger) or B (older) than 5 representing true age of stage boundary. Dates t with measurement error are compared to trial age ($t_e = 4.6$). Inconsistent ages are indicated by asterisks. $Z = -x$ for younger rocks (A) and $z = x$ for older rocks (B). Standard z-value is fractile of probability P. Total of logs of P gives value of log-likelihood function for $t_e = 4.6$ (Source: Agterberg 1990, Table 3.2)

These numbers with uniform frequency distribution can be regarded as true dates (τ) without measurement errors. The stage boundary was set equal to 5 (=mid-point of interval). Results of two runs (1 and 7) are taken for example here. Values of τ less than 5 belong to the younger stage A, and those greater than 5 to the older stage B (Table 3.1). The measurement error was introduced by adding to τ a normal random number with zero mean and standard deviation equal to one. As a result of this, each value of τ was changed into a date t. Some values of t ended up outside the interval [0, 10], like 11.197 for Run 1 (Fig. 3.4 and Table 3.1), and were not used later. In Run 1, a single date for the younger stage (A) has $t > 5$, and a date for B has $t < 5$. Suppose now, for example, that the trial age of the stage boundary t_e is set equal to 4.6. Then there are three inconsistent ages for run No. 1 and these are marked by asterisks in Table 3.1. Each normalized date $x = t - t_e$ was converted into

3.1 Applications of Maximum Likelihood to the Geologic Timescale

a z-value (=fractile of normal distribution in standard form) by changing its sign if it belongs to the younger stage A. The value of z was transformed into a probability $P = \Phi(z)$ for values on the interval $[t_e-3, t_e+3]$. The frequency corresponding to 3 is equal to 0.999 of which the natural logarithm is equal to -0.001. Consequently, values outside the interval $[t_e \pm 3]$ yield probabilities which are approximately 1 (or 0 for the log-likelihood function) and these were not used for further analysis. Thus a natural window is provided screening out dates that are not in the vicinity of the age of chronostratigraphic boundary that is to be estimated. Most probabilities are greater than 0.5. Only inconsistent dates (asterisks in Table 3.1) give probabilities less than 0.5. The value of the log-likelihood function for t_e is the sum of the logs of the probabilities as illustrated for $-t_e = 4.6$ in Table 3.1.

Log-likelihood values for Run No. 1 are shown in Table 3.2 with t_e ranging from 3 to 7 in steps of 0.1. The largest log-likelihood value is reached for $t_e = 5.6$ and this value was selected as the first approximation of the true age of the stage boundary. Ten values before and after 5.6 were used to fit a parabola that is shown in Fig. 3.5. The fitted parabola is more or less independent of the number of values (=21) used to fit it and of width of neighborhood (=2). However, the neighborhood should not be made too wide because of random fluctuations (local minima and maxima near $t_e = 3$ or 7; see e.g. Table 3.2). Such edge effects should be avoided. They are due to the fact that the initial range of simulated time was arbitrarily set equal to 10 in the computer simulation experiment. The peak of the best-fitting parabola provides the second approximation (=m) of the estimated age. The standard deviation (=s) of the corresponding normal distribution can be used to estimate the 95 % confidence interval $m \pm 1.96 s$ also shown in Fig. 3.5.

The sum of squares E^2 for L_a, using inconsistent dates only, also is tabulated in Table 3.1 as a function of t_e. The first approximation of its minimum value is 5.3. The corresponding parabola is shown in Fig. 3.5. Te mean value resulting from L_a is about 0.3 less than the mean based on L and its standard deviation is nearly the same. It is fortuitous that the mean based on L_a is closer to the population mean (-5) than that based on L. On the average, the L method gives better results (see results for 50 runs summarized at the end of this section).

Means and standard deviations obtained for the first ten runs of the computer simulation experiment are listed in Table 3.3. If they could be estimated, results obtained for L_a are close to those for L. The estimated standard deviations tend to be slightly smaller or much greater. It can be seen from the results for Run No. 7 shown in Fig. 3.6 that the large standard deviations are due to a break-down of the L_a method if there are no inconsistent dates. Results obtained by means of the method of scoring (see, e.g., Rao 1973, pp. 366–374) also are shown in Table 3.3. In application of this method, the following procedure was followed. As before, the log-likelihood was calculated for 0.1 increments in t_e and the largest of these values was used as the initial guess. Suppose that this value is written as y. Then two other values x and z were calculated representing log-likelihood values close to y at very small distances -10^{-4} and 10^4 along the t_e-axis. The quantities $D_1 = 0.5(z-x) \cdot 10^4$ and $D_2 = (x-2y+z) \cdot 10^8$ were used to obtain a second approximation of the mean by subtracting D_1/D_2 from the initial guess. The procedure was repeated until the

Table 3.2 Values of log-likelihood functions estimated for Run 1 and predicted values for parabola fitted by method of least squares

Time	Log-likelihood ($\Sigma \log P$)	Sum of squares (E^2)	Predicted LLF	Predicted E^2
3.0	−15.58	10.86		
3.1	−14.41	9.37		
3.2	−13.30	8.00		
3.3	−12.27	6.75		
3.4	−11.31	5.63		
3.5	−16.98	13.54		
3.6	−15.83	12.07		
3.7	−14.75	10.73		
3.8	−13.75	9.52		
3.9	−12.81	8.43		
4.0	−11.94	7.46		
4.1	−11.13	6.59		
4.2	−10.39	5.84		
4.3	−9.72	5.21		5.11
4.4	−9.10	4.69		4.69
4.5	−8.54	4.27		4.32
4.6	−8.04	3.93	−7.98	3.99
4.7	−7.59	3.65	−7.57	3.71
4.8	−7.20	3.44	−7.21	3.47
4.9	−6.87	3.27	−6.89	3.28
5.0	−6.58	3.15	−6.61	3.14
5.1	−6.35	3.06	−6.38	3.04
5.2	−6.16	3.02	−6.19	2.99
5.3**	−6.02	3.01**	−6.05	2.98**
5.4	−5.93	3.05	−5.95	3.01
5.5	−5.88	3.13	−5.89	3.09
5.6*	−5.88*	3.24	−5.88*	3.22
5.7	−5.92	3.40	−5.91	3.39
5.8	−6.00	3.59	−5.98	3.61
5.9	−6.13	3.84	−6.10	3.88
6.0	−6.29	4.15	−6.26	4.18
6.1	−6.49	4.51	−6.46	4.54
6.2	−6.73	4.94	−6.71	4.94
6.3	−7.01	5.42	−7.00	5.38
6.4	−7.33	5.97	−7.33	
6.5	−7.69	6.57	−7.71	
6.6	−8.08	7.23	−8.13	
6.7	−8.50	7.91		
6.8	−8.97	8.65		
6.9	−9.47	9.43		
7.0	−10.01	10.24		

Initial guesses of extreme values are indicated by asterisks (Source: Agterberg 1990, Table 3.3)

3.1 Applications of Maximum Likelihood to the Geologic Timescale

Fig. 3.5 Maximum likelihood methods of Fig. 3.4 used for estimating mean age of stage boundary in Run 1. (**a**) All dates; (**b**) Inconsistent dates only (Source: Agterberg 1990, Fig. 3.16)

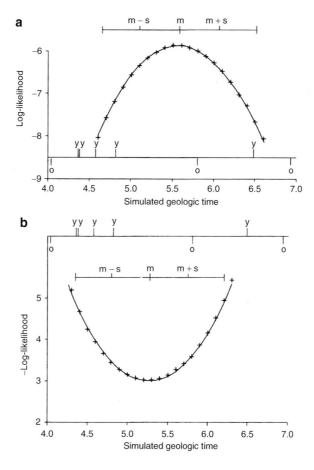

Table 3.3 Summary of results for first 10 runs of computer simulation experiment

	Maximum likelihood method					Gaussian weighting function				
Run		Parabola		Scoring			Parabola		Scoring	
No.	Mid-point	Mean	S.D.	Mean	S.D.	Mid-point	Mean	S.D.	Mean	S.D.
1	5.6	5.582	0.479	5.554	0.481	5.3	5.269	0.470	5.260	0.500
2	5.7	5.632	0.481	5.663	0.489	6.3	6.190	0.480	6.264	0.500
3	5.1	5.153	0.420	5.142	0.423	4.8	4.884	0.335	4.828	0.316
4	4.5	4.506	0.447	4.507	0.452	4.2	4.321	0.395	4.216	0.354
5	5.1	5.070	0.461	5.089	0.466	5.3	5.217	0.482	5.293	0.408
6	4.4	4.419	0.502	4.448	0.505	4.6	4.625	0.749*		
7	5.7	5.710	0.531	5.728	0.542	5.8	5.767	3.924*		
8	5.2	5.205	0.406	5.200	0.411	5.0	5.025	0.364	5.017	0.408
9	5.0	5.022	0.417	5.018	0.419	5.0	4.966	0.614*		
10	4.2	4.231	0.609	4.232	0.623	4.3	4.248	1.001*		

Comparison of estimates obtained by fitting parabola and scoring method, respectively. Standard deviations marked by asterisks are too large (Source: Agterberg 1990, Table 3.4)

Fig. 3.6 Maximum likelihood methods of Fig. 3.4 used for estimating mean age of stage boundary in Run 7. Log-likelihood function $-E^2$ does not provide a good result in this application. (**a**) All dates; (**b**) Inconsistent dates only (Source: Agterberg 1990, Fig. 3.18)

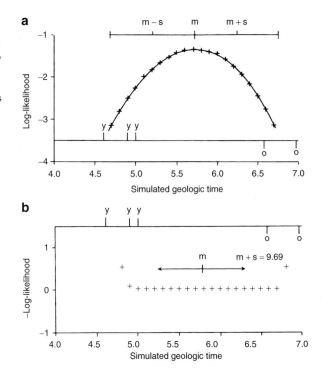

difference between successive approximations became negligibly small, Then the standard deviation of the final estimate is given by $SD = 1/|D_2|$.

For L, the scoring method generally yields estimates of SD which are slightly greater than those resulting from the parabola method. However, the difference is negligible small (Table 3.3). For L_a the scoring method provides an answer only in six of the experiments of Table 3.3. Similar results were obtained in a second type of computer simulation experiment using variable measurement error (see Agterberg 1988, for details). In total, 50 runs were made for each of the two types of experiments. For constant variance of measurement errors, the parabola method for L gave an overall mean of 4.9287 and standard deviation 0.4979 as calculated from 50 means. The corresponding numbers for the second type of experiment were 4.9442 and 0.5160. The Gaussian weighting scheme provided overall means equal to 4.9213 and 4.9414 in the two types of experiments, and corresponding standard deviations equal to 0.5790 and 0.6541, respectively. If the parabola did not provide a good fit to the function E^2, because of zero values around its minimum, the mean was approximated by the mid-point of the range of zero values in these calculations. The results of the 50 runs for the two types of experiments confirm the earlier results described in this section. Additionally, they show that the Gaussian weighting function (using L_a) provides results that are almost as good as the method of maximum likelihood, unless there are no inconsistent dates in the data set.

3.1 Applications of Maximum Likelihood to the Geologic Timescale 87

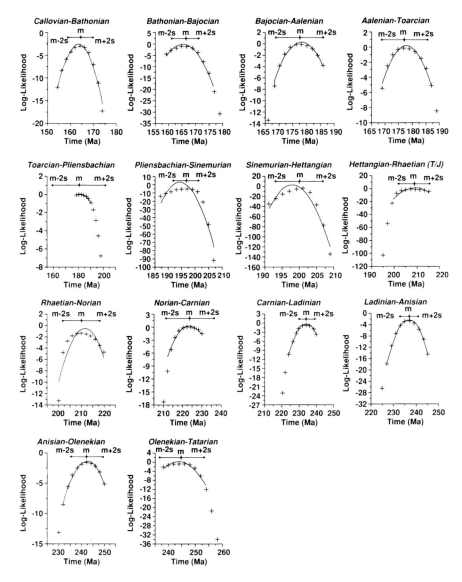

Fig. 3.7 Ages of Mesozoic stage boundaries estimated by maximum likelihood (Modified from Gradstein et al. 1995, Fig 6)

3.1.6 Mesozoic Timescale Example

The maximum likelihood method (L method) was used repeatedly for stage boundary age estimation during the 1990s (*cf.* Agterberg 1994). An example is shown in Fig. 3.7. Estimates obtaining by using the scoring method were refined by splining, a method to be discussed in Chap. 9. Details of how the Mesozoic time was computed can be found in Gradstein et al. (1994, 1995). A general disadvantage of the chronogram and maximum likelihood methods is that the relative stratigraphic position of any rock sample is generalized with respect to stage boundaries that are relatively far apart in time. The relative stratigraphic position of one sample with respect to others within the same stage is not considered. A better approach for estimating the age of stage boundaries currently is to incorporate precisely known stratigraphic positions for which high-precision age determinations are available. More data of this type have become available during the past 20 years resulting in two international geologic time scales (GST2004 and GST2012). Moreover, the establishment of GSSP's (GSSP = Global Stratigraphic Section and Point; see Chap. 9) is allowing more precise estimation of the age from extensive data at or near stage boundaries as in the following example for the age of Maastrichtian-Paleocene boundary.

Precision of Cenozoic epoch boundaries in GTS2012 (Gradstein et al. eds, 2012) has become excellent because of use of the astronomical clock. This improvement applies to Cenozoic stage boundaries as well. Remaining uncertainty of ages of stage boundaries is illustrated by the following example. Currently, the best age estimate of the Maastrichtian-Paleocene boundary is 66.0 ± 0.5 Ma. This 95 % confidence interval extends to the GTS2004 (Gradstein et al. eds, 2004) estimate of 65.5 ± 0.3 Ma. The GTS2004 estimate was close to Renne et al.'s (1998) estimate of 65.46 ± 1.26 Ma, which was based on back-calculation with external errors of Swisher et al.'s (1993) estimate for an ash layer probably coincident with the Cretaceous-Paleogene boundary. New estimates for this boundary include two ages recently proposed by Husson et al. (2011) which are (1) 65.59 ± 0.07 Ma on the basis of 405 ka eccentricity variation resulting from astronomical solution La2010a, and (2) 66 ± 0.07 Ma in coherence with radio-isotope datings. However, Renne et al. (2012)'s preferred estimate of 66.043 ± 0.086 Ma supports the GTS2012 estimate of the Maastrichtian-Paleocene boundary. These authors ascribe earlier estimates, which are closer to 65.5 Ma as in GTS2004, to previous miscalibration by two 405 ka eccentricity cycles (Renne et al. 2012, p. 684).

Undoubtedly, there will be further improvements of the numerical geological time scale in future. These may include the following: it is likely that the astronomical clock will be extended downward from the Cenozoic into the Mesozoic and older periods; zone boundary age estimation may become possible (in addition to further improved stage boundary age estimation); future GSSPs will help to refine the Mesozoic and Paleozoic time scales; and the Precambrian time scale will continue to be further improved. Rapidly increasing numbers of high-precision dates are bound to significantly improve overall precision in future time scales, although GTS2012 already is more accurate and precise than its predecessors. The maximum likelihood method discussed in this chapter may become useful again in future for finer chronostratigraphic sub-divisions.

3.2 Lognormality and Mixtures of Frequency Distributions

A random variable X has lognormal frequency distribution if the logarithms of its values are normally distributed. Theory of the lognormal distribution is explained in detail by Aitchison and Brown (1957). In general, if X represents a random variable

3.2 Lognormality and Mixtures of Frequency Distributions

with cumulative distribution $F_2(x)$, and one knows the distribution $F_1[h(x)]$ where $h(x)$ is a function of x, then $F_2(x) = F_1[h(x)]$. Differentiation of both sides with regard to x gives:

$$\frac{dF_2(x)}{dx} = \frac{dF_1[h(x)]}{dh(x)} \cdot \frac{dh(x)}{dx}$$

Box 3.1: Moments of Lognormal Distribution

If $h(x) = \log_e x$, the lognormal density function becomes:

$$f_2(x) = \frac{1}{x\sigma\sqrt{2\pi}} \exp\left[-\tfrac{1}{2}\left(\frac{\log_e x - \mu}{\sigma}\right)^2\right].$$

In this expression the mean μ and standard deviation σ are for logarithmically (base e) transformed data. If X represents the random variable with lognormal distribution, its moments can be derived as follows. The moment generating function of the normal distribution is: $m(u) = \exp[\mu u + \sigma^2 u^2/2]$. From the equation for the moment generating function of a continuous random variable, it can be derived that for the lognormal distribution: $\mu'_r = \exp[\mu r + \sigma^2 r^2/2]$. Consequently, the moments of the lognormal distribution constitute the moment generating function of the normal distribution. The mean and variance written as α and β^2, are:

$$\mu(X) = \exp\left[\mu + \tfrac{\sigma^2}{2}\right];\ \sigma^2(X) = \exp[2\mu + 2\sigma^2] - \exp[2\mu + \sigma^2].$$

3.2.1 Estimation of Lognormal Parameters

Examples of application of the method of moments to the data of Tables 2.4 and 2.5 are as follows. The 118 zinc values of Table 2.4 produce the estimates $m = 2.6137$ and $s^2 = 0.2851$ after logarithmic transformation. It follows that $m(x) = 15.74$ and $s^2(x) = 81.74$ for the untransformed zinc values. However, direct estimation of mean and variance of the 118 zinc values gives mean and variance equal to 15.61 and 64.13, respectively. The slightly larger value of variance ($s^2(x) = 81.74$) obtained by the method of moments suggests a slight departure from lognormality (large-value tail slightly weaker than lognormal) possibly related to the fact that the largest possible zinc value is significantly less than 66 % (see Sect. 2.4.1). Similar estimates based on the 61 gold values of Table 2.5 are as follows. Mean and standard deviation of original data are 907 and 1,213 in comparison with 897 and 1,088 based on logarithmically transformed values. This is not a large difference indicating that the sample size in this example is sufficiently large so that it is not necessary to apply the following method that is useful for smaller data sets or when positive skewness is larger.

Finney (1941) and Sichel (1952, 1966) have used the maximum likelihood method to estimate the mean α and variance β^2 of the lognormal distribution from the mean variance (s^2) of n logarithmically transformed values. Their results are:

$$\hat{\alpha} = e^{\overline{\log_e(x)}} \cdot \Psi_n\left(\frac{s^2}{2}\right); \hat{\beta}^2 = e^{2 \cdot \overline{\log_e(x)}} \left\{\Psi_n(2s^2) - \Psi_n\left(\frac{n-2}{n-1}s^2\right)\right\}$$

where $\Psi_n(t) = 1 + \frac{n-1}{n}t + \frac{(n-1)^2}{n^2(n+1)}\frac{t^2}{2!} + \frac{(n-1)^5}{n^2(n+1)(n+3)}\frac{t^3}{3!} + \cdots$

The series defining $\Psi_n(t)$ converges only slowly, but its values for variable n and t can be calculated quickly by digital computer.

An example of application of the maximum likelihood method was given in Agterberg (1974). Previously, McCrossan (1969) had shown that probable ultimate recoverable oil reserves of 52 Leduc reef pools in Alberta satisfy a lognormal distribution. The arithmetic mean and variance of the natural logs of these numbers were 15.011 and 6.983. It follows that $t = s^2/2 = 3.491$. The maximum likelihood estimate of the mean becomes $83.5 \cdot 10^6$ barrels. This unbiased estimate exceeds the arithmetic mean of $53 \cdot 10^6$ barrels. Although the arithmetic mean also is unbiased, it underestimates the true mean of values drawn from the lognormal distribution in this example because it is relatively imprecise.

Approximate lognormality is widespread in the geosciences. Examples include small-particle statistics, grain size (Krumbein 1936); thickness of sedimentary layers of different lithologies (Kolmogorov 1951); ore assays; element concentration values in rocks (Krige 1951); size of oil and gas fields (Kaufman 1963); and trace-element concentration values in rocks (Ahrens 1953). Two methods of explaining positive skewness of frequency curves were recognized by Vistelius (1960). The first, better known, explanation is based on the theory of proportionate effect (see next paragraph). The second model (Vistelius model) is based on the theory that one may be sampling a mixture of many separate frequency distributions, which are interrelated in that they share the same coefficient of variation.

The theory of proportionate effect was originally formulated by Kapteyn (1903) and later in a more rigorous manner by Kolmogorov (1941, 1951). It can be regarded as another type of application of the central-limit theorem. Suppose that a random variable X was generated stepwise by means of a generating process that started from a constant X_0. At the i-th step of this process ($i = 1, 2, \ldots, n$) the variable was subject to a random change that was a proportion of some function g (X_{i-1}), or $X_i - X_{i-1} = \varepsilon_i \cdot g(X_{i-1})$. Two special cases are: $g(X_{i-1}) = 1$ and $g(X_{i-1}) = X_{i-1}$. If g remains constant, X becomes normal because of the central-limit theorem. In the second case, it follows that:

$$\sum_{i=1}^{n}\frac{X_i - X_{i-1}}{X_{i-1}} = \sum_{i=1}^{n}\varepsilon_i \approx \int_{X_0}^{X_n}\frac{dx}{x} = \log_e X_n - \log_e X_0$$

This generating process has the lognormal distribution as its end product. There is an obvious similarity between generating processes and the multiplicative cascade processes to be discussed later in this book in the context of fractals and multifractals. The Vistelius model of lognormality will be applied in the next section.

3.2.2 Muskox Layered Intrusion Example

Vistelius (1960, p. 11) argued as follows. Suppose that a geological process, at any time, yields a different population of values. If a given part of the Earth's crust is sampled, a set of values is obtained which may reflect various stages of the same geological process. In some places, the process may have developed further than in other places. The sampled population then consists of a mixture of many separate populations that are, however, interrelated by a single process. Suppose that for each of these populations, the standard deviation σ is proportional to the mean μ. This is equivalent to assuming that the coefficient of variation $\gamma = \sigma/\mu$ is constant. In that situation, the sampled population can have a positively skew frequency distribution because the subpopulations with relatively large means and standard deviations will generate a long tail of large values. The frequency distribution $f(x)$ then satisfies $f(x) = \Sigma\, w_i \cdot f_i$ where each $f_i(x)$ is the distribution representing the process at stage i; w_i is a weighting factor for that stage. It represents the proportion by which the subpopulation $f_i(x)$ occurs in the sampled population.

In order to test this theory, Vistelius (1960) compiled averages and standard deviations for phosphorus (as weight percent P_2O_5) in granitic rock from various areas. He found that average and standard deviation are positively correlated with a correlation coefficient of 0.56. Individual distributions $f(x)$ are approximately normal but the joint distribution $f_i(x)$ with all values lumped together is positively skew.

The Vistelius model was tested for the element copper in various layers of the 1,175 my old Muskox layered ultramafic gabbroic intrusion, District of Mackenzie, northern Canada. The layers were formed by crystallization differentiation of basaltic magma (Smith 1962; Smith and Kapp 1963). Individual layers are approximately homogeneous with regard to the major rock-forming minerals. Layers that are rich in olivine, such as dunite and peridotite, tend to occur near the bottom of the sequence of layers in the Muskox intrusion, whereas gabbroic layers occur closer to the top. The sequence is capped by an acidic layer of granophyric composition.

Figure 3.8a shows the frequency distribution of 116 copper values from rock samples taken at the surface on a series of gabbros with increasing granophyre content. These gabbros are situated between a clinopyroxenite layer (at the bottom) and a mafic granophyre layer (at the top). A logarithmic scale or ratio scale was used in this figure for plotting the trace element copper in ppm (parts per million). A straight line for logarithms of values indicates approximate lognormality in Fig. 3.8a. However, the 116 copper values are for specimens of three different types of gabbros

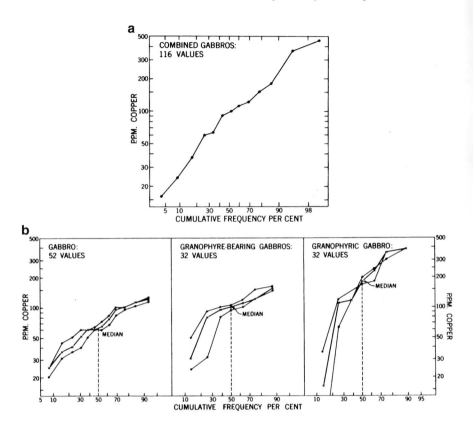

Fig. 3.8 (a). Plot on logarithmic probability paper of histogram for 116 copper determinations from gabbros in the Muskox intrusion. (b). Separate plots for different types of gabbro show that curve in **a** actually represent a mixture of several populations. Precision is indicated by area between curves for two random subsamples, each consisting of 50 % of the data (Source: Agterberg 1974, Fig. 28)

and these subpopulations were plotted separately in Fig. 3.8b. The results prove that the frequency distribution shown in Fig. 3.8a actually can be regarded as a mixture of three separate populations with different properties. Average copper content increases with granophyre content of gabbro in his sequence.

In order to test the stability of the separate frequency curves which are for relatively small samples, a simple procedure for small samples originally suggested by Mahalanobis (1960) was used by Agterberg (1965). Each sample of copper values was randomly divided into two subsamples. For example, if there are 52 values as for gabbro, two subsamples of 26 values were formed and plotted separately (Fig. 3.8b). The area between the two frequency curves for the subsamples represents the so-called "error-area" of the curve for a combined sample. Mahalanobis (1960) had developed an approximate chi-square test based on the error-areas to test for the separation between populations. In this application, the copper distribution for granophyric gabbro differs significantly from those of the other two rock types.

3.2 Lognormality and Mixtures of Frequency Distributions

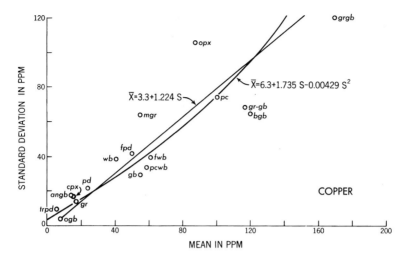

Fig. 3.9 Standard deviation plotted against mean for copper in different rock types of the Muskox intrusion. Two curves (linear and quadratic best fits) are also shown. The pattern in this diagram indicates that, to a first approximation, standard deviation is proportional to mean. Rock types with olivine are: *DN* dunite, *PD* peridotite, *FPI* feldspatic peridotite, *TRPD* troctolitic peridotite, *PC* picrite, *CPX* olivine clinopyroxenite, *PCWB* picritic websterite, *OGB* olivine gabbro, *BGB* bronzite gabbro, *ANGB* anorthositic gabbro; and rock types without olivine are: *OPX* orthopyroxenite, *WB* websterite, *GB* gabbro, *GR-GB* granophyre-bearing gabbro, *GRGB* granophyric gabbro, *MGR* mafic granophyre, *GR* granophyre (Source: Agterberg 1974, Fig. 29)

The Muskox intrusion has been sampled in detail both at the surface and by drilling. Average and standard deviation for copper in 17 different rock types have been plotted in Fig. 3.9. There is positive correlation between these two statistical parameters ($r = 0.9002$). A straight line and a parabola were fitted by least squares, taking the mean as the independent variable. To a first approximation, the standard deviation is proportional to the mean. This implies a constant coefficient of variation. The plots of Fig. 3.9 suggest $\gamma \cong 0.8$.

From the basic properties of variance it follows that $\sigma^2(a+bX) = b^2 \sigma^2(X)$ where a and b are two constants. Further, if $f(x)$ is a function of x that is approximately linear in a range of variation of x, then, according to Taylor's formula: $f(x) \cong f(c) + (x-c) \cdot f'(c)$ where c is another constant. Consequently:

$$\sigma^2\{f(X)\}^2 \cong \{f'(c)\}^2 \cdot \sigma^2(X)$$

Application to $f(X) = \log_e X$ gives:

$$\sigma^2\{\log_e(X)\}^2 \cong \left[\frac{d(\log_e(X))}{dx}\right]^2_{x=x_0} \cdot \sigma^2(X)$$

where x_0 is a constant value for x used in truncated Taylor series approximation. Setting $x_0 = \mu(X)$ it is derived that

Fig. 3.10 Composite histogram of copper values from Muskox intrusion. Original values x_{ik} from different rock types i) were made comparable with one another by replacing them by $x'_{ik} = x_{ik}/\text{ave}(x_i)$ where $\text{ave}(x_i)$ represents rock type mean; horizontal line is for $\log_e (1 + x')$. Best-fitting curve is truncated normal representing maximum-entropy solution showing truncated lognormality of original data (Source: Agterberg 1974, Fig. 30)

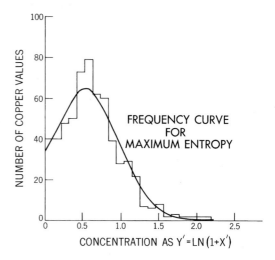

$$\sigma(\log_e(X)) \cong \frac{\sigma(X)}{\mu(X)} = \gamma$$

The coefficient of variation, therefore, is approximately equal to the standard deviation of the logarithmically transformed data. This approximation is good if $\gamma < 0.5$. From these considerations it follows that γ or its estimate from a sample can be used as a criterion to distinguish between normality and lognormality. Hald (1952) has pointed out that if $\gamma < 1/3$, a lognormal curve cannot be distinguished from a normal curve. In general, a sample for which $\gamma > 0.5$ is distinctly non-normal but not necessarily lognormal. Because $\gamma \cong 0.8$ for the rock types in Fig. 3.9, non-normality is indicated.

In total, 622 copper values were used to estimate the means and standard deviations plotted in Fig. 3.9. Suppose that the copper values x_{ik} for rock type i are divided by their rock type mean. This transformation yields 622 new copper values x' with overall average value equal to 1. The frequency distribution of these transformed copper values was studied by Agterberg (1965). It is neither normal nor lognormal but can be fitted by a truncated lognormal distribution for $y' = \log_e (1 + x')$, as shown graphically in Fig. 3.10. This result can be interpreted as an example of validity of the maximum entropy criterion.

The concept of entropy is used in thermodynamics for evaluating the amount of order or disorder in spatial configurations of attributes. For example, the molecules of an ideal gas can occur anywhere within a confined space and their spatial configuration is completely random at any time. The entropy of the system then is at a maximum (complete disorder). Shannon (1948) applied the concept of maximum entropy to ordinary frequency distributions with n classes i). If p_i represents relative frequency of occurrence in the i-th class, then a first constraint is $\Sigma\, p_i = 1$. Suppose that the variance σ^2 of the system is predetermined as a second constraint. The entropy statistic S satisfies:

$$S = -H\sum_{i=1}^{n} p_i \log_e p_i$$

Partial differentiation of S with respect to the p_i subject to the two constraints yields a maximum entropy curve that has the familiar Gaussian shape. Suppose now that in addition to the two constraints there is also the constraint that the mean μ is predetermined as well. Tribus (1962) has shown that the relative frequencies then become in the limit:

$$p(x) = \exp\left[-\left(A_0 + A_1 x + A_2 x^2\right)\right] = c \cdot \exp\left[-\left(x + A_1/2A_2 x^2\right)\right] \text{ with } x > 0$$

where A_0, A_1 and A_2 are Lagrangian multipliers and $c = -A_0 + A_1/4A_2^2$. This represents the equation of a Gaussian curve that is truncated at the origin. The peak of this curve occurs at the point with $x = -A_1/2A_2$. This maximum falls at $y' = 0.624$ in Fig. 2.18. Working backwards from $y' = 0.624$, it follows that $x' = 0.866$. Consequently, all rock types in the Muskox intrusion have maximum frequencies at values that are approximately equal to their arithmetic means multiplied by 0.866. The trace element copper primarily occurs in chalcopyrite crystals which are disseminated through the rocks. The concentration of these crystals (and to a lesser extent their size) varies according to complex spatial patterns of clusters.

3.2.3 Three-Parameter Lognormal Distribution

Suppose that in the generating process $X_i - X_{i-1} = \varepsilon_i \cdot g(X_{i-1})$ it is assumed that $g(X_{i-1}) = X_{i-1} + \alpha$ where α is a positive constant. The result of this modification of the original process that it generates a lognormal distribution with its origin displaced to the point with abscissa equal to $-\alpha$. Of course, negative element concentration values are not possible and this would not be a viable model yielding realistic results. However, suppose that the process was subject to the following constraint: because negative concentration values are not possible, there was an absorbing boundary at zero implying that all concentrations reaching zero remained at zero. This modified generating process would generate a secondary sharp peak at zero concentration.

The absorbing boundary model may offer an explanation for the frequency distribution of gold in the relatively thin conglomerate beds (reefs) of the Witwatersrand gold fields in South Africa (cf. Table 2.5). Until 1960, the general opinion was that these gold concentration values satisfy a lognormal model. However, Krige (1960) has shown that a significant refinement in Witwatersrand gold ore-evaluation methods was obtained by assuming that the underlying frequency distribution is three-parameter lognormal instead of lognormal (see Fig. 3.11).

As mentioned before, the unit of measuring gold in South Africa was inch-dwt or inch-pennyweight which is equivalent to 0.165 cm-g (centimeter-gram) used in

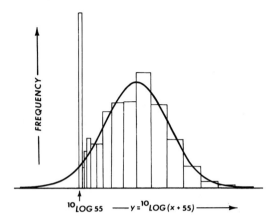

Fig. 3.11 Three-parameter lognormal model for gold values in Merriespruit Mine, Witwatersrand goldfield, South Africa (After Sichel 1961, Fig. 2; original data from Krige 1960). Because of the additional peak at zero, the corresponding cumulative frequency distribution curve is normal for gold values that are positive only. Note that this situation differs from that in Fig. 3.10, although the transformation is of the same type (Source: Agterberg 1974, Fig. 32)

South Africa after 1971. The original inch-dwt value represented the amount of gold present in a column of ore with a base of one square inch and perpendicular to the reef. The best-fitting normal curve shown in Fig. 3.11 was originally derived by Sichel (1961). It was fitted to all values except those at zero which form a sharp, secondary peak.

3.2.4 Graphical Method for Reconstructing the Generating Process

The starting point of the preceding discussion of the generating process also can be written as $dX_i = g(X_i) \cdot \varepsilon_i$ where dX_i represents the difference $X_i - X_{i-1}$ in infinitesimal form. Several systems of curve-fitting are in existence for relating an observed frequency distribution to a normal form. An example of this is Johnson's (1949) system which makes use of $dX_i = g(X_i) \cdot \varepsilon_i$. An excellent review of curve-fitting systems was given by Kendall and Stuart (1949). The so-called Johnson S_B-system results in lognormal distribution of the variable $X/(1+X)$. Jizba (1959) used this approach for modeling geochemical systems. It may be possible to obtain information on the nature of the function $g(X_i)$ by graphical analysis using the normal Q-Q plot. This method is illustrated in Fig. 3.12 for a lognormal situation.

In Fig. 3.12a, a lognormal curve was plotted on normal probability paper. An auxiliary variable z is plotted along the horizontal axis. It has the same arithmetic scale as x that is plotted in the vertical direction. The variable x is a function of z or

3.2 Lognormality and Mixtures of Frequency Distributions

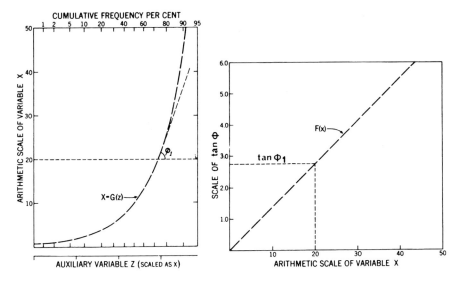

Fig. 3.12 Graphical construction of $F(x)$ which is proportional to the tangent of the slope of $x = G(z)$ representing the theoretical lognormal distribution plotted on arithmetic probability paper (Source: Agterberg 1974, Fig. 33)

$x = G(z)$. The slope of $G(z)$ is measured at a number of points giving the angle φ. For example, if $z = x = 20$ in Fig. 3.12a, $\varphi = \varphi_1$, and:

$$\frac{dG(z)}{dZ} = \tan \varphi$$

If, in a new diagram (Fig. 3.12b), $\tan \varphi$ is plotted against x, we obtain a function $F(x)$ that represents $g(X_i)$ except for a multiplicative constant that remains unknown. This procedure is based on the general solution of the generating process which can be formulated as:

$$Z = \int_0^\infty \frac{dX_1}{g(X_i)} = \sum_0^\infty \varepsilon_i$$

This procedure is based on the assumption that Z is normally distributed.

The function $g(X_i)$ derived in Fig. 3.12b for the lognormal curve of Fig. 3.13a is simply a straight line through the origin. This result is in accordance with the origin theory of proportional effect. However, the method can produce interesting results in situations that the frequency distribution is not lognormal. Two examples are given in Fig. 3.13. The original curves for these two examples (Fig. 3.13a) are on logarithmic probability paper. They are for the 1,000 Merriespruit gold values shown in histogram form in Fig. 3.11, and a set of 84 copper concentration values for rock samples from sulphide deposits that surround the Muskox intrusion as a rim. The results obtained by the graphical method are shown in Fig. 3.13b.

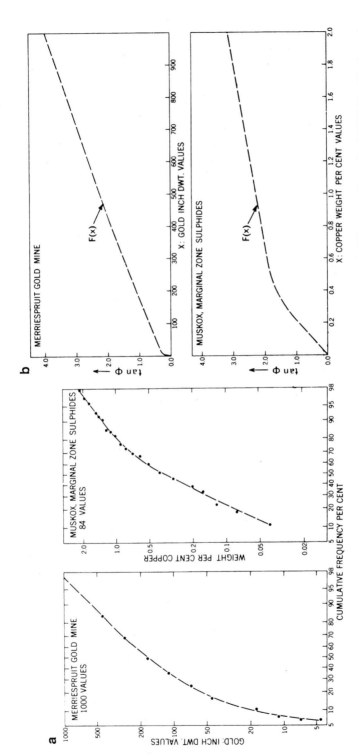

Fig. 3.13 (a). Two frequency distributions that depart from the lognormal model plotted on logarithmic probability paper. Example of 1,000 from Merriespruit gold mine is after Krige (1960). (b). Method of Fig. 3.12 resulted in two types of $F(x)$, the one for the Merriespruit Mine is approximately a straight line that would intersect the X-axis at a point with abscissa of about −55 (value used in Fig. 3.11 for the same data set). Function $F(x)$ for Muskox copper values would suggest a change in the generating process after 0.5 % Cu (Source: Agterberg 1974, Fig. 34)

The generating function $g(X_i)$ for Merriespruit is according to a straight line whose prolongation would intersect the X-axis at a point with negative abscissa $\alpha \cong 55$. The second example (Muskox sulphides) gives a function $g(X_i)$ that resembles a broken line. It would suggest a rather abrupt change in the generating process for copper concentration after the value $X_i \cong 0.5$ %. The influence of X_i on dX_i may have decreased with increasing X_i.

The usefulness of a graphical method of this type is limited, in particular because random fluctuations in the original frequency histograms cannot be considered carefully. However, the method is rapid and offers suggestions with respect to the physico-chemical processes that may underlie a frequency distribution.

3.3 Compound Random Variables

Applications of the theory of compound random variables are helpful in situations that geological entities occurring at points have properties such as size or volume that can be modeled as continuous random variables. This topic is useful in mineral potential evaluation where the targets of interest are ore deposits or oil pools that are randomly distributed in a study area according to a Poisson or negative binomial distribution but have sizes that satisfy, for example, a lognormal or Pareto distribution.

3.3.1 Compound Frequency Distributions and Their Moments

Suppose that K random variables X_i have the same frequency distribution and generating function $g_x(s)$. Their sum Y is a random variable with K terms. If K and the X_i are discrete random variables, the probability that Y is equal to the integer number j is equal to the sum of the probabilities $p_1 \cdot P(X_1 = j), p_2 \cdot P(X_1 = j)$, where the $p_1, p_2, \ldots$, represent the probabilities that K is equal to 1, 2, …, or:

$$P(Y = j) = \sum_{i=0}^{\infty} p_i \cdot P(X_1 + X_2 + \cdots + X_i = j)$$

The summation may start at $i = 0$, for which $P(X_0)$ and $P(Y = j)$ are both equal to zero. If $g_y(s)$ and $g_k(s)$ are the generating functions of Y and K, then:

$$g_y(s) = \sum_{i=0}^{\infty} p_i \{g_{x(s)}\}^i = g_k\{g_x(s)\}$$

This expression is called a compound generating function. Well-known applications of this theory are the compound Poisson and negative binomial distributions.

> **Box 3.2: Moments of Compound Poisson and Negative Binomial Distribution**
>
> If K is Poisson, $g_y(s) = \exp[-\lambda + \lambda \cdot g_x(s)]$. Then, if X satisfies the so-called logarithmic series distribution: $P(X = k) = \alpha \cdot \frac{\theta^k}{k}$ where $\alpha = -\frac{1}{\log_e(1-\theta)}$; $k = 1, 2, \ldots; 0 < \theta < 1$, Y becomes negative binomial (Feller 1968, p. 291). If X is a continuous random variable, whereas K is discrete, the characteristic function of Y satisfies: $g_y(u) = g_k\{g_x(u)\}$ with mean $E(Y) = E(K) \cdot E(X)$ and variance $\sigma^2(Y) = E(K) \cdot \sigma^2(X) + \sigma^2(K) \cdot E^2(X)$. Suppose that K is Poisson and that X is lognormally distributed, then: $E(Y) = \lambda \alpha = \lambda e^{\mu + \sigma^2/2}$ and $\sigma^2(Y) = \lambda(\alpha^2 + \beta^2) = \lambda e^{2\mu + 2\sigma^2}$.
>
> If the negative binomial distribution is used for K instead of the Poisson, whereas X is kept lognormal, the characteristic function of Y becomes: $g_y(u) = \left[\frac{p}{1-qg_{x(u)}}\right]^r$ with: $E(Y) = \frac{rq}{p}e^{\mu+\sigma^2/2}$ and $\sigma^2(Y) = \frac{rq}{p}e^{2\mu+\sigma^2}\left(e^{\sigma^2} + \frac{q}{p}\right)$.

The combination of completely random (Poisson) distribution of mineral deposits with lognormal value distribution originally was used for exploration strategy by Allais (1957). Using the negative binomial instead of the Poisson represents the exploration strategy problem later conceived by Griffiths (1966) and formalized by Uhler and Bradley (1970).

3.3.2 Exploration Strategy Example

One of the objectives of economic geology is to design methods for locating places that are likely to contain hidden mineral deposits. Exploration is expensive mainly because large deposits are rare events and it is difficult to locate them. Most deposits that outcrop at the surface of the Earth may already have been found. Deeper deposits are to be discovered by geological process modeling, drilling and new, mainly geophysical, exploration methods. The drilling of deep boreholes or wells remains essential. In later chapters, several techniques developed for the evaluation of regional mineral potential that provide a starting point for exploration will be discussed. In this section, equations presented in Box 3.2 will be used to predict total value of mineral deposits on the Canadian Shield using historical data. This example originally was presented in Agterberg (1974).

Slichter (1960) compiled information on number of valuable mines per unit area, for 185 units of 1,000 sq. miles each in Ontario. The rocks for the entire area of size 185,000 sq. miles belong to different structural provinces and geological environments in the Canadian Shield. In total, his study area constituted approximately 9 % of the shield. Next he fitted the so-called exponential distribution. Later, Griffiths (1966) showed that it is better to apply the negative binomial model to Slichter's data.

Suppose that the "control" area in Ontario is representative for the entire shield. Because there occur 147 mines in Slichter's area, the total number of mines for the entire shield would be approximately $(100/9) \times 147 \cong 1,600$ mines. Of course, this estimate will be too low if not all mines in the control area were discovered. On the other hand, it will be too large, if the control area is richer in mines (and larger deposits) than the remainder of the Precambrian Shield.

The problem of predicting total value of all orebodies in the Canadian Shield was discussed by De Geoffroy and Wu (1970) who argued as follows. The Canadian Shield occupies an area of about 2,146, 000 sq. miles. Nearly 90 % of commercial mineral deposits known in 1968 occur in volcanic belts and Lower Proterozoic sedimentary belts. These relatively favorable environments, which alternate with other rock types, occupy approximately 341,220 sq. miles or 15.9 % of the total area. An area of 50,000 sq. miles (Timmins-Kirkland Lake area, Ontario; Noranda-Val-d'Or area, Québec) for the favorable environments was treated in detail by De Geoffroy and Wu (1970) with the following results:

Size of area: 50,000 sq. miles
Total number of deposits: 254 orebodies
Average number of deposits per 10×10 sq. miles: Ave $(k) = 0.508$
Variance: $s^2(k) = 2.016$
Average value of deposits: Ave $(x) = 60.5 \cdot 10^6$ US dollars (based on 1968 prices)
Logarithmic mean (base e): Ave $(\log_e x) = 2.858$ (unit of value is 10^6 dollars)
Variance: $s^2(\log_e x) = 3.103$

These statistics can be used for a preliminary prediction of number of deposits and their value in the $341,220 - 50,000 = 291,220$ sq. miles of territory that, for the larger part, has not been explored in the same detail as the 50,000 sq. miles of favorable environment in the control area. Variances of these estimates can be computed to express their uncertainties. According to the statistics compiled by De Geoffroy and Wu (1970), $0.508 \times 341,220/100 = 1,733$ mines are predicted to exist on the Canadian Shield. This number is fairly close to that based on Slichter's data. However, 129 of 254 orebodies of De Geoffroy and Wu fall in Slichter's area, and the control area underlying the two estimated values overlap in part. Also, for the second estimate, deposits outside the more favorable environments were not considered.

De Geoffroy and Wu (1970) fitted the negative binomial to number of orebodies per unit area with satisfactory results. They also fitted a normal distribution to the logarithms of values in dollars for mines. The chi-square test discussed in Sect. 2.4.1 gave a chi-square value of 12.51 for six degrees of freedom (nine classes), which may be compared to $P\{\chi^2(6) < 12.5\} = 94.8$ %. This result indicates that the lognormal model provides a degree of fit that is only moderately good. In fact, the histogram for logarithmically transformed data shows a positive skewness that is probably meaningful. This becomes a practical estimation problem in that, if lognormality is accepted, the risk is taken of the model yielding biased estimates because of a departure from lognormality. The Uhler-Bradley model discussed in the preceding section is fairly sensitive to departures from lognormality. Assuming lognormality, the preceding statistics give an estimated value of $82.18 \cdot 10^6$ dollars. An unbiased variance estimate can be obtained by using $0.985 \cdot s^2 = 3.096$ instead of $s^2 = 3.103$. This results in an estimated value of $80.34 \cdot 10^6$ dollars.

Inasmuch as the original estimates of mean ($=60.5 \cdot 10^6$) and logarithmic mean ($=2.858$) are based on 254 data, both are fairly good estimates. The preceding two estimates of logarithmic variance may be too large because of positive skewness of the value histogram. For this reason, it may be preferable to base the estimate of logarithmic variance on the estimates of mean and logarithmic mean. This gives $s^2 = 2.490$, which is less than the preceding two estimates ($=3.096$ and 3.103).

The number of orebodies in favorable environments outside the control area is estimated at $(219{,}220/100) \times 0.508 = 1{,}478$. With average expected value of $60.5 \cdot 10^6$, their total value would be $1{,}478 \times 60.5 \cdot 10^6 = 89.4 \cdot 10^9$ dollars. The corresponding average value per unit area of $10 \times$ miles size is $E(Y) \cong 0.508 \times 60.5 \cdot 10^6 = 30.73 \cdot 10^6$. Our estimate of the corresponding variance $\sigma^2(Y)$ can be based on the negative binomial distribution model for which $E(K) = rq/p \cong 0.508$ and $\sigma^2(K) = rq/p^2 \cong 2.016$. With $p+q=1$, this yields the estimated values $p' = 0.2520$, $q' = 0.7480$, $r' = 0.1711$. With $\mu \cong 2.858$ and $\sigma^2 \cong 2.490$, it follows that $\sigma^2(Y) \cong 27{,}922 \cdot 10^{12}$ and $\sigma(Y) \cong 167 \cdot 10^6$ dollars.

It seems that an analytical expression for the random variable Y does not exist. However, it can be assumed that its frequency distribution is positively skew and multimodal. Because of the central-limit theorem, the sum of n random variables Y that can be written as $Z = \Sigma\, Y_i$ ($i = 1, 2, \ldots, n$) converges to normal form when n increases. For example, if $n = 400$, we have $E(Z) \cong 12.30 \cdot 10^9$ and $\sigma(Z) \cong 3.34 \cdot 10^9$. The coefficient of variation $\gamma(Z) \cong 0.27$ then is quite small. In that situation it is permissible to determine the 95 % confidence interval that becomes $6.5 \cdot 10^9$ on the predicted total value of $12.30 \cdot 10^9$ dollars. It should be kept in mind that the preceding calculations were based on orebodies located in a control area that had been discovered in 1968, and only the uppermost part of the Earth's crust had been scrutinized for mineral occurrences.

References

Agterberg FP (1965) Frequency distribution of trace elements in the Muskox layered intrusion. In: Dotson JC, Peters WC (eds) Short course and symposium on computers and computer applications in mining and exploration. University of Arizona, Tucson, pp G1–G33

Agterberg FP (1974) Geomathematics. Elsevier, Amsterdam

Agterberg FP (1988) Quality of time scales – a statistical appraisal. In: Merriam DF (ed) Current trends in geomathematics. Plenum, New York, pp 57–103

Agterberg FP (1990) Automated stratigraphic correlation. Elsevier, Amsterdam

Agterberg FP (1994) Estimation of the Mesozoic geological time scale. Math Geol 26:857–876

References

Agterberg FP (in press) Self-similarity and multiplicative cascade models. J S Afr Inst Min Metall
Ahrens LH (1953) A fundamental law in geochemistry. Nature 172:1148
Aitchison J, Brown JAC (1957) The lognormal distribution. Cambridge University Press, Cambridge
Allais M (1957) Method of appraising economic prospects of mining exploration over large territories: Algerian Sahara case study. Manag Sci 3:285–347
Cox AV, Dalrymple GB (1967) Statistical analysis of geomagnetic reversal data and the precision of potassium-argon dating. J Geophys Res 72:2603–2614
DeGeoffroy JG, Wu SM (1970) A statistical study of ore occurrences in the greenstone belts o the Canadian Shield. Econ Geol 65:496–509
Feller W (1968) An introduction to probability theory and its applications. Wiley, New York
Finney PJ (1941) On the distribution of a variate whose logarithm is normally distributed. J R Stat Soc Lond Suppl 7:151–161
Fisher RA (1922) On the mathematical foundations of theoretical statistics. Philos Trans R Soc Lond Ser A 222:209–368
Gradstein FM, Agterberg FP, Ogg JG, Hardenbol J, van Veen P, Thierry T, Huang Z (1994) A Mesozoic time scale. J Geophys Res 99(12):24 051–24 074
Gradstein FM, Agterberg FP, Ogg JG, Hardenbol J, van Veen P, Thierry T, Huang Z (1995) A Triassic, Jurassic and cretaceous time scale. In: Berggren WA, Kent DV, Aubry M-P, Hardenbol J (eds) Geochronology, time scales and global stratigraphic correlations: a unified temporal framework for a historical geology, vol 54. Society of Economic Paleontologists and Mineralogists, Tulsa, pp 95–128
Gradstein FM, Ogg JG, Smith AG (eds) (2004) A geologic time scale 2004. Cambridge University Press, Cambridge/New York
Gradstein FM, Ogg JG, Schmitz M, Ogg G (eds) (2012) The geologic time scale 2012, 2-vol set. Elsevier, Amsterdam
Griffiths JC (1966) Exploration for mineral resources. Oper Res 14:189–209
Hald A (1952) Statistical theory with engineering applications. Wiley, New York
Harbaugh JW, Bonham-Carter GF (1970) Computer simulation in geology: modelling dynamic systems. Wiley, New York
Harland WB, Cox AV, Llevellyn PG, Pickton CAG, Smith AG, Walters R (1982) A geologic time scale. Cambridge University Press, Cambridge/New York
Harland WB, Armstrong RI, Cox AV, Craig LA, Smith AG, Smith DG (1990) A geologic time scale 1989. Cambridge University Press, Cambridge/New York
Husson D, Galbrun B, Laskar J, Hinnov LA, Thibault N, Gardin S, Locklair RE (2011) Astronomical calibration of the Maastrichtian (Late Cretaceous). Earth Planet Sci Lett 305:328–340
Jizba ZV (1959) Frequency distribution of elements in rocks. Geoch Cosmoch Acta 16:79–82
Johnson NL (1949) Systems of frequency curves generated by methods of translations. Biometrika 36:149–176
Kapteyn JC (1903) Sew frequency curves in biology and statistics. Noordhoff, Groningen
Kaufman GM (1963) Statistical decision and related techniques in oil and gas exploration. Prentice-Hall, Englewood Cliffs
Kendall MG, Stuart A (1949) The advanced theory of statistics, (3-vol ed). Hafner, New York
Kendall MG, Stuart A (1961) The advanced theory of statistics, vol 2, 2nd edn. Hafner, New York
Kolmogorov AN (1941) On the logarithmic normal distribution law for the sizes of splitting particles. Dokl Akad Nauk SSSR 31:99–100
Kolmogorov AN (1951) Solution of a problem in probability theory connected with the problem of the mechanism of stratification. Am Math Soc Trans 53:1–8
Krige DG (1951) A statistical approach to some basic valuation problems on the Witwatersrand. J S Afr Inst Min Metall 52:119–139
Krige DG (1960) On the departure of ore value distributions from the lognormal model in South African gold mines. J S Afr Inst Min Metall 61:231–244

Krumbein WC (1936) Application of logarithmic moments to size frequency distributions of sediments. J Sediment Petrol 6:35–47

Mahalanobis PC (1960) A method of fractile graphical analysis. Econometrica 28:325–351

McCrossan RG (1969) Analysis of size frequency distribution of oil and gas reserves of western Canada. Can J Earth Sci 6:201–212

Rao CR (1973) Linear statistical inference and its applications. Wiley, New York

Renne PR, Swisher CC III, Deino AL, Karner DB, Owens TL, DePaolo DJ (1998) Intercalibration of standards, absolute ages and uncertainties in ^{40}Ar/^{39}Ar dating. Chem Geol 145:117–152

Renne PR, Deino AL, Hilgen FJ, Kuiper F, Mark DF, Mitchell WS III, Morgan LE, Mundil R, Smit J (2012) Time scales of critical events around the Cretaceous-Paleogene boundary. Science 339:684–687

Shannon CE (1948) A mathematical theory of communication. Bell Syst Tech J 27:379–423

Sichel HS (1952) New methods in the statistical evaluation from the lognormal model in South African gold mines. Trans Inst Min Metall 61:261–288

Sichel HS (1961) On the departure of ore value distributions from the lognormal model in South African gold mines. J S Afr Inst Min Metall 62:333–338

Sichel HS (1966) The estimation of means and associated confidence limits for small samples from lognormal populations. In: Proceedings of symposium on mathematical statistics and computer applications in ore valuation, The South African Institute of Mining and Metallurgy, Johannesburg, pp 106–122

Slichter LB (1960) The need for a new philosophy of prospecting. Min Eng 12:570–576

Smith CH (1962) Notes on the Muskox intrusion, Coppermine River area, District of Mackenzie. Geological Survey of Canada Paper, Ottawa, pp 61–25

Smith CH, Kapp HE (1963) The Muskox intrusion, a recently discovered layered intrusion in the Coppermine River area, Northwest Territories, Canada. Mineralogical Society of America Special Paper 1, Chantilly, Virginia, pp 30–35

Swisher CC III, Grajales-Nishimura JM, Montanari A, Margolis SV, Claeys W, Alvarez W, Renne P, Cedillo-Pardo E, Maurrasse R, Curtis GH, Smit J, McWilliams MO (1993) Coeval Ar-Ar ages of 65 million years ago from Chixculub crater melt rock and cretaceous-tertiary boundary tectites. Science 257:954–958

Tribus M (1962) The use of the maximum entropy estimate of reliability. In: Machol E, Gray P (eds) Recent developments in information and decision processes. MacMillan, New York

Uhler RS, Bradley PG (1970) A stochastic model for petroleum exploration over large regions. J Am Stat Assoc 65:623–630

Vistelius AB (1960) The skew frequency distribution and the fundamental law of the geochemical processes. J Geol 68:1–22

Chapter 4
Correlation, Method of Least Squares, Linear Regression and the General Linear Model

Abstract The scatterplot in which two variables are plotted against one another is a basic tool in all branches of science. The ordinary correlation coefficient quantifies degree of association between two variables for the same object of study. In some software packages, the squared correlation coefficient (R^2) is used instead of the correlation coefficient to express degree of fit. A best-fitting straight-line obtained by the method of least squares can represent underlying functional relationship if one variable is completely or approximately free of error. When both variables are subject to error, use of other methods such as reduced major axis construction is more appropriate. A useful generalization of major axis construction in which individual observations all have different errors in both variables is Ripley's Maximum Likelihood for Functional Relationship a (MLFR) fitting method. Kummell's equation (*cf.* Agterberg 1974) for linear relationship between two variables that are both subject to error can be regarded as a special case of MLFR.

Multiple regression can be used for curve-fitting if the relationship between two variables is not linear but other explanatory variables have to be considered as well. The general linear model is another logical extension of simple regression analysis. It is useful in mineral resource appraisal studies. Although this approach can be too simplistic in some applications such as estimation of probabilities of occurrence of discrete events, it remains useful as an exploratory tool. During the late 1970s and early 1970s a probabilistic regional mineral potential evaluation was undertaken at the Geological Survey of Canada (*cf.* Agterberg et al. 1972) to estimate probabilities of occurrence of large copper and zinc orebodies in the Abitibi area on the Canadian Shield. These predictions of mineral potential made use of the general linear model relating known orebodies in the area to rock types quantified from geological maps and regional geophysical anomaly maps. About 10 and 40 years later, after more recent discoveries of additional copper ore, two hindsight studies were performed to evaluate accuracy and precision of the mineral potential predictions previously obtained by multiple regression. This topic will be discussed in detail because it illustrates problems encountered in projecting known geological relations between orebodies and geological framework over long distances both horizontally and vertically.

Keywords Correlation • Functional relationship • Least squares • Linear regression • 95 %-confidence belts • Multiple regression • Mineral potential estimation • Abitibi copper hindsight study • Pareto distribution • Abitibi copper deposits

4.1 Correlation and Functional Relationship

It is good to keep in mind that the linear relationship resulting from linear regression only provides an estimate for linear functional relationship between two variables if one of the variables is not a random variable; i.e., free of measurement error. This is illustrated in Fig. 4.1 (based on original data from Krige 1962). Along the X-axis, it shows the average value of amount of gold in inch-dwt values for panel faces in a Witwatersrand gold mine. Along the Y-axis, similar values are plotted for panels which are located 30 ft. (9 m) ahead of the panels whose values are plotted along the X-axis. Each value for a panel is the average of ten single values for narrow channel samples cut across the gold reef. The ten channel samples cover a distance of 150 ft. across the panel along which mining proceeded. Figure 4.1 can be considered as a scattergram. Individual values are not shown; their frequency was counted for small blocks in this diagram. Since panel values are approximately lognormally distributed, a ratio scale is used for the coordinate axes rather than an arithmetic scale. The correlation coefficient amounts to 0.59. The elliptical contour shown in the diagram contains most data points and this suggests that the bivariate frequency distribution is bivariate normal.

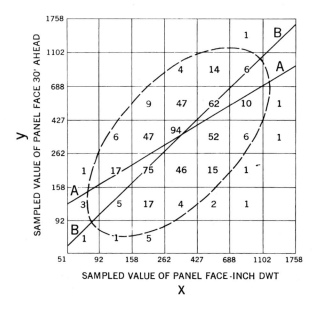

Fig. 4.1 Contour ellipse contains most of gold values (from Krige 1962); frequencies shown for blocks. Krige's regression line (A) is used for prediction; line B represents solution of Kummell's equation for linear functional relationship between two random variables (cf. Deming 1943). Kummell's method represents a special case of MLFR. The resulting line coincides approximately with the axis of the contour ellipse because, on average, y is about equal to x (Source: Agterberg 1974, Fig. 37)

4.1 Correlation and Functional Relationship

The practical problem of gold mining in this example consists of predicting the value of a panel that will be 30 ft. ahead of the panel whose value is known at a given time. In order to solve this problem, one must estimate, as Krige (1962) has done: $E(Y|x) = \beta_0 + \beta_1 x$ using the means, standard deviations and correlation coefficient (see Box 4.1). This linear regression results in the straight line (A) in Fig. 4.1. Krige (1962) pointed out that, in this manner, one avoids making the erroneous assumption that a given amount of gold per panel in a mine is representative for a larger region surrounding it. For example, if a panel values is $x = 1{,}102$, the value predicted from it for a panel that is 30 ft. ahead is not 1,102 but 688 which is considerably less. On the other hand, if $x = 92$, $E(Y|x)$ is estimated to be 158 which is significantly more. Consequently, relatively high values overestimate amount of gold per unit volume in their immediate surroundings, but relatively low values tend to underestimate it. Suppose that each sample value is representative of a larger volume of ore around it and mining would be restricted to only those blocks for which the sample value is relatively large, exceeding a cut-off value and that the remainder of the orebody would be left unmined. In the past, mining engineers found to their surprise that the average for a volume blocked out in this manner systematically overestimates the average amount of metal that is actually mined out later. Problems of this type can be avoided by the use of "kriging". Figure 4.1 provides a simple example of this: if a cut-off grade it set, it should be based on the predicted values (curve A) rather than on the actual values (curve B).

In general, the aim of the various methods of kriging that have been developed in geostatistics is to provide unbiased estimates of values at points with arbitrary coordinates (Sect. 7.2) or for 3-D block averages. If only a single element concentration value for a small block would be available, it could be used as an estimate of a larger block surrounding the small block. The variance of the single value for the larger block then would be a so-called extension variance (Sect. 6.2.5).

Box 4.1: Basic Elements of Covariance and Correlation Coefficient

If X and Y are two random variables, then: $E(XY) = \int \int xy \cdot f(x,y)\, dxdy$ where $f(x, y)$ is the bivariate frequency distribution. Suppose that the double integral is approximated by a double sum, then: $E(XY) = \sum_i \sum_j x_i y_j P(x_i, y_j)$. Mutual independence of two random variables can be defined as: $P(x_i, y_j) = P(x_i) P(y_j)$. Hence: $E(XY) = \sum_i \sum_j [x_i P(x_i)]\, y_j P(y_j)] = (EX) \cdot (EY)$ but only for independent random variables. The covariance of X and Y is: $\sigma(X,Y) = E(X - \mu_x)(Y - \mu_y) = E(XY) - \mu_x \mu_y$. The correlation coefficient is the covariance of two standardized random variables, or $\rho = \rho(X, Y) = \frac{\sigma(X,Y)}{\sigma(X) \cdot \sigma(Y)}$. Conditional dependence can be defined as: $P(Y=y|X=x) = P(X=x, Y=y)$ or $P(y|x) = P(x,y)/P(x)$. Consequently, $P(x,y) = P(y|x) \cdot P(x)$. Suppose X and Y are normal. Standardization to Z_1 and Z_2, then results in the bivariate frequency density: $\varphi(z_1, z_2) = \frac{1}{2\pi\sqrt{1-\rho^2}} \exp\left[-\frac{1}{2(1-\rho^2)}(z_1^2 - 2\rho z_1 z_2)\right]$. After some manipulation, it follows that: $\varphi(z_2|z_1) = \frac{\varphi(z_1, z_2)}{\varphi(z_1)} = \frac{1}{\sqrt{1-\rho^2}} \varphi\left[\frac{z_2 - \rho z_1}{\sqrt{1-\rho^2}}\right]$. This is the

(continued)

Box 4.1 (continued)

frequency distribution of the random variable $(Z_2|z_1)$ which denotes the probability of $Z_2 = z_2$ when it is given that Z_1 has assumed the value z_1. The method of moments yields: $E(Z_2|z_1) = \rho z_1$; $\sigma^2(Z_2|z_1) = 1 - \rho^2$. Transforming back to X and Y: $E(Y|x) = \mu_y + \rho \frac{\sigma_y}{\sigma_x}(x - \mu_x)$; and $\sigma^2(Y|x) = \sigma_y^2(1 - \rho^2)$. The expected value of Y for a given value $X = x$, is commonly written in the form $E(Y|x) = \beta_0 + \beta_1 x$ with $\beta_0 = \mu_y + \rho \frac{\sigma_y}{\sigma_x}\mu_x$; $\beta_1 = \rho \frac{\sigma_y}{\sigma_x}$. It denotes the linear regression of the dependent variable Y on the independent or explanatory variable X. The frequency distribution of the correlation coefficient (r) was derived by Fisher (1915) who also showed that $\frac{1}{2}\log_e\frac{1+r}{1-r}$ is normally distributed with, approximately, mean of $\frac{1}{2}\log_e\frac{1+\rho}{1-\rho} + \frac{\rho}{2(n-1)}$ and variance of $\frac{1}{n-3}$.

There are many situations in which one wishes to determine the relationship between two random variables that are both subject to uncertainty. For example, suppose that a set of rock or ore samples is chemically analyzed in a new laboratory and it is necessary to determine whether or not the newly produced results are unbiased by having the same analyses done in a laboratory known to produce unbiased results. This problem was solved by Ripley and Thompson (1987) who developed a maximum likelihood fitting method for (linear) functional relationship (MLFR) that can be applied in other situations as well. Their method generalizes the original "major axis" method (Agterberg 1974). The major axis differs from the linear regression line in that it assumes that both variables (X and Y) are subject to uncertainty. The two methods have in common that the best-fitting straight line passes through the point with x and y equal to their sample means. It is readily shown that the estimator of slope of the major axis (β) satisfies $\tan 2\beta = \frac{2\sum x'y'}{\sum(x'^2 - y'^2)}$ where x' and y' represent deviations of x and y from their sample means. A disadvantage of the major axis is that it is not independent of scale (unit distances along X- and Y-axes). For this reason, the data for x and y are often standardized before the major axis is constructed. This result is known as the reduced major axis that passes though the origin with slope equal to $\beta^* = \arctan\left\{\frac{s(y)}{s(x)}\right\}$. If more than two random variables are analyzed simultaneously, the correlation coefficients form a correlation matrix that can be subjected to principal component analysis. This is a multivariate extension of constructing the reduced major axis of two variables. If principal component analysis is applied to the variance-covariance matrix instead of to the correlation matrix of more than two variables the result is equivalent to a multivariate extension of estimation of the major axis for two variables. Principal component analysis and a modification of it called factor analysis are well known multivariate methods with useful geoscience applications (see, e.g., Davis 2003).

Box 4.2: Maximum Likelihood for Functional Relationship (MLFR) Method

MLFR (Ripley and Thompson 1987) works as follows: one is interested in the linear relation $v_i = \alpha + \beta \cdot u_i$ where u_i and v_i are observed as x_i ($= u_i + $ error) and y_i ($= v_i + $ error), respectively. If $\kappa_i \approx s^2(x)$ and $\lambda_i \approx s^2(y)$ represent the variances of x_i and y_i, respectively, the problem reduces to minimizing the expression $Q = \sum \left[\frac{(x_i - u_i)^2}{\kappa_i} + \frac{(y_i - \alpha - \beta u_i)^2}{\lambda_i} \right]$ over u_i. First minimizing over u_i and introducing the weights $w_i = 1/(\lambda_i + \beta^2 \kappa_i)$, this minimum is reached when $Q_{min}(\alpha, \beta) = \sum [w_i(y_i - \alpha - \beta u_i)^2]$. The weights w_i depend on β and so does the estimate of α that satisfies $\alpha = \frac{\sum [w_i(y_i - \beta u_i)^2]}{\sum w_i}$. The slope a is found by minimizing $Q_{min}(a, \beta) = \sum [w_i (y_i - a - \beta u_i)^2]$. This, in turn, yields the weights w_i and intercept a. The covariance $s(a,b)$ can be calculated from the relation $s(a,b) = -s^2(b) \cdot \frac{\sum w_i x_i}{\sum w_i}$. So-called scaled residuals r_i can be computed by means of: $r_i = (y_i - \alpha - \beta u_i)\sqrt{\frac{1}{\lambda_i + b^2 \kappa_i}}$. The sum of squares of these scaled residuals should be approximately equal to number of observations -2. If estimates of u_i are written as X_i with $X_i = w_i[\lambda_i x_i + \kappa_i b(y_i - a)]$, a plot of r_i against X_i should not show any noticeable pattern.

The MLFR method was applied by Agterberg (2004) for the purpose of estimating the age of Paleozoic stage boundaries in the GTS-2004 geologic time scale (also see Sect. 9.5).

4.2 Linear Regression

In Box 4.1, bivariate linear regression was written in the form: $E(Y|X) = \beta_0 + \beta_1 X$ with $\beta_0 = \mu_y + \rho \frac{\sigma_y}{\sigma_x} \mu_x$; $\beta_1 = \rho \frac{\sigma_y}{\sigma_x}$. In practical applications, the statistical parameters in this expression can be replaced by sample estimates. The Gauss-Marlov theorem of mathematical statistics (*cf.* Bickel and Doksum 2001) states that the least-squares estimate of $E(Y|X)$ is unbiased, regardless of the nature of the frequency distribution of $Y|X$, even if it is not normal. The only condition to be fulfilled for application is that the residuals for Y are random and uncorrelated.

4.2.1 Degree of Fit and 95 % – Confidence Belts

A useful expression is: TSS = SSR + RSS meaning that the total sum of squares for deviations from the mean (TSS) is sum of squares due to regression (SSR) and

Table 4.1 Analysis of variance table for linear regression

Source	Sum of squares	Degrees of freedom	Mean square
Linear regression	SSR	1	SSR
Residuals	RSS	$n-2$	$RSS/(n-2)$
Total	TSS	$n-1$	

Source: Agterberg (1974, p. 255)

sum residual sum of squares (RSS). The multiple squared correlation coefficient (r^2 but often written as R^2) satisfies $r^2 = SSR/TSS$.

> **Box 4.3: Degree of Fit and Analysis of Variance**
>
> TSS = SSR + RSS signifies that the sum of squares of n deviations of a dependent variable Y from their mean: $TSS = \sum(Y_i - \overline{Y})^2$ can be decomposed as the so-called sum of squares due to regression: $SSR = \sum(\hat{Y}_i - \overline{Y})^2 = \sum(Y_i - \overline{Y})(\hat{Y}_i - \overline{Y})$ plus the residual sum of squares: $RSS = \sum(Y_i - \hat{Y}_i)^2$, or $\sum(Y_i - \overline{Y})^2 = \sum(Y_i - \overline{Y})(\hat{Y}_i - \overline{Y}) + \sum(Y_i - \hat{Y}_i)^2$. If Y has normal distribution, then: $SSR = \sum(\hat{Y}_i - \overline{Y})^2 = \beta^2 \sum(X_i - \overline{X})^2 = \sigma^2 \chi^2(1)$. The so-called residual variance is: $s_{res}^2 = RSS/(n-2)$. Analysis of variance as developed primarily by Fisher (1960) uses the following F-test. Because chi-square statistics can be added: $\sigma^2 \chi^2(n-1) = \sigma^2 \chi^2(1) + \sigma^2 \chi^2(n-2)$. Consequently,
> $$\frac{SSR}{RSS/(n-2)} = \frac{\chi^2(1)}{\chi^2(n-2)/(n-2)} = F(1, n-2).$$

Linear regression results often are summarized in an analysis-of-variance table (Table 4.1). The ratio of the two mean squares in the last column of Table 4.1 provides an estimate of $F(1, n-2)$. If this F-ratio is significantly greater than 1, it may be assumed that there is a significant linear association between X and Y and that the slope of the regression line differs significantly from zero.

It can be shown that the variance of the calculated values satisfies: $s^2(\hat{Y}_i) = s_{res}^2 \cdot R_i$; with $R_i = \frac{1}{n} + \frac{(X_i - \overline{X})^2}{\sum(X_i - \overline{X})^2}$. From this result, the following four types of confidence belt can be derived:

1. $\hat{Y}_i \pm t(n-2)s_{res}\sqrt{R_i}$; This belt consists of two hyperbolas that enclose an area about the best-fitting straight line; $t(n-2)$ is Student's t for $(n-2)$ degrees of freedom. A 95-% confidence belt has $t(n-2) = t_{0.975}(n-2)$. The purpose of this belt is to set confidence intervals on all single values of $\hat{Y}_i$ that could be estimated for given values of X_i.

4.2 Linear Regression

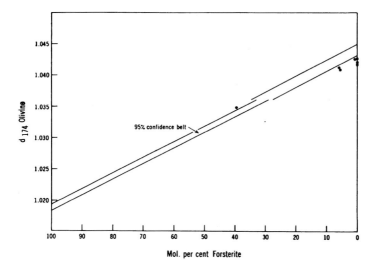

Fig. 4.2 Example of practical use of confidence belts for a regression line. *Straight line* was fitted to data with more than 30 % forsterite; only one of these falls outside the 95-% belt. All observation data with %FO < 30 % fall below the 95-% confidence belt for data not used for estimation (Source: Agterberg 1974, Fig. 38)

2. $\hat{Y}_i \pm \sqrt{2F(2, n-2)} s_{res} \sqrt{R_i}$; The purpose of this belt which is wider than the preceding one is to define a confidence region for the entire calculated regression line.
3. $\hat{Y}_i \pm t(n-2) s_{res} \sqrt{1 - R_i}$; This represents a confidence belt for the results of data which have been used to estimate $\hat{Y}_i$. It allows testing of the residuals for normality. Like belt (4), belt (3) is considerably wider than belts (1) and (2), although all belts are hyperbolas.
4. $\hat{Y}_i \pm t(n-2) s_{res} \sqrt{1 + R_i}$; This belt is for "new" observations that have not been used to estimate $\hat{Y}_i$.

The techniques for linear regression discussed in this section can be extended to (a) testing additional terms (x^2, x^3, etc.) in curvilinear regression for statistical significance, (b) testing other explanatory variables, in addition to x in applications of the general linear model, and (c) additional terms consisting of powers and cross-products of powers in 2-D or 3-D trend analysis. Various generalizations of these types will be discussed later. Use of two of the preceding confidences belts will be illustrated in the following example.

One method of determining the forsterite (FO) content of olivine consists of measuring the d_{174} cell edge. In Agterberg (1964), a linear regression equation was fitted to 20 measurements with %FO > 30 in an experiment for which the olivine crystals were chemically analyzed. The XY-plot for this experiment (Fig. 4.2) showed that these 20 data closely follow a straight line with equation: $\hat{Y}_i = 4,145 - 3,970 \cdot X$. Figure 4.2 shows the 95 % confidence belt for these data. Only one of the 20 points falls slightly outside this Type (3) belt. This indicates that there is a linear relationship between the two variables. However, six olivine crystals consisting entirely or nearly entirely of forsterite that also were treated by both methods (chemical analysis and d_{174} determination) do not seem to fall on the best-fitting (%FO > 30) straight line. This discrepancy was tested by using the slightly wider 95 % confidence belt of Type (4). The conclusion of this statistical analysis was that the best-fitting straight-line relationship could only be used for olivine crystals with more than 30 % forsterite.

The topic of 95 % confidence belts also will be discussed in Chap. 7 in the context of trend surface analysis. In that kind of 2-D application the two surfaces constituting the confidence belt are relatively flat in the area where the observation points are located but the vertical distance between them increases very rapidly near the edge of the area with observation points.

4.2.2 Mineral Resource Estimation Example

Geoscientists are using both facts and concepts to determine the probable occurrence of ore deposits. Examples of basic facts are age and lithological data on rock units, chemical determinations and geophysical measurements. There is a gradual transition from basic facts to conceptual projections of structures and composition of rock formations. Standard conceptual projections normally made by geoscientists; e.g., extrapolations to bedrock partially overlain by regolith or debris, can be taken as facts. A few historical remarks are that, originally, ore was discovered by prospectors but later economic geologists could narrow their search by using genetic models. Bateman (1919) argued convincingly that the old saying "where ore is, there it is" was to be replaced by answering the question "why ore is where it is". He promoted "intelligently directed search for ore or oil". To-day much ore is being found by means of advanced geophysical prospecting techniques. Moreover, as Zhao et al. (2008) point out, increasingly new deposits are being discovered at greater depths with the aid of 2-D and 3-D specialized technologies on the one side (see, e.g., de Kemp 2006) and non-linear modeling on the other (Cheng 2008).

Both geomathematics and conceptual thinking are needed to extrapolate data laterally into less explored areas or vertically from the surface downward into hidden rock formations at greater depths. Such projections remain subject to significant uncertainty that has to be quantified in order to allow valid decision-making. The problem to be considered is how the mineral potential of a region can be assessed systematically by statistical extrapolation from known facts. Because of the complexity of the geological framework, many authors have employed a variety of more subjective methods for mineral potential estimation, often with good results (see, e.g., Harris et al. 2008) but these other, "knowledge-driven" methods (*cf*. Bonham-Carter 1994; Bardossy and Fodor 2004; Carranza 2008; Singer and Menzie 2010) are outside the scope of this book. In this section, some basic principles of quantitative treatment of the geological framework of a region will be considered and demonstrated on the basis of a simple, 2-D mosaic model. Later, in a recently performed case history study (Agterberg 2011), a multivariate prognosis made from 1968 data for copper potential of the Abitibi area on the Canadian Shield will be reviewed and compared with amounts of copper and copper ore discovered in this area during the past 40 years. It turns out that most newly discovered copper occurs in the same favorable environments where deposits were already known to exist closer to the surface of bedrock.

The future of fully automated regional mineral resource estimation is promising because, increasingly, sophisticated geophysical remote sensing techniques are becoming available, while rapid progress is being made in the field of 3-D geological mapping. It should be kept in mind, however, that the geological framework generally is highly heterogeneous. (*cf.* Chap. 1) In addition to continuous spatial variability observed for geophysical fields, there are numerous discontinuities in the upper Earth crust, e.g. at contacts between different rock units and where there are faults. In general, advanced pre-processing techniques are required to produce realistic 3-D images providing the inputs for mineral potential estimation (de Kemp et al. 2013).

For the purpose of this discussion, it is useful to make a distinction between mineral exploration and mineral resource estimation. The objective of mineral exploration is to delineate high-potential target areas. This can be achieved by ranking cells or pixels in a region by means of a probability index for relative prospectivity. In mineral resource estimation, the primary objective is to predict numbers of deposits and their sizes for larger regions. Any probability index has to be converted into a probability that is unbiased. Early on, mineral resource estimation problems were considered by relatively few authors including Allais (1957) who used the Poisson model for completely random spatial distribution of large mineral deposits of any type (*cf.* Sect. 3.3.2). Griffiths (1966) advocating use of "unit regional value" lumping different types of metal and hydrocarbon deposits together, and Harris (1965) who quantified geological maps for cells relating "total dollar value" based on all metals to bedrock variables by means of multivariate statistical analysis. A characteristic feature of these early statistical publications was that natural resources of different types were analyzed simultaneously. Such lumping can be advantageous if statistical models have the property of additivity (e.g., a mixture of two spatial Poisson process models is another Poisson process model) but often it is better to incorporate different genetic models into the mineral resource estimation. Agterberg (1971, 1974) used a commodity-based approach that can be summarized as follows.

Various sources of uncertainty have to be considered in mineral resource estimation, and to some extent in exploration. These different types of uncertainty were considered separately and combined with one another when copper and zinc mineral potential maps were constructed for the Abitibi area on the Canadian Shield in the late 1960s and early 1970s (Agterberg et al. 1972) to be reviewed in more detail later in this chapter. To-day, of course, better answers could be obtained than in the early 1970s, because of both theoretical and computational advances.

However, the basic problems to be solved remain the same. The five principal sources of uncertainty are:

1. The first major source of uncertainty is provided by the probability of occurrence itself. Any point in a study area on a map has probabilities associated with it that a small unit area surrounding it contains mineral deposits of different types. (If depth can be considered as a third dimension, unit volumes can be used in addition to unit areas.)
2. The estimated probabilities have variances to express their uncertainty. Suppose that one is concerned with a single deposit type or commodity in 2-D. For small unit areas, the probabilities of occurrence then are very small. For example, suppose the 10 % largest probabilities are approximately 0.01. This does not only mean that a unit area with probability 0.01 would contain a deposit but also that the variance of this probability is 0.01. This intrinsic variance normally exceeds the estimation variance of the probability itself.
3. Intensity of exploration is a third source of uncertainty. This is a largely unknown variable that is difficult to quantify. Fortunately, uncertainty associated with variable intensity of exploration is much less than the uncertainty intrinsic in the probability itself. However, it should be kept in mind that, from an economic point of view, intensity of exploration can be regarded as the most important variable because it principally determines number of undiscovered ore deposits.
4. A second major source of uncertainty in mineral resource estimation is size distribution of the deposits for which the probabilities of occurrence are being estimated. In general, size of mineral deposits as a random variable covers several orders of magnitude with the largest deposits (supergiants; *cf.* Agterberg 1995) being exceedingly rare but of utmost economic importance. It also should be kept in mind that it is possible that deposit size is positively correlated with probability of occurrence.
5. Metal grades including cut-off grades are to be considered as well although these can often be incorporated in the definition of deposit type. In general, economic data on past production, various types of reserves and grades are of highest quality for the largest deposits with amount of information diminishing and tending to become unavailable for smaller and lower-grade deposits. Two factors to be considered are that mineral deposits for the same metal may occur in different geological settings and that usually more than a single metal is mined from the same deposit suggesting that total amount of ore also is useful as a variable for estimating probabilities of occurrence together with size frequency distribution modeling.

In order to further illustrate uncertainties (1) and (2), let us take a typical weights-of-evidence (*cf.* Bonham-Carter 1994; also see Chap. 5) result for example. The output map with posterior probabilities in weights-of-evidence usually is accompanied by a t-value map. Suppose that the t-value associated with a posterior probability of 0.01 is equal to 4. This would mean that the estimation variance of the probability of 0.01 amounts to (the square of 0.04 =) 0.0016, and this is less than 0.01 representing the intrinsic variance associated with the probability itself.

4.2 Linear Regression

Uncertainties (1) and (2) can be combined with one another by adding the variances associated with them. In the preceding example the combined variance is 0.0116. Suppose now, in the preceding example, that the probability of a (larger) unit cell is 0.1. It would imply that the intrinsic variance is 0.1, with estimation variance of 0.16, and combined variance of 0.26. It illustrates that for larger unit areas and for larger posterior probabilities, relative uncertainty associated with estimation increases significantly. It is noted that probabilities for groups of adjoining pixels can be added. The resulting sums can be interpreted as probabilities if they are less than 1 but must be considered to be expected values if they are greater than 1.

4.2.3 Elementary Statistics of the Mosaic Model

A small-scale geological map of bedrock in a region is a mosaic on which mineral deposits are projected as points. As already discussed in Chap. 2, a simple example of how one can proceed when information of this type is available is as follows: Suppose a study area contains one million pixels of which 20 % are underlain by "favorable" environment A. There are 10 pixels with mineral deposits in this study area of which 8 are on A. The other 2 are on "unfavorable" $A\sim$ where the $\sim$ symbol denotes "not". Therefore, the probability that any pixel contains a deposit is $P(D) = 0.000,01$. The probability that a pixel on A has a deposit can be written as $P(D|A) = 8/200,000 = 0.000,4$; likewise, $P(D|A\sim) = 2/800,000 = 2.5 \cdot 10^{-6}$. If a probability of occurrence map is constructed on the basis of this information, it contains 200,000 pixels with probability 0.000,4, and 800,000 with probability $2.5 \cdot 10^{-6}$.

The second type of uncertainty is related to precision of the statistics. When weights-of-evidence modeling is applied, the positive weight for the preceding example is 0.982 representing the natural log of the ratio $P(A|D)/P(A|D\sim) = 0.75/0.281 = 2.669$, and the negative weight is -1.056 representing the natural log of the ratio $P(A\sim|D)/P(A\sim|D\sim) = 0.25/0.719$. A useful measure of degree of special association between a point pattern and a mosaic layer is the contrast C, which is positive weight minus negative weight (cf. Chap. 5). The contrast for this example is 2.04 with approximate standard deviation equal to 1.18. The corresponding t-value of 1.73 is barely significant at the 95 % level if a one-sided test is used under the normality assumption. It is interesting to apply other resource estimation techniques to this simple mosaic model as well.

For example, one can fit the linear model $Y = a + b \cdot x$ where Y is a random variable assuming the value of 1 at pixels on "A" where $x = 1$, and 0 where $x = 0$. Using the method of least squares, this gives $a = 2.5 \cdot 10^{-6}$ and $b = 37.5 \cdot 10^{-6}$. Obviously, this linear regression model exactly reproduces the two probabilities estimated in the first paragraph of this section. The linear equation also can be used in logistic regression with Y representing the logit of occurrence instead of the probability itself (cf. Sect. 5.2). Application of this technique gave $a = -12.90$ and $b = 2.773$, with variances of 0.50 and 0.624, respectively, and covariance of -0.50. Conversion of logits into probabilities again reproduces $P(D|A) = 0.000,4$

and $P(D| A\sim) = 2.5 \cdot 10^{-6}$. The preceding four methods (probability calculus, weights-of-evidence, linear least squares, and logistic regression) all produce the same estimates of the probabilities (uncertainty type 1). However, they produce slightly different answers for the variances of these probabilities (uncertainty type 2).

Some remarks on other applications pertaining to the mosaic model are as follows. This model was used by Bernknopf et al. (2007) for different rock units with probabilities of occurrence for mineral deposits of different types. Probabilities and expected values were modified according to relative amount of exposure of each rock unit by these authors. In the context of weights of evidence modeling, Carranza (2009) asked the question of what would be the optimum pixel size. For the mosaic model, the answer to this question is simply that pixels should be sufficiently small to allow precise estimates of relative areas of rock units on the map. Further size decrease does not affect estimation results when mineral deposits are modeled as points, because of the dichotomous nature of every rock unit represented by a mosaic model.

4.3 General Model of Least Squares

The linear model discussed in Sect. 4.2 can be generalized by including p additional explanatory variables $X_1, X_2, \ldots, X_p$, so that: $E(Y|X) = \beta_0 + \beta_1 X_1 + \beta_2 X_2 + \ldots + \beta_p X_p$. As before Y is a random variable with the same variance as the random variable with zero mean that distorts the deterministic component $\beta_0 + \beta_1 X_1 + \beta_2 X_2 + \ldots + \beta_p X_p$. It is convenient to use matrix algebra and write:

$$\begin{bmatrix} Y_1 \\ Y_2 \\ \cdot \\ \cdot \\ \cdot \\ Y_n \end{bmatrix} = \begin{bmatrix} 1 & X_{11} & X_{21} & \cdot & \cdot & X_{p1} \\ 1 & X_{12} & X_{22} & \cdot & \cdot & X_{p2} \\ \cdot & \cdot & \cdot & \cdot & \cdot & \cdot \\ \cdot & \cdot & \cdot & \cdot & \cdot & \cdot \\ \cdot & \cdot & \cdot & \cdot & \cdot & \cdot \\ 1 & X_{1n} & X_{2n} & \cdot & \cdot & X_{pn} \end{bmatrix} \cdot \begin{bmatrix} \beta_1 \\ \beta_2 \\ \cdot \\ \cdot \\ \cdot \\ \beta_p \end{bmatrix} + \begin{bmatrix} E_1 \\ E_2 \\ \cdot \\ \cdot \\ \cdot \\ E_n \end{bmatrix}$$

Box 4.4: Multiple Regression

The preceding matrix equation also can be written as $\mathbf{Y} = \mathbf{X}\boldsymbol{\beta} + \mathbf{E}$. Best estimates of the coefficients satisfy: $\hat{\boldsymbol{\beta}} = (\mathbf{X}'\mathbf{X})^{-1} \mathbf{X}'\mathbf{Y}$. The estimated values are $\hat{\mathbf{Y}} = \mathbf{X}\hat{\boldsymbol{\beta}} = \mathbf{X}(\mathbf{X}'\mathbf{X})^{-1} \mathbf{X}'\mathbf{Y}$. In regional mineral resource appraisal, it can be convenient to define a matrix $\mathbf{D} = \mathbf{X}(\mathbf{X}'\mathbf{X})^{-1}\mathbf{X}'$ so that $\hat{\mathbf{Y}} = \mathbf{D}\mathbf{Y}$. The 95-% confidence interval of any individual estimated value is $\hat{\mathbf{Y}}_k \pm t_{0.975} \cdot s \sqrt{\mathbf{X}'_k (\mathbf{X}'\mathbf{X})^{-1} \mathbf{X}_k}$. If all estimated values are considered simultaneously, the wider belt

(continued)

4.3 General Model of Least Squares

Table 4.2 Analysis of variance table to test p explanatory variables for statistical significance

Source of variation	Sum of squares	Degrees of freedom	Mean square
Regression	SSR	p	SSR/p
Residuals	RSS	$n-p-1$	RSS/$(n-p-1)$
Total	TSS	$n-1$	

Table 4.3 Analysis of variance table to test q additional explanatory variables for statistical significance

Source	Sum of squares	Degrees of freedom	Mean square
First regression (p var.)	SSR_1	p	
Difference between 1 and 2	$\Delta SSR = SSR_2 - SSR_1$	q	$\Delta SSR/q$
Second regression ($p+q$ var.)	SSR_2	$p+q$	
Residuals	RSS	$n-p-1$	RSS/$(n-p-q-1)$
Total	TSS	$n-1$	

Box 4.4 (continued)

$\hat{Y}_k \pm \sqrt{X'_k(X'X)^{-1}X_k(p+1)s^2 F_{0.95}(p+1, n-p-1)}$ should be used. The squared multiple correlation coefficient $R^2 = $ SSR/TSS provides a measure of the degree of fit. The residual variance becomes: $s^2_{res} = $ RSS/$(n-p-1)$. The analysis of variance table (Table 4.1) becomes as is shown in Table 4.2 with application of the following F-test: $\hat{F}(p, n-p-1) = \frac{SSR/p}{RSS/(n-p-1)}$.

However, it is more common to apply analysis of variance by adding q new explanatory variables to the p explanatory variables already considered. Then the analysis of variance becomes as is shown in Table 4.3 with: $\hat{F}(p, n-p-q-1) = \frac{\Delta SSR/q}{RSS/(n-p-q-1)}$.

Various techniques of sequential regression analysis are useful. These are forward selection, stepwise regression and backward elimination (Draper and Smith 1966). When there are p explanatory variables, forward selection begins by finding the variable that has the largest squared correlation coefficient with the dependent variable. It is selected first. At the next and later steps the variable that most increases R^2 is included. The forward selection is stopped when none of the remaining explanatory variable significantly increases the degree of fit. Stepwise regression does the same as forward selection except for one refinement. After completing a single step, one goes back one step whereby the variables already included in the equation are again checked for statistical significance. Finally, backward elimination consists of first including all variables in the regression equation and eliminating them one by one until the F-ratio exceeds a predetermined level.

4.3.1 Abitibi Copper Deposits Example

Stepwise regression was used in the study (Agterberg et al. 1972) to relate occurrences of large copper deposits to lithological and geophysical variables in the Abitibi area on the Canadian Shield (Figs. 4.3 and 4.4). This region was selected in 1967 for various projects in the former Geomathematics Program of the Geological Survey of Canada on the systematic quantification of geoscience data for larger regions and to correlate the points of occurrence of mineral deposits to the resulting digitized versions of the geological framework. Initially the grid shown in Fig. 4.4 was used but later a grid of equal-area (10 km × 10 km) cells was projected on various geological maps for the area at scales 1 in. to 4 miles (approx. 1: 250,000) or larger. This grid (see, e.g., Fig. 4.5) corresponds to the Universal Transverse Mercator map projection used for Canadian topographic maps at scales 1:250,000 and larger.

Traditionally, the Abitibi area was important for its gold deposits, but in the 1960s gold-mining was becoming uneconomical because the price of gold had been kept artificially low (at $35.00 US per ounce) and mining shifted from lode gold to volcanogenic massive sulphide deposits that also occur abundantly in this region. Most large sulphide bodies are lenticular, massive to disseminated, stratabound deposits enclosed by volcanic and sedimentary rocks of Archean age. They contain mixtures, in various proportions, of pyrite, pyrrhotite, sphalerite, chalcopyrite, and, in some instances, galena. Copper from chalcopyrite is economically most significant; zinc from sphalerite is recovered from a number of deposits, in some of which significant copper is absent. The volcanogenic massive sulphide deposits also may contain significant (minable) amounts of gold and silver.

The Abitibi area in Fig. 4.4 consists of 814 cells measuring 10 km on a side. For 644 cells, values for the following ten attributes were determined: (1) granitic rocks (acidic intrusives and gneisses); (2) mafic intrusions; (3) ultramafics; (4) acidic volcanics; (5) mafic volcanics; (6) Archean sedimentary rocks; (7) metamorphic rocks of sedimentary origin; (8) combined bedrock surface length of layered iron formations; (9) average Bouguer anomaly; and (10) aeromagnetic anomaly at cell center after removal of effect of Earth's total magnetic field. The geophysical variables were corrected for their overall regional means so that below-average values became negative. More explanatory details on these variables are provided in Agterberg et al. (1972).

A geological environment is not only characterized by individual attributes but, more importantly, by the different types of coexistences of attributes. Some examples of pairwise coexistences are: (1) coexistence of Archean sedimentary rocks and iron formations in a cell may define another sedimentary facies; (2) coexistence of acidic volcanic rocks and relatively high Bouguer anomaly may indicate an ancient volcanic centre where rhyolites and tuffs cap a relatively thick pile of andesites and basalts; and (3) coexistences of different rock types indicate presence of contacts between these rock types in a cell (various types of mineral deposits tend to occur at or near contacts between specific rock types). In order to consider

4.3 General Model of Least Squares

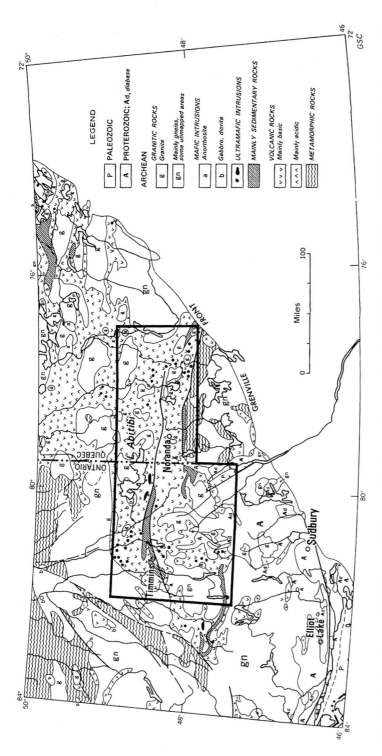

Fig. 4.3 Geological map of Abitibi area, which is part of the Superior Province, Canadian Shield (based on G.S.C. map 1250A by R.J.V. Douglas). Area used in Fig. 4.4 is outlined (From Agterberg 1974, Fig. 79A)

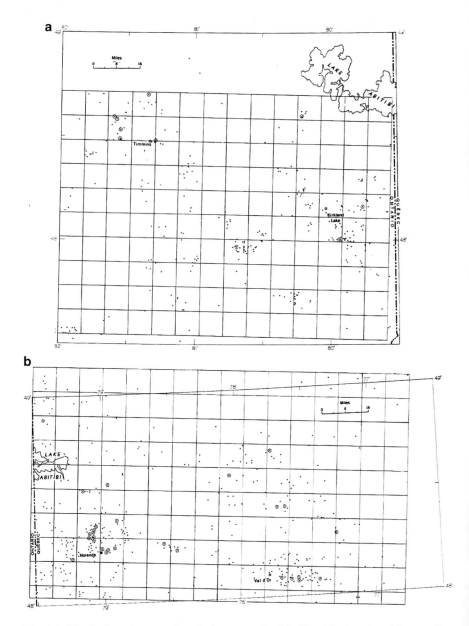

Fig. 4.4 Distribution of copper deposits in area outlined in Fig. 4.3; deposits with more than 1,000 t of copper are *circled*. (**a**) East-central Ontario; (**b**) Western Quebec. *Alternate lines* (8 miles apart) are shown for grid to be used for harmonic analysis in Chap. 7 (Source: Agterberg 1974, Fig. 80)

combinations of features co-occurring in the same cells, 45 additional explanatory variables were formed that are cross-products of the preceding 10 attributes. As is usual in applications of multiple regression, a dummy variable with value equal to 1 in all cells was added to obtain 56 explanatory variables in total.

4.3 General Model of Least Squares

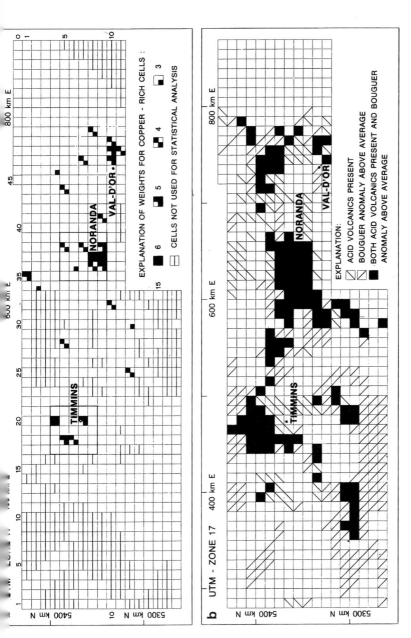

Fig. 4.5 Abitibi area, Canadian Shield; 643 (10 km × 10 km) cells are based on Universal Transverse Mercator grid. (**a**) Weight of 27 copper-rich cells is rounded value of logarithm (base 10) of total tonnage of copper per cell; control areas of 25 cells around Timmins and Noranda will be used for scaling. (**b**) Out of 55 binary variables, the one indicating both acidic volcanics present and Bouguer anomaly above average is most strongly correlated to copper weights shown in **a** (Source: Agterberg 1974, Fig. 119)

Consequently, the data matrix **X** used for the Abitibi area consists of 644 rows and 56 columns. The matrix $\mathbf{D} = \mathbf{X}(\mathbf{X}'\mathbf{X})^{-1}\mathbf{X}'$ for the relation between estimated and observed values of the dependent variable ($\hat{\mathbf{Y}} = \mathbf{DY}$) has 644 rows and 644 columns. Observed values for the dependent variable were based on logarithmically transformed amount of copper in (1968) production and reserves per cell (Fig. 4.5a). Only 27 cells contained one or more large copper deposits (with more than 1,000 short tons of copper in past production and estimated reserves; 1 short ton = 0.907 t). The remaining (644–27 =) 617 cells had zeros for the dependent variable. Rock type variables assume values that are either 0 or positive. Suppose that all explanatory variables are subjected to a Heaviside transformation changing their values that are greater than 0 into 1. Negative values for variables, that involve one of the two geophysical variables, were changed into zeros. Transforming all 55 variables in this manner and correlating them individually with the copper pattern of Fig. 4.5a showed that the single yes-no explanatory variable most strongly correlated with copper is acid volcanics present in a cell and Bouguer anomaly above average. This explanatory variable is shown in Fig. 4.5b. Clearly, there is positive spatial correlation between the patterns of Fig. 4.5a, b. The explanation of this association is that nearly all large copper deposits in the Abitibi area are volcanogenic massive sulphide deposits that are genetically associated with acidic volcanics. Relatively high Bouguer anomaly in this area indicates underground presence of a relatively thick pile of mafic volcanics with above average specific gravity.

Another simplified illustration of use of the general linear model for correlating occurrence of large copper deposits with the geological framework can be based of the **D** matrix. The Kidd Creek Mine near Timmins, Ontario, is the largest volcanogenic massive sulphide deposit in the Abitibi area. Suppose that the dependent variable that is related to all 56 explanatory variables by multiple regression has zero values in all cells except in the cell that contains the Kidd Creek Mine. This means that the 643 other cells are being compared with this single cell. If a full regression is carried out using all values of the dependent variable and the explanatory variables, the estimated values of the dependent variable are related to the corresponding observed values through the **D** matrix that consists of 644 rows of 644 values. The experiment of comparing all cells with the Kidd Creek cell then is equivalent to restricting the input for the dependent variable to the 644 values of the 56 explanatory variables in this single cell only. Figure 4.6 shows the corresponding estimated values for the dependent variable. These predicted values were classified according to three classes: very similar, more similar than average, and less similar than average. This map graphically represents the row of the **D** matrix that corresponds to the Kidd Creek Mine. From Fig. 4.6 it can be seen that comparing all (10 km × 10 km) cells in the Abitibi with the cell that contains the Kidd Creek deposit produces a pattern that has predictive potency.

4.3 General Model of Least Squares

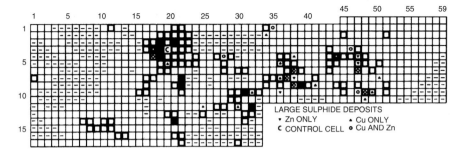

Fig. 4.6 Abitibi area; all cells were compared with a single control cell (C) containing giant (Cu, Zn, Pb, Ag) sulphide deposit (Kidd Creek Mine); 12 *black cells* have geological setting similar to that of C; 123 other cells with *black frame* have similarity index greater than average. Systematic comparisons of this type are useful to problem-solving in exploratory strategy (Source: Agterberg 1974, Fig. 116)

4.3.2 Forward Selection and Stepwise Regression Applied to Abitibi Copper

Standard methods of multiple regression were used for numerical analysis. The main computational step consists of inversion of the $\mathbf{XX'}$ matrix. Because the 55 explanatory variables contain several linear and near-linear relationships, the full matrix cannot be inverted. At a maximum, 51 explanatory variables could be used. Results of forward selection are partially shown as seven contour maps in Fig. 4.7. As illustrated in Fig. 4.8, the contours represent sums of estimated values for overlapping (40 km × 40 km) squares. Each larger unit square contains 16 values estimated for the smaller (10 km × 10 km) cells it contains. Estimated values for the smaller cells are less than 1 and can be interpreted as probabilities of occurrence of one or more copper deposits. Any of the sums of 16 probabilities for (40 km × 40 km) unit areas represents the expected value of number of (10 km × 10 km) cells with one or more copper deposits (also called "control cells"). Each contour map was scaled with respect to a relatively well explored "control area" that contains the mining districts of Timmins, Kirkland Lake, Noranda and Val d'Or. The sum of estimated probabilities in this control area was set equal to the number of cells with copper deposits in this area. It is shown that selection of eight variables (Fig. 4.7d) already nearly produces the final pattern based on 51 variables (Fig. 4.7g).

In another experiment along the same lines, stepwise regression was used to relate the copper cell values to all explanatory variables with individual stepwise selection and final regression coefficients shown in Table 4.4. The Q-value in this table represents level of significance for the F-distribution for every forward or backward step. Note that there occurred only a single backward step in this application. It is customary to terminate the regression run when Q, at a forward

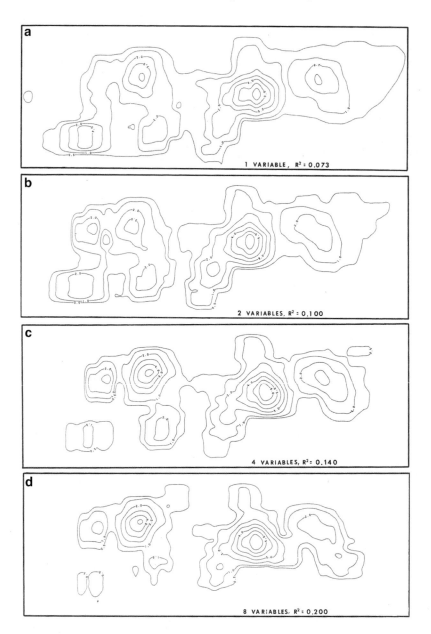

Fig. 4.7 Abitibi area; the copper weights of Fig. 4.5 were regressed on the ten variables for (10 km × 10 km) cells and all possible cross-products of these variables (*cf.* Table 4.4); contour pattern changes when increasingly more of these variables are included using stepwise multiple regression. Number of variables is (**a**) 1; (**b**) 2; (**c**) 4; (**d**) 8; (**e**) 16; (**f**) 32; (**g**) 51 (Source: Agterberg 1974, Fig. 124)

4.3 General Model of Least Squares 125

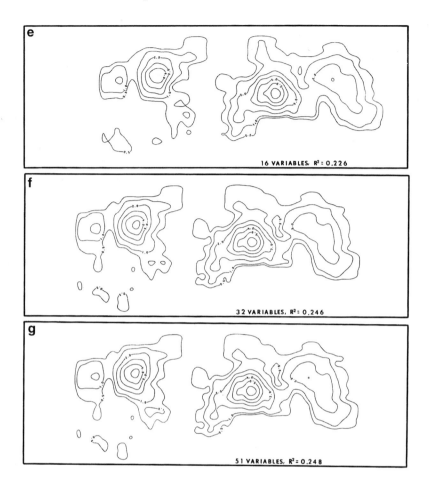

Fig. 4.7 (continued)

step, crosses a predetermined level Q. Setting $Q \leq 0.5$ for the present example resulted in the selection of 24 explanatory variables. Q-values are small (<0.05) for the first nine variables that were selected.

Finally, Fig. 4.9 shows how well smaller sets of known copper deposits in parts of the study area perform in predicting copper potential in the entire study area. Nineteen of the 27 (10 km × 10 km) cells with known copper deposits occurred in western Quebec, mainly in the Noranda and Val d'Or mining districts. The contour map of Fig. 4.9b is based on these 19 control cells only. There is strong similarity between Fig. 4.9a, b that was based on all 27 control cells. Using the eight control cells in Ontario (western part of study area) produced the pattern of Fig. 4.9c, which underestimates Fig. 4.9a.

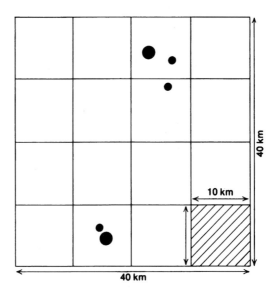

Fig. 4.8 Illustration of occurrence of deposits with respect to 16 (10 km × 10 km) cells within a (40 km × 40 km) unit area. Three smaller cells ($k = 3$) contain one or more deposits. This could be a realization of a random variable with expected value $m = 4$ that is shown on the contour map. Suppose that this random variable has a binomial distribution, then the probability that $k = 3$ amounts to 21 % (Source: Agterberg and David 1979, Fig. 3)

4.4 Abitibi Copper Hindsight Study

The Abitibi area copper potential map constructed in 1971 was based on 1968 statistics for production and reserves (Agterberg et al. 1972). During the 1970s a considerable amount of exploration for additional massive sulphide deposits was undertaken in this region. Agterberg and David (1979) evaluated the prognostic copper potential contours constructed from 1968 data on the basis of the locations and sizes of seven discoveries made between 1966 and 1977 (Millenbach, Louvem, Conigo, Iso-Copperfield, New Insco, Corbet and Montcalm deposits). The first three of these deposits already had been discovered when the original statistical analysis was performed but published figures on production and reserves were not yet available for them at that time. All seven new discoveries occurred either within the vicinity of one or more of the original set of 41 deposits, or within the three relatively high copper potential subareas without known deposits in 1968 mentioned before (also see Wellmer 1983). Together the 41 deposits contained 3.12 Mt of copper at the end of 1968. In 1977, the set of (41 + 7=) 48 deposits contained 5.23 Mt Cu. This increase was largely due to increased production and reserve estimates for the Kidd Creek mine (near Timmins, Ontario). The overall change in geographic distribution of large copper deposits from 1968 to 1977 can be seen by comparing Fig. 4.10b with Fig. 4.10a. Later discoveries (until 2008) are considered in Fig. 4.10c.

4.4 Abitibi Copper Hindsight Study

Table 4.4 Abitibi area; Variables selected by stepwise regression (Q = 0.5) of logarithmically transformed copper per (10× 10 km) cell on 55 geological and geophysical variables

Rank	Name of variable	$\hat{Q}$ – value	Regression coefficient
0	Constant term		−0.14863
1	Acidic volcanics × mafic volcanics	$0.36 \cdot 10^{-9}$	0.00045
2	Granitic rocks × acidic volcanics	$0.44 \cdot 10^{-4}$	0.00147
3	Acidic volcanics × sediments	$0.70 \cdot 10^{-3}$	0.00182
4	Acidic volcanics × iron formations	$0.45 \cdot 10^{-4}$	−0.00334
5	Acidic volcanics × aeromagn. anomaly	$0.63 \cdot 10^{-3}$	−0.01958
6	Mafic intrusions × acidic volcanics	$0.14 \cdot 10^{-3}$	0.00187
7	Metamorphic rocks × iron formations	$0.87 \cdot 10^{-3}$	0.00618
8	Mafic intrusions × metamorphic rocks	$0.91 \cdot 10^{-2}$	0.00129
9	Granitic rocks × sediments	$0.32 \cdot 10^{-1}$	0.00037
10	Sediments × Bouguer anomaly	0.13	0.00027
11	Mafic volcanics × sediments	0.021	0.00017
12	Acidic volcanics	0.13	0.05068
13	Sediments × aeromagnetic anomaly	0.12	0.00560
14	Sediments	0.14	−0.01726
15	Ultramafics × sediments	0.27	0.00049
16	Granitic rocks × iron formations	0.30	
17	Iron formations × Bouguer anomaly	0.10	−0.00490
18	Iron formations	0.19	−0.40777
19	Mafic volcanic × iron formations	0.09	0.00171
–	Backward pass, No. 16 deleted	0.65	
19	Iron formations × aeromagn. anomaly	0.20	0.02133
20	Ultramafics × iron formations	0.29	−0.00332
21	Sediments × metamorphic rocks	0.36	−0.00025
22	Acidic volcanic × methamorphic rocks	0.29	0.00123
23	Mafic intrusions × mafic volcanics	0.41	0.00026
24	Granitic rocks	0.38	0.00128
25	Granitic rocks × mafic volcanics	0.44	0.00003
26	Mafic intrusions × sediments	0.50	−0.00030

Figure 3.8a was based on this run. The rank of the first 16 variables selected is the same as that obtained by forward selection used for Fig. 4.10 (Source: Agterberg 1974, Table L)

4.4.1 Incorporation of Recent Discoveries

Lydon (2007) has published a comprehensive overview of the economic and geological contexts of Canada's major mineral deposit types accompanied by a DVD with production and estimated reserves of Canadian mineral deposits including large copper deposits in the Abitibi area. Between 1977 and 2007, there were five major new discoveries (Ansil, Bouchard-Hebert, Bousquet-Laronde, Amos and Louvicourt deposits), all within the vicinities of the 48 deposits known to exist in 1977. Revising our original 1968 data set (Agterberg et al. 1972, Appendix 1) using Lydon's data set, and including one 2008 estimate for the newly discovered Upper Beaver ore zone near Kirkland Lake, Ontario (cf. www.queenston.ca/news/pdf/080922.pdf), yields a combined set of 66 copper deposits containing 9.50 Mt Cu,

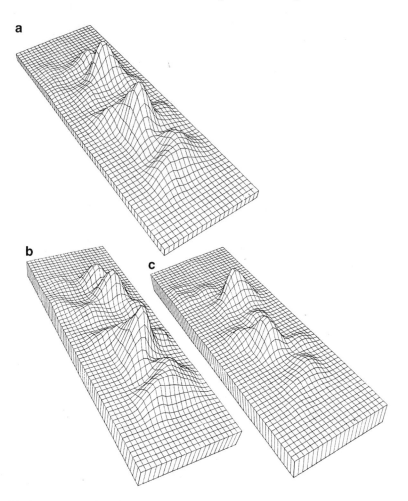

Fig. 4.9 3-D representation of three copper probability maps for same area as depicted in Fig. 4.7. The main peaks near Val d'Or, Noranda and Timmins fall approximately on east-west line pointing upwards to the left in these diagrams. (**a**) Probability index (based on 27 control cells shown in Fig. 4.5), scaled with respect to control area of 50 cells shown in Fig. 4.5; result is based on stepwise regression with $Q = 0.5$; 26 variables were included. (**b**) Ditto, as computed from 19 control cells east of 600 km E line of Fig. 4.7; control area consisted of the 25 cells near Noranda only. (**c**) Ditto, from 8 control cells in western part of area; control area of 25 cells near Timmins only. Note that **b** resembles **a**, but **c** underestimates **a** in the eastern part of the region (Source: Agterberg 1974, Fig. 117)

about three times as much as was contained in the 41 copper deposits in 1968. The 41 copper deposits in the 1968 dataset occurred in 27 "copper cells" measuring 10 km on a side and belonging to the original set of 644 cells for which copper potential was estimated in 1971. Figure 4.10 shows most copper cells, with (A) locations of 1968 copper cells, (B) locations of 1977 copper cells including two new cells with new discoveries, and (C) locations of copper cells in 2008 each

4.4 Abitibi Copper Hindsight Study

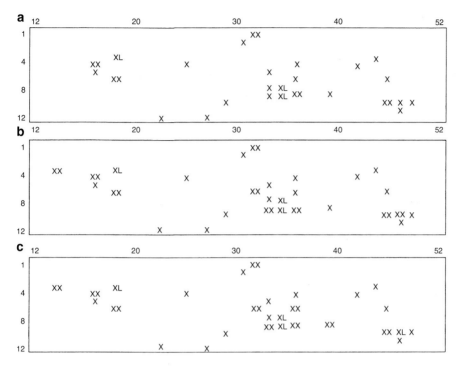

Fig. 4.10 Abitibi copper hindsight study; comparison of copper cells in (**a**) 1968, (**b**) 1977 and (**c**) 2008. Pattern (**a**) is same as pattern of copper weights in Agterberg et al. (1972) but for slightly smaller area; note minor change in areal copper distribution from 1968 to 1977 and no change at all in copper cell locations from 1977 to 2008. Single X denotes one or more deposits with production and reserves (Cu) between 1,000 short tons (st) but less than 50,000 tons (t); XX for cells with $50,000 < Cu < 500,000$ t; XL for cells with $Cu > 500,000$ t (1 st = 0.907 t). In total, the 2008 copper cells contain about three times as much copper as they did in 1968 (Source: Agterberg 2011, Fig. 2)

containing one or more copper deposits in the combined data set. The 1968 data set has three deposits not in Lydon's data base but plotted in Fig. 4.10, whereas the Lydon data base contains three Ni-Cu deposits with 0.1 % copper grade not in Agterberg et al. (1972, Appendix 1) and not plotted in Fig. 4.10. On average, a copper cell shown in Fig. 4.10c contains $(63/35) = 1.80$ large copper deposits but because of localized strong spatial clustering, the frequency distribution of number of deposits per copper cell is highly positively skewed with one cell (37,8) in the Noranda mining district containing as many as 11 large copper deposits. Comparison with Fig. 4.10b shows that nearly all differences between the patterns in Fig. 4.10a, c date from before 1977. The ten copper cells with most copper are listed in Table 4.5.

As mentioned before, new discoveries during the 1970s either were discovered close to known 1968 deposits or within the areas with relatively high copper potential outlined in Fig. 4.7. By 1977, geographical distribution of large copper

Table 4.5 Abitibi copper hindsight study; Amounts of copper for ten cells with most copper in 2008

Rank	Cell #,#	Type	Copper (kT)
1	20,4	1,0,0	4,516.1
2	37,9	11,0,0	1,710.1
3	37,8	1,0,0	1,436.8
4	48,10	4,0,0	576.1
5	41,9	5,0,0	241.2
6	47,10	1,0,1	166.5
7	36,8	2,0,0	114.9
8	38,7	2,0,0	92.6
9	35,7	2,0,0	90.3
10	34,1	1,0,0	79.9

Largest value (rank 1) for Kidd Creek Mine accounts for 47.5 % of total Abitibi production and reserves. Cell numbers as in Fig. 4.10. Three numbers in column for "Type" are for VMS, Ni-Cu and porphyry copper deposits per cell

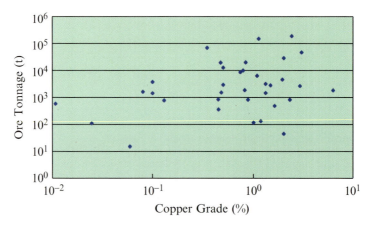

Fig. 4.11 Log-log plot of Ore Tonnage versus Copper grade (2008 data). The three points on the left are probably outliers. When these three points are deleted, the correlation coefficient ($r = 0.079$) is nearly zero suggesting lack of functional relationship between grade and ore tonnage (Source: Agterberg 2011, Fig. 3)

deposits in the Abitibi area had stabilized and further increases in production and reserves (from 5.23 to 9.50 Mt) were for copper within the known deposits and for new discoveries close to (and generally at greater depths than) the known deposits. Average grade of total production and reserves is about 1.6 % copper in the original 1968 data set with 41 copper deposits as well as in the 2008 data set with 66 copper deposits. Figure 4.11 is a log-log plot of copper grade versus amount of ore for the corresponding 35 copper cells that also will be analyzed in more detail in the next section.

Predictions made in Agterberg et al. (1972), such as the one for a "test area" in the surroundings of Timmins, Ontario, were based on the assumption that the frequency distribution of amount of copper per control cell could be used for this purpose. A relatively recent development is that increasingly it is realized that ore deposits, like earthquakes and several other types of natural phenomena, have fractal characteristics and resulted from non-linear processes. Mandelbrot (1983, p. 263) posed a challenge to geoscientists by stating that oil and other natural resources have Pareto distributions and "this finding disagrees with the dominant opinion, that the quantities in question are lognormally distributed. The difference is extremely significant, the reserves being much higher under the hyperbolic than under the lognormal law." This topic will now be investigated in more detail for copper in the Abitibi area. Later, it will be discussed again in the context of fractals (Chap. 10).

4.4.2 Comparison of Weight Frequency Distributions for Copper Metal and Ore

Size frequency distribution studies usually are carried out on populations of mineral deposits of the same type. In this study, it is applied to total amount of copper in the one or more copper deposits per (10 km × 10 km) cell. This procedure has advantages as well as drawbacks. An advantage is that the effect of strong localized clustering of deposits is curtailed, and total number of observations (27 in 1971 versus 35 in the combined data set) is stabilized. A disadvantage is that copper deposits of different types are being combined with one other although frequency distributions for different types of deposits can be different, especially if two or more metals are considered.

Nearly all (86 % of 66) large copper deposits in the combined data set are volcanic massive sulphide type (VMS). There are relatively few (six) magmatic Ni-Cu deposits (and five of these are small), plus three porphyry-type copper deposits. Preliminary statistical analysis was performed on various subsets such as using copper deposits instead of copper cells, VMS deposits only, Lydon's statistics only, but these exercises produced results similar to those to be presented here. However, explicit consideration of average copper grade (= amount of copper/amount of ore) generates somewhat different results. For this reason, the following analysis is for two variables per copper cell: (1) total weight (amount) of copper, and (2) total weight of ore. A comparison will be made between the 2008 and 1968 data.

Figure 4.12 shows log-log plots of copper and ore weight versus rank. A Pareto distribution plots approximately as a straight line on this type of plot as also shown for gold tonnages in lode gold deposits in the Superior Province of the Canadian Shield (Agterberg 1995; *cf.* Fig. 10.10). In each plot of Fig. 4.12, a straight line was fitted by least squares to most data points excluding the smallest copper or copper ore cells for which information is probably incomplete. Also, it can be expected that the Pareto distribution does not provide a good fit for low weight cells because it has the property that frequency of occurrence continues to increase with decreasing

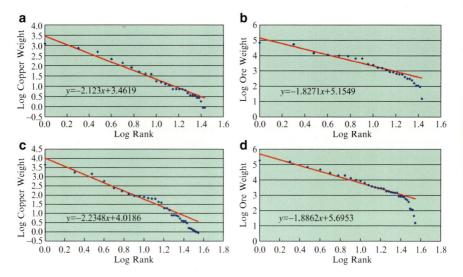

Fig. 4.12 Log-log Weight-Rank plots for 1968 and 2008 data with *straight lines* fitted by least squares. (**a**) 1968 Copper Weight; (**b**) 1968 Ore Weight; (**c**) 2008 Copper Weight; (**d**) 2008 Ore Weight. Base of logarithms = 10; Weight measured in (metric) tons. *Straight lines* approximate Pareto distributions. For 1968 data, first 18 of 27 data points were used to fit straight lines. For 2008 data, first 27 of 35 data points were used (Source: Agterberg 2011, Fig. 4)

weight. Figure 4.13 shows the corresponding four lognormal Q-Q plots. In Fig. 4.13a, c for copper weight, the patterns are not linear but in Fig. 4.13b, d they are, and straight lines were fitted by least squares using all data points. Degree of fit is good in these two diagrams as illustrated by the 95 % confidence interval in Fig. 4.13d.

It may be concluded that six of the eight plots (Figs. 4.12 and 4.13) show straight line patterns. The patterns in Fig. 4.14a, c are not approximately linear, probably because in several deposits copper is not the main metal of economic interest but only mineable as a by-product. For these deposits, total weight of ore fits in with the population of all copper deposits but total weight of copper does not because of the lower copper grades. It seems that both the Pareto and the lognormal are good candidates for modeling total copper and ore weight frequency distributions. The high-value tail of a Pareto frequency distribution is thicker than that of the lognormal. As will be discussed in more detail later (Sect. 11.4), the Pareto and lognormal each can be considered as the end product of a multiplicative cascade model. Cascade models can be regarded as a generalization of the generating process models discussed in the previous chapter. Pareto and lognormal frequency distributions are the end products of a de Wijs cascade and a Turcotte cascade, respectively. Other cascades (*cf.* Lovejoy and Schertzer 2007) can result in frequency distributions that resemble a lognormal except in their high-value Pareto-type tails (cf. Sect. 10.7). It may not be possible to determine whether a high-value tail is lognormal or Pareto-type if the frequency distribution it belongs to

4.4 Abitibi Copper Hindsight Study

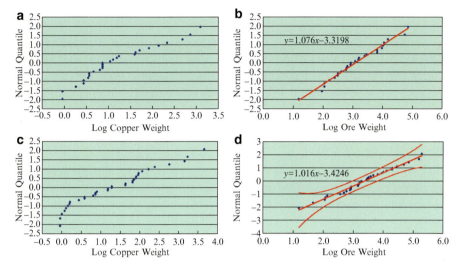

Fig. 4.13 Lognormal Q-Q plots of copper and ore weights for 1968 and 2008 data. (**a**) 1968 Copper Weights; (**b**) 1968 Ore Weights; (**c**) 2008 Copper Weights; (**d**) 2008 Ore Weights. *Straight lines* approximate lognormal frequency distributions with logarithmic standard deviation estimated by inverse of slope. *Curves* in Fig. 4.13d represent 95 % confidence belts for points deviating randomly from *straight line*. All data points were used to fit *straight lines* (Source: Agterberg 2011, Fig. 5)

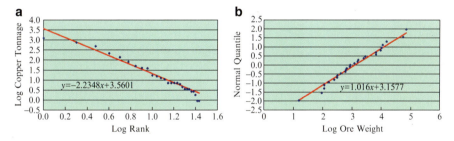

Fig. 4.14 Best-fitting straight lines for 1968 data with slopes set equal to slopes of *straight lines* fitted to 2008 data. (**a**) Log-Log Copper Weight; points same as in Fig. 4.12a; (**b**) Lognormal Q-Q plot of Ore Weights; points same as in Fig. 4.13b. Comparison with Figs. 4.12c and 4.13d shows 1968 and 2008 intercept increases (Source: Agterberg 2011, Fig. 6)

has strong positive skewness like the distributions of Figs. 4.13 and 4.14, because then there are too few very large values for application of standard goodness-of-fit tests (also see Agterberg 1995). Nevertheless, the few largest values contribute much or most of total weight for all deposits in the data set. This difficulty can be avoided by using the following method of comparing the Pareto and lognormal frequency distributions with one another (Agterberg 2011).

Comparison of coefficients of the straight lines fitted in Figs. 4.12 and 4.13 shows that 1968 and 2008 data have approximately the same slope but 2008 intercepts are markedly greater than 1968 intercepts. In Fig. 4.12 the straight lines are based on 18 and 27 data points for 1968 and 2008, respectively. All data points were used for the lognormals of Fig. 4.13. Because slope differences are small, it can be assumed that 1968 slopes are unbiased estimates of 2008 slopes. This is illustrated in Fig. 4.14 for 1968 copper and ore weight data where the best-fitting lines were forced to have the 2008 slopes. The intercept of the Pareto distribution for copper weight in Fig. 4.12a is less than its intercept in Fig. 4.12c. Suppose this difference is written as $\Delta_P = 0.4585$. Equations of the straight lines in Figs. 4.12, 4.13, and 4.14 are of the form $y = bx + a$ indicating that the dependent variable (Y) was regressed on the explanatory variable (x). All uncertainty is assumed to be associated with Y that is plotted in the vertical direction. These equations can be rewritten as $x = b'y + a'$; for example, $x = 0.9843y + 3.3737$ for Fig. 5d and $x = 0.9843y + 3.1080$ for Fig. 4.13b. It is noted that the least squares method used in this section results in slightly biased estimates of the coefficients. If this type of bias cannot be neglected, a different method of fitting the Pareto distribution can be used (Sect. 10.2.3).

The intercept (a') of the lognormal distribution in Fig. 4.14b is only $\Delta_L = 0.2627$ less than the intercept in Fig. 4.13d. Each intercept difference Δ corresponds to a factor of 10^Δ for increase in average weight per cell between 1968 and 2008. These factors are 2.875 for copper and 1.831 for copper ore, respectively. Incorporating a 6.1 % correction related to the relatively slight increase in total number of cells with known deposits due to new discoveries (cf. Fig. 4.10), the factors of increase in total weight become 3.049 for the copper-weight Pareto model, and 1.943 for the ore-weight lognormal model. Observed factors of increase are 3.026 for total copper weight and 3.030 for total ore weight, respectively. Consequently, the copper-weight Pareto agrees better with observed increase in total copper weight than the ore-weight lognormal, which significantly underestimates observed overall change in total ore weight. As an additional test, it was determined from the straight lines in Fig. 4.12b, d, that $\Delta_P = 0.4885$ for total ore weight, resulting in an increase factor of 3.266, slightly overestimating the observed value of 3.030.

4.4.3 Final Remarks on Application of the General Linear Model to Abitibi Copper

The large copper deposits in the Abitibi are almost all of the volcanogenic massive sulphide type. These deposits are associated with geological and geophysical variables that can be mapped. Other types of deposits may not so clearly associated with mappable variables. For example, in the Abitibi area there are many lode gold deposits that occur in places that are not clearly different from other places in the region with respect to the geological and geophysical variables used in the application of the general linear model to Abitibi copper.

4.4 Abitibi Copper Hindsight Study

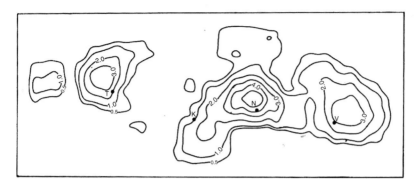

Fig. 4.15 Copper potential map for Abitibi area based on combined model with polynomial terms added to the geological and geophysical explanatory variables. Its contour values near Val d'Or exceed those in Fig. 4.7e–g (Source: Agterberg 1974, Fig. 126)

The geomathematical mineral resource prediction project in the Abitibi area which was commenced in 1967 (cf. Sect. 4.3.1) on the premise that the large amounts of data available for many hundreds of gold deposits including a few hundred gold mines (almost exclusively past producers) could be fruitfully used to develop methodology that would be applicable to other types of deposits. The Kidd Creek mine, which is on a very large volcanogenic massive sulphide deposit, had already been discovered and prospecting for new deposits in the Abitibi area had shifted almost entirely to large copper and zinc deposits instead of lode gold deposits that could no longer be profitably mined because the price of gold had been kept at 35 dollars per troy ounce. It turned out that spatial frequency of occurrence (number of deposits per unit of area) and sizes of the lode gold deposits are only partially controlled by mappable geological and geophysical variables. In situations of this kind it can be assumed that the regional mineralization processes were deep-seated and not mainly controlled by the depositional environments.

The problem of uneven geographical distribution of amounts of mineable metal across a region can be partially overcome by including terms in the regression equations that are functions of geographic location in the Abitibi area as described in Agterberg (1970, 1971), and Agterberg and Kelly (1971). An example for the large copper deposits used for example earlier in this chapter is shown in Fig. 4.15. The stepwise regression resulting in Table 4.4 and Fig. 4.7 was repeated after adding 44 new variables to the 55 explanatory variables used before. These additional variables were functions of the geographical coordinates of all cells in the region. Together these functions form an octic trend surface (cf. Chap. 7). The contour pattern of Fig. 4.15 is similar to the patterns obtained previously (e.g., Fig. 4.7) except in the southern Val d'Or-Noranda area. This would suggest that, relatively, there occurs more copper in this vicinity than elsewhere in the Abitibi area. Reddy et al. (1991) used logistic regression to predict base-metal potential in the Snow Lake area, Manitoba, using explanatory variables that included geographic locations. A two-stage least-squares model for the relationship between

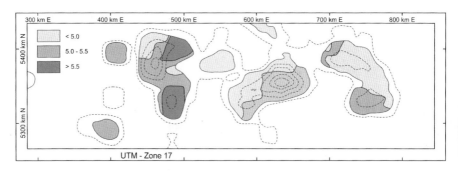

Fig. 4.16 Expected total amounts of copper for ore-rich cells predicted by Fig. 12.2a. Contour values are logarithms (base 10) for (30 km × 30 km) unit cells (Source: Agterberg 1973, Fig. 5A)

mappable geological variables with variable regional mean was developed by Agterberg and Cabilio (1969).

In most mineral potential research studies it is assumed that the size and grade distribution of the deposits is independent of location within areas considered favorable for their occurrence (see, e.g., Singer and Menzie 2010). However, a hypothesis worthy of investigation is that frequency of occurrence in more favorable environments is positively correlated with amounts of metal contained in the deposits. Examples of positive correlations of this type will be given later for gold deposits in Meguma Terrane, Nova Scotia (Sect. 5.2.1) and worldwide occurrence of porphyry copper deposits (Sect. 10.2.3). The following example taken from Agterberg (1973, 1977) applies to volcanogenic massive sulphide deposits in the Abitibi area (Fig. 4.16). This example will be discussed in more detail in an application of the jackknife method for bias elimination (Sect. 12.1.2). The Abitibi study area contained 35 (10 km × 10 km) cells with one or more large copper deposits. Two types of multiple regressions were carried out with the same explanatory variables. First the dependent variable was set equal to 1 in the 35 control cells, and then it was set equal to a logarithmic measure (base 10) of short tons of copper per control cell. Suppose that estimated values for the first regression are written as P_i and those for the second regression as Y_i. Both sets of values were added for overlapping square blocks of cells to obtain estimates of expected values (30 km × 30 km) unit cells. In Fig. 4.16 the ratio $Y_i^* = Y_i/P_i$ is shown as a pattern that is superimposed on the pattern for the P_i values only in Fig. 12.22 (see later). The values of Y_i^* cannot be estimated when Y_i and P_i are both close to zero. Little is known about the precision of Y_i^* for $P_i \geq 0.5$. These values (Y_i^*) should be transformed into estimated amounts of copper per cell here written as X_i. Because of the extreme positive skewness of the size-frequency distribution for amounts of copper per cell (X_i), antilogs (base 10) of the values of Y_i^* as observed in the 55 control cells were multiplied by the constant $c = \sum X_i / \sum 10^{Y_i^*}$ in order to reduce bias under the assumption of approximate lognormality (*cf.* Box 3.1). The pattern of Fig. 4.16 is useful as a suggested outline of subareas where the largest volcanogenic massive sulphide deposits are more likely to occur.

References

Agterberg FP (1964) Statistical analysis of X-ray data for olivine. Mineral Mag 33:742–748
Agterberg FP (1970) Multivariate prediction equations in geology. J Int Assoc Math Geol 2:319–324
Agterberg FP (1971) A probability index for detecting favourable geological environments. Can Inst Min Metall Spec 12:82–91
Agterberg FP (1973) Probabilistic models to evaluate regional mineral potential. In: Proceedings of the symposium on mathematical methods in geoscience, Přibram, Czechoslovakia, Oct 1973, pp 3–38
Agterberg FP (1974) Geomathematics. Elsevier, Amsterdam
Agterberg FP (1977) Probabilistic approach to mineral resource evaluation problems. In: Yu V (ed) Application of mathematical methods and computers for mineral search and prospecting. Computing Centre, Novosibirsk, pp 3–19 (in Russian)
Agterberg FP (1995) Multifractal modeling of the sizes and grades of giant and supergiant deposits. Int Geol Rev 37:1–8
Agterberg FP (2004) Geomathematics. In: Gradstein JM, Ogg JG, Smith AG (eds) A geologic time scale 2004. Cambridge University Press, New York, pp 106–125 & Appendix 3, pp 485–486
Agterberg FP (2011) Principles of probabilistic regional mineral resource estimation. J China Univ Geosci 36(2):189–200
Agterberg FP, Cabilio P (1969) Two-stage least-squares model for the relationship between mappable geological variables. J Int Assoc Math Geol 2:137–153
Agterberg FP, David M (1979) Statistical exploration. In: Weiss A (ed) Computer methods for the 80's. Society of Mining Engineers, New York, pp 90–115
Agterberg FP, Kelly AM (1971) Geomathematical methods for use in prospecting. Can Min J 92(5):61–72
Agterberg FP, Chung CF, Fabbri AG, Kelly AM, Springer J (1972) Geomathematical evaluation of copper and zinc potential in the Abitibi area on the Canadian Shield. Geological Survey of Canada, Paper 71-11
Allais M (1957) Method of appraising economic prospects of mining exploration over large territories: Algerian Sahara case study. Manag Sci 3:285–347
Bardossy G, Fodor J (2004) Evaluation of uncertainties and risks in geology. Springer, Heidelberg
Bateman AM (1919) Why ore is where it is. Econ Geol 14(8):640–642
Bernknopf R, Wein A, St-Onge M (2007) Analysis of improved government geological map information for mineral exploration incorporating efficiency, productivity, effectiveness and risk considerations, Bulletin (Geological Survey of Canada) 593. Geological Survey of Canada, Ottawa
Bickel PJ, Doksum KA (2001) Mathematical statistics, vol 1, 2nd edn. Prentice-Hall, Upper Saddle River
Bonham-Carter GF (1994) Geographic information systems for geoscientists: modelling with GIS. Pergamon, Oxford
Carranza EJM (2008) Geochemical anomaly and mineral prospectivity mapping in GIS, vol 11, Handbook of exploration and environmental geochemistry. Elsevier, Amsterdam
Carranza EJM (2009) Objective selection of suitable unit cell size in data driven modeling of mineral prospectivity. Comput Geosci 35(10):2031–2046
Cheng Q (2008) Non-linear theory and power-law models for information integration and mineral resources quantitative assessments. In: Bonham-Carter GF, Cheng Q (eds) Progress in geomathematics. Springer, Heidelberg, pp 195–225
Davis JC (2003) Statistics and data analysis in geology, 3rd edn. Wiley, New York
De Kemp E (2006) 3-D interpretative mapping: an extension of GIS technologies for the earth scientist. In: Harris JR (ed) GIS for the earth sciences. Geological Association of Canada, Special Publication 44, pp 591–612

De Kemp EA, Monecke T, Sheshpari M, Girard E, Lauzière K, Grunsky EC, Schetselaar EM, Goutier JE, Perron G, Bellefleur G (2013) 3D GIS as a support for mineral discovery. Geochem Explor Environ Anal 11:117–128

Deming WE (1943) Statistical adjustment of data. Wiley, New York

Draper NR, Smith H (1966) Applied regression analysis. Wiley, New York

Fisher RA (1915) Frequency distribution of the values of the correlation coefficient in samples from an indefinitely large population. Biometrika 10:507–521

Fisher RA (1960) The design of experiments. Oliver and Boyd, Edinburgh

Griffiths JC (1966) Exploration for mineral resources. Oper Res 14:189–209

Harris DP (1965) An application of multivariate statistical analysis in mineral exploration. Unpublished PhD thesis, Penn State University, University Park

Harris JR, Lemkov D, Jefferson C, Wright D, Falck H (2008) Mineral potential modelling for the greater Nahanni ecosystem using GIS based analytical methods. In: Bonham-Carter G, Cheng Q (eds) Progress in geomathematics. Springer, Heidelberg, pp 227–269

Krige DG (1962) Economic aspects of stopping through unpayable ore. J S Afr Inst Min Metall 63:364–374

Lovejoy S, Schertzer D (2007) Scaling and multifractal fields in the solid earth and topography. Nonlinear Process Geophys 14:465–502

Lydon JW (2007) An overview of economic and geological contexts of Canada's major mineralogical deposit types. In: Goodfellow WD (ed) Mineral deposits in Canada. Geological Survey of Canada, Special Publication 5, pp 3–48

Mandelbrot B (1983) The fractal geometry of nature. Freeman, San Francisco

Reddy RKT, Agterberg FP, Bonham-Carter GF (1991) Application of GIS-based logistic models to base-metal potential mapping in Snow Lake area, Manitoba. In: Proceedings of the 3rd international conference on geographical information systems, Canadian Institute of Surveying and Mapping, Ottawa, pp 607–619

Ripley BD, Thompson M (1987) Regression techniques for the detection of analytical bias. Analyst 112:377–383

Singer DA, Menzie WD (2010) Quantitative mineral resource assessments. Oxford University Press, New York

Wellmer FW (1983) Neue Entwicklungen in der Exploration (II), Kosten, Erlöse, Technologien. Erzmetall 36(3):124–131

Zhao P, Cheng Q, Xia Q (2008) Quantitative prediction for deep mineral exploration. J China Univ Geosci 19(4):309–318

Chapter 5
Prediction of Occurrence of Discrete Events

Abstract Many geological bodies or events including various types of mineral deposits, earthquakes and landslides can be represented as points on small-scale maps. Various methods exist to express probability of occurrence of such events in terms of various map patterns based on geological, geophysical and geochemical data (Agterberg 1989a). The machine-based approach was greatly facilitated by the development of Geographic Information systems (GIS, *cf.* Bonham-Carter 1994). Weights-of-Evidence modeling and weighted logistic regression are two powerful methods useful for estimating probabilities of occurrence of an event within a small unit area. Weights-of-Evidence (WofE) consists of first assuming that the event can occur anywhere within the study area according to a completely random Poisson distribution model. This equiprobability assumption provides the prior probability that only depends on size of an arbitrarily small unit area. Various indicator map patterns commonly reduced to binary (presence-absence) or ternary (presence-absence-unknown) form are used to update this prior probability by means of Bayes' rule in order to create a map of posterior probabilities that is useful for selecting target areas for further exploration for undiscovered mineral deposits or for the prediction of occurrence of other discrete events such as earthquakes or landslides. If probabilities are transformed into logits, Bayes' rule is simplified: the posterior logit simply is equal to the sum of the prior logit and the weights of which there is only one for each map layer at the same point. These weights are either positive (W^+) or negative (W^-) depending on presence or absence of the indicator, or zero for missing data. An important consideration in WofE applications is that the indicator patterns should be approximately conditionally independent (CI). WofE will be illustrated by applications to gold deposits in Meguma Terrain, Nova Scotia, and to flowing wells in the Greater Toronto area. Weighted logistic regression (WLR) also can be used to estimate probabilities of occurrence of discrete events. Both WofE and WLR are applied to gold occurrences in the Gowganda area on the Canadian Shield, northern Ontario, and to occurrences of hydrothermal vents on the East Pacific Rise. Indicator patterns used include favorable rock types, proximity to anticlinal structures or contacts between rock units, indices representing various geochemical elements, proximity to lineaments and igneous intrusives, aeromagnetic data, relative age, and topographic elevation. The Kolmogorov-Smirnov test is used for testing goodness of fit.

Keywords Discrete event prediction • Weights-of-evidence method • Contrast • Missing data • Weighted logistic regression • Meguma terrain gold deposits • Greater Toronto area flowing wells • Gowganda gold occurrences • Pacific rise volcanic vents • Conditional independence test

5.1 Weights-of-Evidence Modeling

In the preceding chapter, stepwise multiple regression analysis was used to estimate a probability index for occurrence of large copper deposits in the Abitibi area on the Canadian Shield. The input for explanatory variables primarily consisted of information on rock types systematically quantified for cells measuring 10 km on a side. Geophysical field data at cell centers were used for gravity (Bouguer) and regional aeromagnetic anomaly. Cross-products of variables provided better results than scores for individual variables. The dependent variable used in the multivariate linear model was logarithmically transformed total amount of copper in one or more copper deposits per input cell. Neither of the two geophysical input variables made a significant contribution to the magnitudes of the probabilities that were being estimated. However, in a separate computer experiment using binary (presence-absence) input data only, the variable most strongly correlated with occurrence of large copper deposits was a combination of presence of felsic volcanics at the surface of bedrock and higher than average Bouguer anomaly. This result could be interpreted in terms of a mineral deposit model, because nearly all large copper deposits in Abitibi are of the volcanogenic massive sulphide (VMS) type and were formed near volcanic centers in association with acidic (felsic) volcanics, while a relatively high Bouguer anomaly on the Canadian Shield indicates relatively large amounts of mafic volcanic rocks with above average specific gravity at greater depths.

The probabilities of occurrence estimated for (10 km × 10 km) cells in the last chapter were added for larger (40 km × 40 km) unit areas to produce a regionally based prognostic contour map for expected numbers of known and unknown copper deposits (Agterberg et al. 1972). Later, Assad and Favini (1980) conducted a statistical mineral exploration study for the eastern part of the Abitibi region using localized geophysical (aeromagnetic, gravity and terrain elevation) anomalies. It could be assumed that localized geophysical and geochemical anomalies can be superimposed on broader regionally based probabilities of mineral occurrence. This prompted an experiment of using the probabilities previously estimated in the Abitibi case study as priors in weights-of-evidence (WofE) modeling followed by later incorporating Favini and Assad's map layers that had been based on local geophysical anomalies only, in order to obtain updated posterior probabilities of occurrence (Agterberg 1989b). A somewhat similar method of updating prognostic

5.1 Weights-of-Evidence Modeling

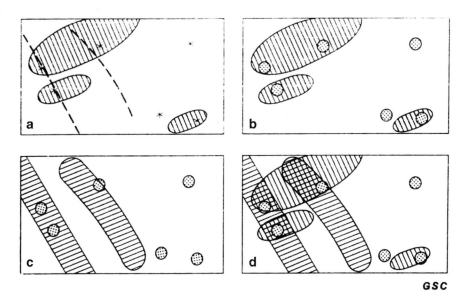

Fig. 5.1 Artificial example to illustrate concept of combining two binary patterns related to occurrence of mineral deposits; (**a**) outcrop pattern of rock type, lineaments, and mineral deposits; (**b**) rock type and deposits dilatated to unit cells; (**c**) lineaments dilatated to corridors; (**d**) superposition of three patterns (Source: Agterberg et al. 1990, Fig. 1)

maps by incorporating new data originally had been used by Dowds (1969) who updated probabilities of oil well occurrence maps by including later discoveries. The method of weights of evidence modeling was invented by Good (1950). In its original application to mineral resource potential mapping (Agterberg 1989a), extensive use was made of medical applications (Spiegelhalter and Knill-Jones 1984).

5.1.1 Basic Concepts and Artificial Example

Figure 5.1 illustrates the concept of combining two binary patterns for which it can be assumed that they are related to occurrences of mineral deposits of a given type. Figure 5.1a shows locations of six hypothetical deposits, the outcrop pattern of a rock type (B) with which several of the deposits may be associated (Fig. 5.1b), and two lineaments that have been dilatated in Fig. 5.1c. Within the corridors around the lineaments, the likelihood of locating deposits may be greater than elsewhere in the study area. In Fig. 5.1b–d, the deposits are surrounded by a small unit area. This permits estimation the unconditional "prior" probability $P(D)$ that a unit area with random location in the study area contains a deposit, as well as the conditional "posterior" probabilities $P(D|B)$, $P(D|C)$ and $P(D|BC)$ that unit areas located on the rock type, within a corridor and both on the rock type and within a corridor contain

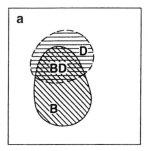 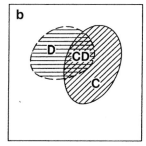 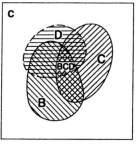

Fig. 5.2 Venn diagrams corresponding to areas of binary patterns of Fig. 5.1; (**a**) is for Fig. 5.1b; (**b**) is for Fig. 5.1c; (**c**) is for Fig. 5.1d (Source: Agterberg et al. 1990, Fig. 2)

a deposit. These probabilities are estimated by counting how many deposits occur within the areas occupied by the polygons of their patterns. The relationships between the two patterns (B and C), and the deposits (D) can be represented by Venn diagrams as shown schematically in Fig. 5.2.

Operations such as creating corridors around line segments on maps and measuring areas can be performed by using Geographic Information Systems (GIS's). The Spatial Data Modeller (SDM) (Sawatzky et al. 2009) is an example of a system that provides tools for weights of evidence, logistic regression, fuzzy logic and neural networks. The availability of excellent software for WofE and WLR has been a factor in promoting widespread usage of these methods. Examples of applications to mapping mineral prospectivity can be found in Carranza (2004), Cassard et al. (2008), Porwal et al. (2010), Coolbaugh et al. (2007) and Lindsay et al. (2014). Applications of these techniques in other fields include Cervi et al. (2010), Cho et al. (2008), Gorney et al. (2011), Neuhäuser and Terhorst (2007), Ozdemir and Altural (2013), Regmi et al. (2010) Romero-Calcerrada et al. (2010) and Song et al. (2008).

Box 5.1: Bayes' Rule for Single Map Layer

When there is a single pattern B, the odds $O(D|B)$ for occurrence of mineralization if B is present is given by the ratio of the following two expressions of Bayes' rule (*cf.* Sect. 2.2.1): $P(D|B) = \frac{P(B|D)P(D)}{P(B)}$; $P(\overline{D}|B) = \frac{P(B|\overline{D})P(\overline{D})}{P(B)}$ where the set $\overline{D}$ represents the complement of D. Consequently, $\ln O(D|B) = \ln O(D) + W^+$ where the positive weight for presence of B is: $W^+ = \ln \frac{P(B|D)}{P(B|\overline{D})}$. The negative weight for absence of B is: $W^- = \ln \frac{P(\overline{B}|D)}{P(\overline{B}|\overline{D})}$.

5.1 Weights-of-Evidence Modeling

Relationships between probabilities, odds and logits previously were discussed in Sect. 2.2. The result of application of Bayes' rule applied to a single map layer can be extended by using it as prior probability input for a second map layer. This process can be repeated by further adding additional map layers provided that there is approximate conditional independence (CI) of map layers. The order in which new patterns are added is immaterial.

Box 5.2: Bayes' Rule for Two Map Patterns

When there are two map patterns as in Fig 5.1: $P(D|B \cap C) = \frac{P(B \cap C|D)P(D)}{P(B \cap C)}$;

$P(\overline{D}|B \cap C) = \frac{P(B \cap C|\overline{D})P(\overline{D})}{P(B \cap C)}$. Conditional independence of D with respect to B and C implies: $P(B \cap C|D) = P(B|D)P(C|D)$; $P(B \cap C|\overline{D}) = P(B|\overline{D})P(C|\overline{D})$. Consequently, $P(D|B \cap C) = P(D)\frac{P(B|D)P(C|D)}{P(B \cap C)}$;

$P(\overline{D}|B \cap C) = P(\overline{D})\frac{P(B|\overline{D})P(C|\overline{D})}{P(B \cap C)}$. From these two equations it follows that: $\frac{P(D|B \cap C)}{P(\overline{D}|B \cap C)} = \frac{P(D)P(B|D)P(C|D)}{P(\overline{D})P(B|\overline{D})P(C|\overline{D})}$. This expression is equivalent to $\ln O$ $(D|B \cap C) = \ln O(D) + W_1^+ + W_2^+$.

The posterior logit on the left side of the final result shown in Box 5.2 is the sum of the prior logit and the weights of the two map layers. The posterior probability follows from the posterior logit. Similar expressions apply when either one or both patterns are absent. Cheng (2008) has pointed out that, since it is based on a ratio, the underlying assumption is somewhat weaker than assuming conditional independence of D with respect to B_1 and B_2. If there are p map layers, the final result is based on prior logit plus the p weights for these map layers. A good WofE strategy is first to achieve approximate conditional independence by pre-processing. A common problem is that final estimated probabilities usually are biased. If there are N deposits in a study area and the sum of all estimated probabilities is written as S, WofE often results in $S > N$. The difference $S-N$ can be tested for statistical significance (Agterberg and Cheng 2002). The main advantage of WofE in comparison with other methods such as WLR is transparency in that it is easy to compare weights with one another. On the other hand, the coefficients resulting from logistic regression generally are subject to considerable uncertainty (Sect. 5.2).

The contrast $C = W^+ - W^-$ is the difference between positive and negative weight for a binary map layer. It is a convenient measure for strength of spatial correlation between a point pattern and the map layer (Bonham-Carter et al. 1988; Agterberg 1989b). It is somewhat similar to Yule's (1912) "measure of association" $Q = \frac{\alpha-1}{\alpha+1}$ with $\alpha = e^C$. Both C and Q express strength of correlation between two binary variables that only can assume the values 1 (for presence) or −1 (for absence) (cf. Bishop et al. 1975, p. 378). Like the ordinary correlation coefficient, Q is confined to the interval $[-1, 1]$. If the binary variables are uncorrelated, then $E(Q) = 0$.

If a binary map layer is used as an indicator variable, the probability of occurrence of a deposit is greater when it is present than when it is absent, and W^+ is positive whereas W^- is negative. Consequently, C generally is positive. However, in a practical application, it may turn out that C is negative. This would mean that the map layer considered is not an indicator variable. In a situation of this type, one could switch presence of the map layer with its absence, so that W^+ and C both become positive (and W^- negative). An excellent strategy often applied in practice is to create corridors of variable width x around linear map features that are either lineaments as in Fig. 5.1 or contacts between different rock types (e.g., boundaries of intrusive bodies) and to maximize $C(x)$. From $\frac{dQ}{d\alpha} = \frac{2}{(\alpha+1)^2}$ being positive it follows that $Q(x)$ and $C(x)$ reach their maximum value at the same value of x (*cf.* Agterberg et al. 1990).

5.1.2 Meguma Terrane Gold Deposits Example

The first application of WofE was concerned with Meguma Terrane gold deposits in Nova Scotia (Wright 1988; Bonham-Carter et al. 1988). The study area of 2,591 km^2 is shown in Fig. 5.3. In total, there were 68 gold deposits. Gold production and reserves figures were available for 32 gold mines. If unit cell area is made sufficiently small, maps of posterior probabilities become similar in appearance except that, like the prior probability, all posterior probabilities are directly proportional to unit cell area selected. In this application, unit cell area arbitrarily was set at 1 km^2.

Three examples of map patterns considered are shown in Figs. 5.4, 5.5, and 5.6. Figure 5.4 is a ternary pattern for geochemical signature for Au. This pattern has three states instead of two because it is based on information from lake sediments that are missing in parts of the area close to rivers or the coastline. Each lake is surrounded by a catchment basin. In order to avoid strong violations of conditional independence of map layers, the 16 chemical elements for with measurements were available were combined into a single index or signature by multiple regression. If data are missing, the weight is set equal to 0. The positive weight for the geochemical signature is relatively large but the negative weight is small. Multiple regression is one way to combine variables with similar map patterns into a single index; another good method for accomplishing this is to perform principal component analysis on all variables and then use scores for the first principal component to create a single new map layer (Agterberg and Bonham-Carter 2005; Wang et al. 2012). Figure 5.5 shows corridors constructed around Acadian anticlines. Both positive and negative weights are significant. Figure 5.6 is for a corridor constructed around contacts of Devonian granite. Some gold deposits could be genetically related with these granites. This hypothesis can be tested statistically by application of the following procedure.

5.1 Weights-of-Evidence Modeling

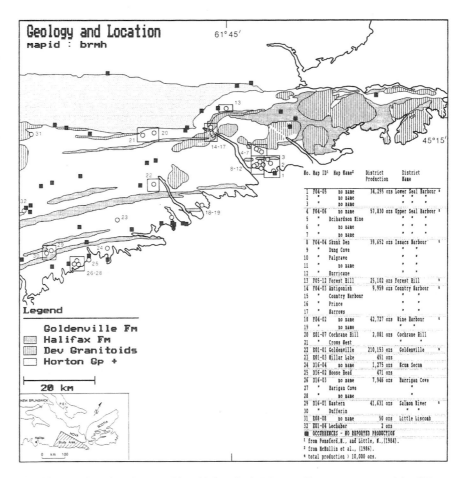

Fig. 5.3 Location of study area with gold deposits in Meguma Terrane, eastern mainland Nova Scotia (Source: Bonham-Carter et al. 1990, Fig. 1)

Approximate variances of the weights can be obtained from:

$$\sigma^2(W^+) = \frac{1}{n(bd)} + \frac{1}{n(b\overline{d})}; \sigma^2(W^-) = \frac{1}{n(\overline{b}d)} + \frac{1}{n(\overline{b}\,\overline{d})}$$

where the denominators of the terms on the right sides represent numbers of unit cells with map layer and deposit present or absent. Spiegelhalter and Knill-Jones (1984) had used similar asymptotic expressions for variances of weights in their GLADYS expert system. The following contingency table can be created:

$$\boldsymbol{P} = \begin{bmatrix} p(bd) & p(b\overline{d}) \\ p(\overline{b}d) & p(\overline{b}\,\overline{d}) \end{bmatrix} \equiv \begin{bmatrix} n(bd)/n & n(b\overline{d})/n \\ n(\overline{b}d)/n & n(\overline{b}\,\overline{d})/n \end{bmatrix}$$

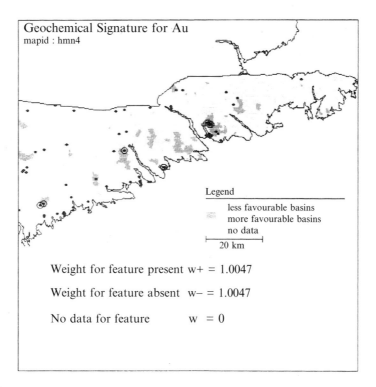

Fig. 5.4 Ternary pattern for geochemical signature (binary pattern for presence or absence of favorable geochemistry plus binary pattern unknown) (Source: Bonham-Carter et al. 1988, Fig. 2a)

If there is no spatial correlation between deposit points and map pattern:

$$P = \begin{bmatrix} p(b)p(d) & p(b)p(\overline{d}) \\ p(\overline{b})p(d) & p(\overline{b})p(\overline{d}) \end{bmatrix}$$

where $p(b)$ is proportion of study area occupied by map layer B, $p(d)$ is total number of deposits divided by total area, and with similar explanations for the other proportions. A chi-square test for goodness of fit can be applied to test the hypothesis that there is no spatial correlation between deposit points and map pattern. It is equivalent to using the z-test to be described in the next paragraph. The variance of the contrast $C = W^+ - W^-$ is:

$$\sigma^2(C) = \frac{1}{n(bd)} + \frac{1}{n(b\overline{d})} + \frac{1}{n(\overline{b}d)} + \frac{1}{n(\overline{b}\,\overline{d})}.$$

The subject of asymptotic estimates of variances will be discussed in more detail in Sect. 5.1.4.

Table 5.1 shows weights and contacts with their standard deviations for nine binary or ternary map layers used in a subsequent application of WofE to Meguma Terrain gold deposits (Bonham-Carter et al. 1990). Halifax Formation and Devonian granite were not used in this WofE application because almost no occurrences

5.1 Weights-of-Evidence Modeling

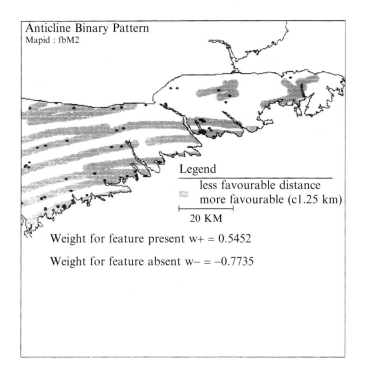

Fig. 5.5 Binary pattern for proximity to axial traces of Acadian anticlines (Source: Bonham-Carter et al. 1988, Fig. 3a)

are known on these two formations in the study area. Bonham-Carter et al. (1990) added a ternary map pattern for biogeochemical measurements of gold in balsam fir to the six map patterns used in earlier applications. Inspection of the standard deviations of the contrasts in Table 5.1 shows that spatial correlation of the 68 gold deposits with NW lineaments and granite contact is probably not statistically significant because the ratio of the contrast C and its standard deviation $\sigma(C)$ is less than 1.96 representing the 95 % confidence limit based on a two-sided z-test (cf. Sect. 2.3.3). Figures 5.7 and 5.9 show posterior probability maps obtained without and with use of the ternary pattern for gold in balsam fir that is shown in Fig. 5.8. The effect of adding the Au in balsam fir data is quite pronounced as shown by comparing the two posterior probability maps (Fig. 5.9 versus Fig. 5.7).

Table 5.2 illustrates how the pattern shown in Fig. 5.8 was selected. Weights, contrasts and their standard deviations were calculated for a series of different Au levels. Although the maximum value of W^+ occurs by thresholding at the 90th percentile, the maximum contrast C occurs for the 80th percentile. At this level, 24 out of 68 gold deposits fall within the balsam fir Au anomaly, which occupies 435 km^2 out of a total of 2,591 km^2.

The Kolmogorov-Smirnov test (Fig. 5.10) can be used to test the hypothesis of conditional independence of the seven map layers used for Fig. 5.9. This hypothesis

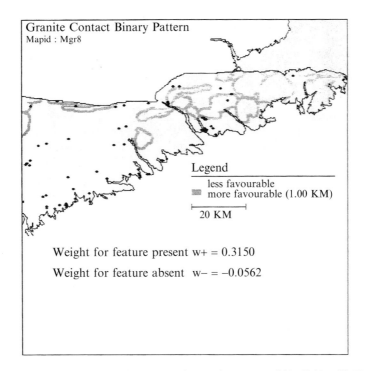

Fig. 5.6 Binary pattern for proximity to Devonian granite contact within Goldenville Formation (Source: Bonham-Carter et al. 1988, Fig. 3d)

Table 5.1 Weights, contrasts and their standard deviations for predictor maps

	W^+	$\sigma(W^+)$	W^-	$\sigma(W^-)$	C	$\sigma(C)$	$C/\sigma(C)$
Goldenville Fm	0.3085	0.1280	−1.4689	0.4484	1.7774	0.4663	3.8117
Anticline axes	0.5452	0.1443	−0.7735	0.2370	1.3187	0.2775	4.7521
Au, biogeochem.	0.9045	0.2100	−0.2812	0.1521	1.1856	0.2593	4.5725
Lake sed. signature	1.0047	0.3263	−0.1037	0.1327	1.1084	0.3523	3.1462
Golden-Hal contact	0.3683	0.1744	−0.2685	0.1730	0.6368	0.2457	2.5918
Granite contact	0.3419	0.2932	−0.0562	0.1351	0.3981	0.3228	1.2332
NW lineaments	−0.0185	0.2453	0.0062	0.1417	−0.0247	0.2833	0.0872
Halifax Fm.	−1.2406	0.5793	0.1204	0.1257	−1.4610	0.5928	2.4646
Devonian granite	−1.7360	0.7086	0.1528	0.1248	−1.8888	0.7195	2.6253

Source: Bonham-Carter et al. (1990, Table 1)
The last column is the "studentized" value of C, for testing the hypothesis that $C = 0$. Values greater than 1.96 indicate that the hypothesis can be rejected at $\alpha = 0.05$. Note that the hypothesis cannot be rejected for the granite contact and NW lineaments

is approximately satisfied because nowhere does the observed curve break the 95 % confidence envelope surrounding the predicted curve. An interesting result is that if the gold occurrences with known production are plotted on a graph of posterior probability versus cumulative area, there appears to be a positive correlation

5.1 Weights-of-Evidence Modeling

Fig. 5.7 Map of posterior probabilities based on weights shown in Table 5.1 (without use of balsam fir) (Source: Bonham-Carter et al. 1990, Fig. 2a)

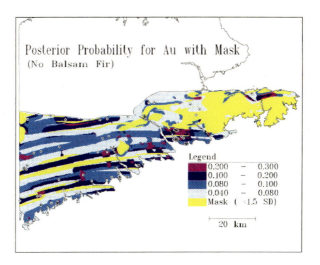

Fig. 5.8 Au in balsam fir map (Source: Bonham-Carter et al. 1990, Fig. 2b)

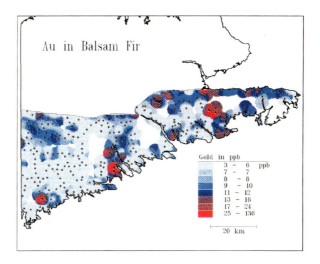

between production and posterior probability (Fig. 5.11). In other words, the larger gold districts are associated with higher predictions of gold potential. Figure 5.12 shows an enlargement of the region roughly centered on the Sherbrooke pluton. The masked areas are where posterior probability divided by its standard deviation is less than 1.5, i.e. where the posterior probability is not significantly greater than 0 in a one-tailed significance test. The geological contacts are superimposed in black, and the masked areas are either granite or Halifax Formation, or where the biogeochemical Au map is uncertain. The rectangular areas A to E with posterior

Fig. 5.9 Posterior probability map including balsam fir data based on weights shown in Table 5.1 (Source: Bonham-Carter et al. 1990, Fig. 2c)

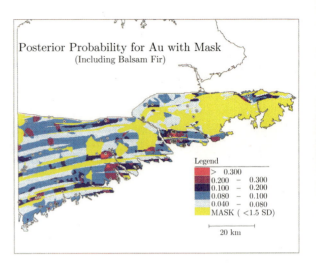

probabilities greater than 0.3 would be of interest, because probability of a gold occurrence within a 1 km² area is about 1–3. Two of the areas (B an E) are known gold districts. A, C and D, on the other hand, have no reported occurrences, yet they contain essentially the same signatures as B and E.

5.1.3 Flowing Wells in the Greater Toronto Area

Cheng (2004) has given the following application of Weights-of-Evidence modeling. Figure 5.13 shows surficial geology of Oak Ridge Moraine (ORM) for a study on assessment of flowing water wells in the Greater Toronto area, Ontario. The ORM is a 150 km long east-west trending belt of stratified glaciofluvial-glaciolacustrine deposits. It is 5–15 km wide and up to 150 m thick. For more detailed discussion of the geology of ORM, see Sharpe et al. (1997). It is generally recognized that the ORM is the main source of recharge in the area. Figure 5.14 shows flowing wells together with ORM and distances between wells and ORM. Cheng (2004) used Weights-of-Evidence to test the influence of the ORM on locations of flowing wells. A number of binary patterns were constructed by maximizing the contrast C. For distance from ORM, C reaches its maximum at 2 km (inset on Fig. 5.14). it means that a YES-NO binary pattern with YES on points belonging to ORM plus all points that occur less than 2 km from ORM and NO for the remainder of the study area with points that are more than 2 km away from ORM provides positive and negative weight that would be best to use for a binary pattern of the type because $C = W^+ - W^{--}$ has been maximized.

Other binary patterns constructed in the same way by Cheng (2004, Table 1) were distance from buffer zone constructed around ORM (Fig. 5.15), distance from a relatively steep slope zone (Fig. 5.16), and distance from a relatively thick glacial

5.1 Weights-of-Evidence Modeling

Table 5.2 Calculation of optimal cut-off of gold in balsam fir, to maximize the contract C with known gold occurrence points

Cut off		Cumulative								
%ile	ppb	Area km²	Occurrences#	W^+	$\sigma(W^+)$	W^-	$\sigma(W^-)$	C	$\sigma(C)$	$C/\sigma(C)$
98	137	42	0	–	–	–	–	–	–	–
95	24	93	3	0.3438	0.58690	−0.0133	0.1254	0.3571	0.6002	0.2090
90	16	227	13	0.9439	0.2856	−0.1349	0.1362	1.0788	0.3164	3.4095
80	12	435	24	0.9045	0.2100	−0.2812	0.1521	1.1856	0.2593	4.5724
70	10	848	31	0.4733	0.1830	−0.2745	0.1659	0.7479	0.2470	3.0279
60	8	1,070	35	0.3582	0.1719	−0.2771	0.1756	0.6353	0.2457	2.5853
50	7	1,360	45	0.3701	0.1516	−0.4732	0.2100	0.8433	0.2590	3.2560
<50	3–6	2,591	64	0.0695	0.1266	−0.7295	0.5028	0.7900	0.5185	1.5237
Outside		2,945	68	–	–	–	–	–	–	–

Source: Bonham-Carter et al. (1990, Table 2)

Fig. 5.10 Test of overall conditional independence, using the Kolmogorov-Smirnov statistic. Note that the observed curve (*open circles*) stays within the 95 % confidence envelope surrounding the predicted curve (*solid line*) (Source: Bonham-Carter et al. 1990, Fig. 3)

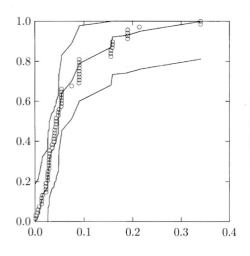

Fig. 5.11 Posterior probability plotted against cumulative area, with producing gold mines shown as *circles* whose radii reflect magnitudes of reported production (Source: Bonham-Carter et al. 1990, Fig. 4)

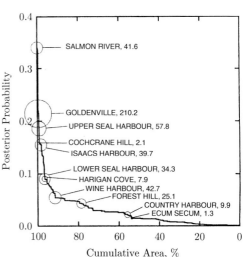

drift layer. The posterior probability map shown in Fig. 5.17 is based on buffer zone around ORM and steep slope zone only. It not only quantifies the relationship between the known artesian aquifers and these two binary map layers, it also outlines areas with no or relatively few aquifers that have good potential for additional aquifers.

5.1.4 Variance of the Contrast and Incorporation of Missing Data

For estimation of the variances of weights and contrasts in Weights-of-Evidence, use was made of asymptotic likelihood expressions (*cf.* Bishop et al. 1975,

Fig. 5.12 Map of posterior probabilities (enlarged subarea of Fig. 5.9) showing areas for follow-up exploration. Area *A* is at the head of Gegogan Harbour; *B* is the Goldenville district including the Goldenville mine working; *C* is north of the Sherbrooke pluton; *D* is an area almost 6 km north of Holland Harbour, through which Indian River flows; and *E* is the area around Isaacs Harbour inlet (Source: Bonham-Carter et al. 1990, Fig. 5)

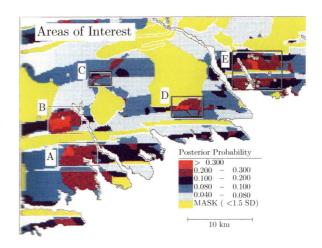

Chap. 14). Such expressions are valid only if a number of conditions are satisfied including the condition that the probabilities in the **P**-matrix (Sect. 5.1.2) are neither large (=close to one) nor small (=close to zero). The latter condition has probably been violated for some of the rock types with few gold deposits in Table 5.1. For example, there are only two occurrences on Devonian granite contributing 0.5 to the variance of their positive weight ($W^+ = -1.7360$) so that $\sigma(W^+) = 0.7086$ in Table 5.1 is probably too large. As mentioned before, presence of Devonian granite could be switched with its absence, which would be a better indicator than its presence. If weights of map layers are very small, including them for calculation of posterior probabilities does not significantly affect final results.

The standard deviation of a posterior probability can be estimated as follows. The variance $\sigma^2(p)$ of a prior probability p satisfies approximately p/n. For $p = 68/2{,}945 = 0.0231$, this yields the standard deviation $\sigma(p) = 0.0028$. The corresponding standard deviation of the prior logit $\log_e [p/(1-p)] = -3.7450$ is approximately $\sigma(p)/p = 0.1213$. This follows from the approximate identity for any variable x with mean $\bar{x}$:

$$\frac{\sigma(\log_e x)}{\sigma(x)} \approx \left| \frac{d(\log_e x)}{dx} \right|_{x=\bar{x}} = \frac{1}{\bar{x}}.$$

Suppose, for example, that a unit cell in Fig. 5.7 has the following features. Its geochemical signature is unknown; it occurs in the Goldenville Formation and not near a granite contact, but in the proximity of an anticline axis, NW lineament and Goldenville/Halifax contact. Then its posterior logit is -2.598 as can be seen when the appropriate weights are used. The variance of the log posterior odds is derived by adding variances of weights to the variance of the log prior odds. It follows that

154   5   Prediction of Occurrence of Discrete Events

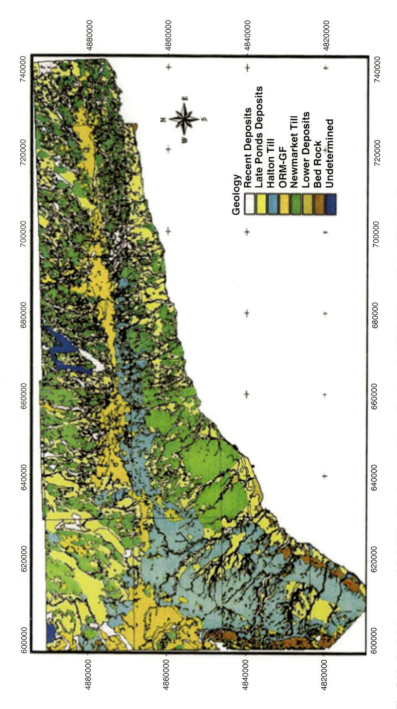

Fig. 5.13 Surficial geology of Oak Ridge Moraine according to Sharpe et al. (1997) (Source: Cheng 2004, Fig. 2)

5.1 Weights-of-Evidence Modeling

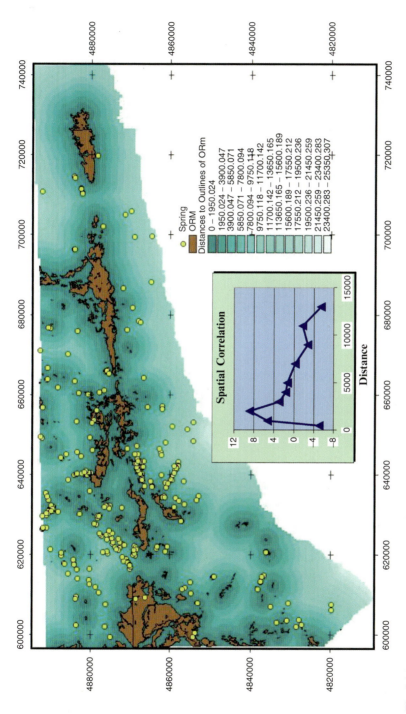

Fig. 5.14 Flowing wells versus distance from Oak Ridge Moraine in study area shown in Fig. 5.13 (Courtesy of Q. Cheng)

156   5 Prediction of Occurrence of Discrete Events

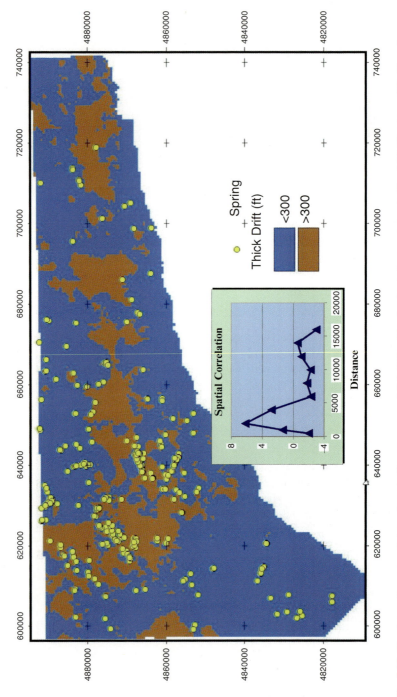

Fig. 5.15 Flowing wells versus distance from buffer zones around Oak Ridge Moraine in study area shown in Fig. 5.13 (Courtesy of Q. Cheng)

5.1 Weights-of-Evidence Modeling

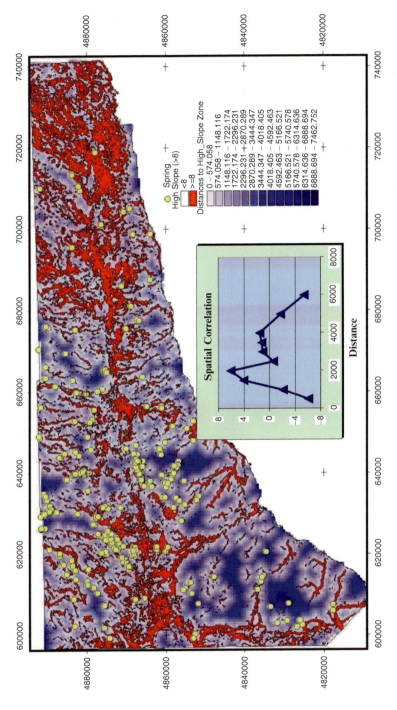

Fig. 5.16 Flowing wells versus distance from relatively steep slope zone in study area shown in Fig. 5.13 (Courtesy of Q. Cheng)

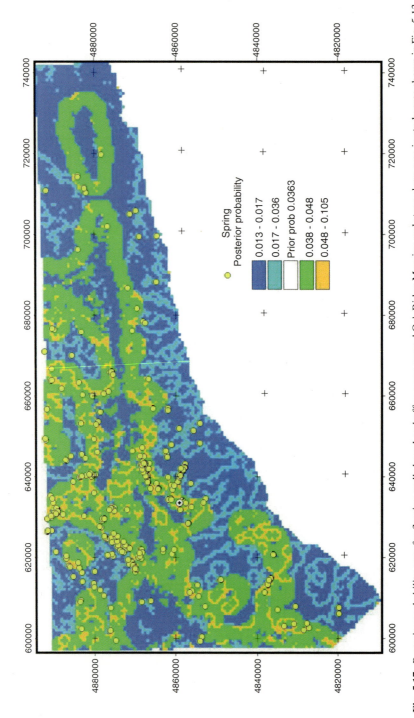

Fig. 5.17 Posterior probability map for flowing wells based on buffer zones around Oak Ridge Moraine and steep slope area in study area shown in Fig. 5.13 (Courtesy of Q. Cheng)

5.1 Weights-of-Evidence Modeling

the standard deviation of the posterior logit is 0.401. The posterior probability of the unit cell containing a deposit becomes 0.069 with approximate standard deviation equal to $0.069 \times 0.401 = 0.028$. In this way, a standard deviation can be estimated for each posterior probability on the final integrated pattern. However, it will be shown later that if one or more patterns are missing, the standard deviation of the posterior probability should be increased due to lack of knowledge. Because no information on geochemical signature is available for the unit cell in the preceding example, the final standard deviation becomes 0.042 instead of 0.028. It is customary to produce $t \approx p/\sigma(p)$ maps that accompany the p maps with posterior probabilities. Use of Student's t instead of standard normal z indicates that the 95 % confidence limit of 1.96 for z is too small. However, the exact number of degrees of freedom for Student's t is not known for this application. In practice, the 95 % confidence limit for t is probably some value between 2.0 and 2.5 for a two-sided test. For a one-sided test the 95 % confidence interval for z is 1.645 and the corresponding value for Student's t would be somewhat greater.

The equation used for estimating the standard deviation of the contrast C is based on the following asymptotic result for large n (see Bishop et al. 1975, p. 377):

$$\sigma_\infty^2(\hat{\alpha}) = \frac{\hat{\alpha}^2}{n}\left[\frac{1}{p(bd)} + \frac{1}{p(\bar{b}d)} + \frac{1}{p(b\bar{d})} + \frac{1}{p(\bar{b}\bar{d})}\right]$$

Extrapolation of this variance to $C = \log_e \alpha$ is only valid if it is small compared to α.

Box 5.3: Incorporation of Uncertainty Because of One or More Missing Patterns

This refinement is based on a proposal by Spiegelhalter (1986, p. 37) to regard any prior probability $p(d)$ as the expectation of the possible final probabilities $p(d|x)$ that may be obtained on observing data x. In general, $p(d) = E_x[p(d|X)] = \int p(d|x) p(x)dx$. In the situation that there are three map layers as in Fig. 5.1: $p(d) = \sum_{ij} p(d|b_i c_j) p(b_i c_j) = p(d|bc)p(bc) + (d|\bar{b}c)p(\bar{b}c) + (d|b\bar{c})p(b\bar{c}) + (d|\bar{b}\bar{c})p(\bar{b}\bar{c})$ with corresponding variance $\sigma_2^2[p(d)] = \sum_{ij} [p(d|b_i c_j) - p(d))]^2 p(b_i c_j)$. If only B is unknown, the information on C can be added to the prior probability in order to obtain updated prior probabilities $p_b(d)$ with variance: $\sigma_1^2[p(d)] = [p(d|b) - p(d)]^2 p(d) + [p(d|\bar{b}) - p(d)]^2 p(\bar{b})$ as follows from $\sum_j p(d|b_i c_j) p(b_i c_j) = p(d|b_i) p(b_i)$.

The expressions for the variances σ_1^2 (one pattern missing) and σ_2^2 (two patterns missing) derived in Box 5.3 are independent of any other patterns for which data were available and used to change the prior probability. In the example that resulted in Fig. 5.7, the ternary pattern for geochemical signature has parts of the area for which no data were available (Fig. 5.4). Uncertainty due to missing data becomes zero in places where all patterns including the geochemical signature are available. The term σ_1^2 is only added to the variance in places where presence or absence of the feature could not be determined.

The weights $W^+ = 1.0047$ and $W^{--} = -0.1037$ for geochemical signature were calculated from likelihood ratios for the entire area. For example, $W^+ = \log_e \frac{p(b|d)}{p(b|\bar{d})} = 1.0047$ was based on (1) $p(b|d) = \frac{p(bd)}{p(d)} = n(bd)/n(d)$ with $n(bd) = 10$ and $n(d) = 68$; and (2) $p(b|\bar{d}) = \frac{p(b\bar{d})}{p(\bar{d})} = n(b\bar{d})/n(\bar{d})$ with $n(b\bar{d}) = 164.9 - 10 = 154.9$ and $n(\bar{d}) = 2,945.0 - 68 = 2,877.0$. The weight W^+ can be regarded as independent of the prior probability. For this example, approximately the same value of W^+ is obtained when (1) the calculation is based on the subarea (=1,765.8 km^2) with known geochemical signature, and (2) the prior probability within the area with known geochemical signature is set equal to the prior probability for the total study area (=2,945 km^2). The second condition would imply that there are 41 deposits within the area with known geochemical signature. In reality, this subarea contains only 24 known gold occurrences. Revised weight based on the subarea only would amount to 1.5444 which is greater than $W^+ = 1.0047$, because the subarea would contain a larger proportion of the deposits. The lesser weight ($W^+ = 1.0047$) was used in Fig. 5.7 and will now be employed for estimating σ_1^2.

For this example, the modified prior probability, which is based on all patterns except geochemical signature, will be set equal to 0.05 and 0.10 within the area without definable geochemical signature. The logits of these values are -2.9444 and -2.1972, respectively. Addition of W^+ and W^- provides the required estimates of $p(d|b)$ and $p(d|\bar{b})$. For $p(d) = 0.05$, the conditional probabilities are equal to 0.1257 and 0.0453, respectively. For $p(b)$, which also is necessary to determine σ_1^2, the ratio of favorable area (=164.9 km^2) to known area (=1,765.8 km^2) can be used. This gives $p(b) = 1 - p(\bar{b}) = 0.9066$. Consequently, $\sigma_1(0.05) = 0.024$. By the same method, it follows that $\sigma_1(0.10) = 0.042$. In a previous example, it was pointed out that a unit cell in the Goldenville Formation with unknown geochemical signature in the proximity of the linear features except granite contact has posterior probability of 0.070 with standard deviation equal to 0.028. Addition of the uncertainty due to missing pattern results in the larger standard deviation of 0.042.

A question that is often asked when WofE is used to define target areas representing areas with relatively high posterior probabilities but with no or few known deposits is related to bias arising when undiscovered deposits probably exist in the study area. All posterior probabilities in a study area systematically underestimate the "true" posterior probabilities if there are undiscovered deposits in the study area. Such bias could be eliminated only if it would be exactly known how many undiscovered deposits there are in a region and this quantity remains unknown. Because the estimated weights depend only on presence or absence of features at the actual sites of the deposits, the weights estimated for known and unknown deposits in a region can be assumed to be independent of the prior probability. In order to hypothetically incorporate unknown events in the study area, one might change the prior probability by increasing it by a factor based on intensity of exploration (Agterberg 1992). If a reasonable guess can be made on number of undiscovered deposits, for example by assuming that all deposits have been discovered in "control" area consisting of mining districts, then the prior probability can be enlarged accordingly. It should be kept in mind, however, that most unknown metal resources may well occur at greater depths within mining districts as was shown for copper in the Abitibi area in Sect. 4.4.2.

Another question asked with respect to WofE applications is that reduction of map layers for indicator variables to binary or ternary form seems to be crude approximation. WofE software generally allows for approximations with more than two or three states. An example of application without missing data in which more than two states are needed are aeromagnetic data in relation to occurrences of earthquakes (Goodacre et al. 1993). Agterberg and Bonham-Carter (1990) have shown that it is possible to construct variable weight functions in an application to the relationship between occurrences of gold deposits and proximity to Devonian anticlines in Meguma Terrain, Nova Scotia. Variable weights also can be calculated when the input layer is obtained by 2-D kriging in which the kriging variance strongly depends on density of observation points as shown by Bonham-Carter and Agterberg (1999) in a study of relating the Meguma Terrain gold deposits to Au in balsam fir trees.

5.2 Weighted Logistic Regression

If it is assumed that there are no undiscovered or new events in a study area, the sum of the posterior probabilities should be, at least approximately, equal to the number of known events if the conditional independence (CI) assumption is satisfied. Weighted logistic regression (WLR) yields unbiased posterior probabilities if undiscovered events are not considered. This technique is equivalent to ordinary logistic regression except that the input values for the explanatory variables are weighted according to their areal extents within the study area. Logistic regression coefficients are equivalent to WofE contrasts ($C = W^+ - W^-$) that measure strength of correlation between point pattern and map layers. Logistic regression will be applied to occurrences of volcanogenic massive sulphide and magmatic nickel-copper deposits in the Abitibi area on the Canadian Shield for comparison with results as obtained in the previous chapter by means of the general linear model of least squares.

Suppose that X' is a row vector $\{1, X_1, X_2, \ldots, X_p\}$ with values of p explanatory variables while Y is a binary variable that can only assume the values 1 or 0. The explanatory variables can be binary like the variables used for WofE earlier in this chapter but they can assume any other value on the real line. The event in a cell or at a location is given by the following two equations:

$$P(Y=1|X) = \pi(X) = \frac{e^{\beta_0+\beta_1 X_1+\ldots+\beta_p X_p}}{1 - e^{\beta_0+\beta_1 X_1+\ldots+\beta_p X_p}}$$

$$P(Y=0|X) = 1 - \pi(X)$$

This represents a non-linear model; the unknown coefficients then can be estimated using the scoring method of maximum likelihood with Newton-Raphson iteration (Agterberg 1992). In applications to presence or absence of deposits, after reaching convergence, the sum of all estimates of the n probabilities π_j, which can be written as S, should be equal to the total number of known mineral deposits, N. Weighted Logistic Regression (WLR) is a variant of logistic regression. It was originally developed for maps with mosaic patterns consisting of numerous small polygons with yes-no data for different map layers. Polygons with the same characteristic features belong to the same "unique condition". The unique conditions can be regarded as separate observations to be weighted according to the areas they occupy on the map of the study area. Because the observations have different weights according to the areas occupied by the unique conditions, WLR differs slightly from ordinary logistic regression.

Logistic regression is used in many branches of science. Cox (1966) has provided a detailed account of the logistic qualitative response model, its multivariate extension employing several explanatory variables, and its relation to discriminant analysis. An elementary introduction to the method with illustrative examples is provided by Hosmer and Lemeshow (1989). There is a close connection between logistic regression and linear discriminant analysis (*cf.* Agterberg 1974). Recently, discriminant analysis was used by Grunsky et al. (2013) in a study of lake sediment geochemistry of the Melville Peninsula. It provides a basis for distinguishing between different map units which are assigned probabilities of occurrence on the basis of the lake sediment geochemistry.

Chung (1978) published LOGIST, a computer program for logistic regression. Pregibon (1981) used the approach to estimate frequencies, which are independent binomial responses, thus expanding the method to deal with multiple qualitative responses. Agterberg (1989c) published LOGDIA, which is a generalization of LOGIST in that frequencies of more than one discrete event could be estimated and logistic regression diagnostics were provided. The further extension (LOGPOL program) to make the method applicable to observations for polygons with different areas is described in Agterberg (1992).

Box 5.4: Newton-Raphson Iteration

The logistic regression model can be written in the form: Logit $\pi(X) = \beta_0 + \beta_1 X_1 + \ldots + \beta_p X_p$. Suppose that Y with elements Y_j is a column vector consisting of n ones and zeros denoting presence or absence in n very small unit cells (e.g. $n = 10^6$) in a study area with N unique conditions. Suppose further that the weights w_i with $\sum w_i = n$ represent numbers of unit cells for the i-th unique condition. The $(N \times N)$ diagonal matric V with non-zero elements $V_{ii} = w_i \cdot \sum \{\hat{Y}_j(1 - \hat{Y}_j)\}$ where $\hat{Y}_j$ is an estimated value of $\pi_j(X)$. If the maximum likelihood method is used for estimation, a column vector of scores S for differences between observed and estimated values of Y is made to converge until $X'S = 0$. Newton-Raphson iteration results in successive estimates: $\beta(t+1) = \beta(t) + \{X^T V(t) X\}^{-1} X^T V(t) \{X\beta(t) + V^{-1}(t)S(t)\}$, $t = 1, 2, \ldots$ At the beginning of the process an arbitrary vector of coefficients (e.g., with all coefficients set equal to 0) is used. After convergence, the estimated logits are converted into probabilities. The LOGPOL program, which is based on Newton-Raphson iteration with weights w_i, was incorporated in the Spatial Data Modeller (SDM, Sawatzky et al. 2009).

5.2.1 Meguma Terrane Gold Deposits Example

Weights of evidence (WofE) modeling and weighted logistic regression (WLR) are different types of application of the loglinear model (*cf.* Agterberg 1992). In WLR, the patterns are not necessarily conditionally independent as in WofE. WLR can also be used in situations where the explanatory variables have many classes or are continuous. In the Meguma Terrane gold deposits example, the map patterns that were selected are approximately conditionally independent. The conditional independence (CI) hypothesis can be tested in various ways. It was already shown by means of the Kolmogorov-Smirnov test (Fig. 5.10) that CI is approximately satisfied for the seven map layers used for Fig. 5.9. One simple way for testing CI is to compare the sum S of all posterior probabilities with the total number of deposits (N). For Fig. 5.9, the total number of deposits predicted by the posterior probabilities is $S = 75.2$ exceeding $N = 68$ representing the actual number of gold deposits. If all patterns would have been conditionally independent, their predicted total would have been 68 as well. Minor violations of the CI hypothesis account for WofE overestimating the total number of gold occurrences (=68) by about 10 %. Bonham-Carter (1994) introduced the so-called omnibus test stating that CI is approximately satisfied if over-estimation is less than 15 %. The hypothesis $S = N$ can be tested statistically (Agterberg and Cheng 2002). For a systematic comparison of various CI testing methods, see Thiart et al. (2006).

Figure 5.18 (top) shows results of WLR applied to the data set previously used for the WofE result shown in Fig. 5.9. Figure 5.18 (bottom) is the corresponding t-value map. The t-values are the posterior probabilities divided by their standard

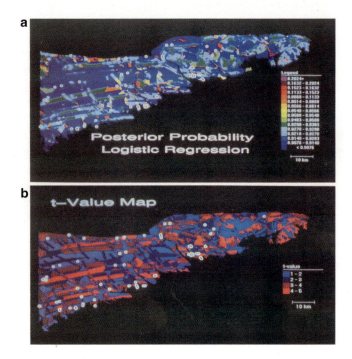

Fig. 5.18 Weighted logistic regression applied to gold deposits (*circles*) in Meguma Terrane, Nova Scotia. (*Top*) Posterior probability map with 91 unique conditions for seven binary patterns without missing data, unit cell size = 1 km^2; (*Bottom*) *t*-value map for Fig. 5.18a (Source: Agterberg et al. 1994, Plate 5a, b)

deviations. Table 5.3 shows regression coefficients in comparison with WofE weights. Estimated standard deviations are shown as well. Finally, Fig. 5.19 is an evaluation of the goodness of WLR fit. The difference between expected and observed relative frequencies is plotted against posterior probability. The absolute value of the largest difference of this type is 0.0775. This is less than the Kolmogorov-Smirnov statistic (=0.1426; 95 % two-tailed test) from which it may be concluded that the fit of the logistic model is good.

It can be seen in Table 5.3 that the seven contrasts show similarity with the seven regression coefficients. Theoretically in discrete multivariate analysis (*cf.* Andersen 1990; Christensen 1990), it can be shown that if, asymptotically, explanatory variables are conditionally independent, then logistic regression and the procedure used in WofE produce identical results (cf. Agterberg 1992).

5.2.2 Comparison of Logistic Model with General Linear Model

Applications of the general linear model to study the occurrence of large copper deposits in the Abitibi Volcanic Belt were discussed in the previous chapter. In a discussion of a paper by Agterberg and Robinson (1972), Tukey (1972) suggested

5.2 Weighted Logistic Regression

Table 5.3 Weights and contrasts (with standard deviations) for seven binary patterns related to gold deposits in Meguma Terrane

Pattern No.	W^+	$S(W^+)$	W^-	$S(W^-)$	C	S(C)	B	S(B)
0							−6.172	0.501
1	0.563	0.143	−0.829	0.244	1.392	0.283	1.260	0.301
2	0.836	0.210	−0.293	0.160	1.129	0.264	1.322	0.267
3	0.367	0.174	−0.268	0.173	0.635	0.246	0.288	0.266
4	0.311	0.128	−1.448	0.448	1.787	0.466	1.290	0.505
5	0.223	0.306	−0.038	0.134	0.261	0.334	0.505	0.343
6	1.423	0.343	−0.375	0.259	1.798	0.430	0.652	0.383
7	0.041	0.271	−0.010	0.138	0.051	0.304	0.015	0.309

Source: Agterberg et al. (1994, Table 4)
Regression coefficients for logistic model (B) and their standard deviations are shown in the last two columns. First row (pattern No. 0) is for constant term in weighted logistic regression

Table 5.4 Summary of WofE results for Experiments 1–5

Experiment number	1	2	3	4	5
Number of cells (Training)	13,964	14,570	14,177	14,012	14,353
Number of cells (Testing)	42,133	41,528	41,920	42,086	41,727
Observed number of deposits (Training)	23	15	19	24	43
W1 (Geology)	−1.7756	−2.2753	−1.6337	−0.8486	−2.0326
W2 (Geology)	1.3301	0.9772	0.6636	1.6187	1.1406
W1 (PC3)	−1.3151	−1.1538	−0.2909	−1.1480	−0.9356
W1 (PC3)	0.8527	0.463	0.3622	1.2633	0.7601
W1 (Magnetics)	−0.5031	−1.2549	−0.0891	−0.6888	−1.5514
W2 (Magnetics)	1.2494	0.4874	0.4297	0.5245	0.2001
Sum of PPs	37.38	18.79	20.04	46.72	63.40
CI-test probability	0.903	0.676	0.569	0.941	0.886
Adjustment factor	0.6153	0.7983	0.9482	0.5137	0.677
Observed number of deposits (Testing)	67	75	71	66	47
Estimated number of deposits (Testing)	56	43	20	64	66
Smallest PP	0.0000	0.0000	0.0002	0.0002	0.0000
Largest PP for cell with deposit	0.0054	0.0011	0.0032	0.00078	0.0030
Largest PP for cell without deposit	0.0299	0.0053	0.0054	0.0253	0.0162

Source: Agterberg and Bonham-Carter (2005, Table 1)
Note: *W1*, *W2* negative and positive weight, *PP* posterior probability, *CI* conditional independence

the use of logits in studies of this type because, basically, the objective is to estimate probabilities of occurrence of discrete events. Berkson (1944) had introduced logits as an alternative approach to probit analysis in bioassay (see Fig. 12.8 for a comparison of logits and probits). Probabilities cannot be negative; neither can they exceed 1, and these conditions can be violated when the linear least squares method is used for estimating probabilities. In a later study (Agterberg 1974), the Abitibi area was made part of a larger study area (Fig. 5.20) and a distinction was

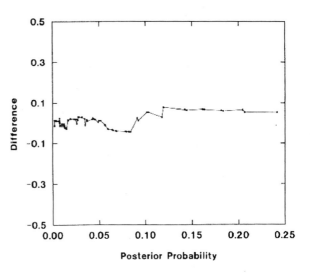

Fig. 5.19 Goodness-of-fit test applied to weighted logistic regression model for gold deposits, Meguma Terrane (Fig. 5.18a). The difference between observed and estimated relative frequencies is plotted against the posterior probability. The absolute value of the largest difference (=0.0775) is less than the Kolmogorov-Smirnov statistic (=0.1426) indicating a good fit (Source: Agterberg et al. 1994, Fig. 3)

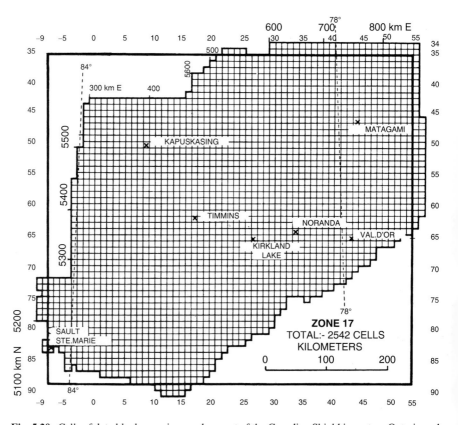

Fig. 5.20 Cells of data block superimposed on part of the Canadian Shield in eastern Ontario and western Quebec. Subset of the (10 km × 10 km) UTM cells was previously used for Abitibi subarea study of Figs. 4.5, 4.6, and 4.7. Heavy frame is boundary of contour maps in Figs. 5.21, 5.22, 5.23, and 5.24 (Source: Agterberg 1974, Fig. 2)

5.2 Weighted Logistic Regression

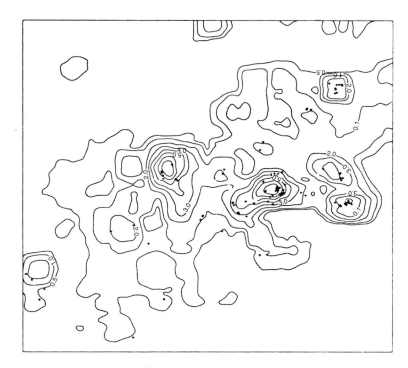

Fig. 5.21 Automatic contour map for occurrence of copper-zinc deposits; linear model applied to 38 lithological variables; contour value represents expected number of (10 km × 10 km) cells containing one or more Cu-Zn deposits per (40 km × 40 km) unit area; crosses represent known Cu-Zn deposits; see Fig. 5.20 for location (Source: Agterberg 1974, Fig. 3)

made between volcanogenic massive sulphide deposits and magmatic nickel-copper deposits associated with mafic and ultramafic intrusions.

Following up on Tukey's (1972) suggestion to use logits, both the linear and nonlinear model were applied to these two deposit types in the larger study area with the results shown in Figs. 5.21, 5.22, 5.23, and 5.24. As before, probabilities estimated for (10 km × 10 km) UTM cells were combined into overlapping unit cells measuring 40 km on a side to produce estimates of expected values that were contoured. Comparison of the logistic model pattern of Fig. 5.23 with the linear model pattern of Fig. 5.21 shows that the two methods gave approximately the same results for the relatively abundant volcanogenic massive sulphide deposits. On the other hand, there are significant differences between Figs. 5.22 and 5.24, which are for the magmatic nickel-copper deposits that occur rarely in the Precambrian rocks of the Canadian Shield.

The two types of deposits exhibit different types of geographic distribution patterns. Ordinary multiple regression could be employed to estimate probability of occurrence of the massive sulphide deposits but not for the magmatic nickel-copper deposits because relatively many estimated probabilities were outside the [0,1] interval. For these relatively rare deposits, logistic regression gave decidedly better results.

Fig. 5.22 Occurrences of nickel-copper deposits as predicted by linear model as in Fig. 5.21 (Source: Agterberg 1974, Fig. 4)

5.2.3 Gowganda Area Gold Occurrences Example

The purpose of the following experiments on 90 gold deposits in the Gowganda Area (1,405.51 km^2) located between Timmins and Sudbury in the Abitibi Subprovince on the Canadian Shield is to compare WofE with the more flexible weighted logistic regression method. These experiments were originally conducted by Agterberg and Bonham-Carter (2005). Other GIS applications from this region in east-central Ontario that have been published include a WofE study by Thiart et al. (2006). The Gowganda area was divided into small cells (pixels). The gold occurrences are modeled as single pixels representing discoveries at points. In most other WofE and WLR applications used for example in this chapter, the study area was not subdivided into gridded pixels but unique conditions measured by GIS were used. However, both methods (gridded pixels or unique conditions) work equally well when resolution is adequate.

The Gowganda study area was subdivided into 560,976 small (100 m × 100 m) cells. In Experiments 1 and 2, about 25 % of the cells were randomly selected to provide two training sets. In Experiments 3–5, geographically coherent 'blocks' each measuring about 25 % of the study area, were used as training sets. In each experiment, the remaining 75 % of the study area was used as testing area. Model parameters (coefficients or weights) estimated on the training set were used to estimate favorability of all cells. In some experiments, training and testing areas were kept separate for evaluation purposes.

5.2 Weighted Logistic Regression

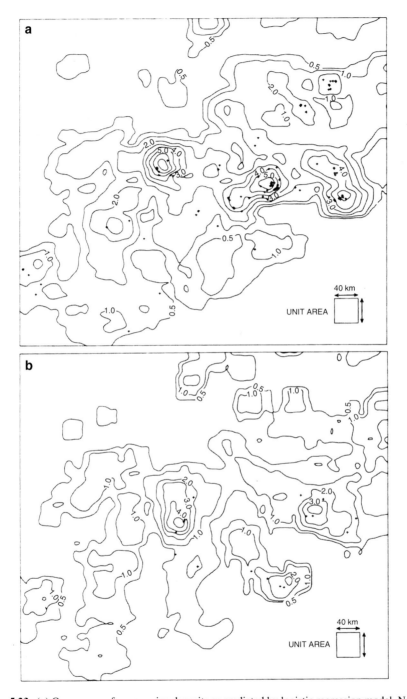

Fig. 5.23 (**a**) Occurrence of copper-zinc deposits as predicted by logistic regression model. Note resemblance with contour map of Fig. 5.21 in Abitibi subarea also shown in Fig. 4.7; (**b**) Occurrence of nickel-copper deposits as predicted by logistic regression model (Source: Agterberg 1984, Fig. 3)

170   5 Prediction of Occurrence of Discrete Events

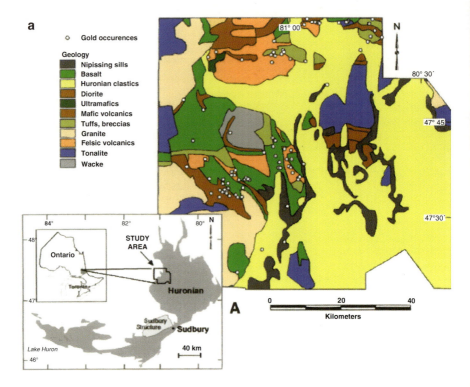

Fig. 5.24 (a) Gowganda area, east-central Ontario. Locations of 80 gold occurrences in relation to bedrock geology; (b) third principal component (PC3) of lake geochemistry data, and (c) aeromagnetics (Source: Agterberg and Bonham-Carter 2005, Fig. 1)

Normally, a sample of geoscientific data compiled for one area is not representative of another area, especially if some suitable form of discretization is not used. Because of strong spatial variability, patterns including geophysical and geochemical contour maps differ from place to place not only locally but also at regional scales. Discretization by reduction of these patterns to binary or ternary form often helps to prevent adverse effects on mineral potential mapping resulting from regional variability. This positive aspect of discretization, which is not necessarily restricted to WofE, will be illustrated in this section by geographically separating training areas from testing areas in Experiments 3–5.

Location and patterns for gold deposits, geology, geochemistry and aeromagnetics of the study area are shown in Fig. 5.24. These map patterns were used as input in the experiments. On the generalized geological map it can be seen that the 80 gold occurrences are spatially associated with felsic and mafic metavolcanics, which were selected to form the binary map pattern for bedrock geology.

Figure 5.24a is for the third principal component (PC3) derived from geochemical elements for lake sediments in the area (original data obtained from the Ontario Geological Survey "Treasure Hunt" database). PC3 is primarily determined by Cd, Zn, Mo, Br, Cu and Pb concentration values. Its pattern shows

5.2 Weighted Logistic Regression

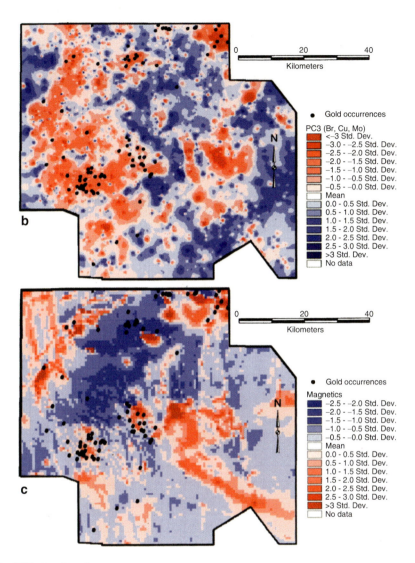

Fig. 5.24 (continued)

relatively strong spatial association with the Au occurrences. Class values are relative with limits determined by regional mean and standard deviation. These PC3 values were transformed to define a WofE binary pattern using a multiclass WofE analysis to select a threshold. Original classes as shown in Fig. 5.24a were used for WLR. In a similar way, the pattern of Fig. 5.24b was derived from aeromagnetic maps. The Au occurrences are mostly associated with magnetic highs. All classes shown were used for WLR and a binary map layer for WofE was formed by reclassification, again using multiclass WofE to determine a threshold.

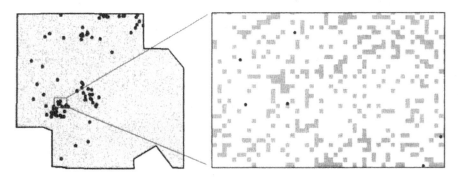

Fig. 5.25 Training set of 139,634 (100 m × 100 m) randomly selected from grid on Gowganda area (Fig. 5.24a) used in Experiment 1. Enlargement of small part of pattern (*right*) to show relations between randomly selected cells and gold occurrences. Different selection of cells, also randomly selected was used in Experiment 2 (Source: Agterberg and Bonham-Carter 2005, Fig. 2)

As has been explained before, the WofE method consists of updating the prior probability for a small unit cell or pixel. Posterior probabilities are derived by applying Bayes' rule separately to every map layer considered. Results are independent of the order of map layers during updating. In Experiments 1–5, unit cell area was set equal to 0.1 km^2. This implies that the prior probability for a unit cell anywhere in the study area is equal to (90/56,097.6=) 0.001604. WofE is a GIS-based statistical method. The study area is divided into polygons (or pixels) characterized by different strings of code numbers for map layers. The area of a "unique condition" is equal to the sum of areas (polygons or pixels) with the same string of code numbers. In the simplest kind of application, map layers are binary with two code numbers denoting presence and absence, respectively. In the Gowganda WofE applications, the three binary map layers result in eight unique conditions only.

In the WLR applications, the input data also are for unique conditions but there are many more of these, because all patterns delineated by contours in Fig. 5.24b, c were assigned separate integer code numbers increasing according to value. For example, the training set for WLR Experiment 1 has 850 unique conditions. Every unique condition is weighted according to the sum of areas of its polygons (or grid cells) and according to the number of mineral deposits it contains (nearly always equal to 1 in Experiments 1–5 because there are relatively few mineral deposits; e.g., 23 in Experiment 1).

The WofE posterior probability map for the Gowganda area (not shown here), which is based on the eight unique conditions, shows stronger spatial correlation with the 80 Au occurrences than the original map patterns of Fig. 5.24a. The posterior probabilities range from 0.00074 to 0.17880. An accompanying confidence map indicates that the smallest and largest posterior probabilities are either significantly lower or higher than the prior probability.

Figure 5.25 shows the training set of randomly selected cells for Experiment 1. As mentioned before, the Gowganda study area was subdivided into 560,976

5.2 Weighted Logistic Regression

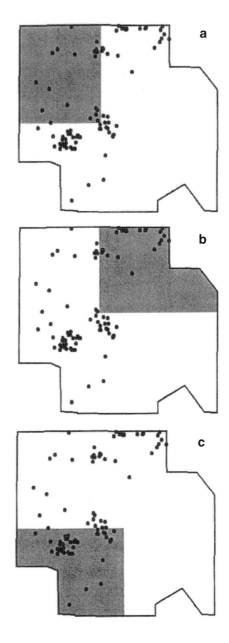

Fig. 5.26 Diagrams showing the training areas (spatially coherent blocks) used in (**a**), Experiment 3, (**b**) Experiment 4, and (**c**) Experiment 5 (Source: Agterberg and Bonham-Carter 2005, Fig. 3)

small (100 m × 100 m) cells. Only 139,643 (25 % of total area) of these were used for training. The inset is an enlargement of part of the complete pattern of Fig. 5.25. In total, 23 of the 90 Au occurrences belong to the training set for Experiment 1. This experiment was repeated: the training set for Experiment 2 (not shown as a figure) consists of 145,701 cells (26 % of total area) containing 15 Au occurrences. Figure 5.26 shows the three geographically coherent subareas or 'blocks' used for training in Experiments 3–5.

Table 5.5 Summary of WLR results for Experiments 1–5

Experiment number	1	2	3	4	5
Constant term, B0	−8.8639	−7.3394	−9.8428	−5.7877	−9.0784
SD (B0)	1.3241	0.7746	1.9138	1.2414	1.1576
B (Geology)	1.6059	1.6620	2.2080	1.3413	1.2101
SD (Geology)	0.4468	0.2543	0.7606	0.4537	0.3193
B (PC3)	−0.1201	−0.1492	−0.0496	−0.1697	−0.1081
SD (PC3)	0.0490	0.0281	0.0652	0.0444	0.0423
B (Magnetics)	0.1657	0.0504	0.0064	−0.0148	0.3010
SD (Magnetics)	0.0554	0.033	0.0674	0.0523	0.0732
Sum of PPs	23	15	19	24	43
Observed number of deposits (Testing)	67	75	71	66	47
Estimated number of deposits (Testing)	69	43	22	99	198
Smallest PP (Training)	0.0000	0.0002	0.0001	0.0001	0.0001
Largest PP for cell with known deposit (Training)	0.0303	0.0035	0.0037	0.0296	0.0400
Largest PP for cell without known deposit (Training)	0.0788	0.0129	0.0047	0.0344	0.0953
Smallest PP (Testing)	0.0000	0.0092	0.0045	0.0303	0.1633
Largest PP for cell with known deposit (Testing)	0.0312	0.0092	0.0045	0.0303	0.1633
Largest PP for cell without known deposit (Testing)	0.0788	0.0142	0.0047	0.0349	0.3217

Source: Agterberg and Bonham-Carter (2005, Table 2)
Note: *B* WLR coefficient, *SD* standard deviation, *PP* posterior probability

5.2.4 Results of the Gowganda Experiments

Weights-of-Evidence (WofE) and Weighted Logistic Regression (WLR) results for the five experiments are summarized in Tables 5.5 and 5.6. For all five experiments, the sum of WofE posterior probabilities (the product of probability and area in unit cells summed over all unique conditions) exceeds the number of deposits in the training area. This suggests that the assumption of conditional independence may be violated. However, application of the conditional independence (CI) test of Agterberg and Cheng (2002) yields a test statistic with probabilities ranging from 0.569 (Experiment 3) to 0.941 (Experiment 4). Normally, the CI hypothesis would be rejected only if the probability for this one-tailed z-test exceeds 95 % or 99 %, corresponding to level of significance set equal to 0.05 or 0.01, respectively. However, it is unlikely that in five successive experiments none of the CI probabilities are less than 0.5, and some CI violation probably exists. For this reason, all WofE posterior probabilities were reduced by a factor set equal to the quotient of number of deposits and sum of WofE posterior probabilities in the training area. This factor ranges from 0.514 (Experiment 4) to 0.948 (Experiment 3).

Theoretically, the WLR sum of posterior probabilities is exactly equal to number of deposits in the testing area. In practice, this sum may differ somewhat from

5.2 Weighted Logistic Regression

Table 5.6 Seafloor example. Number of volcanic vents per unique condition (unit area = 100 m^2)

Age	Topo	Contact	Rocktype	Fissures	# of vents	Area
0	0	0	0	0	0	10,052
0	0	0	0	1	1	3,363
0	0	0	1	0	0	3,268
0	0	0	1	1	0	1,074
0	1	0	0	0	0	5,455
0	1	0	0	1	0	25
0	0	1	0	0	0	3,482
0	1	0	1	0	0	2,518
0	0	1	0	1	0	1,474
0	1	0	1	1	0	1,371
1	0	0	0	0	0	5
1	0	0	0	1	0	705
0	0	1	1	0	0	5
0	0	1	1	1	0	744
1	0	0	1	0	0	422
1	0	0	1	1	0	58
0	1	1	0	0	0	12
0	1	1	0	1	0	179
1	1	0	0	0	2	1,766
1	1	0	0	1	0	119
0	1	1	1	0	1	1,055
0	1	1	1	1	0	33
1	0	1	0	0	0	10
1	1	0	1	0	0	146
1	0	1	0	1	1	623
1	1	0	1	1	0	145
1	0	1	1	0	2	504
1	0	1	1	1	0	1
1	1	1	0	0	2	317
1	1	1	0	1	1	277
1	1	1	1	0	3	348
1	1	1	1	1	0	295

Source: Agterberg (2011, Table 1)

number of deposits because of numerical precision restrictions depending on choice of critical parameters that control the iterative process by which a WLR solution is obtained. In the experiments, these parameters were set such that WLR sum of posterior probabilities became equal to number of deposits after rounding off to the nearest integer. In both WofE and WLR, it can be expected that the sum of posterior probabilities in the testing area is more or less equal to the number of deposits in the testing area if the training area is a random sample of the study area as in Experiments 1 and 2. When training area and testing area are geographically distinct blocks (as illustrated in Fig. 5.26 for Experiments 3–5), expected number of deposits can be either greater or less than sum of posterior probabilities.

Estimated numbers of deposits in testing area in Tables 5.5 and 5.6 are nearly equal to one another for Experiments 2 and 3. One reason for this similarity may be that the three binary patterns used are probably conditionally independent of occurrence of Au deposits in the training areas selected for these two experiments as indicated by relatively low CI-test probabilities (0.676 and 0.569, respectively). This CI-test result should be considered together with the fact that, with the exception of the WLR coefficient for PC3 in Experiment 3, WLR coefficients for non-binary variables in Experiments 2 and 3 are not significantly different from zero.

In Experiments 4 and 5, the sums of WLR posterior probabilities exceed numbers of deposits in the testing areas. In these two situations, WLR results are probably worse than the corresponding WofE results. The main reason is that WLR was performed on non-binary variables. Because of regional changes in the spatial PC3 and aeromagnetic variability patterns, the testing areas include cells with values that are outside the ranges of these variables in the training areas. As a rule, posterior probabilities should not be used for prediction of mineral occurrence unless in the special situation satisfied in Experiments 1 and 2 that the testing area is the complement of a training area consisting of many small randomly selected cells so that both training and testing area are representative of the entire study area.

The preceding conclusion also is reached when individual posterior probabilities are considered. Smallest and largest WLR posterior probabilities are shown in the bottom rows of Table 5.5 keeping training and testing areas separate for all five experiments. A further separation is made by distinguishing between largest posterior probabilities for cells with and without deposits, respectively. Smallest and largest posterior probabilities for WofE in Table 5.4 were the same because there were only eight unique conditions per experiment (and no deposits for the unique condition with maximum posterior probability but relatively small area in each experiment). In general, the largest posterior probability clearly exceeds the largest probability for unique condition with one or more deposits. Worst-case scenario is for WLR in Experiment 5 where the relatively unconstrained posterior probability of 0.3217 in the testing area is nearly 20 times greater than its WofE counterpart ($=0.0162$). It is mainly because of discretization that WofE results then are more realistic than corresponding WLR results.

5.2.5 *Training Cells and Control Areas*

The best strategy in mineral potential mapping is to base a mineral potential map on all data available for a study area without geographical separation of training and testing areas. In general, however, degree of knowledge about existence of mineral deposits varies from place to place within the same study area according to patterns that often cannot be quantified explicitly. Under-explored subareas become targets of new exploration only if their probability index matches that in places where mineral deposits are known to occur. Discretization, which is not necessarily restricted to WofE, may help to stabilize extrapolation to relatively poorly explored subareas by constraining ranges and magnitudes of the values of the variables.

5.2 Weighted Logistic Regression

One of the basic rules of WofE is that the binary patterns selected for further work should be approximately conditionally independent. A number of conditional independence tests are in existence (cf. Thiart et al. 2006). An overview of how one should proceed in the application of WofE is provided in the book by Bonham-Carter (1994). Basically, there are two CI tests: one for pair-wise comparisons of map layers, and the other (the omnibus test) for comparing sum of posterior probabilities to total number of mineral deposits. These two totals are approximately equal when the assumption of conditional independence is satisfied. The statistical CI test of Agterberg and Cheng (2002), which was applied earlier in this section, will be discussed in more detail in the next section. As an artificial example to illustrate lack of CI, these authors presented duplication of a map layer with positive weight of 2. This creates posterior probabilities that are more than seven times too large, because this map layer then receives a weight of 4 instead of 2 (and $e^2 = 7.39$). In situations of this type, WLR automatically compensates for lack of conditional independence.

Suppose that WofE is applied using a 0.01 km^2 unit cell and color the resulting posterior probability map using red for the 2 % largest posterior probabilities. If resolution does not present a problem, one could equally well select a 100 m^2 unit cell. All posterior probabilities then would become approximately 100 times smaller but the colored map would stay approximately the same as long as the colors correspond to percentage values for ranked posterior probabilities. Although there now are now 100 times as many cells in the maximum 2 % probability class, the area that is colored red remains approximately the same in this artificial example.

In WofE or WLR it is not necessary to randomly select a set of training cells. If the unit cell is sufficiently small, any large number of randomly selected training cells would produce a posterior probability map that is approximately the same as the map of posterior probabilities for the cells not used in training. The three probability maps for training, testing and total area are approximately the same. Consequently, it is not necessary to use a grid as is used in these experiments for the Gowganda area. Map patterns are either spatially continuous (e.g., Fig. 5.24a, b), or discontinuous (geology in Fig. 5.24). Continuous patterns may be subjected to minor discretization if their values are to be used for statistical analysis. Significant differences, between values of cells for the same variable, are likely to adversely affect statistical analysis results if training and testing area are geographically distinct as in Experiments 3–5. This is because other values as well as combinations of values of variables are likely to exist outside the training area.

For example, suppose that a geophysical variable is restricted to 0–10 range in the training area where it receives a weighting coefficient, but that in part of the testing area its values exceed 100. Consequently, its contribution to expected value in this part of the testing area becomes more than 10 times as large when the same weighting coefficient is used, and predictions probably are meaningless. Regional trends in other continuous variables may aggravate the situation. Discretization to ternary or binary form then would improve results because the possibility of

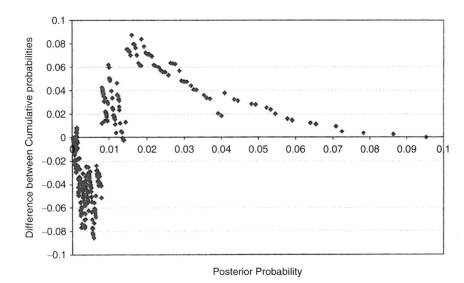

Fig. 5.27 Kolmogorov-Smirnov plot for WLR Experiment 5 (Source: Agterberg and Bonham-Carter 2005, Fig. 6)

extremely high values of variables would either be eliminated or severely restricted. Although discretization implies loss of information within a given study area, reduction to binary form has the advantage of transportability in that weights derived for one study area may become applicable in other areas. Coloring the posterior probability map resulting from WofE (or another method), e.g. by using red for the largest 2 % class as before. Such a colored map retains validity if there are undiscovered deposits in the study area that have the same weights as the known deposits. Consequently, the posterior probability provides target areas for further exploration where the colors (e.g. red) indicate relatively high probability of occurrence but where no or few deposits have been discovered.

A suitable method for measuring the performance of mineral potential maps consists of attempting to predict occurrences of deposits that are deleted from the data set. A jackknife type approach of successively omitting individual deposits was proposed by Chung and Agterberg (1980). A different jackknife-based approach will be discussed in Sect. 14.1. Alternatively, mineral potential maps can be constructed without use of deposit data in parts of the study area and evaluation of how well mineral potential is predicted in these parts (*cf*. Chap. 4). Ideally, a historical hindsight study can be performed in which a mineral potential map is evaluated against new discoveries, after a number of years of subsequent mineral exploration (Sect. 4.4.2).

A difference plot as shown for WLR in Fig. 5.27 can be subjected to an approximate Kolmogorov-Smirnov test for goodness of fit. This test checks for the largest difference (in absolute value) between observed and expected frequencies. In this example, there are 43 deposits and 430 unique conditions. The maximum difference of 0.088 is within the 95 % confidence interval that amounts to ($\pm 1.36/\sqrt{43}=$) ± 0.207 in the two-tailed Kolmogorov-Smirnov test. Thus the WLR model provides a good fit for the training area of Experiment 5.

A mineral potential map can be viewed as the contour map of a landscape covered by disks representing mineral deposits. If the fit is good, the sum of posterior probabilities for any subarea is equal to observed number of deposits in that subarea. It is possible to improve the fit; e.g., by using non-linear functions of the variables. However, the end product then may be equivalent to an empirical contour map of deposit density derived without use of geoscience information and without predictive potency. In this situation, superimposing a grid with very small cells on the study area and random sampling (see Fig. 5.25) produces a representative sample of this empirical contour map. Another way to illustrate this concept is to imagine maintaining a constant proportion of cells that are sampled at random (e.g., 25 %) but to steadily decrease grid spacing. In the limit (infinitely small cells), both training and target area exactly duplicate the study area. A better fit then does not prove that the method used provides a better predictive tool.

If the concept of random sampling from a study area is applied, one could argue as in Agterberg (1992) that the known mineral deposits in a region constitute a random subset of a larger population of discovered plus undiscovered deposits that are of the same type in that they relate in the same way to the variables from which the indicator map layers are formed. In WofE, one then simply can increase the prior probability by a factor equal to total number of deposits divided by number of known deposits. Of course, in practice the problem is that, generally, it is not possible to estimate this ratio accurately. However, in a relative sense, the patterns on a mineral potential map are not affected by such lack of knowledge.

5.3 Modified Weights-of-Evidence

The approach to be discussed in this section was originally introduced by Spiegelhalter and Knill-Jones (1984) as a refinement of weights of evidence as used in the GLADYS expert system. Agterberg (1992) had suggested applicability of this indirect method to GIS-based regional mineral resources estimation but this modification had not yet been tested until the method was applied in Agterberg (2011) and Zhang et al. (2013). Suppose that a study area is digitized as a number (n) of pixels and that X_i ($i=1, 2, \ldots, p$) are a number of binary explanatory variables used to predict a dichotomous random variable Y representing presence ($Y_k = 1$) or absence ($Y_k = 0$) of mineralization at the k-th pixel. Provided that n is very large, we can redefine the situation in terms of binary sets B_i corresponding to the X_i and a set D corresponding to Y. In most WofE applications, the B_i's are binary with or without missing data, although the method also could be used with multistate explanatory variables.

5.3.1 East Pacific Rise Seafloor Example

If there are N discrete events in a study area and the sum of all estimated probabilities is written as S, WofE generally results in $S > N$. The difference S-N can be tested for statistical significance. The main advantage of WofE in comparison with WLR is transparency in that it is easy to compare weights with one another. Although WLR yields $S = N$, WLR coefficients generally have relatively large variances. By preprocessing it is usually possible to obtain WofE weights that approximately result in $S = N$. It is also possible to first perform WofE modeling and to follow this by WLR applied to the weights. This method results in modified weights with unbiased probabilities satisfying $S = N$. An additional advantage of this approach is that it automatically copes with missing data on some layers because weights of unit areas with missing data can be set equal to zero as is generally practiced in WofE applications.

Box 5.5: Proof That the WLR Likelihood Function Results in $S = N$

Suppose that the probability of occurrence or non-occurrence of an event is written in the form: $P(Y=1|x) = \pi(x) = e^{f(x)}/\{1+e^{f(x)}\}; P(Y=0|x) = 1 - \pi(x)$. The likelihood function then becomes: $l(\beta) = \prod \pi(x_i)^{y_i} \{1 - \pi(x_i)^{1-y_i}\}$. When there would be a single explanatory variable with $\beta = [`\beta_0\ \beta_1]$: Logit$(\pi_i) = \beta_0 + \beta_1 x_i$, and the Log likelihood function is $L(\beta) = \log_e\{l(\beta)\} = \sum y_i \cdot \log_e\{\pi(x_i)\} + (1-y_i) \cdot \log_e\{1 - \pi(x_i)\}$. Differentiation with respect to β_0 and β_1 gives: $\sum \{y_i - \pi(x_i)\} = 0$; $\sum x_i \{y_i - \pi(x_i)\} = 0$. Consequently, the total number of discrete events (N) is equal to the sum the estimated probabilities (S), or $\sum y_i = \sum p(x_i)$. The relation $S = N$ also applies when there are p explanatory variables. It is noted here that the $S = N$ also is useful as a final test on posterior probabilities obtained by the LOGPOL program (Box 5.3). Although various calculations in this FORTRAN program are carried out in double precision, it is possible, for very large databases, that the sum of the final posterior probabilities is slightly less than N due to lack of complete convergence for some posterior probabilities The logistic regression coefficients on which these probabilities are based then can be used as input for a new LOGPOL run to check for the $S = N$ requirement.

The problem of obtaining unbiased posterior probabilities in Bayesian approaches to regional mineral resource evaluation has been considered by several authors including Caumon et al. (2006). These authors proposed a cross-validation technique to cope with violation of conditional independence of explanatory variables in weights-of-evidence modeling. Their approach is a modification of a method originally proposed by Journel (2002), Krishnan (2008) and Krishnan et al. (2004). Several WofE-based methods to obtain unbiased posterior

5.3 Modified Weights-of-Evidence

Fig. 5.28 Seafloor Example. Two of patterns used for to correlate occurrences of 13 hydrothermal vents on the seafloor (East Pacific Rise, 21°N; based on Fig. 5 of Ballard et al. 1981). *Top*: litho-age units and vents (*dots*); relative age classes were based on measuring relative amounts of sediments deposited on top of the volcanics with 1.0–1.4 (youngest), 1.4–17 (intermediate), and 1.7–20 (oldest); *Bottom*: topography (depth below sea level) (Source: Agterberg et al. 1994, Plate 1a, b)

probabilities were used by Bonham-Carter et al. (2009). Schaeben (2012) has provided a comprehensive review of all these methods used for mineral resource appraisal. Cheng (2011) has developed a boosting version of WofE in which $S = N$ is nearly satisfied. Boosting (Freund and Schapire 1997) consists of sequentially a classification algorithm to reweighted versions of a set of training data followed by taking the weighted majority of the sequence of classifiers produced (*cf.* Friedman et al. 2000). Deng (2009) had introduced a conditional dependence adjusted weights of evidence method. In an application, it was shown this method does not eliminate bias because it results in $S < N$ but bias is reduced in comparison with ordinary WofE (Agterberg 2011). This existence of bias also has been pointed out by Schaeben and van den Boogaart (2011). Schaeben (2014) has made a detailed comparison of weights of evidence with logistic regression in terms of Markov random fields expanding on work by Sutton and McCallum (2007). However, mit should be kept in mind that "interpreting regression coefficients can be very tricky" (Christensen 1990, p. 259). Agterberg's (1989c) LOGDIA computer program contains logistic regression diagnostics as developed by Pregibon (1981). However, in practical WLR applications, usually is best to evaluate statistical significance by using the Student's t-map that accompanies the posterior probability map (*cf.* Fig. 5.18) and the Kolmogorov-Smirnov test (Fig. 5.19).

The example of application to be used for illustrating modified WofE here is the Seafloor Example consisting of 13 volcanic vents on the East Pacific Rise near 21° N (Fig. 5.28). It was previously analyzed by means of both WofE and WLR (Agterberg et al. 1994). The datasets used for the current study are slightly different from those used previously but the new WofE and WLR results are similar to those derived before and more detailed explanations of input and output patterns can be found in the earlier papers.

Table 5.6 shows input data for the Seafloor Example. There are five binary patterns (relative age of basaltic rocks; depth below sea-level; vicinity to contact between youngest basaltic rocks; type of basaltic rock; and vicinity of fissures) and $2^5 = 32$ unique conditions. Unit of area is 10 m × 10 m. Ordinary WofE weights and contrasts are shown in the first three columns of Table 5.7. WLR was applied to the dataset of Table 5.6 after replacing every 1 for a map layer by its positive weight and every 0 by its negative weight. Because presence of fissures initially resulted in a negative value of W^+, its 0s for absence were replaced by this negative value, because it is absence (instead of presence) of fissures that is weakly positively correlated with vent occurrence. The resulting logistic regression coefficients and their standard deviations are shown in columns 4 and 5 of Table 5.7. Multiplying them by the weights yields the modified weights shown in the next two columns. The modified WofE weights in the last columns of Table 5.7 yield $S = 13$ so that bias due to lack of conditional independence is avoided.

Agterberg and Cheng (2002) proposed their conditional independence test using original WofE results for the Seafloor Example. Figure 5.29 show estimated and observed numbers of vents in the 5-layer model. The sum of posterior probabilities based on all five map layers was $S = 37.59$ and much greater than $N = 13$. Clearly, there is significant violation of the conditional independence assumption. In a separate experiment (Table 5.9) for a slightly larger study area of 3,985 km^2 a three-layer model was analyzed. The first column of Table 5.9 shows unique conditions with "1" for presence and "0" for absence. The area of the unit cell was set at 0.01 km^2 in this experiment. With weights and standard deviations similar to those listed for WofE in Table 5.7 this resulted in the posterior probabilities P_f with standard deviations $s(P_f)$ shown in columns 3 and 4 of Table 5.9. Multiplication of each P_f by area (number of unit cells) of its unique condition results in the eight predicted vent frequencies $N_{IJK} P_f$. Their sum provides the estimate $T = 14.05$ which is only slightly larger their $n = 13$. The corresponding variance $s^2(T)$ (=41.6511) is the sum of the eight values listed in the last column of Table 5.9. The square root of this number gives $s(T) = 6.45$. The z-test can be applied to test the standardized difference $(T − n)/s(T) = 1.05$ for statistical significance. The 95 % confidence level for a one-tailed test to see if T is significantly greater than n is 1.645. Consequently, the hypothesis that the three layers in the model of Table 5.9 are conditionally independent can be accepted. On the contrary, if the same z-test is applied to the five-layer model (cf. Fig. 5.29), the assumption of conditional independence is rejected. For more detailed explanations of this test, see Agterberg and Cheng (2002).

It is interesting to compare the newly derived WLR results with those resulting from application of WRL directly to the input data shown in Table 5.6. In Table 5.8 it is shown that indirectly estimated coefficients divided by their standard deviations are exactly equal to direct estimates divided by their standard deviations. Repetition of the preceding experiment using the Meguma Terrain Example yielded the results shown in Table 5.10. In this application, the seven map layers used were approximately conditionally independent of the 68 gold deposits, and the results obtained by either direct or indirect application of WLR are not greatly different as they were for the first example.

5.3 Modified Weights-of-Evidence

Table 5.7 Seafloor example. Weights and adjusted weights

Variable	W⁺	W⁻	Contrast	B	s(B)	New W⁺	New W⁻	New contrast
Age	1.77205	−1.71651	3.48856	0.83716	0.23350	1.48349	−1.43699	2.92048
Topography	0.67433	−0.74367	1.41800	0.52821	0.44554	0.35619	−0.39281	0.74901
Contact	1.18719	−1.19888	2.38607	0.84137	0.29766	0.99886	−1.00870	2.00757
Rock type	0.42831	−0.26130	0.68961	−0.00164	0.86058	−0.00070	0.00043	−0.00113
No fissures	0.04299	−0.13127	0.17426	5.05353	3.95312	0.21727	−0.66338	0.88065

Source: Agterberg (2011, Table 2)
Logistic regression coefficients in column B

Table 5.8 Seafloor example. Comparison of WLR results

Variable	W⁺	s(W⁺)	W⁻	s(W⁻)	W_R⁺	s(W_R⁺)	W_R⁻	s(W_R⁻)
Age	1.77205	0.30180	−1.71651	0.70713	1.96126	0.37914	−1.89979	0.81868
Topography	0.67433	0.33344	−0.74367	0.50004	−0.36013	0.41223	0.34885	0.60241
Contact	1.18719	0.31640	−1.19888	0.57738	−0.28949	0.35277	0.28041	0.63265
Rock type	0.42831	0.40835	−0.26130	0.37801	0.04663	0.44083	−0.04517	0.41710
No fissures	0.04299	0.57743	−0.13127	0.31628	−0.19919	0.62718	0.19294	0.37332

Source: Agterberg (2011, Table 3)

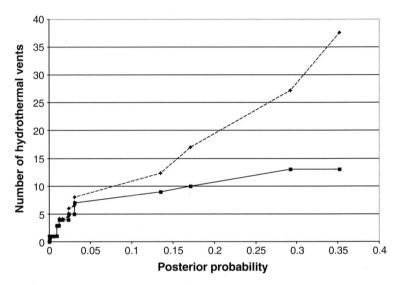

Fig. 5.29 Seafloor Example: Hydrothermal vents on East Pacific Rise. Estimated and observed numbers of vents in 5-layer model (Source: Agterberg and Cheng 2002, Fig. 1)

Table 5.9 Estimation of T and $s^2(T)$ for 3-layer model

IJK	Area (km²)	P_f	$s(P_f)$	$N_{IJK} P_f$	$N_{IJK}^2 s^2(P_f)$
222	1.344	0.0003	0.0005	0.0403	0.0045
212	0.9007	0.0057	0.0065	0.5134	0.3428
221	0.4351	0.0008	0.0013	0.0348	0.0032
122	0.4187	0.0111	0.0126	0.4648	0.2783
112	0.3415	0.1709	0.0907	5.8362	9.5939
211	0.2223	0.0154	0.0176	0.3423	0.1531
111	0.1771	0.3604	0.153	6.3827	7.3421
121	0.1456	0.0297	0.336	0.4324	23.9332
Sum	3.985			14.0470	41.6511

Source: Agterberg and Cheng (2002, Table 3)
I age, J topography, K rock type; $N = 100 \times$ Area

Table 5.10 Meguma Terrane example. Weights and adjusted weights

Variable	W⁺	W⁻	Contrast	B	New W⁺	New W⁻	New contrast
Anticlines	0.54559	−0.81603	1.36162	0.91524	0.49935	−0.74686	1.24621
Au in balsam fir	0.65006	−0.25920	0.90927	1.16490	0.75726	−0.30194	1.05920
GH contact	0.35704	−0.26256	0.61961	0.45987	0.16419	−0.12074	0.28494
Goldenville	0.30231	−1.45564	1.75795	0.73623	0.22257	−1.07168	1.29425
Devonian granite	0.21771	−0.03706	0.25476	1.80401	0.39274	−0.06686	0.45960
Geochemistry	0.85240	−0.08388	0.93628	0.75700	0.64527	−0.06350	0.70877
NW lineaments	0.03998	−0.01011	0.05009	−0.65798	−0.02631	0.00665	−0.03296

Source: Agterberg (2011, Table 6)

The new approach advocated in this section results in unbiased estimates of the posterior probabilities. An additional advantage of this approach is that it automatically copes with missing data on some layers because weights of unit areas with missing data can be set equal to zero as is generally practiced in WofE applications. As usual, the method presented here was applied using known events in the study area only.

Because logistic regression results in posterior probabilities that are unbiased in the sense that the condition $S = N$ (sum of posterior probabilities = total number of events) is satisfied, it is possible to modify the WofE further by assigning other values to the pixels or unique conditions that contain events as discussed and applied by Zhang et al. (2013).

References

Agterberg FP (1974) Automatic contouring of geological maps to detect target areas for mineral exploration. J Int Assoc Math Geol 6:373–395

Agterberg FP (1984) Use of spatial analysis in mineral resource evaluation. Math Geol 16:565–589

Agterberg FP (1989a) Computer programs for mineral exploration. Science 245:76–81

Agterberg FP (1989b) Systematic approach to dealing with uncertainty of geoscience information in mineral exploration. In: Proceedings of the 21st international symposium on Computers in the mineral industry, Las Vegas, March 1989. Society of Mining Engineers, AIME, Littleton, pp 165–178

Agterberg FP (1989c) LOGDIA, Fortran 77 program for logistic regression with diagnostics. Comput Geosci 15:599–614

Agterberg FP (1992) Combining indicator patterns in weights of evidence modeling for resource evaluation. Nonrenew Resour 1:39–50

Agterberg FP (2011) A modified weight-of-evidence method for regional mineral resource estimation. Nat Resour Res 20(1):95–101

Agterberg FP, Bonham-Carter GF (1990) Deriving weights of evidence from geoscience contour maps for the prediction of discrete events. In: Proceedings of the 22nd APCOM symposium, Technical University of Berlin, vol 2, pp 381–396

Agterberg FP, Bonham-Carter GF (2005) Measuring the performance of mineral potential maps. Nat Resour Res 14(1):1–18

Agterberg FP, Cheng Q (2002) Conditional independence test for weights-of-evidence modeling. Nat Resour Res 11:249–255

Agterberg FP, Robinson SC (1972) Mathematical problems in geology. In: Proceedings of the 38th session on Bulletin of the International Statistical Institute, The Hague, vol 38, pp 567–596

Agterberg FP, Chung CF, Fabbri AG, Kelly AM, Springer J (1972) Geomathematical evaluation of copper and zinc potential in the Abitibi area on the Canadian Shield. Geological Survey of Canada Paper 71-11

Agterberg FP, Bonham-Carter GF, Wright DF (1990) Statistical pattern integration for mineral exploration. In: Gaal G, Merriam DF (eds) Computer applications in resource estimation, prediction and assessment for metals and petroleum. Pergamon, Oxford, pp 1–21

Agterberg FP, Bonham-Carter GF, Cheng Q, Wright DF (1994) Weights of evidence and weighted logistic regression for mineral potential mapping. In: Davis JC, Herzfeld UC (eds) Computers in geology – 25 years of progress. Oxford University Press, New York, pp 13–32

Andersen EB (1990) The statistical analysis of categorical data. Springer, Berlin

Assad R, Favini G (1980) Prévisions de minerai cuprofere dans le Nord Ouest Québecois. Min Energ Ress Qué. DVP-470

Ballard RD, Francheteau J, Juteau T, Rangan C, Norwark W (1981) East Pacific Rise at 21° N; the oceanic, tectonic and hydrothermal processes of the central axis. Earth Planet Sci Lett 55:1–10

Berkson J (1944) Application of the logistic function to bio-assay. J Am Stat Assoc 39:357–365

Bishop YMM, Fienberg SE, Holland PW (1975) Discrete multivariate analysis. MIT Press, Cambridge, MA

Bonham-Carter GF (1994) Geographic information systems for geoscientists: modelling with GIS. Pergamon, Oxford

Bonham-Carter GF, Agterberg FP (1999) Arc-WofE – a GIS tool for statistical integration of mineral exploration data sets. Bull Int Stat Inst 58(2):497–500

Bonham-Carter GF, Agterberg FP, Wright DF (1988) Integration of geological data sets for gold exploration in Nova Scotia. Photogramm Eng Remote Sens 54:1585–1592

Bonham-Carter GF, Agterberg FP, Wright DF (1990) Weights of evidence modelling: a new approach to mapping mineral potential. Geological Survey of Canada Paper 89-9, pp 171–183

Bonham-Carter GF, Agterberg FP, Cheng Q, Behnia P, Raines G, Kerswill J (2009) Correcting for bias in weights-of-evidence applications to assessing mineral potential. In: Proceedings of the IAMG annual meeting, Stanford University (CD-ROM)

Carranza EJ (2004) Weights of evidence modeling of mineral potential: a case study using small number of prospects, Abra, Philippines. Nat Resour Res 13(3):173–185

Cassard D, Billa M, Lambert A (2008) Gold predictivity mapping in French Guiana using an expert guided data-driven approach based on a regional-scale GIS. Ore Geol Rev 34:471–500

Caumon G, Ortiz JM, Rabeau O (2006) A comparative study of three data-driven mineral potential mapping techniques. IAMG-2006 Proc CD, 523-05 (4 p)

Cervi F, Berti M, Borgatti L, Ronchetti F, Manenti F, Corsini A (2010) Comparing predictive capability of statistical and deterministic methods for landslide susceptibility mapping: a case study in the northern Apennines (Reggio Emilia Province, Italy). Landslides 7:433–444

Cheng Q (2004) Application of weights of evidence method for assessment of flowing wells in the Greater Toronto Area, Canada. Nat Resour Res 13(2):77–86

Cheng Q (2008) Non-linear theory and power-law models for information integration and mineral resources quantitative assessments. In: Bonham-Carter GF, Cheng Q (eds) Progress in geomathematics. Springer, Heidelberg, pp 195–225

Cheng Q (2011) Integration of AdaBoost and weights of evidence model or mineral potential probabilistic mapping. In: Marschallinger R, Zobl F (eds) Mathematical geosciences at the crossroads of theory and practice. Proceedings of IAMG 2011 conference. GIScience, Salzburg, p 923

Cho SH, Poudyal NC, Roberts RK (2008) Spatial analysis of the amenity value of green open space. Ecol Econ 66(2–3):403–416

References

Christensen R (1990) Loglinear models. Springer, Berlin

Chung CF (1978) Computer program for the logistic model to estimate the probability of occurrence of discrete events. Geological Survey of Canada Paper 78-11

Chung CF, Agterberg FP (1980) Regression models for estimating mineral resources from geological map data. Math Geol 12:473–488

Coolbaugh MF, Raines GL, Zehner RE (2007) Assessment of exploration bias in data-driven predictive models and the estimation of undiscovered resources. Nat Resour Res 16(2):199–207

Cox DR (1966) Some procedures connected with the logistic qualitative response curve. In: David FN (ed) Research papers in statistics. Wiley, London, pp 55–71

Deng M (2009) A conditional dependence adjusted weights of evidence model. Nat Resour Res 18:249–258

Dowds JP (1969) Statistical geometry of petroleum reservoirs in exploration and exploitation. J Pet Technol 7:841–852

Freund Y, Schapire (1997) A decision-theoretic generalization of online learning and an application to boosting. J Comput Syst Sci 55(1):119–139

Friedman J, Hastie T, Tibshirani R (2000) Additive logistic regression: a statistical view of boosting. Ann Stat 28(2):337–407

Good H (1950) Probability and the weighing of evidence. Griffin, London

Goodacre A, Bonham-Carter GF, Agterberg FP, Wright DF (1993) A statistical analysis of the spatial association of seismicity with drainage patterns and magnetic anomalies in western Quebec. Tectonophysics 217:285–305

Gorney RM, Ferris DR, Ward AD, Williams LR (2011) Assessing channel-forming characteristics of an impacted headwater stream in Ohio, USA. Ecol Eng 37:418–430

Grunsky EC Mueller UA Corrigan D (2013) A study of the lake sediment geochemistry of the Melville Peninsula using multivariate methods. doi: 10.1016/j.gexplo.20130098

Hosmer DW, Lemeshow S (1989) Applied logistic regression. Wiley, New York

Journel AJ (2002) Combining knowledge from diverse sources; an alternative to traditional conditional independence hypothesis. Math Geol 34(5):573–596

Krishnan S (2008) The tau model for data redundancy and information combination in earth sciences. Math Geosci 40(6):705–727

Krishnan S, Boucher A, Journel A (2004) Evaluating information redundancy through the Tau model. In: Leuangthong O, Deutsch C (eds) Geostatistics Banff 2004. Springer, Heidelberg, pp 1037–1046

Lindsay MD, Betts PG, Ailleres L (2014) Data fusion and porphyry copper prospectivity models southeastern Arizona. Ore Geol Rev 61:120–140

Neuhäuser B, Terhorst B (2007) Landslide susceptibility assessment using "weights-of-evidence" applied to a study area at the Jurassic escarpment (SW-Germany). Geomorphology 86:12–24

Ozdemir A, Altural T (2013) A comparative study of frequency ratio, weights of evidence and logistic regression methods for landslide susceptibility mapping: Sultan Mountains, SW Turkey. J Asian Earth Sci 64:180–197

Porwal A, González-Álvarez I, Markwitz V, McCuaig TC, Mamuse A (2010) Weights-of-evidence and logistic regression modeling of magmatic nickel sulfide prospectivity in the Yilgarn Craton, Western Australia. Ore Geol Rev 38:184–196

Pregibon D (1981) Logistic regression diagnostics. Ann Stat 9(3):705–724

Regmi NR, Giardino JR, Vitek JD (2010) Modeling susceptibility to landslides using the weight of evidence approach: Western Colorado, USA. Geomorphology 115(1–2):172–187

Romero-Calcerrada R, Barrio-Parra F, Millington JD, Novillo CJ (2010) Spatial modelling of socioeconomic data to understand patterns of human-caused wildfire ignition risk in the SW of Madrid (central Spain). Ecol Model 221(1):34–45

Sawatzky DI, Raines GL, Bonham-Carter GF, Looney CG (2009), Spatial Data Modeller (SDM): ArcMap 9.3 geoprocessing tools for spatial modelling, logistic regression, fuzzy logic and neural networks. http//arcscripts.esri.com/details/.asp/dbid=15341

Schaeben H (2012) Comparison of mathematical methods of potential modeling. Math Geosci 44:101–129

Schaeben H (2014) A mathematical view of weights-of-evidence, conditional independence, and logistic regression in terms of Markov random fields. doi: 10.1007/s11004-013-9513-y

Schaeben H, van den Boogaart KG (2011) Comment on "A conditional dependence adjusted weights of evidence model" by Minfeng Deng in natural resources research. Nat Resour Res 20:401–406

Sharpe DR, Barnett PJ, Brennand TA, Finley D, Gorrel G, Russell HAJ (1997) Surficial geology of the Greater Toronto sand Ridges Moraine areas, compilation map sheet. Geol Surv Can Open File 3062, map 1:200,000

Song H, Hiromu D, Kazutoki A, Usio K, Sumio M (2008) Modeling the potential distribution of shallow-seated landslides using the eights of evidence method and a logistic regression model: a case study of the Sabae Area, Japan. Int J Sediment Res 23:106–118

Spiegelhalter DJ (1986) Uncertainty in expert systems. In: Gale WA (ed) Artificial intelligence and statistics. Addison-Wesley, Reading, pp 17–55

Spiegelhalter DJ, Knill-Jones RP (1984) Statistical and knowledge-based approaches to clinical decision-support systems, with an application in gastroenterology. J R Stat Soc A 147(1):35–77

Sutton C, McCallum A (2007) An introduction to conditional random fields for relational learning. In: Getoor L, Taskar B (eds) Introduction to statistical relational learning. MIT Press, Cambridge, pp 93–127

Thiart C, Bonham-Carter GF, Agterberg FP, Cheng Q, Pahani A (2006) An application of the new omnibus test for conditional independence in weights-of-evidence modeling. In: Harris J (ed) GIS in the earth sciences. Geological Association of Canada, Special Volume, pp 131–142

Tukey JW (1972) Discussion of paper by F. P. Agterberg and S. C. Robinson. Int Stat Inst Bull 38:596

Wang W, Zhao J, Cheng Q, Liu J (2012) Tectonic-geochemical exploration modeling for charactering geo-anomalies in southeastern Yunnan District, China. J Geochem Explor 122:71–80

Wright DF (1988) Data integration and geochemical evaluation of Meguma Terrane, Nova Scotia, for gold mineralization. Unpublished PhD thesis, University of Ottawa

Yule GU (1912) On the methods of measuring association between two attributes. J R Stat Soc 75:579–642

Zhang D, Agterberg FP, Cheng Q, Zuo R (2013) A comparison off modified fuzzy weights of evidence, fuzzy weights of evidence and logistic regression for mapping mineral prospectivity. doi: 10.1007/s11004-013-9496.8

Chapter 6
Autocorrelation and Geostatistics

Abstract Time series analysis has a long history of useful applications in many branches of science. Estimation of the autocorrelation function and power spectrum of serial data are important tools that are complementary to one another. In general, it is assumed that a series is "stationary" with constant mean and variance. Analysis of glacial varve-thickness data will be presented as an example of establishing periodicities in temperature related to climate change. The cross-spectrum and coherence are time series equivalents generalizing the correlation coefficient between two variables. In geostatistics, the semivariogram is favored as a principal tool for analysis of space series consisting of mining assays in ore deposits or chemical element concentration values in rock units. In general, there is a simple linear relationship between autocorrelation function and semivariogram. The most frequently used geostatistical semivariograms including the spherical and exponential models allow for a nugget effect at their origin representing non-zero variance and assume asymptotic convergence to a constant value representing regional variance at large sampling intervals exceeding a range. However, a semivariogram model without finite variance for modeling of short-distance autocorrelation as originally proposed by Matheron (Traité de géostatistique appliquée. Mém BRGM 14, Paris, 1962, Estimating and choosing, an essay on probability in practice (trans: Hasover AM). Springer, Heidelberg, 1989) can be more appropriate in some types of applications. Geometrical considerations related to shapes and volumes of blocks of rocks are important in geostatistical modeling. Average values estimated for blocks can be extrapolated into their surroundings with use of their extension variances. In addition to the time-series applications to glacial varve-thickness data, practical examples in this chapter are mainly for copper, zinc and gold concentration values obtained by channel sampling in various ore deposits in Bolivia, Canada and South Africa, and for copper in chip samples from along the 7-km deep KTB borehole drilled into the Bohemian Massif in southern Germany. Stationary discrete random variables, for example, those related to successions of different KTB lithologies also can be modeled geostatistically.

Keywords Time series analysis • Autocorrelation • Power spectrum • Spatial statistics • Semivariogram • Nugget effect • Geometric probability • Extension variance • Pulacayo zinc values • Whalesback copper deposit • KTB copper • KTB geophysics

6.1 Time Series Analysis

Time series analysis is a well-established topic of mathematical statistics. It is closely related to Fourier analysis. A good review of the history of this topic can be found in Bloomfield (2000). It is of interest to mathematical geoscientists for analysis of geoscientific time series and space series. The latter arise when a rock unit is sampled at regular intervals along a line. Obviously, there is then in a 3-D situation. Geostatistics was originally developed by Matheron (1962) for 3-D domains. It was adopted by mathematical statisticians under the name "spatial statistics" (*cf.* Sect. 2.1.2)

Suppose that a given series $\{x_k\}$ is part of an infinite series which, in turn, is a realization of an ordered set of random variables $\{X_k\}, k = -\infty, \ldots, -1, 0, 1, \ldots, \infty$. The theoretical set can be seen as a population that is doubly infinite, because any point out of an infinite number of points along the line can assume any one value in an infinite set of possible values. Suppose the random variables $\{X_k\}$ all have the same probability distribution with constant mean and variance. However, they can be correlated with one another. In the example of Fig. 4.1, the two sets of gold values from panels that are ($h =$) 30 ft. apart have approximately the same mean and variance but are autocorrelated with $r(h) = 0.59$. The serial correlation coefficient would become larger for shorter distance and less when h is increased. A series is "weakly" stationary if the autocorrelation function for the $\{X_k\}$ (with mean μ and variance σ^2) is constant and does not depend on location along the series. A property of stationary series is that $r(h)$ goes to zero when h approaches infinity. Serial correlation $r(h)$ can be negative even if a series is stationary. Many series are not stationary in that the mean of the $\{X_k\}$ changes systematically along the series.

Discussions of the concept of "stationarity" of time or spatial series can be found in Bloomfield (2000, Sect. 9.3) or Cressie (1991). "Strong" stationarity involves time- or space-independence of frequency distributions and implies "weak" stationarity also simply referred to as "stationarity". Cressie (1991, p. 40) defines "intrinsic" stationarity through first differences between successive values with the properties $E(X_{k+h} - X_k) = 0$ and $\sigma^2 (X_{k+h} - X_k) = 2 \cdot \gamma_h$ where γ_h is the semivariogram. This assumption is weaker than "weak" stationarity because it allows for infinitely large variance and non-existence of the mean μ, a topic to be discussed in more detail in Sect. 6.2.1. A time series is called "ergodic" if the time average of a quantity is equal to its ensemble average (Brillinger 1981, Sect. 2.11) where "ensemble" comprises all possible realizations of the series. Ergodicity implies stationarity. In this chapter, stationarity and ergodicity will be assumed to exist if possible. However, intrinsic stationarity with infinitely large variance will be assumed to hold true as an alternative approach to describe strictly local variability of element concentration values in rocks (Sect. 6.2.6).

> **Box 6.1: Autocovariance and Power Spectrum**
>
> The autocovariance of an ordered sequence of random variables is defined as $\Gamma_h = E[(X_k - \mu)(X_{k+h} - \mu)]$. The sample autocovariance for n values can be estimated by: $C_h = \frac{1}{n-h} \sum_{k=1}^{n-h} (x_h - \bar{x})(x_{k+h} - \bar{x})$ where $\bar{x} = \frac{1}{n} \sum_{k=1}^{n} x_k$. The autocorrelation function satisfies $\rho_h = \Gamma_h/\Gamma_0 = \Gamma_h/\sigma^2$ and the sample autocorrelation coefficient is $r_h = C_h/C_0$. Together the r_h values form a so-called "correlogram". Wiener's (1933) theorem of autocorrelation states that the power spectrum satisfies: $\varphi(\omega) = \int_{-\infty}^{\infty} \rho_h e^{-i\omega h} dh$. Since ρ_h is even, $\varphi(\omega) = \int_{-\infty}^{\infty} \rho_h \cos \omega h \, dh$. In the next section the power spectral density $P(f)$ will be computed as: $P(f) = C_0 + 2 \sum_{k=1}^{m} W(k) C_k \cos 2\pi k f$ with $W(k) = \frac{1}{2}(1 + \cos \pi k/m)$. The weighting function $W(k)$ defines a cosine-shaped lag window for the autocovariance function whose use is called "hanning".

The purpose of hanning is that individual values $P(f)$ can be considered to estimate a smoothed version of the underlying true spectrum. For more explanations of why this procedure can be useful, see Blackman and Tukey (1959). The power spectrum $P(f)$ can be regarded as a decomposition of all variability in a series in terms of components of the variance for narrow frequency bands. Smoothing operations such as hanning significantly improve their estimation from the autocovariances by eliminating distortions. An alternative approach of constructing the power spectrum (Sect. 6.2.7) consists of averaging adjoining values in the periodogram over equal intervals.

6.1.1 Spectral Analysis: Glacial Lake Barlow-Ojibway Example

Glacial varves, because of their presumed annual nature have attracted the attention of geologists as a geochronological tool. Anderson and Koopmans (1963) were among the first to apply spectral analysis to varves. Theory of cross-correlation and cross-spectral analysis was treated by Goodman (1957), Amos and Koopmans (1963), and Kendall and Stuart (1966), with applications by Hamon and Hannan (1963), Koopmans (1967), Anderson (1967) and Agterberg and Banerjee (1969).

Lake Barlow-Ojibway is a late-glacial water body that formed approximately 11,000 years ago during the retreat of the Late Wisconsin ice-sheet in northern Ontario and western Quebec. The lake, in its maximum extent, measured about 960 km in the east-west and 240 km in the north-south direction (Fig. 6.1 inset). The most extensive deposit formed in it was the sheet-like body of varves which

192   6 Autocorrelation and Geostatistics

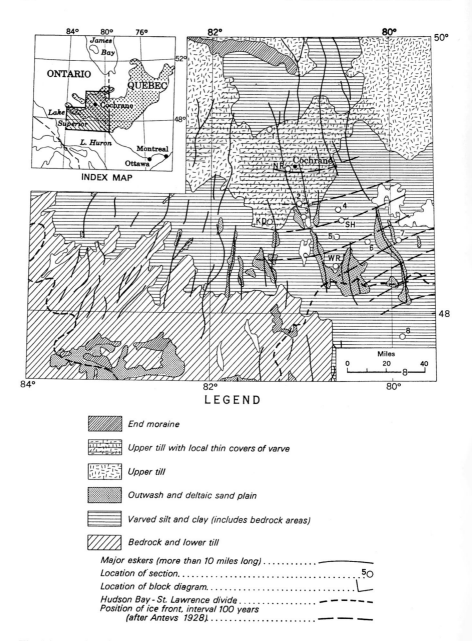

Fig. 6.1 Location of study area with outline of Lake Barlow-Ojibway (*inset*) and surficial geology (Source: Agterberg and Banerjee 1969, Fig. 1)

extends uninterruptedly across the Hudson Bay – St. Lawrence divide (Fig. 6.1). More detailed geological background with references to earlier work is given in Agterberg and Banerjee (1969). It is mentioned here that individual varves can be traced laterally from their proximal (northern) end to distal (southern) end successively changing from sandy through silty to diamictic facies (Fig. 6.2). In the

6.1 Time Series Analysis

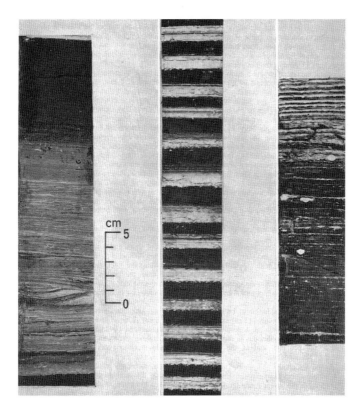

Fig. 6.2 Typical examples of the three facies of varves (from *left* to *right*: sandy, silty and diamictic facies) (Source: Agterberg and Banerjee 1969, Fig. 4)

vertical sequence too, individual varves were successively overlain by more distal facies as the ice-front and with it the proximal end of the varves receded northward with time. Kuenen (1951) proposed a mechanism of varve deposition by annual turbidity currents with silt deposition during the summer followed by slow clay deposition during the winter.

In total, 4,310 thickness measurements from eight sections (locations shown in Fig. 6.1) were statistically analyzed. In the longest series (No. 4 with 537 silt-clay couplets), silt layers near the bottom are about ten times as thick as those near the top. The clay layers near the top are about three times thinner than those near the bottom. A logarithmic transformation was desirable in order to ensure that variations in thickness are of the same magnitude in parts of the series that are relatively thick and relatively thin, respectively. For example, the sum of squared silt thickness data on which measures of variability such as the variance are based, amounts to 1,037.2 cm^2 for series 4. The first 100 values in this series contribute 91.99 % and the last 100 values only 0.02 % to this total sum of squares that is based on all 537 data. If no logarithmic transformation would be applied to the observations, results from a statistical study would be largely determined by variations in the thicker parts of the series, whereas variations in the thinner parts of the series would have a negligibly small effect on the

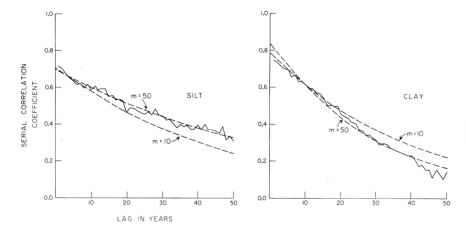

Fig. 6.3 Correlograms of log-thickness data, silt and clay, series 4. Best-fitting negative exponential curves with $m = 10$ and $m = 50$ are also shown (Source: Agterberg and Banerjee 1969, Fig. 6)

end results. The logarithmic transformation stabilizes the variance. For the example, the ratio of the variances for the first and last sets of 100 silt values for series 4 is 33.6, but after logarithmic transformation it becomes 1.5.

Box 6.2: Signal-Plus Noise Model

Suppose a series of observed values (x_k) is the sum of two series: signal (s_k) and white noise (n_k) with $k = 1, 2, \ldots, n$. The autocovariance functions of these series written as $\Gamma_x(h)$, $\Gamma_s(h)$ and $\Gamma_n(h)$ satisfy the relations: $\Gamma_n(h) = 0$ if $h \neq 0$ and $\Gamma_x(h) = \Gamma_s(h) + \Gamma_n(h)$. Suppose $\Gamma_x(0) = 1$ and $\Gamma_n(0) = c$. Let $F(h)$ denote a filter to which the record must be subjected in order to obtain the signal: $s_k = \int_{-\infty}^{\infty} F(h) x(k+h) dh$. Then, $\Gamma_s(h) = \int_{-\infty}^{\infty} F(h+\tau) \Gamma_s(\tau) d\tau + \int_{-\infty}^{\infty} F(h+\tau) \Gamma_n(\tau) d\tau$ or $\Gamma_s(h) = \int_{-\infty}^{\infty} F(h+\tau) \Gamma_s(\tau) d\tau + (1-c) F(h)$. Fourier transformation of both sides gives: $G(\omega) = \Phi(\omega) \cdot G(\omega) + (1-c) \Phi(\omega)$ where $G(\omega) = \int_{-\infty}^{\infty} \Gamma_h \cos \omega h \, dh$ and $\Phi(\omega) = \int_{-\infty}^{\infty} F_h \cos \omega h \, dh$. It follows that: $\Phi(\omega) = \frac{G(\omega)}{G(\omega) + (1-c)}$. The filter $F(h)$ can be found by taking the inverse Fourier transform: $F(h) = \int_{-\infty}^{\infty} \Phi(\omega) \cos \omega h \, d\omega$. When $\Gamma_x(h) = \exp(-a|h|)$, then: $G(\omega) = \frac{2ac}{a^2 + c^2}$ and $F(h) = \frac{ac}{\pi p^2 (1-c)} \int_{-\infty}^{\infty} \frac{1}{\frac{\omega^2}{p^2} + 1} \cos \omega h \, d\omega$ where $p = [a^2 + 2ac/(1-c)]^{0.5}$. After some manipulation it follows that: $F(h) = q \cdot \exp(-p|h|)$ where $q = ac/[(1-c) \cdot p]$ (cf. Yaglom 1962; Agterberg 1967, 1974).

Correlograms for log-thickness data, silt and clay, for series 4 are shown in Fig. 6.3. Best-fitting semi-exponential curves with $r_h = c \cdot \exp(-a|h|)$ where

6.1 Time Series Analysis

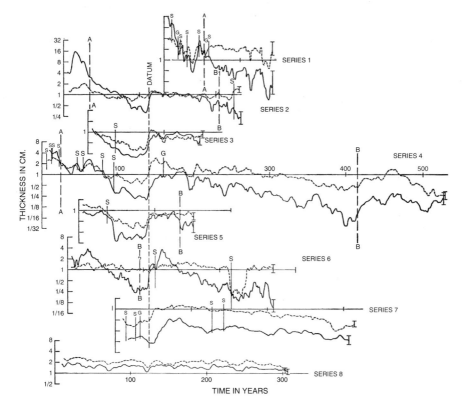

Fig. 6.4 Filtered curves for thickness data, series 1–8. *Solid lines* denote silt thickness, *broken lines* for clay thickness. A-A is boundary between sandy and silty facies, B-B is boundary between silty and diamictic facies, s slumped intervals, G gaps in record with 5–10 varves missing. Width of 95 % confidence belt is shown at end of each series (Source: Agterberg and Banerjee 1969, Fig. 7)

c and a are constants are shown for $m = 10$ and $m = 50$. Deming's (1948) method of least squares for exponentials was used for the curve-fitting. Clearly, the correlogram curves (original data and lines of best fit) intersect the vertical axis at points that are less than one. This indicates the presence of random (uncorrelated) noise with variance of standardized data equal to $(1 - c)$. This white noise can be removed from the data by using the filter of Box 6.2 as follows.

Application of the semi-exponential filter to series 4, silt, (with $m = 10$) yielded, $a = 0.022$, $c = 0.72$, $p = 0.33$ and $q = 0.17$ (*cf*. Agterberg and Banerjee 1969) indicating that the bilateral filter used to derive the relatively smooth "signal" in Fig. 6.4 is restricted to a relatively narrow neighborhood. Six of the eight series shown in Fig. 6.4 were aligned with respect to one another on the basis of the "datum" which is a relatively abrupt increase in thickness of the varves. This datum coincides with the Cochrane readvance of the land-ice about 8,000 years ago (Antevs 1925; Hughes 1955).

The varve time series are not stationary in that, for example, both mean silt and clay layer thickness tend to decrease with time. As a rule, such decreases are stronger in the silt than in the clay. Observed varve series are incomplete because thicknesses could not be measured in places where slumped layers occur. These places are shown as "s" in Fig. 6.4. Some gaps where a larger number of varves is probably missing are identified as "G" in this figure. In total as much as 10 % of total number of varves may be missing in some series including series 4.

Varve-thickness data, like tree-rings, provide a sensitive tool to measure very small temperature fluctuations on Earth. Power spectra with $m = 50$ are shown in Fig. 6.5 for the four longest series ($n > 250$). All spectra commence with a strong peak close to the origin for low-frequency waves representing long-term variations or "trends". For 20-year and lesser periods, the spectra tend to flatten out. It is likely that the peaks for shorter periods are not exclusively caused by random variations but are partly due to weak periodical phenomena. The first peak for silt (S_1 in Fig. 6.5) occurs at a period of about 14 years in all four spectra. Because some varves are missing, this indicates that a cyclical variation of approximately 15 years existed in the silt thickness variation over large parts of Lake Barlow-Ojibway. A periodicity of this type is not known to occur in glacial deposits elsewhere in the world. Two possible explanations have been offered. Agterberg and Banerjee (1969) suggested that the phenomenon could be related to the existence of ridges in the bedrock profile known to occur at about 8 km intervals underneath some eskers. If the average rate of retreat of the ice-front was about 500 m per year, occurrence of the annual turbidity currents during the summer months would have been about 15 years. Later, another explanation was offered by Schove (1972). This author suggested climatological variations due to soli-lunar cycles as an explanation. Brier (1968) had shown that soli-lunar cycles produce significant tidal effects in the atmosphere at 13.5 and 27 years. According to Schove (1972), silt components of varves are suitable for study of soli-lunar cycles as affect the summer months of June and July. The 13.5 year (163 calendar months) cyclicity is the beat period between the calendar month (30.44 days) and the synodic month (29.5 days). These two cycles are in phase with one another every 163 months, and both then are approximately in phase with the cycle arising from the anomalistic month (27.55 days). The synodic cycle is the period from one new moon to the next and the anomalistic cycle that from one perigee to the next. These two cycles are most important in determining the magnitude of the soli-lunar gravitational tide, which has significant effects on monthly weather conditions according to Brier (1968). There is no consensus among climatologists that the moon significantly affects climate.

Peaks in power spectra often are followed by secondary peaks that tend to be equally spaced along the frequency scale. These may represent "harmonics" reflecting deviations from sinusoidal shape of the periodic phenomenon (*cf.* Schwarzacher 1967). This phenomenon is most conspicuous for clay in series 4 where there are nine peaks, all of which are multiples of frequency 0.05 or a value

6.1 Time Series Analysis

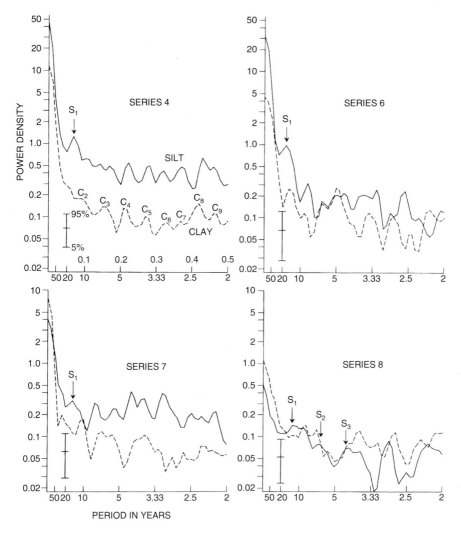

Fig. 6.5 Power spectra with $m = 50$ for four longest series ($n > 250$); Frequency scale is shown in spectrum for series 4 only. Confidence belts also are shown (Source: Agterberg and Banerjee 1969, Fig. 5)

slightly larger than 0.05. This probably reflects the 22-year and 11-year sunspot cycles that have been found in varves elsewhere in the world (Schove 1983; Berry 1987) and in other laminated sediments, e.g. in marl-limestone couplets belonging to the Late Campanian *Radotruncana calcarata* Zone (Wagreich et al. 2012). Sunspot cycles have variable duration but, on average, they last 11 years. The 22-year cycle is due to alternation of stronger and weaker 11-year cycles.

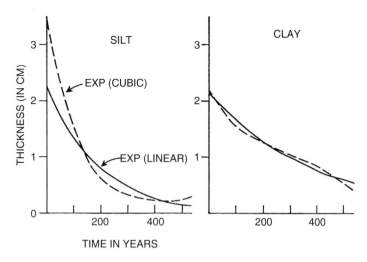

Fig. 6.6 Exponential trends, silt and clay, series 4 (Source: Agterberg and Banerjee 1969, Fig. 8)

6.1.2 Trend Elimination and Cross-Spectral Analysis

A more detailed time series analysis has been performed of series 4 which, with 537 varves, covers the longest time interval. First polynomial curve-fitting was performed on the logarithmically (base e) transformed data. Percentages of explained sums of squares due to linear fits were 43.0 and 38.0 % for silt and clay thickness data, respectively. These %ESS values were increased to 48.2 and 38.6 %, after addition of quadratic terms. Further improvements due to cubic fits were small with %ESS values of 48.5 and 39.5 %, respectively. The linear and cubic exponential trends are shown in Fig. 6.6. These curves were obtained by elimination of the effects of the logarithmic transformation on the trend. In the linear case, this procedure yields the following exponential curve for thickness in cm: $H(t) = \exp\{a + bt + \frac{1}{2}s^2\}$ where $H(t)$ represents the exponential thickness decrease with t measured in years, a and b are constants, and s^2 is the residual variance (Agterberg 1968; Heien 1968). It represents the solution of the deterministic differential equation: $dH(t)/dt = H(t)$. In Agterberg and Banerjee (1969), the dimension of time (t) is replaces by that of distance (x) so that the exponential trend curves represent thickness profiles of individual varves. The preceding two equations then can be written as: $H(x) = \exp(-cx)$ and $dH(x)/dx = -cH(x)$. It follows that: $\Delta H(x) = -cH(x)\Delta x$ where $\Delta H(x)$ represents the decrease in varve thickness away from the source over a short distance Δx. Therefore, this model would mean that thickness away from the source is everywhere proportional to thickness. It provides a fair approximation for the clay, but for the silt relatively more material was deposited close to the source (Fig. 6.6).

Refined spectral analysis and cross-spectral analysis can be applied to the residuals from the linear trends for the log-thickness data (base e) of both silt and clay. The resulting new correlograms are shown in Figs. 6.7 and 6.8, respectively. Shifting the series of clay residuals with respect to the series of silt residuals resulted

6.1 Time Series Analysis   199

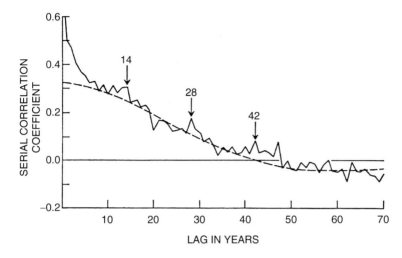

Fig. 6.7 Correlogram for residuals from linear trend, silt, series 4. The 14-year periodicity that shows as a weak oscillation is indicated by *arrows*. Fitted curve satisfies stochastic differential equation of second order Markov process with $c = 0.32$, $\alpha = 0.03$, and $\omega_0 = 0.05$ (Source: Agterberg and Banerjee 1969, Fig. 9)

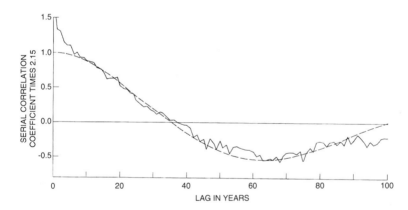

Fig. 6.8 Correlogram for residuals from linear trend, clay, series 4. Fitted curve satisfies stochastic differential equation of second order Markov process with $c = 1$, $\alpha = 0.01$, and $\omega_0 = 0.05$ (Source: Agterberg and Banerjee 1969, Fig. 10)

in the 201 cross-correlation coefficients shown in Fig. 6.9. The three figures based on residuals instead of original thickness data confirm results obtained by spectral analysis in the preceding section but produce additional information as well. The new silt correlogram (Fig. 6.7) clearly shows the 14-year periodicity that could not be perceived in the correlogram of original silt data shown in Fig. 6.3. The 10-year (and 20-year) sunspot cycle that is evident in Fig. 6.5 (series 4) cannot be seen in the clay residual correlogram of Fig. 6.6. As pointed out before, all periodicities in the power spectrum for series 4 are underestimated by about 10 % due to missing varves. The 11-year

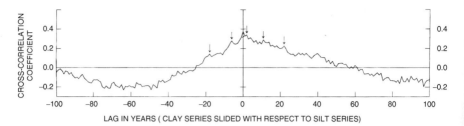

Fig. 6.9 Cross-correlation function for residuals from linear trend, clay, series 4. Weak oscillations with period of 10 years indicated by *arrows*. Note that the correlation coefficient for zero lag is only slightly larger than its neighboring values, indicating that the noise components for silt and clay are nearly uncorrelated (Source: Agterberg and Banerjee 1969, Fig. 11)

sunspot cycle is visible in the cross-correlogram of Fig. 6.9. In this diagram, it can be seen that the clay leads the silt by about 2 years. Such phase differences are better studied by using cross-spectral analysis. The fitted curves in Figs. 6.7 and 6.8 satisfy an equation for a stochastic model to be explained later (Sect. 6.1.3).

> **Box 6.3: Cross-Spectrum, Coherence and Phase**
>
> The cross-spectrum consists of the co-spectrum $C_{sc}(f)$ and the quadrature spectrum $Q_{sc}(f)$ with: $C_{sc}(f) = r_{sc}(0) + \sum_{k=1}^{m}[W(k)\cos 2\pi k f \{r_{sc}(k) + r_{cs}(k)\}]$; and $Q_{sc}(f) = \sum_{k=1}^{m}[W(k)\sin 2\pi k f \{r_{sc}(k) - r_{cs}(k)\}]$. In these expressions, $W(k)$ is the same weighting function as before. The data were standardized and $r_{sc}(k)$ and $r_{cs}(k)$ together form the cross-correlation function shown in Fig. 6.9. The coherence $R(f)$ and phase $\varphi(f)$ satisfy: $R(f) = \sqrt{\frac{C_{sc}^2(f) + Q_{sc}^2(f)}{P_s(f)P_c(f)}}$ and $\varphi(f) = \arctan \frac{Q_{sc}(f)}{C_{sc}(f)}$. $R(f)$ is a measure of the strength of linear relationship between the two series for frequency bands around f. It is equivalent to the correlation coefficient between two variables as a function of frequency (Koopmans 1967).

As mentioned before, the power spectrum $P(f)$ represents a decomposition of total variance of a series in terms of variance components for narrow frequency bands. Likewise, the coherence is the decomposition of the total correlation coefficient between two variables. For example, Anderson (1967) has shown that two time series can be uncorrelated when time is not considered as a variable whereas, in reality, the long-term fluctuations are negatively correlated and the short-term fluctuations positively correlated (or vice versa). Partial correlation with trend elimination can give a solution to problems of this type but cross-spectral analysis may provide a more refined answer. The coherence $R(f)$ is positive for all frequencies and should not be interpreted separately from the phase $\varphi(f)$ that can be either positive or negative and falls between $-180°$ and $180°$. When the phase is close to $180°$ or $-180°$, the two variables are nearly $180°$ out of phase implying negative correlation.

Fig. 6.10 Power spectra with $m = 100$ for standardized residuals from linear trend, silt (*solid line*) and clay (*broken line*), series 4. The coherence (COH) is also shown (Source: Agterberg and Banerjee 1969, Fig. 12)

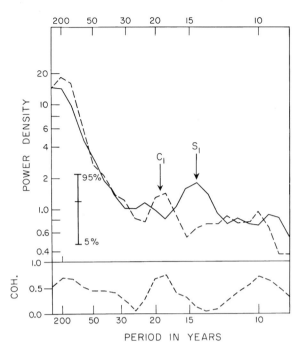

From Fig. 6.4 it can be concluded that the lowest frequency waves that form sharp peaks near the origin of the power spectra in Fig. 6.5 are positively correlated. Figure 6.10 shows power spectra of the silt and clay residuals. Both have a large long-period peak that will be discussed later in this section. Clay shows a peak just below 20 years that was not visible in Fig. 6.5 (series 4). The 14-year peak in the silt power spectrum also is seen in the silt residual spectrum. The coherence is shown in Fig. 6.10 as well. It shows local maxima near 20 and 10 years for the sunspot cycle showing that both clay and silt thicknesses were influenced by this cyclicity. Phase differences for periods of 9.5 and 10 years were $-91°$ and $-72°$, respectively, that both represent a 2-year lag of silt with respect to clay. The 2-year lead of clay could already be seen in the cross-correlogram of Fig. 6.9. It is interesting that clay did not participate in the 14-year silt cycle according to the coherence diagram. It would suggest that, every 15 years, coarser grained material predominated in the annual turbidity currents that probably took place toward the end of summer. The fine-grained clay was slowly deposited during the fall and winter. The fact that it led the silt by about 2 years probably reflects slower response of the coarser grained material. A greater rate of retreat of the land ice due to an increase in temperature would produce more melt water and consequently more sediments to be deposited. The clay was transported across the lake almost immediately, but the silt was delayed. Initially, the silty material was dumped close to the ice front leaving the thickness profile of varves in more distal areas relatively unaffected. However, later the steeper profile of underlying varves, acting as a floor to turbidity currents for later years, helped to spread the silt farther with the thicker parts moving to more distal areas.

A useful technique for the study of cross-correlation, which had not yet been developed at the time of the varve analysis project of Agterberg and Banerjee (1969), is wavelet analysis (see Hubbard 1996). Applications of wavelet analysis to sedimentary sequences include Prokoph and Agterberg (1999) and Prokoph and Bilali (2008).

6.1.3 Stochastic Modeling

Agterberg and Banerjee (1969) discuss how the theoretical autocorrelation function of a second order stochastic process was fitted experimentally to the silt and clay thickness correlograms shown in Figs. 6.7 and 6.8, respectively. The corresponding theoretical power spectra are shown in Fig. 6.11. The corresponding theoretical phase difference is given in Fig. 6.12. For periods greater than 130 years, the clay leads the silt. This corresponds to the frequency bands where most of the power in concentrated in Fig. 6.11 for both silt and clay. The zero crossing points in the cross-correlogram of Fig. 6.9 occur at 56½ and −25½ years indicating that the major oscillations for the silt are not in phase with those of the clay, but that, on the average, clay leads silt by about 15½ years. This phase lag of the silt is explained, at least in a qualitative manner by the second-order stochastic model (Fig. 6.12).

Box 6.4: Time-Dependent Stochastic Processes

The continuous m-th order autoregressive process is represented by the stochastic differential equation: $a_m \frac{d^m x}{dt^m} + a_{m-1} \frac{d^{m-1} x}{dt^{m-1}} + \ldots + a_0 x(t) = \epsilon(t)$ where $\epsilon(t)$ is a white-noise function driving the stationary random variable $x(t)$. Its autocorrelation function is of the type: $\rho(x) = A_1 e^{-\lambda_1 |x|} + A_2 e^{-\lambda_2 |x|} + \ldots + A_m e^{-\lambda_m |x|}$ where $A_1, A_2, \ldots, A_m$ are constants, and $\lambda_1, \lambda_2, \ldots, \lambda_m$ represent the real or complex roots of a polynomial. A pair of conjugate complex roots can be combined into a single term of the type $e^{k_1 |x|} \cos(k_2 + \varphi)$ where k_1, k_2 and φ are constants. The first order process results in $\rho(x) = \exp(-a \cdot |x|)$. The second order can be written in the form: $\frac{d^2 x}{dt^2} + 2\alpha \frac{dx}{dt} + (\omega_0^2 + \alpha^2) x(t) = \epsilon(t)$. If $\omega_0^2 > 0$, λ_1 and λ_2 form a pair of conjugate roots, and: $\rho(x) = e^{-\alpha |x|} \left[\cos \omega_0 x + \frac{\alpha \sin \omega_0 |x|}{\omega_0} \right]$. The latter result can only be applied to discrete data if the sampling interval is sufficiently small (cf. Yaglom 1962; Jenkins and Watts 1968; Kendall and Stuart 1958). Fourier transformation of the second order autocorrelation function gives: $f(\omega) = \frac{2\alpha (\omega_0^2 + \alpha^2)/\pi}{(\omega^2 - \alpha^2 - \omega_0^2)^2 + 4\alpha^2 \omega^2}$ where ω represents angular frequency. The corresponding phase lag satisfies: $\varphi(\omega) = \frac{-2\alpha \omega}{\omega_0^2 + \alpha^2 - \omega^2}$ (cf. Parzen 1962, p. 112; Sommerfeld 1949, p. 101).

6.1 Time Series Analysis

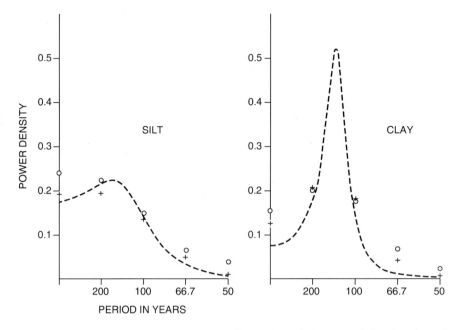

Fig. 6.11 Theoretical power spectra corresponding to theoretical autocorrelation functions of Figs. 6.7 and 6.8. *Crosses* denote smoothed theoretical power density values for frequency bands using the hanning response function. Observed values (o) from Fig. 6.10 are also shown and can be compared to crosses (Source: Agterberg and Banerjee 1969, Fig. 13)

Fig. 6.12 Theoretical phase angle for clay and silt according to which the response $h(t)$ lags behind the random process $\eta(t)$. Note that the clay leads the silt for $T > 130$ years where most of the power density occurs according to Fig. 6.11 (Source: Agterberg and Banerjee 1969, Fig. 14)

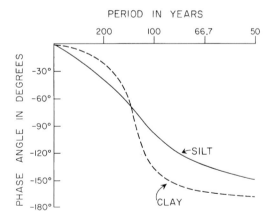

Summarizing it can be concluded that the non-random thickness pattern of varves in Lake Barlow-Ojibway primarily consists of an exponential decrease away from the retreating ice front. In individual sections this decrease shows as gradual thickness decreases in both silt and clay. Elimination of this overall trend shows that additional changes in thickness were mainly controlled by a second-order stochastic process with phase differences between silt and clay. Finally, there are weak cyclicities for the 22-year and 11-year sunspot cycles involving both silt and clay and a 15-year cyclicity mainly restricted to the silt. The exponential thickness decrease in most varve time series suggests a rapid linear retreat of the land-ice. Superimposed on this trend there occurred two relatively rapid increases in thickness. One of these falls at the "datum" in Fig. 6.4 coinciding with the so-called Cochrane readvance; the other one occurred near the end of series 4. On average, the stochastic model indicated a 126-year cyclicity superimposed on the linear retreat. Thus, the rate of land ice retreat was accelerated and decelerated periodically. It is possible that at the end of a deceleration, the ice-sheet not only came to a stand-still but readvanced relatively rapidly during a short period of time. This phenomenon is known as "surging".

Beginning with the theoretical models of Weertman (1969) and Lliboutry (1969), there has been much progress in the deterministic modeling of land-ice retreat. A comprehensive review of these theoretical developments can be found in Fowler (2011). Surges of glaciers and land-ice are modeled by Fowler (2011). They can indeed be periodic. There may even have existed mega-surges (Heinrich events including the Hudson Strait mega-surge) with a periodicity of about 7,000 years.

6.2 Spatial Series Analysis

With respect to space series, two geostatistical topics of practical interest are existence of "sill" and "nugget effect" (see e.g. Journel and Huijbregts 1978; Isaaks and Srivastava 1989; Cressie 1991; or Goovaerts 1997). Suppose $\gamma(h)$ represents the semivariogram, which is half the variance of the difference between values separated by lag distance h. Semivariogram values normally increase when h is increased until a sill value is reached for large distances. If element concentration values are subject to second-order stationarity, $\gamma(h) = \sigma^2 (1 - \rho_h)$ where σ^2 represents variance and ρ_h is the autocorrelation function. The sill is reached when there is no spatial autocorrelation or $\gamma(h) = \sigma^2$. If regional trends can be separately fitted to, for example, element concentration values, the residuals from the resulting regional, systematic variation may become second-order stationary because the overall mean in the study area then is artificially set equal to zero. Within most rock types such as granite or sandstone, randomness of chemical concentration is largely restricted to microscopic scale and sills for compositional data are reached over very short distances. The nugget effect occurs when extrapolation of $\gamma(h)$ towards the origin ($h \to 0$) from observed element concentration values yields estimates with $\gamma(h) > 0$ (or $\rho_h < 1$). A pseudo-nugget effect arises when there is strong local autocorrelation that cannot be detected because locations of samples subjected to chemical analysis are too far apart to describe it adequately.

If a segment of the Earth's crust is sampled and element concentration values are determined on the resulting rock samples, the spatial variability of the chemical determinations generally can be subdivided into a number of separate components. In some applications the original data are stochastic in that they can be described by random functions. However, often the main component of spatial variability is deterministic, either because it is related to differences between rock units separated by discontinuities (contacts), or because there are regional trends. The latter can be extracted from the data by a variety of methods; e.g., by trend surface analysis (Chap. 7), calculation of moving averages with or without weights that are powers of the inverse of distance, by various methods of kriging, by using splines, or by means of other methods of signal extraction. After extraction of a deterministic component, the residuals generally are stochastic in that they can be described by means of spatial random functions. In the simplest case, these residuals are uncorrelated and their correlogram is a Dirac delta function representing white noise. Measurement errors would create white noise. If extrapolation towards the origin by means of a function fitted to the correlogram results in a variance that significantly exceeds variance due to measurement errors, this would create a pseudo-nugget effect hiding strong autocorrelation over short distances.

6.2.1 Finite or Infinite Variance?

A problem of considerable interest for spatial series (and for time series as well) is whether or not the random variable used for the modeling is stationary in that it would have a definite mean and finite variance. Stationarity implies intrinsic stationarity but intrinsic stationarity does not imply stationarity. Jowett (1955) was among the first to assume intrinsic stationarity rejecting the commonly made automatic assumption of existence of a constant or variable mean μ. Matheron (1962) initially introduced the variogram under the assumption of intrinsic stationarity. His original approach recently was summarized by Serra (2012).

Box 6.5: Pseudo-Parabolic Behavior at the Origin

Like Matheron, Serra (2012, p. 59) defines the variogram as: $2\gamma(h) = E\,[f(x) - f(x+h)]^2$ where $\gamma(h)$ is the semivariogram as used in this chapter, and $f(x)$ denoted the value of a regionalized random variable at location x. In an example, Serra uses a Poisson process with parameter λ to generate values along a line. At every next equidistant location a value drawn from the Poisson population is added to the value at the preceding location. This artificial series does not have a finite variance but the increments have variance equal to λ, and values for a finite segment of length L along the series can be assumed to have finite variance σ^2. Suppose now that a pseudo-covariance is estimated in the usual way as:

(continued)

Box 6.5 (continued)

$$\text{Cov}'(h) = \frac{1}{L-h}\int_0^{L-h}[f(x+h)-f_L][f(x)-f_L]dx \text{ where } f_L = \frac{1}{L}\int_0^L f(x)dx.$$

It can be shown that the expected value of this pseudo-covariance is $E[\text{Cov}'(h)] = \lambda\sigma^2\left(\frac{L}{3}-\frac{4}{3}h+\frac{2h^2}{3L}\right)$ $(0 \leq h \leq L)$. Such pseudo-parabolic behavior completely distorts the real behaviour of the random variable at the origin.

Behavior of covariance and semivariogram near the origin and at very great distances ($h \to \infty$) generally is difficult to determine in practice, because it must be based on extrapolation from observations at sampling intervals that, for practical reasons, cannot be very small nor very large. Additional problems may arise in practice when the data have frequency distributions that are positively skewed with relatively few very large values that strongly affect estimations of covariance and semivariogram. The purpose of the following example is to consider extrapolations for $h \to 0$ and $h \to \infty$ before and after logarithmic transformation.

Suppose that the semivariogram of X and $\log_e X$ are written as $\gamma^*(h)$ and $\gamma(h)$ with $\gamma^*(h) = \frac{1}{2}E\ (X_i - X_{i+h})$ and $\gamma(h) = \frac{1}{2}E\ (\log_e X_i - \log_e X_{i+h})$, respectively. If it can be assumed that the mean $EX = \mu$ exists and the variance (to be written as var) is finite, the covariance satisfies $cov\ (h) = E\ (X_i \cdot X_{i+h}) - \mu^2 = var - \gamma^*(h)$. This covariance is related to the semivariogram for logarithmically transformed values by means of (cf. Matheron 1974; Agterberg 1974, p. 339) as $cov\ (h) = var \cdot \exp[-\gamma(h)]$ provided that $\log_e X$ is approximately normal (Gaussian) with variance $\sigma^2 \gg \gamma(h)$. An example is as follows.

Estimated covariances for the 118 Pulacayo zinc values of Table 2.4 are shown in Fig. 6.13a. The scale used for the covariance is logarithmic. Also shown is the straight line corresponding to the signal-plus-noise model for $cov\ (h) = c \cdot var \cdot \exp(-ch)$ ($a = 0.1892$; $c = 0.5157$) previously shown in Fig. 2.10. The exponential covariance model of Fig. 6.13a corresponds to a linear semivariogram for logarithmically transformed zinc values as follows immediately from substitution of $\gamma(h) = 3A \cdot h$ into $cov\ (h) = var \cdot \exp[-\gamma(h)]$. Consequently, the semivariogram of logarithmically transformed zinc values plotted on log-log paper should be according to a straight line with slope equal to one. In Fig. 6.13b it is shown that this model is approximately satisfied. It can be concluded that the logarithmically transformed zinc values approximately have a linear semivariogram. This conclusion is in accordance with the linear semivariogram originally fitted by Matheron (1964) to the logarithmically transformed zinc values.

6.2 Spatial Series Analysis

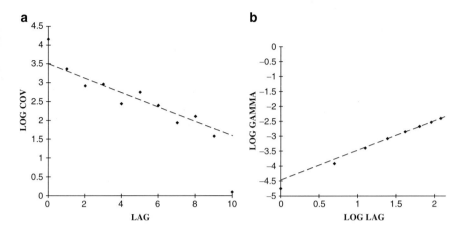

Fig. 6.13 Comparison of covariance models fitted to 118 Pulacayo zinc values under finite and infinite variance assumption. (**a**) Exponential function (straight line with slope $-a = 0.1892$). Nearly half of the variance of the zinc values can be attributed to white noise; (**b**) Semivariogram for logarithmically transformed zinc signal values (previously shown in Fig. 2.10) with logarithmic distance scale. *Straight line* has unit slope. This is in agreement with exponential covariance model applied to original zinc values of Table 2.4 (Source: Agterberg 1994, Fig. 2)

6.2.2 Correlograms and Variograms: Pulacayo Mine Example

De Wijs (1951) assumed that, if a block of ore is divided into halves, the ratio of average element concentration values for the halves is equal to the same constant regardless of the size of the block that is divided into halves. If greater value is divided by lesser value, this ratio can be written as $\eta > 1$. Matheron (1962) generalized this original model by introducing the concept of "absolute dispersion" written as $\alpha = (\log_e \eta)^2 / \log_e 16$. This approach is equivalent to what is now better known as scale invariance. It leads to the more general equation $\sigma^2 (\log_e x) = A \times \log_e V/v$ where $\sigma^2 (\log_e x)$ represents logarithmic variance of element concentration values x in smaller blocks with volume v contained within a larger block of ore with volume V. The corresponding semivariogram along a line then satisfies: $\gamma_h = 3A \cdot \log_e h$. This model does not have a sill but is useful for modeling spatial correlation over very short distances.

Matheron (1989) pointed out that in rock-sampling there are two possible infinities if number of samples is increased indefinitely: either the sampling interval is kept constant so that more rock is covered, or size of study area is kept constant whereas sampling interval is decreased. These two possible sampling schemes provide additional information on sample neighbourhood, for sill and nugget effect, respectively. In practice, the exact form of the nugget effect usually remains unknown because extensive sampling would be needed at a scale that exceeds microscopic scale but is less than scale of sampling space commonly used for ore deposits or other geological bodies. Nevertheless, there are now several methods by means of which the nugget effect can be studied. The de Wijs zinc data set used for example is rather small (118 values). Because of this, larger data sets also will be analyzed: a series of 2,796 copper concentration values for chip samples taken at 2-m intervals along the Main KTB borehole shows a persistent nugget effect that will be analyzed separately. As an example taken from another geoscience field, it will be discussed in Sect. 6.2.6 that alternating, detrended lithologies over a length of about 7 km in the KTB borehole (Goff and Hollinger 1999) show a small-scale nugget effect as well.

The following remarks pertain to the effect of logarithmic transformation on autocorrelation functions and semivariograms. Matheron (1962) applied geostatistical methods to logarithmically transformed assay values for the Pulacayo Mine (Table 2.2). This can have advantages with respect to using untransformed element concentration values. He assumed that "effective length" of each channel sample could be set equal to $L=0.5$ m, representing the average width of the Tajo vein on the 446-m level. Obviously, the 118 zinc values of Fig. 2.1 systematically underestimate true zinc content of the massive sulphide vein filling because the original sample length was 1.30 m for the massive vein augmented by lower zinc grade wall rocks. The effective length can be assumed to be a variable parameter. In the absence of more complete information, it is not unreasonable to assume, as Matheron did, that all massive sulphide zinc concentration values were underestimated by the same factor during the channel sapling. The logarithmic variance σ^2 ($\log_e x$) is not affected if this bias factor is constant. For our example, σ^2 ($\log_e x$) is estimated to be 0.2851. One relatively simple geostatistical sampling method can be illustrated as follows. Suppose that the 118 values for channel samples that are 2 m apart together provide an estimate of average zinc content (=15.61 %) of an elongated rod-shaped mining block with a length of 238 m. Dividing this number by $L=0.5$ m and raising the quotient to the power 3 then yields $V/v=476$. Combining this number with the estimate of logarithmic variance and using Matheron's equation σ^2 ($\log_e x$) $= A \cdot \log_e\{V/v\}$ then yields the absolute dispersion estimate $A=0.015$, which would apply to other block sizes as well.

If the logarithmic variance of element concentration values is relatively large, it may not be easy to obtain reliable estimates of statistics such as mean, variance, autocorrelation function and power spectrum by using untransformed element concentration values. However, lognormality of the frequency distribution often can be assumed. This is the main reason for using logarithmically transformed values instead of original values. Suppose that element concentration values can be described by X_i and $Y_i = \log_e X_i$ has normal, Gaussian frequency distribution with mean μ and variance σ^2. Transformations have been studied by Grenander and Rosenblatt (1957) and Granger and Hatanaka (1964). Representing the autocorrelation functions for X_i and Y_i as $\rho_x(h)$ and $\rho_y(h)$, respectively, the following relationship applies:

$$\sigma^2 \rho_y(h) = \log_e\left[1 + \gamma^2 \rho_x(h)\right]$$

where $\gamma^2 = \sigma^2(X)/\mu^2(X)$ (cf. Agterberg 1974, Eq. 10.40). If γ^2 is sufficiently small, $\rho_x(h)$ and $\rho_y(h)$ are approximately equal. For our example, this condition is

6.2 Spatial Series Analysis

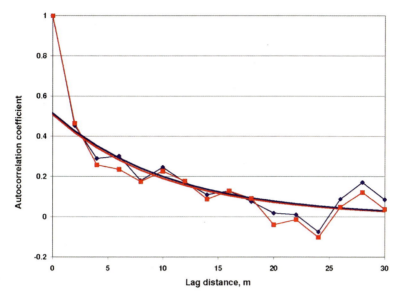

Fig. 6.14 Estimated autocorrelation coefficients for original data (*diamonds*) and logarithmically transformed Pulacayo zinc values (*squares*), shown together with best-fitting negative exponential autocorrelation functions. The curves for original and transformed data coincide approximately illustrating that logarithmic transformation of the original data does not significantly affect autocorrelation in this application (Source: Agterberg 2012, Fig. 5)

satisfied as demonstrated in Fig. 6.14. Approximate equality of results shown in Fig. 6.14 applies to both the estimated autocorrelation coefficients and negative exponential functions fitted by non-linear least squares to data points with $h > 0$. Consequently, variograms of zinc values and logarithmically transformed zinc values also are approximately the same. Later the variogram of logarithmically transformed zinc values will be used. Substituting fitted values from Fig. 6.14 into $\gamma(h) = \sigma^2 (1 - \rho_h)$ yields a variogram (Table 6.1, see later) that is close to estimates originally obtained by Matheron (1962).

Box 6.6: Whittle's Space-Time Model

Whittle (1962) considered a variable $\xi(x,t)$ that adopts a value at every point of a space with Cartesian co-ordinates $x = (x, y, z)$ for time t with $\frac{\partial \xi}{\partial t} + \alpha \xi = \frac{1}{2} \nabla^2 \xi + \epsilon$. This is the standard diffusion equation "driven" by $\epsilon = \epsilon(x, t)$ with $\nabla^2 \xi$ representing the spreading of, for example, a chemical element through the medium and $\alpha \xi$ as a spatial "trend" term. The solution is:

(continued)

Box 6.6 (continued)

$$\xi(x,t) = \int_{-\infty}^{\infty} dy \int_0^{\infty} G(x,y,\tau)\,\epsilon(x,t-\tau)d\tau \quad \text{where } G(x,y,\tau) = (2\pi\tau)^{-\frac{n}{2}}$$

$\exp\{-\alpha\tau - \frac{|x-y|^2}{2\tau}\}$. The spatial covariance function is isotropic and satisfies:

$$\Gamma(s) = \text{cov}[\xi(x,t), \xi(x+s,t)], \text{ and } \Gamma(s) = \int_{-\infty}^{\infty} dy \int_0^{\infty} G(x,y,\tau)\, G(x+s,y,\tau)$$

$d\tau = \frac{1}{(4\pi\tau)^{n/2}} \int_0^{\infty} e^{-2\alpha\tau - s^2/4\tau} d\tau$ where s is distance along a line in any direction.

This expression can be evaluated as follows: $\Gamma(s) = \frac{e^{-s\sqrt{2\alpha}}}{2\sqrt{2\alpha}}$ ($n = 1$) representing the semi-exponential also generated from the first-order stochastic differential equation; $\Gamma(s) = \frac{1}{2\pi}K_0(s\sqrt{2\alpha})$ ($n = 2$) representing the well-known 2-D result first independently derived in 1948 by von Kármán (1948) and Matérn (1981; English version of book published in Swedish in 1948); $\Gamma(s) = \frac{e^{-s\sqrt{2\alpha}}}{2\pi s}$ ($n = 3$) representing Whittle's relatively unknown 3-D solution.

Table 6.1 Pulacayo Mine variogram model

h, m.	Experimental	Exponential	$f(L,h)$	$\beta(h)$	σ_h^2	Deviation
2	0.303	0.325	2.891	0.105	0.286	0.017
4	0.402	0.367	3.580	0.112	0.354	0.048
6	0.436	0.401	3.985	0.109	0.394	0.042
8	0.465	0.429	4.273	0.109	0.422	0.043
10	0.408	0.452	4.496	0.091	0.444	−0.036
12	0.412	0.471	4.678	0.088	0.462	−0.050
14	0.464	0.486	4.832	0.096	0.477	−0.013
16	0.452	0.499	4.966	0.091	0.491	−0.039
18	0.472	0.510	5.083	0.093	0.502	−0.030
20	0.545	0.518	5.189	0.105	0.513	0.032

Source: Agterberg (2012, Table 1)
Experimental values from Matheron (1962, p. 180); Lag distance (h) in m; Experimental values from model of Fig. 6.16; $f(\theta)$ as in text; $\beta(h) = $ Experimental value/$f(\theta)$; $\sigma_h^2 = \beta \times f(\theta)$ is extension variance of 50 cm line segments; Deviation is difference between columns 2 and 6. The small deviations indicate good fit of Matheron's variogram model

As explained in Box 6.5, Whittle (1962) derived another theoretical equation for a space series along a line in 3-D. It differs from the semi-exponential correlogram shown in Fig. 6.14 in that distance (lag) occurs as a linear term of the denominator. Suppose that both sides of the equation are multiplied by distance. Then the term on the left-hand side (called xy in Fig. 6.15a) becomes semi-exponential. However, application of this simple curve-fitting method to the Pulacayo zinc values results in a line that is approximately horizontal suggesting $\alpha = 0$. Consequently, the resulting

6.2 Spatial Series Analysis

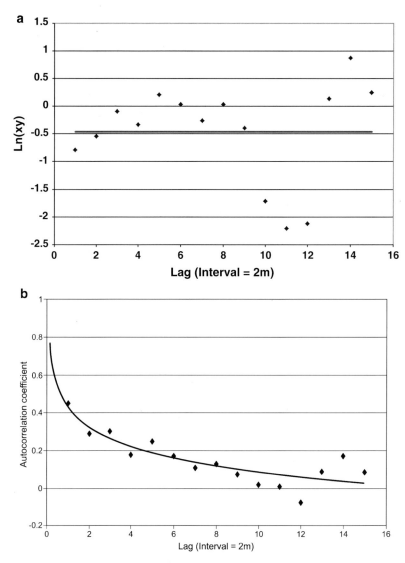

Fig. 6.15 Whittle's 3-D correlogram fitted to Pulacayo zinc values. (**a**) Log *XY* plotted against Lag distance (*X*); (**b**) Horizontal line of Fig. 6.15a in comparison with serial correlation coefficients

theoretical correlogram (Fig. 6.15b) is approximately hyperbolic. It primarily differs from the semi-exponential model of Fig. 6.14 in the vicinity of the origin where the autocovariance approaches ∞. Whittle's model produces a result that resembles Matheron's linear semivariogram model and also two other models to be discussed in Sects. 6.2.5 and 11.6.3, respectively.

6.2.3 Other Applications to Ore Deposits

The question can be asked of how representative the relatively small, historical data set of 118 Pulacayo zinc values is of ore deposits in general. Matheron (1962) used several other mineral deposits exemplifying his extension of the model of the Wijs that resulted in linear variograms. His primary examples were from the Mounana uranium deposit, Gabon, and the Mehengui bauxite deposit, Guyana. These two deposits occur relatively close to the Earth's surface and were explored by means of subvertical boreholes drilled on regular grids. His other examples included the Bou-Kiama, Montbelleux, Laouni, Mpassa, and Brugeaud orebodies. In all these situations, the model of de Wijs proved to be satisfactory. Some of these examples and others also were discussed in geostatistical textbooks including David (1977) and Journel and Huijbregts (1978). Later, however, this type of modeling became de-emphasized, probably because the model of de Wijs does not allow for sills that occur generally and problems associated with working with logarithmically transformed concentration values instead of original data. However, as pointed out by Matheron (1974), lognormality is an issue that must be considered generally. Multifractal modeling (*e.g.*, use of multiplicative cascades, see Chap. 12) confirms the validity of several aspects of Matheron's original approach. The multifractal autocorrelation function of Cheng and Agterberg (1996) has a sill as well as a nugget effect with exceptionally strong autocorrelation over very short distances (*cf.* Sect. 12.2.3).

Agterberg (1965) estimated autocorrelation coefficients for the original de Wijs zinc data and obtained similar results for titanium data from adjoining borehole samples in a magnetite deposit, Los Angeles County, California, originally described by Benson et al. (1962). Figure 6.16a (modified from Agterberg 1974, Fig. 56) shows average autocorrelation coefficients and best-fitting negative exponential function derived from logarithmically transformed element concentration values for copper from the Whalesback copper deposit, Newfoundland, and Fig. 6.16b, c are for two relatively long series of gold assays from the Orange Free State Mine, Witwatersrand goldfields, South Africa (data from Krige et al. 1969). In these three examples, the negative exponential function with significant noise component provides a good fit. In each situation, there is finite variance (existence of sill) and a de Wijsian variogram can only be fitted for the copper and gold examples over relatively short distances (over approximately the first six values from the origin in the three examples of Fig. 6.16).

A typical sample of 1,090 copper concentration values from the Whalesback deposit (*cf.* Agterberg 1974, p. 301) had mean value of 1.57 % Cu and logarithmic variance of 1.21. Converting these values back into copper concentration values using the method of Sect. 3.3 yields $\mu = 0.857$ and $\sigma^2(X) = 43.84$. The positive skewness of the copper concentration is so large that it is not possible to obtain reliable statistics from original data without use of a more efficient estimation method involving logarithmic transformation (Aitchison and Brown 1957; Sichel 1966). The logarithmic variance of the gold values in the other example is approximately 1.03. Krige et al. (1960) did not report the corresponding mean value but the following statistics can be derived from the relatively small data set of 61 gold values in Table 2.6: $\mu(X) = 906.6$; $\sigma^2(X) = 1{,}470{,}410$; $\mu = 6.144$; and $\sigma^2 = 0.929$.

6.2 Spatial Series Analysis

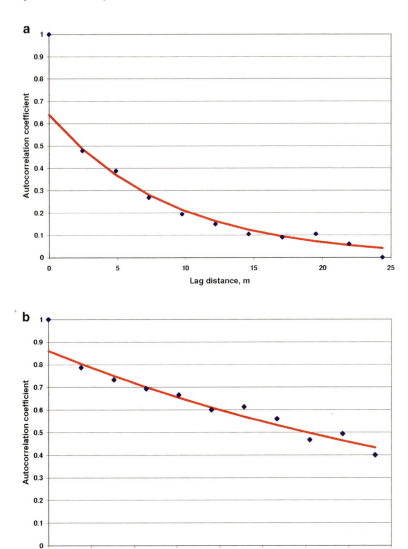

Fig. 6.16 Estimated autocorrelation coefficients and best-fitting negative exponential autocorrelation functions (*curves*) derived from logarithmically transformed element concentration values: (**a**) Average correlograms for 24 copper channel sample series from drifts at various levels of Whalesback copper deposit, Newfoundland (After Agterberg 1974); (**b**) Series of 462 gold assays from the Orange Free State Mine, Witwatersrand goldfields, South Africa (Modified from Krige et al. 1969); (**c**) Other series with 540 values from same gold mine. In each diagram the fitted exponential intersects the vertical axis at a point with autocorrelation coefficient less than 1 indicating existence of "nugget" effect (From Agterberg 2012, Fig. 6)

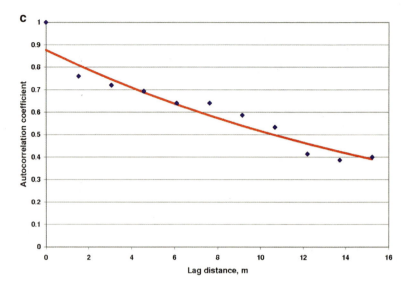

Fig. 6.16 (continued)

This yields new estimates of $\mu(X) = 879.1$; $\sigma^2(X) = 1,183,972$. In this application, the new estimates are probably better than those obtained from the original gold values without use of an appropriate transformation.

The comparison of the Pulacayo zinc example with the Whalesback copper and Witwatersrand gold examples illustrates that there are similarities in that the frequency distributions of channel samples in all three examples are positively skew and approximately lognormal. Also, in all three cases, the autocorrelation function can be approximated by a negative exponential function with value less than unity at the origin indicating existence of a noise component superimposed on the spatial random variable representing more continuous variability at larger distances. Negative exponential autocorrelation functions are closely related to Markov chain analysis and to scaling properties of sequences of mineral grains in igneous rocks. For example, Xu et al. (2007) demonstrated existence of small-scale scaling in "ideal granite" grain sequences previously modeled as Markov chains (Vistelius et al. 1983). Wang et al. (2008) applied multifractal and Markov chain analysis to sphalerite banding at the microscopic scale in the Jinding lead-zinc deposit, Yunnan Province, China.

In Fig. 2.10 for the Pulacayo zinc example, the noise component was filtered out to retain a "signal" with approximately unity autocorrelation function value at the origin (cf. Agterberg 1974). The nugget effect can be modeled as random noise at lag distances greater than 2 m. Existence of a sill is not obvious in the Pulacayo zinc example. However, as originally realized by Matheron (1971), a nugget effect of this type may reflect strong autocorrelation so close to the origin that it cannot be seen in semivariograms or correlograms because its spatial extent is less than the sampling interval used in practice. The frequency distribution of the Pulacayo zinc example has less positive skewness than those of copper and gold in the other examples.

6.2.4 Geometrical Probability Modeling

In principle, Matheron (1962)'s semivariogram $\gamma(h) = 3\alpha \cdot \log_e h$ also can be applied to untransformed data. However, the following applications of geometrical probability are for logarithmically transformed distance.

Returning to the geometry of channel sampling (Fig. 2.9): suppose that AA'BB' represents a rectangle with sides $AA' = BB' = h$, $AB = A'B' = L$ and $\tan\theta = L/h$. If the concentration value for a small volume at a point is taken to be the concentration value of another volume of rock that either contains the small volume or is located elsewhere, this results in uncertainty expressed by means of the "extension variance". In Matheron (1962, Sect. 39) or Agterberg (1974, Sect. 10.11) it is discussed in detail that the variogram value of parallel line segments of length L that are distance h apart along a straight line can be interpreted as an extension variance $\sigma_h^2 = \beta \bullet f(\theta)$ with $\beta = 6A$ and:

$$f(\vartheta) = -\ln\frac{L}{\sqrt{L^2+h^2}} + \frac{2h}{L}\tan^{-1}\frac{L}{h} + \frac{h^2}{L^2}\ln\frac{h}{\sqrt{L^2+h^2}}$$

Table 6.1 shows the first ten Pulacayo Mine variogram values as estimated by Matheron (1962, p. 180) applying this equation to log-transformed (base e) zinc values. For comparison, theoretical variogram values for the exponential model (derived from autocorrelation model graphically shown in Fig. 2.10) are listed as well, illustrating that this model with a sill also provides a good fit. For other theoretical autocorrelation functions fitted to the Pulacayo zinc values, see Sect. 6.2.1 and Chen et al. (2007).

Use of Matheron's original variogram model resulted in multiple estimates of $\beta(h)$ for different lag distances (h) in Table 6.1. A better estimate is obtained by using constrained least squares estimation as follows. The theoretical variogram values in the second last column of Table 6.1 are based on a single estimate ($\beta = 0.0988$) representing the slope of a line of best fit (Fig. 6.17) forced through the point where $f(\theta) = 0$ and $h = 0$. This additional point receives relatively strong weight in the linear regression because it is distant from the cluster of the other ten points used. The constraint can be used because, for decreasing h:

$$\lim_{h\to 0} f(\vartheta) = -\lim_{h\to 0}\left\{\ln\frac{L}{\sqrt{L^2+h^2}}\right\} + \lim_{h\to 0}\left\{\frac{2h}{L}\tan^{-1}\frac{L}{h}\right\} + \lim_{h\to 0}\left\{\frac{h}{\sqrt{L^2+h^2}}\right\} = 0$$

The new estimate of absolute dispersion $A = \beta/6 = 0.0165$ not only produces theoretical variogram values, which are nearly equal to the estimates based on the logarithmically transformed zinc values, it also is nearly equal to $\alpha = 0.015$ previously derived from the logarithmic variance in the previous section, confirming the applicability of Matheron's original method within a neighbourhood extending from about 2 to 400 m.

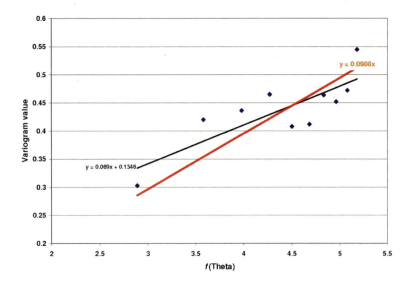

Fig. 6.17 Straight line with equation $y = 0.0988 \cdot x$ fitted by constrained least squares to 10 variogram values taken from Matheron (1962). Horizontal axis is for $f(\theta)$. This line was forced through the point with $f(\theta) = 0$ and $h = 0$. Best-fitting line without this constraint has two coefficients and is significantly different (Source: Agterberg 2012, Fig. 8)

The preceding experiment illustrates (a) different variogram models applied to the same data sets can produce similar estimates of extension variances; and (b) extension variance estimates are too large if there is a "nugget effect" incorporating strong autocorrelation over very short distances. In the remainder of this section it will be attempted to model this type of nugget effect by (a) extrapolation from the original variogram values, (b) multifractal modeling, and (c) spectral analysis. The Pulacayo zinc example will be re-analyzed. Because this series is based on 118 values only, the estimated autocorrelation (or semivariogram) values have limited precision as previously shown by Agterberg (1965, 1967). For this reason, autocorrelation for a very large data set was studied as well. It will be shown (Sect. 6.2.7) that there is a nugget effect in copper concentration values from along the deep KTB borehole with short-distance extent that is similar in consecutive series of 1,000, 1,000 and 796 values, respectively.

6.2.5 Extension Variance

Matheron's geometrical approach can be used for several other purposes. Basic geostatistical theory (Box 6.7) results in equations for the extension variance σ_E^2 for the uncertainty associated with using the element concentration value of a small block as the concentration of a larger block that surrounds it. For example, in

6.2 Spatial Series Analysis

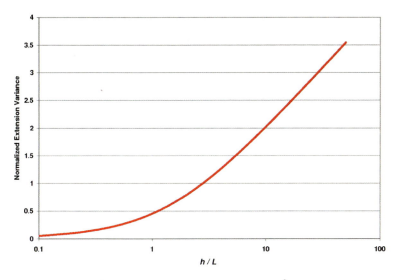

Fig. 6.18 Relationship between normalized extension variance ($\sigma_E{}^2$) and h/L (Source: Agterberg 2012, Fig. 9)

applications of multifractal modeling to the Pulacayo Mine (Cheng and Agterberg 1996; Chen et al. 2007; Lovejoy and Schertzer 2007), it is assumed that the zinc concentration values can be converted into measures of amounts of zinc in adjoining 2-m wide samples along a line parallel to the drift on the 446-level. It implies that every zinc concentration value for a channel sample at a point along this line is taken as representative for a width of 2 m. Associated uncertainty then is given by the extension variance σ_E^2. Figure 6.18 shows that normalized extension variance $\sigma_E^2/3A$ depends on h/L. From our estimate $A = 0.0165$, it follows that $\sigma_E^2 = 0.0622$ for $h = 2$ m wide samples that are $L = 50$ cm long. It probably significantly overestimates true value because absolute dispersion is less than 0.0165 over very short distances due to the nugget effect (see later). If $A < 0.0165$, the normalized extension variance is greater than $\sigma_E^2 = 0.0622$ as derived for the same value of h/L from the curve in Fig. 6.18, that is for $A = 0.0165$.

Box 6.7: Geostatistics of Mean Block Values

Suppose that $f(x)$ represents concentration of a chemical element at a point x, which may be a 3-D vector (x_1, x_2, x_3). The average m for any volume v is $m = \frac{1}{v} \cdot \int_v f(x)dx$. Suppose M is an assemblage average for a large volume V, then $M = \frac{1}{V} \cdot \int_V f(x)dx = \sum m_i$ where the m_i are for smaller volumes v_i. On the basis of V and $f(x)$ the intrinsic functions can be defined for the covariance

(continued)

Box 6.7 (continued)

$C(y)$ and semivariogram $\gamma(y)$ as $C(y) = \frac{1}{V} \cdot \int_V [f(x) - M][f(x) - M]dx$ and $\gamma(y) = \frac{1}{2V} \cdot \int_V [f(x+y) - f(x)]^2 dx$. The covariance of the averages m_1 and m_2 for two volumes v_1 and v_2 can be determined as: $\sigma(m_1, m_2) = \int_V [m_1 - M][m_2 - M]dx = \frac{1}{Vv_1v_2} \int_{v_1} dx_1 \int_{v_2} dx_2 \int_V [f(x+x_1) - M][f(x+x_2) - M]dx$. When y represents distance between all possible pairs of points, it follows that : $\sigma(m_1, m_2) = \frac{1}{v_1v_2} \int_{v_1} dx_1 \int_{v_2} C(y)dx_2$. When v_1 and v_2 coincide it reduces to $\sigma^2(m) = \frac{1}{v^2} \int_v dx_1 \int_v C(y)dx_2$ representing the variance of average concentration m for a block with volume v. By defining an auxiliary function $F(m) = \frac{1}{v^2} \int_v dx_1 \int_v \gamma(y)dx_2$ this variance can be rewritten as $\sigma^2(m) = F(M) - F(m)$. Definition of $G(m_1, m_2) = \frac{1}{v_1v_2} \int_{v_1} dx_1 \int_{v_2} \gamma(y)dx_2$ results in $\sigma(m_1, m_2) = F(M) - G(m_1, m_2)$. Using m_1 of v_1 to estimate m_2 of v_2 results in the "extension" variance $\sigma_E^2 = \sigma^2(m_1 - m_2)$ that can also be written as $\sigma_E^2 = \sigma^2(m_1) - 2\sigma(m_1, m_2) + \sigma^2(m_2)$ or $\sigma_E^2 = 2G(m_1, m_2) - F(m_1) - F(m_2)$.

Matheron (1964) has shown that the average of n adjoining channel sample concentration values has variance equal to σ^2_E/n. This is another important result: In Chap. 11, average values with n equal to 3, 5, 7, and 9 will be used extensively. The extension variance $\sigma_E^2 = 0.0622$ is for logarithmically transformed zinc concentration values. As discussed in Sect. 2.2, it can be assumed that the zinc values (X_i with $i = 1, \ldots, 118$) for the original channel samples systematically underestimate zinc values for the massive sulphide (Fig. 2.9). By setting $\sigma^2 = \sigma_E^2$ and $\mu(X) = X_i$, it is possible to estimate the variances $\sigma^2(X_i)$ of the original zinc values. These variances can then be used to calculate approximate 95 % confidence limits for zinc concentration values of 1.3 m × 2 m plates formed by extending the 1.3 m long channel samples by 1 m on both sides. Table 6.2 shows ±1.96 $\sigma(X_i)$ error bars for 11 original zinc values and for averages of adjacent values for wider plates at the same locations. These sets of overlapping plates, that are 20 m apart, were selected for example so that both low and high zinc concentrations are represented. The error bars in Table 6.2 for plates wider than 2 m are relatively narrow. Uncertainty is greatest for the 1.3 m × 2 m plates but this is probably because $\alpha = 0.0165$ is underestimated over very short distances resulting in error bars that are too wide.

6.2 Spatial Series Analysis

Table 6.2 Zinc concentration values (in %) with 95 % confidence intervals for thin plates in the direction of the mining drift with channel samples at their centers

Plate size	1.3 m × 2 m	1.3 m × 6 m	1.3 m × 10 m	1.3 m × 14 m	1.3 m × 18 m
#10	24.1 ± 12.2	19.9 ± 5.6	19.4 ± 4.2	17.5 ± 3.2	17.0 ± 2.8
#20	15.1 ± 7.7	13.8 ± 3.9	14.0 ± 3.1	13.3 ± 2.4	13.2 ± 2.1
#30	9.5 ± 4.8	12.1 ± 3.4	15.2 ± 3.3	13.2 ± 2.4	14.7 ± 2.4
#40	10.6 ± 5.4	15.6 ± 4.4	17.0 ± 3.7	15.5 ± 2.9	14.2 ± 2.3
#50	27.4 ± 13.9	18.6 ± 5.3	17.4 ± 3.8	17.4 ± 3.2	17.2 ± 2.8
#60	4.7 ± 2.4	9.0 ± 2.5	8.7 ± 1.9	8.1 ± 1.5	9.0 ± 1.5
#70	9.7 ± 4.9	9.2 ± 2.6	10.5 ± 2.3	10.3 ± 1.9	10.2 ± 1.6
#80	10.6 ± 5.4	11.1 ± 3.2	10.8 ± 2.3	9.3 ± 1.7	9.6 ± 1.6
#90	30.8 ± 15.6	31.6 ± 9.0	30.8 ± 6.7	30.7 ± 5.7	29.2 ± 4.7
#100	22.6 ± 11.5	16.4 ± 4.6	18.6 ± 4.1	20.8 ± 3.8	21.4 ± 3.5
#110	7.9 ± 4.0	17.8 ± 5.0	17.2 ± 3.8	15.9 ± 2.9	14.6 ± 2.4

From Agterberg (2012, Table 2)
Results are shown for every 10th value in the original series of 118 values. Error bars for 1.3 m × 2 m are too wide because small-scale spatial correlations are not being considered

The problem of overestimation of extension variances of average element concentration values for small plates due to local strong autocorrelation was previously considered by Matheron (1989, pp. 73–75) as follows. Ten professional geostatisticians were provided with a set of variogram values with unit of lag distance equal to 180 m. Independently the participants in this experiment were asked to (a) fit a variogram, and (b) calculate the corresponding extension variance for a square plate measuring 180 m on a side. Each variogram fitted by a participant had a nugget effect, with, in addition, an exponential or (third-order polynomial) "spherical" variogram curve. The corresponding average of ten estimated extension variances was 0.4019 ± 0.0127 indicating excellent agreement between participants. Next, the same ten people were provided with additional variogram values for shorter unit lag distance interval of 20 m. Again they were asked (a) fit a variogram, and (b) calculate the corresponding extension of the 180 m × 180 m square plate. The variogram models used during the second stage of the experiment were "richer" becoming either: nugget + spherical + spherical, or nugget + exponential + spherical, or nugget + exponential + exponential. A few other answers were given as well. The revised average extension variance became 0.3686 ± 0.0062. Clearly this revised estimate of the extension variance is less than the first estimate and outside the 95 % confidence of the first estimate. Similar results were obtained during a third stage of this experiment using an even shorter unit lag distance. Although Matheron (1989) did not report the equation of the model used to generate the variograms for longer lag distances, this model obviously had no or very small nugget effect that is overestimated by extrapolation toward the origin by means of the standard models.

It is noted that geostatisticians often use the spherical semivariogram with $\gamma(h)/\sigma^2 = 3h/2a - (h/a)^3/2$ for $0 \leq h < a$ where a is a constant called the range; and $\gamma(h) = \sigma^2$ for $h \geq a$. This model also arises in the following situation. Suppose that in 3-D two identical copies of a sphere with radius a and volume equal to

1 initially coincide but one of the two spheres is shifted in the h direction. Then the amount of overlap of overlap of the two spheres then decreases from its maximum value (=1) at zero shift to its minimum (=0) at lag a. The cubic semivariogram function describes 1 minus the amount of overlap. This spherical model often provides a good fit and is relatively easy to use in geometrical extrapolations. In situations of asymmetry as frequently occur in Nature; the sphere can be replaced by an ellipsoid. The spherical-ellipsoidal model of diminishing autocorrelation formalizes the idea of a probable influence cell ("prince"; Agterberg 1965) according to which strong local spatial continuity diminishes with distance until spatial variability becomes totally dominated by other, mainly deterministic, genetic processes.

6.2.6 Short-Distance Nugget Effect Modeling

In Sect. 2.5.1, it was pointed out that there is uncertainty associated with the definition of effective length $L = 0.5$ m of the channel samples in the Pulacayo Mine. This is because these samples were taken across the entire width (=1.30 m) of drift whereas the Tajo vein has (horizontally measured) thickness of 0.50 m on the 446-m level. This thickness value was used by Matheron and earlier in this chapter as a best estimate of L. It has been shown that the choice of $L = 0.5$ results in estimates of A that are satisfactory for lag distances greater than 2 m (up to 400 m). For shorter lag distances, however, it is useful to generalize Matheron's concept of absolute dispersion by defining $A(L)$, which depends on the value of L. Consequently, $A = A(0.5)$ for the applications described in Sect. 6.2.1. Theoretically, the method used to estimate $\beta(0.5) = 6\alpha(0.5)$ in Fig. 6.17 can be used to optimize our choice of L. In Agterberg (2012, Fig. 10) estimates of $A(L)$ are shown that would be obtained for effective channel sample lengths less than 0.5 m. For $L > 3$ cm, $A(L)$ increases slightly from about 0.01 to 0.0165 at $L = 0.5$ m; for $L < 0.03$ m, there is rapid decrease to $A(L) = 0$. Sums of squared deviations from lines of best fit for different values of L. showed that the optimum solution ($\alpha(L) = 0.021$) is obtained at $L = 13$ cm (Agterberg 2012, Fig. 12). The de Wijsian variogram model that best fits the 10 observed values of Table 6.1 is for linear samples that are not only shorter than the channel samples on which zinc concentration was measured ($L = 1.3$ m) but also shorter than the thickness of the Tajo vein ($L = 0.5$ m). This result probably reflects small-scale clustering of the chalcopyrite crystals. In Agterberg (2012) it was tentatively suggested that the very narrow optimum effective vein width may reflect the fact that the Tajo vein was originally formed along a fissure.

It should be kept in mind that the preceding conclusions remain subject to uncertainty because of limited precision of the variogram values of Table 6.1. Also, anisotropy may have played a role because zinc concentration value variability perpendicular to the Tajo vein could well differ from variability parallel to the vein. However, the best explanation is that over short lag distances h (e.g. within the

6.2 Spatial Series Analysis

domain 0.003 m $< h <$ 2 m) there exists a strong nugget effect that is not readily detectable at distances of $h \geq 2$ m. At the microscopic level α would be expected to increase rapidly again, because of measurement errors and the fact that the zinc occurs in sphalerite crystals only (cf. Fig. 2.8). The crystal boundary effect may have become negligibly small because channel sample length greatly exceeded crystal dimensions.

The preceding considerations imply that the negative exponential autocorrelation function previously used (see, e.g., Fig. 2.10) is too simple for short distances ($h <$ 2 m). The true pattern is probably close to that shown in Fig. 6.19, which differs from the earlier model in that strong autocorrelation is assumed to exist over very short distances. It is probably caused by clustering of ore crystals, although at the microscopic scale there remains rapid decorrelation related to measurement errors and crystal shapes. The model of Fig. 6.19 is an example of a nested semi-exponential autocorrelation function as previously used by Matérn (1981) and Serra (1966). The graph in Fig. 6.19a satisfies the equation:

$$\rho(h) = c_1 e^{-a_1 h} + c_2 e^{-a_2 h}$$

The coefficients in the first term are $c_1 = 0.5157$ and $a_1 = 0.1892$ as in Fig. 2.10. The second term represents the strong autocorrelation due to clustering over very short distances. The decorrelation at microscopic scale is represented by a small white noise component (probably augmented with a random measurement error) with variance equal to $c_0 = 0.0208$ as will be determined in Sect. 11.3.1. The coefficient c_2 in the second term on the right side satisfies $c_2 = 1 - c_0 - c_1 = 0.4635$. Because of lack of more detailed information on autocorrelation over very short distances, it is difficult to choose a good value for the coefficient a_2. Choosing $a_2 = 2$ provides a good fit over the entire observed correlogram (Fig. 2.10). It affects extrapolation toward the origin with $h <$ 2 m only. Figure 6.19b shows that the second term on the right side of the preceding equation cannot be detected in the correlogram for sampling intervals greater than 2 m. Other types of evidence for existence of strong autocorrelation over very short distances in the Pulacayo orebody have already been discussed (e.g., Fig. 6.15) and will be presented later. In the Sect. 11.6.3, a multifractal autocorrelation function will be derived on the basis of self-similarity assumptions. It results in a curve that resembles the curve in Fig. 6.19 (with $a_2 = 2$). For example, for lag distance equal to 60 cm, the theoretical value according to Fig. 6.19a is 0.6, while the curve in Sect. 11.6.3 yields 0.7.

Independent evidence that a model similar to the one shown in Fig. 6.19 also applies to copper in the Whalesback deposit (see single exponential fit in Fig. 6.16a) is as follows. Figure 6.16a represents an average correlogram based of 24 series of channel samples taken at 8 ft. intervals perpendicular to drifts on various levels of the Whalesback Mine. Agterberg (1966) had obtained separate results for a series of 111 channel samples along a single drift on the 425-ft. level of this mine. These 8-ft. long channel samples had been divided into 4-ft. long halves that were separately analyzed. The correlation coefficient for copper in the two halves amounted to 0.80

222　　　　　　　　　　　　　　　　　　　　　　　6 Autocorrelation and Geostatistics

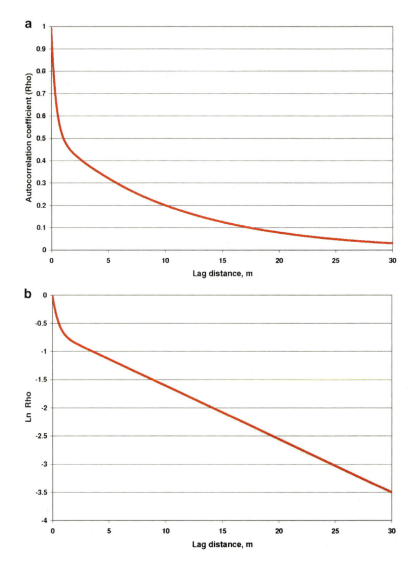

Fig. 6.19 Hypothetical autocorrelation function consisting of two negative exponentials to incorporate nugget effect over short lag distances h with $h < 2$ m. Autocorrelation function for nugget effect is superimposed on negative exponential curve for distances $h \geq 2$ m. (**a**) Vertical scale is linear; (**b**) Vertical scale is logarithmic (Source: Agterberg 2012, Fig. 12)

and this is significantly larger than 0.56 predicted by the negative exponential curve shown in Fig. 6.16a. Although this result is based on relatively few data and the two half-samples were adjoining, it strongly suggests that autocorrelation close to the origin is not adequately described by the curve shown in Fig. 6.16a.

6.2 Spatial Series Analysis

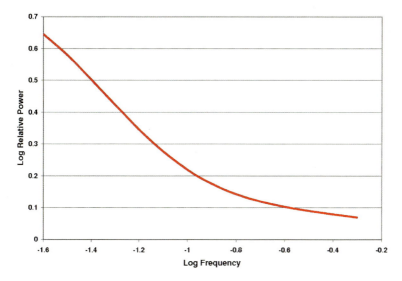

Fig. 6.20 Relative power spectrum for autocorrelation function of Fig. 6.19. Decrease in slope at higher frequency side is caused by the superimposed nugget effect (Source: Agterberg 2012, Fig. 24)

6.2.7 Spectral Analysis: Pulacayo Mine Example

The normalized power spectrum corresponding to the nested design of Fig. 6.19 is:

$$P(f) = \sum_{i=1}^{2}\left[1 - c_i + \frac{c_i/\pi f_{ci}}{1 + (f/f_{ci})^2}\right]$$

where $f_{ci} = a_i/2\pi$. A log-log plot of this spectrum is shown in Fig. 6.20 adopting the coefficients previously used for the nested design plotted in Fig. 6.19. The curve in Fig. 6.20 is approximately a straight line for lower frequencies but for high frequencies there is a marked decrease of slope reflecting the nugget effect.

Box 6.8: Periodograms

If n is even and the origin is chosen in the center of a series y_k ($k=1, 2, \ldots, n$), it can be written as: $y_k = a_0 + 2\sum_{i=1}^{\frac{1}{2}n-1}\left[a_i \cos\left(\frac{2\pi i k}{n}\right) + b_i \sin\left(\frac{2\pi i k}{n}\right)\right] + a_{\frac{1}{2}n}$ $\cos\pi k$; its amplitude and phase representation is: $y_k = R_0 + 2\sum_{i=1}^{\frac{1}{2}n-1}\left[R_i \cos\left(\frac{2\pi i k}{n} + \varphi_i\right)\right] + R_{\frac{1}{2}n}\cos\pi k$ where $R_i = \sqrt{a_i^2 + b_i^2}$; $\varphi_i = \arctan\left(\frac{b_i}{a_i}\right)$. Also:

(continued)

Box 6.8 (continued)

$$y_k = \sum_{i=\frac{1}{2}n}^{\frac{1}{2}n-1} S_i \exp\left(\frac{2\pi I k}{n}\right); \quad S_i = \frac{1}{n}\sum_{i=\frac{1}{2}n}^{\frac{1}{2}n-1} y_i \exp\left(-\frac{2\pi I k}{n}\right) \text{ where } I^2 = 1.$$ The periodogram is a plot of $P_i = R_i^2$ against i. P_i is distributed as χ^2 with 2 degrees of freedom (see e.g. Jenkins 1961). The average of q consecutive periodogram values provides an estimate of the power spectrum $E(P_i)$ that is distributed as χ^2 with approximately $2q$ degrees of freedom. This method to construct power spectra commonly makes use of the Fast Fourier Transform (FFT) method introduced by Cooley and Tukey (1965). The power spectrum of a continuous series is: $P_f = s^2(x) \int_{-\infty}^{\infty} r_h \cos 2\pi f h \, dh$ where r_h is the autocorrelation function. For example, if $r_h = c \cdot \exp(-a|h|)$, $P(f) = (1-c)s^2(x) + \dfrac{cs^2(x)/\pi f_c}{1 + (f/f_c)^2}$.

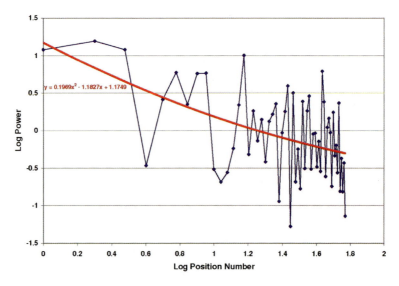

Fig. 6.21 Periodogram of 118 zinc values with quadratic curve fitted by least squares (Logarithms base 10). The flattening of the curve toward higher frequencies is believed to be due to the nugget effect (Source: Agterberg 2012, Fig. 25)

Figure 6.21 shows the power spectrum $E(P_i)$ estimated by averaging pairs of consecutive values of the periodogram for the 118 zinc values together with a quadratic curve fitted by least squares. A best-fitting straight line for the same values results in $\beta = 0.72$, but by means of an F-test it can be shown that the quadratic fit of Fig. 6.21 is significantly better than the linear fit (for level of significance $\alpha = 0.01$). The slope of the curve at the origin in Fig. 6.21 gives $\beta = 1.18$ with gradually decrease to 0.49 at maximum log wave number on the right. A log-log plot of the two-point moving average of the periodogram of Fig. 6.21 produces a pattern that is close to Lovejoy and Schertzer's (2007, Fig. 3b) spectrum for the de Wijs data. This topic will be discussed in more detail later (Fig. 12.35). A straight line fit to the first 20 points of this two-point moving average gives $\beta = 1.03$, which is close to $\beta = 1.18$ at the origin of Fig. 6.21 and close to $\beta \approx 1.12$. A possible explanation is that spectral analysis confirms validity of the Lovejoy-Schertzer universal multifractal model (Sect. 12.7) but with superimposed noise that tends to flatten the spectrum at higher frequencies.

For comparison, the preceding method also was applied in Agterberg (2012) to a sequence of 132 titanium concentration values from the Black Cargo titaniferous magnetite deposit, Los Angeles County, California (Benson et al. 1962). This sequence, previously analyzed in Agterberg (1965), is a composite of four sub-sequences obtained from four different boreholes. All samples were 5 ft. in length except for three 10 ft. samples at the subsequence meeting points. Mean and standard deviation of the 132 numbers are 2.73 and 1.65 % TiO_2, respectively. The resulting periodogram (Agterberg 2012, Fig. 26) is similar to Fig. 6.21 in that the best-fitting quadratic trend line has a slope that decreases toward higher frequencies. At the origin ($x = 0$) its value is -1.088 and at maximum frequency ($x = 1.8195$) the slope is -0.6186. Other results for this example also were similar to those obtained for the 118 Pulacayo zinc values.

The curves in Figs. 6.20 and 6.21 indicate (1) the log-log plots of the three power spectra are not straight lines but curves with slopes that decrease toward higher frequencies; and (2) at their maximum frequency or highest position number the curves are probably not horizontal indicating that the nugget effect is not white noise with Dirac delta autocorrelation function. The sampling intervals of two data sets used for example in this section are too wide to allow a better description of the effect of the nugget effect on the power spectra.

6.2.8 KTB Copper Example

The second example of detection of nugget effect is for a long series consisting of 2,796 copper (XRF) concentration values for cutting samples taken at 2 m intervals along the Main KTB borehole of the German Continental deep Drilling Program (abbreviated to KTB). These data are in the public domain (citation: KTB, WG Geochemistry). Depths of first and last cuttings used for this series are 8 and 5,596 m, respectively. Locally, in the database, results are reported for a 1-m sampling interval; then, alternate copper values at the standard 2 m interval were included in the series used for example. Most values are shown in Fig. 6.22 together with a 101-point average representing consecutive 202-m long segments of drill-core. The data set was divided into three series (1, 2 and 3) with 1,000, 1,000 and 796 values, respectively. Mean copper values for these three series are 37.8, 33.7 and 39.9 ppm Cu, and corresponding standard deviations are 20.3, 11.0 and 20.6 ppm Cu, respectively. Figure 6.23 shows correlograms of the three series. Each series shows a nugget effect that, for series 2 and 3, is accompanied by a

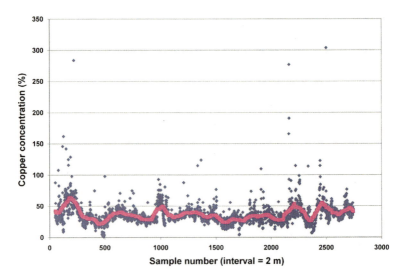

Fig. 6.22 Copper concentration (ppm) values from Main KTB bore-hole together with mean values for 101 m long segments of drill-core. Locally, the original data deviate strongly from the moving average (Source: Agterberg 2012, Fig. 15)

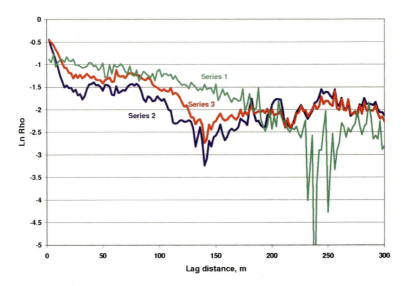

Fig. 6.23 Correlograms for three consecutive series of copper concentration values from Main KTB bore-hole. Series 2 (for depths between 2 and 4 km) and Series 3 (for depths between 4 and 5.54 km) show similar autocorrelations that differ from autocorrelation function for Series 1 (for depths between 0.05 and 2 km) (Source: Agterberg 2012, Fig. 16)

6.2 Spatial Series Analysis

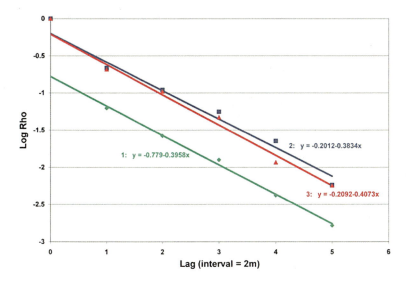

Fig. 6.24 Correlograms (first five lag distances only) for three consecutive series of differences between original copper concentration values and mean values shown in Fig. 6.22. Results are for same series as used for Fig. 6.23. Best-fitting semi-exponentials were obtained by ordinary least squares (Logarithms base 10). The slopes of the three best-fitting straight lines are nearly equal. This indicates existence of nugget effect with same spatial extent along entire KTB borehole (Source: Agterberg 2012, Fig. 17)

relatively steeply increasing curve near the origin. Because the autocorrelation coefficients are logarithmically transformed, random fluctuations for near-zero autocorrelation values are amplified. It is noted, however, that all three series only had positive autocorrelations for the first 150 lag distances. Also, the patterns for series 2 and 3 are strikingly similar.

It can be expected that series of element concentrations over a vertical distance of about 5.5 km will exhibit deterministic trends reflecting systematic changes in rock compositions. It is assumed here that these trends are largely captured by the moving average curve of Fig. 6.22. Figure 6.24 shows autocorrelation coefficients for the three series after subtracting the trend values from the original data. All three series of deviations have autocorrelation functions that are approximately negative exponential in shape over distances less than 10 m. Each can be regarded as representing a nugget effect with equation $\rho_h = c \cdot \exp(-ah)$. The slope coefficients (a) of the three curves are nearly equal to one another (0.40, 0.38 and 0.41 for series 1, 2 and 3, respectively). The spatial extent of this nugget effect is much less than the small scale binary lithology variation for the same borehole discussed in the previous two sections. It is interesting that the parameter (a) that determines the spatial extent of the nugget effect remains the same over a vertical distance of nearly 6 km. The corresponding variance components (c) of the copper nugget effect are 0.46, 0.82 and 0.81, indicating that the white noise component is relatively strong for series 1.

Quantitative modeling of the nugget effect in KTB copper determinations has yielded better results than could be obtained for our examples from mineral deposits including the Pulacayo Mine. This is not only because the series of chemical determinations is much longer but also because the nugget effect remains clearly visible over lag distances between 2 m (= original sampling interval) and 10 m.

6.3 Autocorrelation of Discrete Data

In order to visualize theoretical autocorrelation of discrete data, the example of random succession of lithologies in Sect. 2.2.5 can be considered again. For a point in a specific state (e.g., shale), there is a fixed probability that the next point along the section will be in the same state or that another state will occur. Suppose that instead of using letters to indicate the states, presence of a given state is coded as +1 and its absence as −1. The result is a series of discrete data that can be plotted as in Fig. 6.25. The main characteristic of this situation is that a transition from +1 to −1 or from −1 to +1 can occur any time with a probability that is independent of place of occurrence along the series.

Assume that Fig. 6.25 represents a process that started at T = −∞. For any discrete point k, the probabilities of being in state +1 or −1, are equal to constants adding to one. For simplicity, it is assumed that $P(X_k = 1) = P(X_k = -1) = \frac{1}{2}$. Consequently, $E(X_k) = 0$. This series is stationary and has expectation equal to zero. The autocovariance function satisfies: $\Gamma_h = E(X_k \cdot X_{k+h})$. This is the sum of two terms: (1) the probability that the number of changes in state is even during the interval [k,k + 1] multiplied by +1; and (2) probability of an odd number of changes multiplied by −1, or:

$\Gamma_h = \sum_{k=0}^{\infty} (P_{2k} - P_{2k+1}) = e^{-2\lambda|h|}$ where $P_k = \frac{e^{-\lambda|h|}(\lambda|h|)^k}{k!}$ denotes the Poisson-type probability of having exactly k changes in an interval of length h. The parameter λ can be interpreted as the number of changes (both from +1 to −1 and −1 to +1) expected per unit of time or distance along the axis when $|h| = 1$. The absolute value of lag h indicates that proceeding in the positive or negative

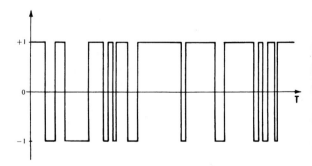

Fig. 6.25 Random telegraph signal (After Jenkins and Watts 1968). Zero crossings between two states +1 and −1 occur at random along the line (Source: Agterberg 1974, Fig. 57)

6.3.1 KTB Geophysical Data Example

The following geophysical example of spatial covariance modeling of velocity and lithology logs provides another example of application of the model of Fig. 6.25. An autocovariance function consisting of two superimposed negative exponentials with different scaling constants originally was obtained by Goff and Holliger (1999) for binary lithology values derived from velocity and lithology logs for the main borehole of the German Continental deep Drilling Program (KTB). In Fig. 6.19, $a_1 = 0.1892$ for larger scale variability and $a_2 = 2$ was assumed for the nugget effect. In Goff and Holliger's Fig. 7, $a_1 = 0.001$ for the "large scale" and $a_2 = 0.019$ for the "small scale" model. The dimensionless ratio a_2/a_1 for KTB binary lithology is 19 and somewhat greater than the ratio of 11 in Fig. 6.19. Lithology in the main KTB borehole was determined at points that are 1 m apart over a length of about 7 km. In general, significant pre-processing is required for the analysis of long series of this type. Goff and Hollinger (1999) commenced this process by plotting raw compressional velocity (V_p) averaged within more or less homogeneous lithological sections against depth. A deterministic component derived from this plot was extracted for the purpose of detrending followed by conversion of the lithology log into a binary residual V_p profile for which the spatial covariance in (km/s)^2 was estimated. The two rock types retained in the binary plot are mainly metabasite ($V_p = +0.2$ km/s) and mainly gneiss ($V_p = -0.2$ km/s).

The von Kármán autocovariance model has been used extensively by geophysicists to characterize crustal heterogeneity properties not only for velocity log properties (e.g., Wu and Aki 1985; Wu et al. 1994; Goff and Hollinger 1999, 2003) but also for geological maps of crustal exposures (e.g., Goff et al. 1994; Goff and Levander 1996), seafloor morphology (Goff and Jordan 1988), and in field simulations (Goff and Jennings 1999). This model was proposed by von Kármán (1948) (but independently by Matérn, in 1948; *cf.* Box 6.6) and can be written as:

$$\rho(h) = \frac{(ah)^\nu K_\nu(ah)}{2^{\nu-1}\Gamma(\nu)}$$

where ν is the Hurst number (*cf.*, Mandelbrot 1983; Chemingui 2001; Klimeš 2002), and K_ν is the modified Bessel function of order ν. Fitting of the two-parameter von Kármán model to an estimated covariance function can be performed using the inversion methodology of Goff and Jordan (1988). If $\nu = 0.5$, the preceding equation reduces to $\rho(h) = \exp(-ah)$. Goff and Hollinger's (1999) best von Kármán model fit for the KTB binary residual V_p profile has $\nu = 0.21$ and $a = 0.00072$. However, a better fit for the autocovariance of this series was obtained by these authors using the nested semi-exponential design model with $c_0 = 0$,

$c_1 = 0.684$, $c_2 = 0.316$, $a_1 = 0.001$ and $a_2 = 0.019$. The Hurst numbers for both negative exponentials are equal to 0.5, more than twice the Hurst number of best fit using the von Kármán equation.

Because the series considered in the preceding paragraphs is binary, it is possible to interpret the scaling constants a_i ($i = 1, 2$) along the lines presented at the beginning of this section. Suppose the two binary states along the borehole are written as +1 and −1. If the mean can be set equal to zero, the autocorrelation $\rho(h)$ is equal to the sum of the probability that number of state changes over the interval h is even minus the sum of the probability that it is odd. If P_k represents the Poisson-type probability that there exist k state changes over h:

$$\rho(h) = \sum_{k=0}^{\infty} (P_{2k} - P_{2k-1}) \, \rho(h) = e^{-2\lambda h}; \quad P_k = e^{-\lambda h} \frac{(\lambda h)^k}{k!}$$

where λ is number of state changes per unit of distance. A similar result is obtained when the mean is not equal to zero.

For the Goff-Hollinger KTB example, the fact that there are two separate negative exponentials illustrates that, over short distances, there are rapid lithology changes or a "nugget effect" for $i = 2$, but changes at larger scale are controlled by the other negative exponential ($i = 1$) function. Thus alternation between mostly metabasite and mostly felsic gneisses in KTB is subject to two separate random processes. The alternation either has high or low frequency with probabilities controlled by the c_i ($i = 1, 2$) coefficients. This type of modeling only applies to the binary residual V_p profile for KTB. For example, Marsan and Bean (1999, 2003) have demonstrated that the KTB sonic log can be modeled using a multifractal approach. Also, Goff and Hollinger (2003) have developed a generic model for the so-called $1/f$ nature of seismic velocity fluctuations. In that paper, these authors modeled the autocovariance function of KTB depth-detrended sonic log through the superposition of four von Kármán autocovariances using negative exponentials with Hurst numbers $\nu = 0.5$ for large, medium, and intermediate scales but $\nu = 0.99$ for the small scale.

References

Agterberg FP (1965) The technique of serial correlation applied to continuous series of element concentration values in homogeneous rocks. J Geol 73:142–162

Agterberg FP (1966) Trend surfaces with autocorrelated residuals. In: Proceedings of the symposium on computers and operations research in the mineral industries. Pennsylvania State University, State College, Pennsylvania, pp 764–785

Agterberg FP (1967) Mathematical models in ore evaluation. J Can Oper Res Soc 5:144–158

Agterberg FP (1968) Application of trend analysis in the evaluation of the Whalesback Mine, Newfoundland. Can Inst Min Metall 9:77–88

Agterberg FP (1974) Geomathematics. Elsevier, Amsterdam

Agterberg FP (1994) Fractals, multifractals, and change of support. In: Dimitrakopoulos R (ed) Geostatistics for the next century. Kluwer, Dordrecht

References

Agterberg FP (2012) Sampling and analysis of element concentration distribution in rock units and orebodies. Nonlinear Process Geophys 19:23–44

Agterberg FP, Banerjee I (1969) Stochastic model for the deposition of varves in glacial Lake Barlow-Ojibway, Ont., Canada. Can J Earth Sci 6:625–652

Aitchison J, Brown JAC (1957) The lognormal distribution. Cambridge University Press, Cambridge

Amos DE, Koopmans LH (1963) Tables of the distribution of the coefficient of coherence for stationary bivariate Gaussian processes, Sandia Corporation monograph SCR-483. Sandia Corporation, Albuquerque

Anderson RY (1967) Sedimentary laminations in time-series analysis. Kans Geol Surv Comput Contrib 18:68–72

Anderson RY, Koopmans LH (1963) Harmonic analysis of varve time series. J Geophys Res 68:877–893

Antevs E (1925) Retreat of the last ice-sheet in eastern Canada. Geol Surv Can Mem 146:1–142, Ottawa

Benson WT, Engel AL, Heinen HJ (1962) Titaniferous magnetite deposits, Los Angeles County, Calif, Report of investigations (United States. Bureau of Mines), no. 5962. Bureau of Mines, Washington, DC

Berry PAM (1987) Periodicities in the sunspot cycle. Vistas Astron 30(2):87–108

Blackman RB, Tukey JW (1959) The measurement of power spectra. Dover, New York

Bloomfield P (2000) Fourier analysis of time series. Wiley, New York

Brier GW (1968) Long-range prediction of the zonal westerlies and some problems in data analysis. Rev Geophys 6:525–551

Brillinger DR (1981) Time series: data analysis and theory, expanded edn. Holt, Rinehart and Winston, New York

Chemingui N (2001) Modeling 3-D anisotropic fractal media. Stanford Exploration Project, Report 80, May 15, pp 1–13

Chen Z, Cheng Q, Chen J, Xie S (2007) A novel iterative approach for mapping local singularities from geochemical data. Nonlinear Process Geophys 14:317–324

Cheng Q, Agterberg FP (1996) Multifractal modeling and spatial statistics. Math Geol 28:1–16

Cooley JW, Tukey JW (1965) An algorithm for the machine calculation of complex Fourier series. Math Comput 19:297–307

Cressie NAC (1991) Statistics for spatial data. Wiley, New York

David M (1977) Geostatistical ore reserve estimation. Elsevier, Amsterdam

Deming WE (1948) Statistical adjustment of data. Wiley, New York

Fowler A (2011) Mathematical geoscience. Springer, London

Goff JA, Hollinger K (1999) Nature and origin of upper crustal seismic velocity fluctuations and associated scaling properties; combined stochastic analyses of KTB velocity and lithology. J Geophys Res 104:13169–13182

Goff JA, Hollinger K (2003) A generic model for the $1/f$-nature of seismic velocity fluctuations and associated scaling properties. In: Goff JA, Hollinger K (eds) Non-linear processes in geophysics. Kluwer Academic, New York, pp 131–154

Goff JA, Jennings TW (1999) Improvement of Fourier-based unconditional and conditional simulations for band limited fractal (von Kármán) statistical models. Math Geol 31:627–649

Goff JA, Jordan TH (1988) Stochastic modeling of seafloor morphology: inversion of sea beam data for second-order statistics. J Geophys Res 93:13589–13608

Goff JA, Levander A (1996) Incorporating "sinuous connectivity" into stochastic models of crustal heterogeneity: examples from the Lewisian gneiss complex, Scotland, the Franciscan Formation, San Francisco, and the Halifax gneiss complex, Egypt. J Geophys Res 101:8489–8501

Goff JA, Hollinger K, Levander A (1994) Model fields: a new approach for characterization of random seismic velocity heterogeneity. Geophys Res Lett 21:493–496

Goodman NR (1957) On the joint estimation of the spectra, cospectrrum and quadrature spectrum of a two-dimensional stationary Gaussian process. GTS-US Commerce Publication, Washington DC

Goovaerts P (1997) Geostatistics for natural resources evaluation. Oxford University Press, New York

Granger CWJ, Hatanaka M (1964) Spectral analysis of economic time series. Princeton University Press, Princeton

Grenander U, Rosenblatt M (1957) Statistical analysis of stationary time series. Wiley, New York

Hamon BV, Hannan EJ (1963) Estimating relations between time series. J Geophys Res 68:6033–6041

Heien DM (1968) A note on log-linear egression. J Am Stat Assoc 63:1034–1038

Hubbard BB (1996) The world according to wavelets. The story of a mathematical technique in the making. Peters, Wellesley

Hughes OL (1955) Surficial geology of Smooth Rock and Iroquois Falls map area, Cochrane District, Ontario. Unpublished PhD thesis, University of Kansas, Lawrence

Isaaks EH, Srivastava RM (1989) An introduction to applied geostatistics. Oxford University Press, New York

Jenkins GM (1961) General considerations in the analysis of spectra. Technometrics 3:133–166

Jenkins GM, Watts DG (1968) Spectral analysis and its applications. Holden-Day, San Francisco

Journel AG, Huijbregts CJ (1978) Mining geostatistics. Academic, New York

Jowett GH (1955) Least-squares regression analysis for trend reduced time series. J R Stat Soc Ser B 17:91–104

Kendall MG, Stuart A (1958) The advanced theory of statistics, 3rd edn. Griffin, London

Klimeš L (2002) Correlations functions in porous media. Pure Appl Geophys 159:1811–1831

Koopmans LH (1967) A comparison of coherence and correlation as measures of association for time or spatially indexed data. Kans Geol Surv Comput Contrib 18:1–4

Krige DG, Watson MI, Oberholzer WJ, du Toit SR (1969) The use of contour surfaces as predictive models for ore values. In: Proceedings of the 8th symposium on application of computers and operations research in the mineral industry. American Institute of Mining and Metallurg, Baltimore, pp 127–161

Kuenen PH (1951) Mechanics of varve formation and the action of turbidity currents. Geol Före Förh 73:69–84

Llibroutry LA (1969) Contribution à la théorie des ondes glaciaires. Can J Earth Sci 6:943–954

Lovejoy S, Schertzer D (2007) Scaling and multifractal fields in the solid Earth and topography. Nonlinear Process Geophys 14:465–502

Marsan D, Bean C (1999) Multiscaling nature of sonic velocities and lithologies in the upper crystalline crust: evidence from the KTB main borehole. Geophys Res Lett 26:275–278

Marsan D, Bean C (2003) Multifractal modeling and analyses of crustal heterogeneity. In: Goff JA, Hollinger K (eds) Non-linear processes in geophysics. Kluwer Academic, New York, pp 207–236

Matérn B (1981) Spatial variation, 2nd edn. Springer, Berlin

Matheron G (1962) Traité de géostatistique appliquée, Mémoires du Bureau de recherches géologiques et minières 14. Éditions Technip, Paris

Matheron G (1971) The theory of regionalized random variables and its applications. Cah Cent Morph Math 5:1–211

Matheron G (1974) Effet proportionnel et lognormalité: Le retour du serpent de mer. Note Géostat. 124, Fontainebleau

Matheron G (1989) Estimating and choosing, an essay on probability in practice (trans: Hasover AM). Springer, Heidelberg

Parzen E (1962) Stochastic processes. Holden-Day, San Francisco

Prokoph A, Agterberg FP (1999) Detection of sedimentary cyclicity and stratigraphic completeness by wavelet analysis: application to Late Albian cyclostratigraphy of the Western Canada Sedimentary Basin. J Sediment Res 69:862–875

Prokoph A, Bilali HE (2008) Cross-wavelet analysis: a tool for detection of relationships between paleoclimate proxy records. In: Bonham-Carter G, Cheng Q (eds) Progress in geomathematics. Springer, Heidelberg, pp 477–489

Schove DJ (1972) A varve teleconnection project. Proc VIIIe Congr INQUA, Paris 1969, pp 928–935

Schove DJ (1983) Sunspot cycles. Hutchison Ross, Stroudsburg

Schwarzacher W (1967) Some experiments to simulate the Pennsylvanian rock sequence of Kansas. Kans Geol Surv Comput Contrib 18:5–14

Serra J (1966) Remarques sur une lame mince de minerai lorrain. Bull BRGM 6:1–6

Serra J (2012) Revisiting 'estimating and choosing'. In: Pardo-Igúzquiza E, Guardiola-Albert C, Heredia J, Moreno-Merino L, Durán JJ, Vargas-Guzmán JA (eds) Mathematics of planet earth. Springer, Heidelberg, pp 57–60

Sichel HS (1966) The estimation of means and associated confidence limits for small samples from lognormal populations. In: Proceedings of the symposium on mathematical statistics and computer applications in ore valuation. South African Institute of Mining and Metallurgy, Johannesburg, pp 106–122

Sommerfeld A (1949) Vorlesungen über theoretische Physik, B 1. Dieterich'sche Verlagbuchhandlung, Wiesbaden

Vistelius AB, Agterberg FP, Divi SR, Hogarth DD (1983) A stochastic model for the crystallization and textural analysis of a fine grained stock near Meech Lake, Quebec. Geological Survey Paper 81-21

Von Kármán T (1948) Progress in the statistical theory of turbulence. J Mar Sci 7:252–264

Wagreich M, Hohenegger J, Heuhuber S (2012) Nannofossil biostratgraphy, strontium and carbon isotope stratigraphy, cyclostratigraphy and an astronomically calibrated duration of the Late Campanian. Cretac Res 38:80–96

Wang Z, Cheng Q, Xu D, Dong Y (2008) Fractal modeling of sphalerite banding in Jingling Pb-Zn deposit, Yunnan, southwestern China. J China Univ Geosci 19(1):77–84

Weertman J (1969) Water lubrication of glacier surges. Can J Earth Sci 6:929–942

Whittle P (1962) Topographic correlation, power-law covariance functions and diffusion. Biometrika 49:305–314

Wiener N (1933) The Fourier integral and certain of its applications. Dover, New York

Wu S-S, Aki K (1985) The fractal nature of heterogeneities in the lithosphere evidenced from seismic wave scattering. Pure Appl Geophys 123:805–818

Wu S-S, Xu Z, Li X-P (1994) Heterogeneity spectrum and scale anisotropy in the upper crust revealed by the German continental deep drilling (KTB) holes. Geophys Res Lett 21:911–914

Xu D, Cheng Q, Agterberg FP (2007) Scaling property of ideal granitic sequences. Nonlinear Process Geophys 14:237–246

Yaglom AM (1962) An introduction to the theory of stationary random functions. Prentice-Hall, Englewood Cliffs

Chapter 7
2-D and 3-D Trend Analysis

Abstract One of the early applications of the general linear model is trend surface analysis (Krumbein and Graybill, An introduction to statistical models in geology. McGraw-Hill, New York, 1965). In the late 1960s, this technique was competing with universal kriging originally developed by Huijbregts and Matheron (Can Inst Min Metall 12:159–169, 1971). To-day, both techniques remain in use for describing spatial trends or "drifts" in variables with a mean that changes systematically in two- or three-dimensional space. Simple moving averaging as practiced by Krige (Two-dimensional weighted moving average trend surfaces for ore valuation. In: Proceedings of the symposium on mathematical statistics and computer applications in ore valuation, Johannesburg, pp 13–38, 1966) or inverse distance weighting methods can be equally effective when there are many observations.

Trend surface analysis was one of the first computer-based methods widely applied in geophysics, stratigraphy and physical geography in the 1960s. Initially, it was assumed that the residuals from a best-fitting trend surface should be independent and normally distributed but Watson (J Int Assoc Math Geol 3:215–226, 1971) clarified that polynomial trend surfaces are unbiased if the residuals satisfy a stationary random process model. Examples of 2-D trend surface analysis include variations in mineral composition in the Mount Albert Peridotite Intrusion, eastern Quebec. A 3-D extension of the method applied to specific gravity data shows that, geometrically, serpentinization of this peridotite body occurred along a northward dipping inverted pyramid. 2-D and 3-D polynomial trends of copper in the Whalesback Deposit, Newfoundland, illustrate how numbers of degrees of freedom are affected by autocorrelation of residuals in statistical significance tests. A useful approach to regional variability of variables subject to both deterministic regional trends and local variability that can be characterized by stationary variability of residuals is to extract the trend by polynomial-fitting and to subject the residuals from the trend to ordinary kriging using 2-D autocorrelation functions. This approach is illustrated by application to (1) depths of the top of the Arbuckle Formation in Kansas, (2) the bottom of the Matinenda Formation in the Elliott Lake area, central Ontario, and (3) variability of sulphur in coal, Harbour seam, Lingan Mine, Cape Breton, Nova Scotia. Use of double Fourier series expansions instead of polynomials to describe regional trends is illustrated on copper in exploratory drill-holes originally drilled from the surface into the Whalesback deposit. Use of 2-D harmonic analysis is also illustrated by application to density of gold and copper deposits in the Abitibi area on the Canadian Shield, east-central Ontario. An advantage of using double Fourier series instead of ordinary polynomials in trend surface analysis is that many geological features are to some extent characterized by similarity of patterns along equidistant straight lines. Periodicities of this type are accentuated by harmonic analysis.

Keywords Trend surface analysis • 2-D and 3-D polynomials • Kriging • Harmonic trend analysis • Arbuckle formation • Mount Albert peridotite intrusion • Whalesback copper deposit • Matinenda uranium deposits • Lingan mine sulphur in coal • Harmonic trend analysis

7.1 2-D and 3-D Polynomial Trend Analysis

Trend analysis is a relatively simple technique that is useful when (1) the trend and the residuals (observed values – trend values) can be interpreted from a spatial geoscience point of view, and (2) the number of observations is not very large so that interpolations and extrapolations must be based on relatively few data. In practical applications, care should be taken not to rely too heavily on statistical significance tests to decide of polynomial degree (analysis of variance) or on 95 % confidence belts that can be calculated for trend surfaces. These statistical tests would produce exact results only if the residuals are uncorrelated (Krumbein and Graybill 1965). If the residuals themselves show systematic patterns of variation, they are clearly not uncorrelated. Then the test statistics would be severely biased. However, often the residuals can be modeled as a zero-mean, weakly stationary random variable. In that situation the trend surface (or hypersurface in 3-D) is unbiased (Agterberg 1964; Watson 1971). Trend surfaces are fitted to data by using the method of least squares that is also used in multiple regression analysis. Consequently, it is also an application of the general linear model.

Suppose that the geographic co-ordinates of a point on a map are written as u and v for the east-west and north-south directions, respectively (in 3-D applications, w is added for the vertical direction). An observation at a point k can be written as $Y_k = Y(u_k, v_k)$ to indicate that it represents a specific value assumed by the variable $Y(u, v)$. One can write: $Y_k = T_k + R_k$ where $T_k = T(u_k, v_k)$ represents the trend and $R_k = R(u_k, v_k)$ the residual at point k. T_k is a specific value of the variable $T(u, v)$ with: $T(u, v) = b_{00} + b_{10}u + b_{01}v + b_{20}u^2 + b_{11}uv + b_{02}v^2 + \ldots + b_{pq}u^p v^q$. In general, u and v form a rectangular co-ordinate system. However, latitudes and longitudes also have been used (Vistelius and Hurst 1964).

In specific applications, $p + q \leq r$ where r denotes the degree of the trend surface. Depending on the value of r, a trend surface is called linear ($r = 1$), quadratic ($r = 2$), cubic ($r = 3$), quartic ($r = 4$), or quintic ($r = 5$). Higher degree trend surfaces also can be used. For example, Whitten (1970) employed octic ($r = 8$) surfaces. Coons et al. (1967) fitted polynomial surfaces of the 13th degree. A trend surface of degree r has $m = \frac{1}{2}(r + 1)(r + 2)$ coefficients b_{pq}. These can only be estimated if the

7.1 2-D and 3-D Polynomial Trend Analysis

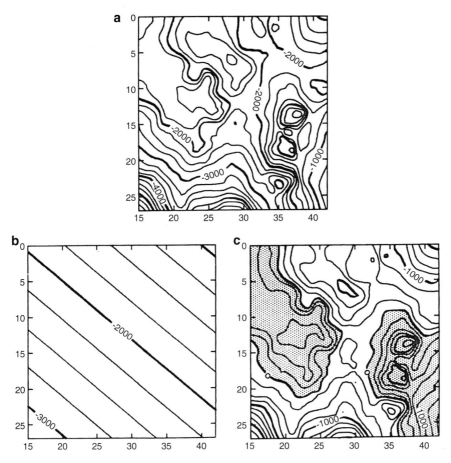

Fig. 7.1 Top of Arbuckle Formation in central Kansas. Contours are in feet below sea level (After Davis 1973). Geographic co-ordinates are in arbitrary units (area shown measures 320 km on a side.) (**a**) Contour map of original data; (**b**) Linear trend surface; (**c**) Residuals from linear trend surface (Source: Agterberg 1984, Fig. 1)

number of observation points (n) satisfies the condition $n \geq m$. If $n = m$, a special type of surface would be fitted with the property $T_k = Y_k$ and no residuals. In most applications, n is several times larger than the number of coefficients (m).

7.1.1 Top of Arbuckle Formation Example

A typical example of trend surface analysis is shown in Fig. 7.1 (after Davis 1973). Figure 7.1a gives the contours (in feet below sea level) of the top of Ordovician rocks (Arbuckle Formation) in central Kansas. The linear trend surface is shown in Fig. 7.1b and the corresponding residuals in Fig. 7.1c. The sum of the values R_k contoured in Fig. 7.1c and the values T_k (Fig. 7.1b) gives the original pattern (Fig. 7.1a).

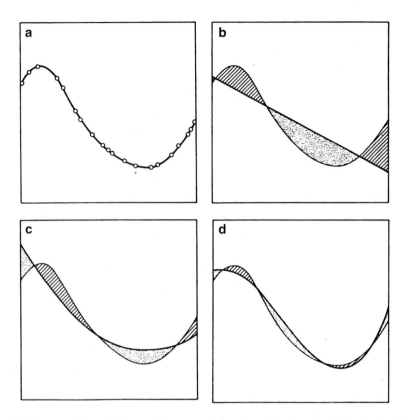

Fig. 7.2 Concept of trend illustrated by means of two-dimensional graphs (After Davis 1973). Value (vertical scale) is plotted against distance (horizontal scale). (**a**) Collection of original data points and smooth curve on which they lie; (**b**) Straight-line trend fitted to the observations; (**c**) Quadratic trend; (**d**) Cubic trend. *Shadings* represent positive and negative residuals from the trends (Source: Agterberg 1984, Fig. 2)

Some problems of trend surface analysis can be illustrated by the fitting of curves to data collected along a line. An example is shown in Fig. 7.2 (after Davis 1973). The observations fall on a smooth curve which, however, is not a low-degree polynomial. The magnitudes of the residuals decrease when the degree of the polynomial is increased. This artificial example illustrates two important points:

1. A good fit is not necessarily the object of trend analysis. Instead of this, the aim usually is to divide the data into a regional trend component and local residuals which can be linked to separate spatial processes of regional and local significance, respectively. Examples of interpretation of trends and residuals will be given later in this chapter.
2. Each curve-fitting of Fig. 7.2 yields residuals that form a continuous curve. Adjoining residuals along a line therefore are not statistically independent. Instead of this, the residuals are spatially correlated.

7.1.2 Mount Albert Peridotite Example

The Mount Albert intrusion is the largest ultramafic mass (approx. 44 km^2) in the Gaspé (Quebec) portion of the so-called Appalachian ultramafic belt. It is probably 530 my old. The intrusion was mapped and sampled in 1959 by C.H. Smith and I.D. MacGregor, who made their data available to the author for performing trend surface analysis (Agterberg 1964). The petrography of the body is relatively simple. Prior to serpentinization, it consisted of from 80 to 90 % olivine, the remainder being primarily orthopyroxene with some chrome spinel (up to 1 %) and diopsidic clinopyroxene. It was attempted to collect rock specimens from the nodes of a rectangular grid with 1,000 ft. spacing. The actual number of specimens that could be collected was less because of overburden. The number of mineralogical determinations was further reduced by serpentine alteration. The following four variables were determined for as many collected specimens as possible: (1) cell edge d_{174} of olivine; (2) refraction index N_z of orthopyroxene; (3) unit cell dimension of chrome spinel; and (4) specific gravity of the whole rock. The first two variables were converted into percentage magnesium in olivine and orthopyroxene, respectively, and the results reported as mol. percent forsterite (Mg-olivine, cf. Fig. 4.2) and enstatite (Mg-orthopyroxene). Mineralogical and thermodynamic implications of variations in the chrome spinel unit cell dimension were reviewed in MacGregor and Smith (1963) and Agterberg (1974). The relationship between percent serpentine and rock density in samples from the Mount Albert intrusion is approximately linear. Serpentinization was a process that continued afterwards, when the ultramafic body was already in place.

Trend surfaces for percentage forsterite in olivine, percentage enstatite in orthopyroxene, and the chrome spinel cell edge are shown in Fig. 7.3. The trends for these three variables are similar in that they show an elongated minimum at approximately the same location. This phenomenon was called cryptic zoning by MacGregor and Smith (1963). The trend pattern is most pronounced for percentage enstatite in orthopyroxene. The percentage explained sum of squares satisfies ESS = 39.2 % for the quadratic enstatite trend surface versus 6.3 and 16.0 % for the quadratic olivine and chrome spinel trend surfaces, respectively. The origin of the coordinate system with U pointing westward and V southward was set at a point outside the study area. Unit of distance was 10,000 ft. Then, for example, the quadratic equation of the surface shown in Fig. 7.3b is: % Mg = 79.116 + 5.806u − 1.245v − 0.0496u^2 − 2.2289uv + 1.9661v^2. Geometrically, this surface is a hyperbolic paraboloid. By using elementary methods of matrix algebra, it can be shown that this is equivalent to % Mg = 88.46 + 2.4608u'^2 − 0.5443v'^2 with the V' pointing WSW and U' NWN from the geometrical center of this quadratic surface.

The cubic trend surface for percentage enstatite in orthopyroxene suggests that there occurs a %Mg maximum in the southeastern corner of the intrusion. Whether or not this peak is real can be evaluated experimentally in several ways: Fig. 7.4 shows the 95 % confidence belt on the entire cubic surface. The shape of this belt mirrors the boundary of the intrusion because all observation points are contained within this

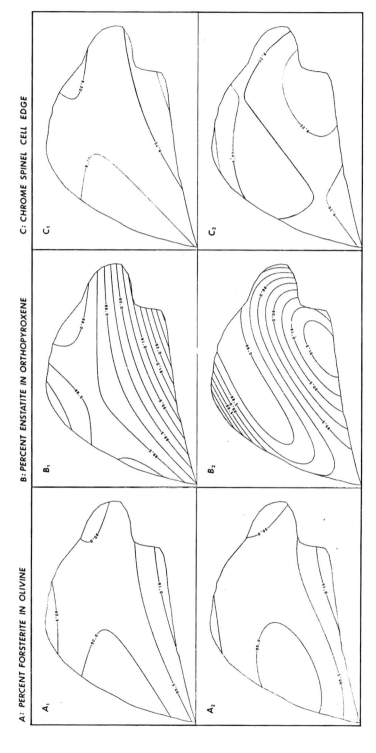

Fig. 7.3 Quadratic (subscript 1) and cubic (subscript 2) trend surfaces for (**a**) percent forsterite in olivine, (**b**) percent enstatite in orthopyroxene, and (**c**) cell edge, chrome spinel. Patterns for Mg content in olivine and orthopyroxene are similar in shape but variation is more pronounced in orthopyroxene (steeper gradients). Area measures 10.6 km in east-west direction (for scale see Fig. 7.5) (Source: Agterberg 1974, Fig. 42)

7.1 2-D and 3-D Polynomial Trend Analysis

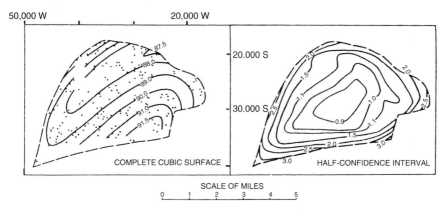

Fig. 7.4 Cubic trend surface for percent enstatite in orthopyroxene shown together with locations of specimens and 95 % half-confidence interval (Source: Agterberg 1974, Fig. 43)

boundary. It confirms the existence of the elongated ENE-WSW trending %Mg minimum but not necessarily the maximum in the southeastern corner of the intrusion. As a further experiment, the original data set that consisted of 174 orthopyroxene determinations was randomly divided into two interpenetrating subsets. The results of trend surface analysis applied separately to these two subsets are shown in Fig. 7.5. The difference between the two cubic trend surfaces in Fig. 7.5 is less than 1 % Mg in most of the Mount Albert peridotite intrusion.

There were 359 observation points for specific gravity. The quadratic surface is not adequate for representing the trend of this variable. Cubic (ESS = 39.9 %) and quintic (ESS = 56.0 %) trend surfaces are shown in Fig. 7.6. The difference in pattern between these two surfaces is caused mainly by the occurrence of a pocket of practically unaltered peridotite in the eastern part of the body. The transition from high-density to lower density material is relatively rapid and is poorly approximated by the cubic.

A schematic contour map for elevation is given in Fig. 7.7a. A cross-section CD was constructed and in Fig. 7.7b all observations within 2,500 ft. from the section line were projected onto it. Then, average values for blocks measuring 5,000 ft. on a side were calculated (Fig. 7.7e). These block averages almost exactly coincide with the intersection of the quintic trend surface with the cross-section CD. This shows that: (1) the quintic trend surface provides a good fit to the specific gravity trend; and (2) a trend surface also can be obtained by the relatively simple method of moving averages. The method of moving or running averages consists of calculating arithmetic averages for a large number of overlapping blocks and contouring the results. Obviously, this method can only be applied when there are many observations. If few observations are available, trend-surface analysis is to be preferred.

Because as many as 359 data points were available for specific gravity, 3-D trend analysis also can be attempted by incorporating the elevation of each observation point. The 3-D cubic trend equation contains 19 explanatory variables (u, v, w, u^2, uv, uw, v^2, uw, w^2, u^3, u^2v, u^2w, uv^2, uvw, uw^2, v^3, v^2w, vw^2, and w^3) and 20 coefficients. Its ESS-value amount to 55.1 % which is close to the previously mentioned

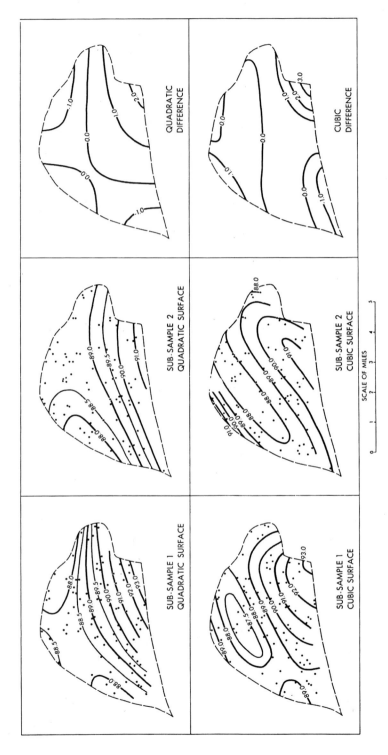

Fig. 7.5 Estimation of statistical significance of trend surfaces for enstatite by computing differential surfaces based on two interpenetrating subsamples (Source: Agterberg 1984, Fig. 3)

7.1 2-D and 3-D Polynomial Trend Analysis

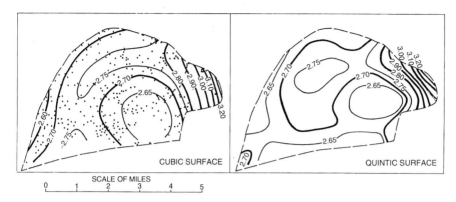

Fig. 7.6 Cubic and quintic trend surfaces for specific gravity data, Mount Albert Peridotite intrusion. *Dots* denote locations of specimens (Source: Agterberg 1974, Fig. 45)

ESS = 56.0 % for the quintic trend surface. Calculated values for the 3-D cubic along the topographic surface are shown for the CD profile in Fig. 7.7e. Because all observation points lie in the topographic surface, the elevation w could be approximated by a 2-D polynomial in u and v. Consequently, a 3-D trend $f(u, v, w)$ becomes a 2-D trend $g(u,v)$ if w, as contained in $f(u, v, w)$, is replaced by its polynomial in u and v. The reliability of the 3-D trend $f(u, v, w)$ decreases rapidly outside the topographic surface. Figure 7.7c, d show extrapolations of the 3-D cubic trend in the vertical direction obtained by setting w equal to 2,000 and 3,000 ft., respectively. This, of course, results in two ordinary cubic trend surfaces for specific gravity.

The results shown in Fig. 7.7c, d can be compared with the contour map for the topographic surface (Fig. 7.7a). The following interpretation is suggested. In Mount Albert, specific gravity is, by approximation, linearly related to volume percentage serpentine. Thus, a low value indicates a soft rock relatively sensitive to erosion. Two of the three rivers originating on Mount Albert follow zones of weakness at the 3,000-ft. level rather than at the 2,000-ft. level (Fig. 7.7). Further comparison with the contour map for elevation suggests present-day topography was controlled by the distribution of less weathering resistant rocks at higher levels. It can be concluded that the pipe of maximum serpentinization moved northward as well as upward.

7.1.3 Whalesback Copper Mine Example

The Whalesback copper deposit near Springdale, Newfoundland, provides an example of 2-D and 3-D trend analysis in a situation that the values are autocorrelated (Fig. 6.15a) and have a frequency distribution that is positively skewed with a long tail of large values. The orebody which has largely been mined out consists of chloritic altered volcanics containing pyrite and chalcopyrite mainly in disseminated form. However, these minerals also tend to cluster and form

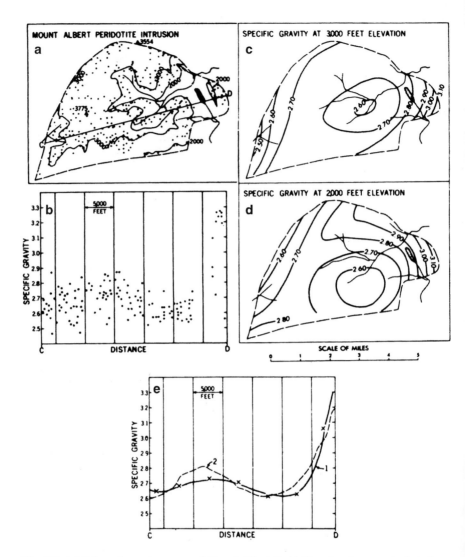

Fig. 7.7 (a) Schematic contour map of Mount Albert Peridotite; elevations are in feet above sealevel. (b) Variations in specific gravity along section CD (see **a** for location); all values within 2,500 ft. Source: section line were perpendicularly projected to it. (**c**) and (**d**). Cubic trend surfaces for 300-ft. and 2,000-ft. levels obtained by contouring three-dimensional cubic hypersurface in two horizontal planes. In most of the area, to-day's topography resembles the 3,000-ft. pattern more closely than the 2,000-ft. pattern. (**e**) *Crosses* represent averages fort values within blocks measuring 5,000 ft. on a side (see **b**). *Curve 1* is intersection of quintic trend surface (Fig. 7.6) with CD; *curve 2* represents variation along topographic surface (see **a**) according to three-dimensional cubic hypersurface (Source: Agterberg 1974, Fig. 46)

7.1 2-D and 3-D Polynomial Trend Analysis

stringers, and occur in well-defined narrow veins with maximum width of 1.2 m. The deposit coincides with a shear zone. It is about 400 m long at the surface. Average width of the central copper zone which averages more than 1 % copper is approximately 20 m, and vertical depth is 270 m. The copper zone is enveloped by a rather strongly altered chlorite zone with 0.35 % Cu on the average. The plate-shaped mineralized zone dips about 70° south-southwest. Before mining (by the British Newfoundland Exploration Company) commenced in 1965, ore reserves were estimated at four million short tons averaging 1.48 % copper.

Trend surface analysis (Agterberg 1974) was applied to copper concentration values from the 425 ft. level that occurs 425 ft. below the surface. The core for underground drill-holes, which were 50 ft. apart, was divided into pieces of 5 ft. length and assay values (percentages copper) were determined from these. About 15 vol.% of the rocks in the deposit consist of dikes that do not contain copper. This yields zero assay values which were omitted from the data to be used for statistical analysis and trends will be projected across the dikes. Therefore, if average grade values are to be determined for larger blocks of ore, these values, after calculation, should be reduced by 15 % for dilution caused by barren dikes. Of course, more precise corrections can be made in places where the precise location of dikes is known.

Because trend surfaces are imprecise at the edges of clusters of observation points, overlapping sets of six drill-holes, or cross-cuts which has been sampled in the same manner, were used and quadratic and cubic trend surfaces $S(u,v)$ were fitted to logarithmically transformed copper values. The fitted values were converted back to ordinary copper values by using the transformation $T(u,v) = \exp[S(u,v) + s^2/2]$ where s^2 is the residual variance for the $S(u,v)$ surfaces. The central part for each converted surface $T(u,v)$ for a 150 ft. wide zone situated between the second and fifth hole in each situation, is shown in the mosaics of Figs. 7.8 and 7.9. This transformation is equivalent to the one applied to lognormally distributed data (Sect. 3.3) and assumes that at any location the copper values are lognormally distributed with mean $T(u,v)$ and logarithmic variance s^2.

Autocorrelation of residuals affects the results of trend surface analysis. This topic will be discussed in more detail in the next section. Here it can be assumed that the autocorrelation function of the residuals is approximately as shown in Fig. 6.15a which is an average correlogram based on 24 separate correlograms for series of logarithmically transformed copper values (8 ft. apart) along drifts at various levels of the mine that were approximately within the mineralized zone. Watson (1971) has shown that if the residuals satisfy a weakly stationary process with an autocorrelation function that is independent of geographic location, then the best fitting trend surfaces are unbiased (*cf.* Cressie 1991, p. 164). However, analysis of variance to help decide on the optimum degree of a trend surface then are not applicable because a set of n autocorrelated data does not contain the same amount of information as n uncorrelated (iid) data. Some authors including Matalas (1963) have developed methods to determine n' representing a hypothetical sample of uncorrelated (iid) values containing the same amount of information as n autocorrelated data (see Box 7.1). It is usually assumed that the underlying autocorrelation function is exponential. The topic is fraught with difficulties and,

246   7 2-D and 3-D Trend Analysis

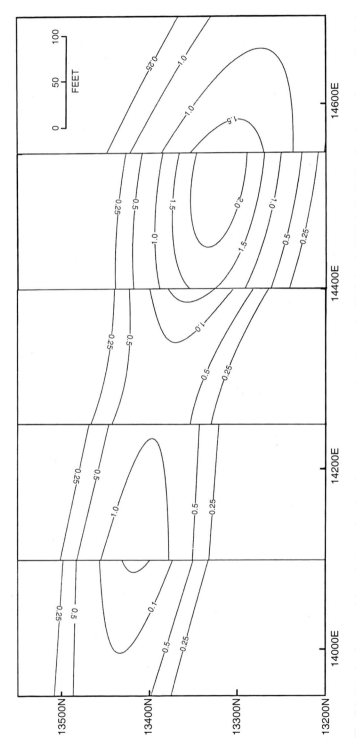

Fig. 7.8 Mosaic of exponential quadratic trend surfaces for copper concentration value on the 425-ft. level of Whalesback Mine; based on underground borehole data; drill-holes were north-south oriented, approximately 50 ft. apart; western boundary coincides with drill-hole No. 2 (Source: Agterberg 1974, Fig. 48)

7.1 2-D and 3-D Polynomial Trend Analysis

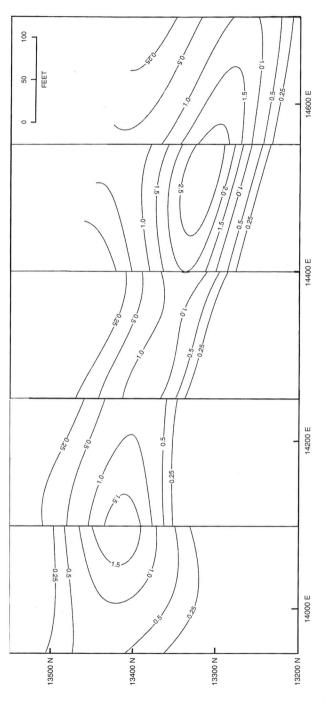

Fig. 7.9 Same as Fig. 7.8 using cubic trend surfaces (Source: Agterberg 1974, Fig. 49)

Table 7.1 Comparison of quadratic and cubic trend surfaces for four sets of six bore-holes at 425-ft. level; logarithmically transformed data

	Source of variation	Sum of squares	Degrees of freedom	$\hat{F}$ · ratio	Corrected df	Corrected $\hat{F}$	Confidence limits 95 %	99 %
Holes 1–6	Complete quadratic	100.68	6	16.4	6	5.5	2.3	3.2
	Residuals	175.01	170		57			
	Total	275.69	176					
	Cubic minus quadratic	21.39	4	5.8	4	1.9	2.6	3.7
Holes 7–12	Complete quadratic	100.46	6	14.5	6	4.8	2.3	3.2
	Residuals	163.76	148		49			
	Total	264.21	154					
	Cubic minus quadratic	41.82	4	11.8	4	3.9	2.6	3.8
Holes 13–18	Complete quadratic	119.89	6	20.0	6	6.7	2.2	3.1
	Residuals	206.71	198		66			
	Total	326.60	204					
	Cubic minus quadratic	16.88	4	4.3	4	1.4	2.5	3.6
Holes 19–24	Complete quadratic	74.88	6	10.2	6	3.4	2.3	3.2
	Residuals	196.52	159		53			
	Total	271.40	165					
	Cubic minus quadratic	14.29	4	3.1	4	1.0	2.6	3.7

Source: Agterberg (1974, Table XXV)

in general, scientists active in spatial statistics do not make use of "degrees of freedom" as are required in statistical analysis based on iid data.

However, Agterberg (1966) argued as follows: Any F-ratio is the ration of two variances for independent random variables. The sum of squares of n' independent data would be distributed according to $\sigma^2\chi^2(n')$ with n' degrees of freedom. In the situation of redundancy due to autocorrelation, one would obtain $n/n'\,\sigma^2\chi^2(n')$. Suppose that the sum of squares due to regression is not seriously affected by the redundancy. This suggests that the F-ratios should be multiplied by n/n' in order to compensate for the redundancy. The number of degrees of freedom for the denominator also should be corrected by this factor. Application of this correction to comparison of the cubic surfaces (Fig. 7.9) with the quadratic surfaces shown in Fig. 7.8 resulted in the modified analysis of variance shown in Table 7.1. Contrary to what would be suggested if independence of residuals would be assumed, the results in Table 7.1 suggest that the step from quadratic to cubic fit should not be made except perhaps for holes 7–12 where the corrected F-ratio is significant at the 99-% level. The area of Figs. 7.8 (quadratics) and 7.9 (cubics) is situated between hole 2 to the west and hole 17 to the east. The difference between these two mosaics is greatest in the central part (between 14,250E and 14,400E) computed from holes 7–12 (between 14,200E and 14, 450E).

It is noted that the analysis of variance results shown in Table 7.1 differ from those shown in the original table of Agterberg (1966, Table 4) that was based on the same data with one exception. The original table shows separate estimates of the first serial correlation coefficient (r_1) for each set of holes. These estimates vary from 0.24 to 0.70 and F-ratios corrected using n' based on these original estimates of r_1. Later (in Agterberg 1974), it was decided to use the same estimate ($r_1 = 0.50$) based on the correlogram of Fig. 6.15a for each set of holes. This single estimate is probably more precise than the single smaller- sample estimates, because it is an average for 24 drifts on different levels of the Whalesback Mine including those on the 425-ft. level. This revision does not significantly change the original conclusions drawn in Agterberg (1966).

Much information on the spatial distribution of copper in the Whalesback

> **Box 7.1: Degrees of Freedom for Autocorrelated Data**
>
> The variance of the mean of n autocorrelated data from a series with the first order Markov property satisfies: $\text{var}(\overline{X}) = n^{-2} \sum_{i=1}^{n} \sum_{j=1}^{n} \text{cov}(X_i, X_j) = \sigma^2 \left[1 + 2\frac{\rho}{1-\rho}\left\{(1-n^{-1}) - \frac{\rho}{1-\rho}(1-\rho^{n-1})/n\right\}\right]/n$ where ρ is the autocorrelation coefficient for adjoining values (cf. Cressie 1991, p. 14). Suppose an equivalent number of independent data (n') is defined as $\text{var}(\overline{X}) = \sigma^2/n'$. For $n > 10$, the preceding equation then gives approximately $n/n' = 1 + 2r_1[n/(1-r_1) - 1/(1-r_1)^2]/n$ (cf. Agterberg 1974, p. 302). For large n this gives approximately $n' = n(1-r_1)/(1+r_1)$ illustrating that autocorrelation strongly affects statistical inference even in large samples (also see Sect. 2.1).

deposit is for points outside subhorizontal levels such as the 425-ft. level used for example earlier in this section. As an experiment, a 3-D analysis was performed on all copper values for samples taken within a relatively small block extending from the surface to the 425-ft. level and situated between the 14,000E and 14,500E sections. In total, 516 values from both core samples and channel samples along drifts were used. Figure 7.10a shows a histogram of the 516 logarithmically transformed copper values. It is clearly bi-modal. 3-D trend analysis gave the following ESS-values: linear 18.0 %, quadratic 32.8 % and cubic 43.3 %. The logarithmic variance of original data is 1.865. 3-D trend analysis reduced this variance as follows: linear residuals 1.531, quadratic residuals 1.2070, and cubic residuals 1.093. In Agterberg (1968) it is discussed in more detail that the cubic fit is better than the quadratic fit in this example. A chi-square test for goodness of fit was applied to test the cubic residuals (Fig. 7.10b) for normality. It gave an estimated chi-square of 17.3 for 17 degrees of freedom, well below the 27.6 representing the

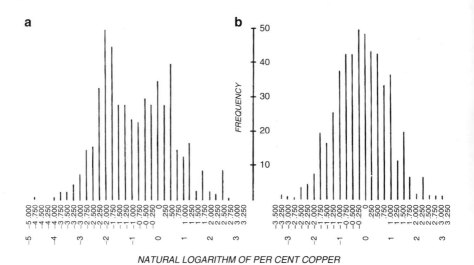

Fig. 7.10 (**a**) Histogram of natural logs of 516 copper values; (**b**) ditto for residuals from cubic hypersurface; this curve is approximately Gaussian (Source: Agterberg 1974, Fig. 51)

95-% confidence level. Consequently, it can be assumed that the cubic residuals of the log-transformed copper values are indeed normally distributed.

A final example of 3-D trend analysis is for 335 Whalesback copper values from 20 holes drilled from the surface during an early stage of development of the orebody. They are for an 800 ft. wide zone between 14,100 E and 14,900 E which is 625 ft. deep. Figure 7.11 shows contours from the exponential cubic hypersurface for copper on two levels together with an outline of the orebody based on later information. Residuals for this hypersurface were shown in Fig. 7.10b. Although the position of the central plane of maximum mineralization in Fig. 7.11 is in close agreement with the outline of the orebody better results could be obtained from the development holes drilled from the surface by using harmonic trend analysis of the copper data as will be discussed in Sect. 7.4.2.

7.2 Kriging and Polynomial Trend Surfaces

During the late 1960s, there was a considerable amount of discussion among earth scientists and geostatisticians regarding the question of which technique is better: trend surface analysis or kriging (see, e.g., Matheron 1967)? The technique of kriging (named after the South African geostatistician Danie G. Krige who first applied regression-type techniques for gold occurrence prediction) is based on the assumption of a homogeneous spatial autocorrelation function. If such a function can be established independently, it can be used to solve the type of problem exemplified in Fig. 7.12. Within a neighborhood, values of a variable are known

7.2 Kriging and Polynomial Trend Surfaces

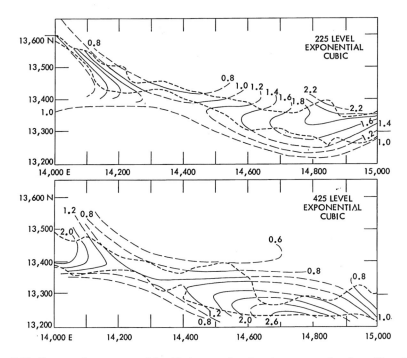

Fig. 7.11 Contours from exponential cubic hypersurface based on copper data from 20 exploratory holes drilled downward from the topographic surface. Outline of orebody is approximately 1 % copper contour based on later underground information (Source: Agterberg 1968, Fig. 6)

at $n = 5$ points (P_1–P_5). These five values are to be used to estimate values at any other point in the neighborhood; e.g., at P_0 in Fig. 7.12. Suppose, firstly, that the five known values are corrected for a hypothetical mean value, which may be a constant estimated as the average of all values in a larger neighborhood. Then application of the general linear model provides estimates of the coefficients (usually called "weights") to be assigned to the n (=5 in Fig. 7.12) known values. The weights can written as a column vector $\mathbf{W}$ with solution $\mathbf{W} = \mathbf{R}_1^{-1} \mathbf{R}_0$ where $\mathbf{R}_1$ is an ($n \times n$) matrix consisting of the autocorrelation coefficients for all possible pairs of the n known points in the neighborhood, and $\mathbf{R}_0$ is a column vector of the autocorrelation coefficients for distances between the locations of the n known values and the point at which the kriging value is to be estimated. In a slightly different version of this simple kriging technique, it is required that the sum of the weights to be assigned to the known values is unity. This constraint can be incorporated by using a Lagrangian multiplier (Sect. 7.2.3). A useful generalization of simple kriging is to correct all values for a trend value instead of for a constant regional mean. The final estimated value at a point P_0 with arbitrary coordinates then is the sum of the trend value at P_0 and the positive or negative value resulting from the simple kriging.

Fig. 7.12 Typical kriging interpolation problem; values are known at five points. Problem is to estimate value at point P_0 from values at points P_1–P_5 (Source: Agterberg 1974, Fig. 64)

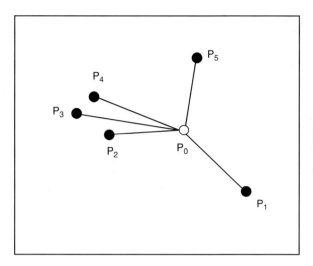

7.2.1 Top of Arbuckle Formation Example

As a contribution to the discussion of which technique is better: trend surface analysis or kriging, Agterberg (1970) performed the following experiment. A set of 200 elevation data for the top of the Arbuckle Formation in Kansas (Fig. 7.13) was randomly divided into three samples: two "control" samples (No. 1 and No. 2) each consisting of 75 data and a test sample (No. 3) of 50 data. Three different techniques were applied to samples No. 1 and No. 2 and the results used to make predictions of the elevations at the top of the Arbuckle Formation at the 50 points of sample No. 3. These three techniques were: (1) Trend surface analysis; (2) Kriging, and (3) Trend surface analysis plus kriging of residuals. Linear, quadratic, cubic and quartic trend surfaces were fitted to the entire data set and the two smaller control samples. Details of how 2-D autocorrelation functions were constructed for kriging applied to residuals from these surfaces are given in Agterberg (1970). Estimates of 2-D autocorrelation functions based on residuals from the linear, quadratic, and cubic surfaces fitted to the entire data set are shown in Fig. 7.14.

Results of the experiments involving the subsamples are given in Tables 7.2 and 7.3. Analysis of variance to decide on the best trend surface cannot be used because the residuals from the surfaces are autocorrelated. From the results shown in Table 7.2 it can be concluded that the quadratic, cubic and quartic fits perform equally well. Note the drop in percentage explained sum of squares from quadratic (74 %) to cubic (60 %) for predictions made by control sample 1. Table 7.3 contains results for sample 3 on the basis of the control samples after subjecting the deviations from the fitted surfaces to kriging. Kriging on its own, without trend surface analysis, is about as good as the fitting of a quadratic trend surface. Because there are strong trends in the data, universal kriging (of. Huijbregts and Matheron 1971) is preferable to simple or ordinary kriging. Kriging of residuals improves the overall degree of fit in all experiments.

7.2 Kriging and Polynomial Trend Surfaces

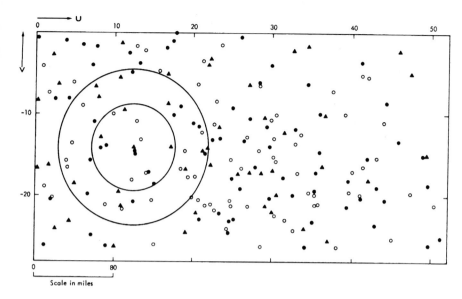

Fig. 7.13 Location of 200 wells in Kansas. Good (1964) fitted trend surfaces to elevation on top of Arbuckle Group for points. (Linear trend surface was shown in Fig. 7.1.) Observations have been randomly divided into three samples. *Solid circles* are for control sample 1; *open circles* for control sample 2; *triangles* for sample 3. *Smaller circle* represents area for estimating autocorrelation function used for kriging (Source: Agterberg 1970, Fig. 2)

The preceding experiments have been discussed in publications by statisticians including those by Tukey (1970), Haining (1987) and Cressie (1991). In the 1950s and 1960s, most earth scientists took a trend surface approach to their mapping problems but later the extra advantages in taking a random-field approach (i.e., universal kriging) became clearly established. As pointed out by Cressie (1991, p. 164) it can be said that comparison of results with or without the random-field approach is easy because the trend-surface model is a special case of a random field model. Consequently, when the spatial-covariance structure is known, universal kriging generally gives more precise predictions than trend-surface analysis, because universal kriging chooses optimal weights to be applied to the data. However, as Watson (1971) has shown it is possible for trend-surface prediction to be just as precise as universal kriging even when the residuals do not satisfy a pure white noise model as would be required for analysis-of-variance applications to decide on optimal degree. Cressie also points out that, in practice, there is a price to pay for using universal kriging because: "One must obtain (efficient) estimators of variogram parameters, whose effects on mean-squared prediction errors should be assessed." The following two sections contain further examples of universal kriging. The objective of the modeling will be to separate the spatial variability into three components: (1) regional trend; (2) localized "signals" based on the assumption that the residuals from the regional trends are weakly stationary and possess the same regional autocorrelation fun; and (3) strictly local white noise.

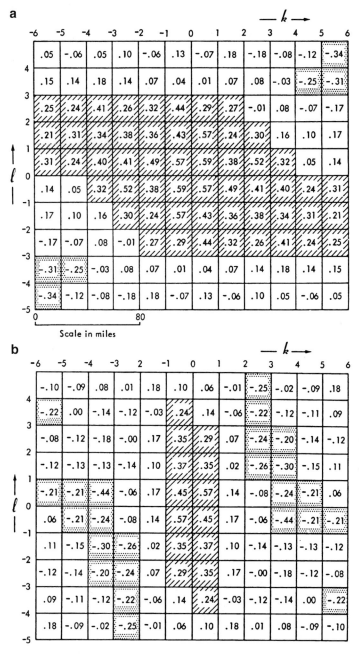

Fig. 7.14 Autocorrelation functions for Arbuckle elevations at locations shown in Fig. 7.13. (**a**) Original data; (**b**) residuals from quadratic trend surface; (**c**) residuals from cubic trend surface. About half of the estimated autocorrelation coefficients are based on samples of more and less than 100 pairs of values, respectively (Source: Agterberg 1970, Fig. 4)

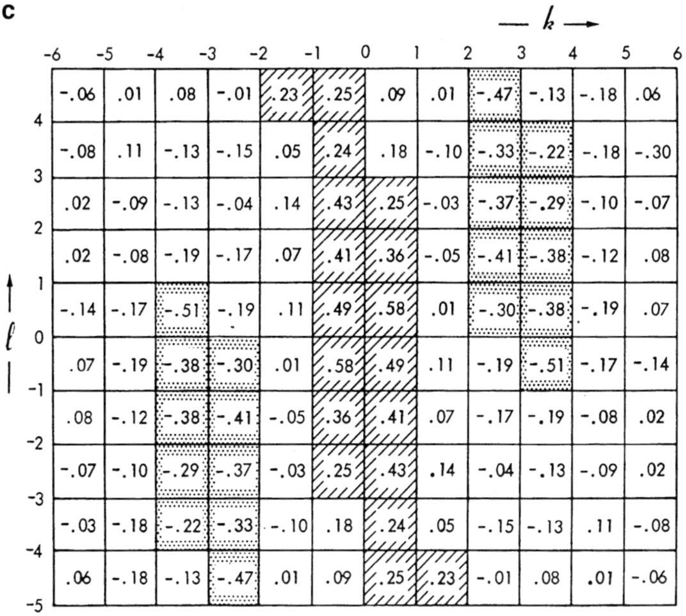

7.2 Kriging and Polynomial Trend Surfaces   255

c

	-6	-5	-4	-3	-2	-1	0	1	2	3	4	5	6
4	-.06	.01	.08	-.01	.23	.25	.09	.01	-.47	-.13	-.18	.06	
3	-.08	.11	-.13	-.15	.05	.24	.18	-.10	-.33	-.22	-.18	-.30	
2	.02	-.09	-.13	-.04	.14	.43	.25	-.03	-.37	-.29	-.10	-.07	
1	.02	-.08	-.19	-.17	.07	.41	.36	-.05	-.41	-.38	-.12	.08	
0	-.14	-.17	-.51	-.19	.11	.49	.58	.01	-.30	-.38	-.19	.07	
-1	.07	-.19	-.38	-.30	.01	.58	.49	.11	-.19	-.51	-.17	-.14	
-2	.08	-.12	-.38	-.41	-.05	.36	.41	.07	-.17	-.19	-.08	.02	
-3	-.07	-.10	-.29	-.37	-.03	.25	.43	.14	-.04	-.13	-.09	.02	
-4	-.03	-.18	-.22	-.33	-.10	.18	.24	.05	-.15	-.13	.11	-.08	
-5	.06	-.18	-.13	-.47	.01	.09	.25	.23	-.01	.08	.01	-.06	

— h →

↑
l 0
↓

Fig. 7.14 (continued)

Table 7.2 Values of sample 3 as predicted by trend surfaces for the two control sample results

		Mean	Linear	Quadratic	Cubic	Quartic
Sum of squares	1	0.46	0.32	0.12	0.18	0.11
	2	0.40	0.27	0.13	0.10	0.11
Percentage ESS	1	0.0	30.2	73.8	60.3	76.5
	2	0.0	34.2	68.5	74.6	72.6

Source: Agterberg (1970, Tables 1 and 3)
Unit for sum of squares is 10^8

Table 7.3 Values of sample 3 predicted by kriging on deviations from mean and residuals

		No. kriging	Mean	Linear	Quadratic	Cubic	Quartic
Sum of squares	1	0.46	0.11	0.21	0.09	0.14	0.09
	2	0.40	0.19	0.16	0.08	0.07	0.09
Percentage ESS	1	0.0	76.7	53.5	80.9	68.9	81.3
	2	0.0	53.6	61.4	79.5	82.6	78.3
Ditto for residuals (Table 3)	1			33.3	27.0	21.4	20.4
	2			41.4	34.6	31.1	20.0

Source: Agterberg (1970, Table 5)

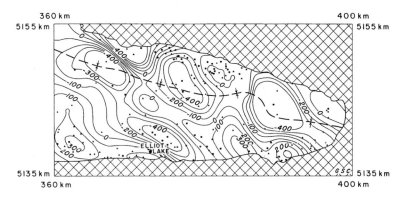

Fig. 7.15 "Signal" for bottom of Matinenda Formation, Elliott Lake area, Ontario, obtained by contouring kriging values (in feet) computed at points with 1 km-spacing. Archaean basement rocks are shown by pattern. Trace of axial plane of Quirke Syncline is shown also. *Squares* and *triangles* represent locations of drill-holes with and without significant mineralization, respectively. Universal Transverse Mercator grid locations of corner points of map area are shown in km (Source: Agterberg 1984, Fig. 7)

7.2.2 Matinenda Formation Example

In the Elliot Lake area (Fig. 7.15), the Archean rocks of the Canadian Shield are unconformably overlain by Proterozoic rocks which have been subjected to folding. This yielded the Quirke Syncline (trough-shaped fold) of which the axial plane is shown as the trace in Fig. 7.15. The bottom of the Matinenda Formation which contains uraniferous conglomerates coincides with the top of the Archean basement except in some places where it is separated from it by a sequence of earlier Proterozoic volcanic rocks of variable thickness. Locations of boreholes are given in Fig. 7.15 where the squares indicate boreholes in which significant uranium mineralization was encountered and the triangles represent other boreholes. The bottom of the Matinenda Formation in feet above or below sealevel was available for all boreholes of Fig. 7.15. Quadratic trend surfaces were fitted to these data from overlapping clusters of observation points (Fig. 7.16). Figure 7.17 shows the sum of this trend and the signal that was already shown in Fig. 7.15.

Robertson (1966) had attempted to reconstruct the topography of the top of the Archean basement as it was before the later structural deformation. He assumed that the Matinenda Formation was deposited in topographical depressions in the top of the basement. This geological model provides a good reason for application of the trend + signal + noise model introduced in the previous section. Ideally, the trend in this model would represent the structural deformation and the signal would represent the original topographical surface on which the Matinenda Formation was deposited. Thus, it would be meaningful to extract the signal from the data by elimination of (1) the trend, and (2) the noise representing irregular local variability which is unpredictable on the scale at which the sampling (by drilling) was carried out. Although the concepts of "trend" (captured by a deterministic polynomial

7.2 Kriging and Polynomial Trend Surfaces

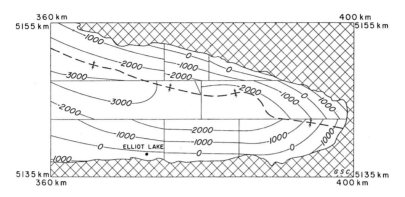

Fig. 7.16 Contours (in feet above sealevel) of parts of eight quadratic trend surfaces fitted to bottom of Matinenda Formation, Elliot Lake area (Source: Agterberg 1984, Fig. 8)

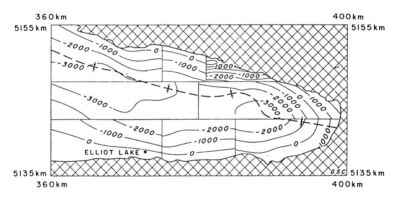

Fig. 7.17 Sum of trend (see Fig. 7.16) and signal (see Fig. 7.15) for bottom of Matinenda Formation. Note strong resemblance between patterns of Figs. 7.16 and 7.17 illustrating that the fluctuations (signal) contoured in Fig. 7.15 have relatively small magnitudes (Source: Agterberg 1984, Fig. 9)

equation), "signal" (characterized by a homogeneous autocorrelation function), and stochastically independent "noise" have different statistical connotations, the differences between the features they represent usually are less distinct. It will be seen that some of the signal in the present example represents structural deformation. It also is known that these concepts depend on the sampling density. In general, if more measurements are performed, some of the signal becomes trend and some of the noise becomes signal.

Although it can be assumed that the major uranium-producing conglomerates in the Elliot Lake area were deposited in channels, the geometrical pattern of the channels remains a subject of speculation, at least in places removed from the mining areas. Bain (1960) assumed a single "uraniferous river channel" (see also Stanton 1972, Figs. 12–16) winding its way through the basin so that it fits the approximately NW trending channels at the Quirke, Nordic and Pronto mineralized

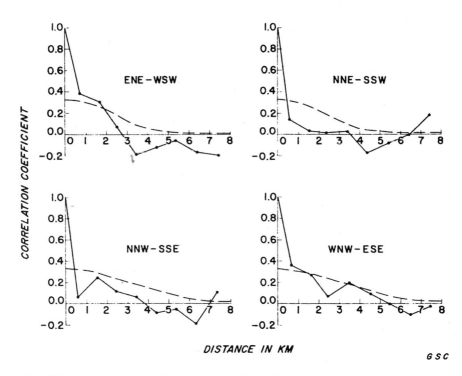

Fig. 7.18 Autocorrelation coefficients computed for residuals. Source: quadratic trend surfaces (see text for method used) computed with profiles (*broken lines*) across theoretical two-dimensional model used for kriging (Source: Agterberg 1984, Fig. 10)

zones. On the other hand, Derry (1960) assumed separate, subparallel channels. It will be seen that the results obtained by our trend + signal + noise model fit in with Derry's hypothesis and not with Bain's hypothesis.

The following procedure was followed to obtain a 2-D autocorrelation function for residuals from the trend shown in Fig. 7.17. By using polar coordinates, distances between pairs of boreholes were grouped according to (1) distance (1-km spacing) and (2) direction of connecting line (45° wide segments). This gave domains of different sizes. The autocorrelation coefficients for pairs of values grouped according to this method are shown in Fig. 7.18. These values were assigned to the centers of gravity of their domains. A 2-D quadratic exponential (Gaussian) function with superimposed noise component (nugget effect) was fitted by trend surface analysis of logarithmically positive autocorrelation coefficients obtained from the grouped data. Both the input pattern of autocorrelation coefficients and the fitted function are symmetrical with respect to the origin where the observed autocorrelation coefficient is equal to 1 because it includes the noise component. A profile through the origin across the quadratic trend surface fitted to logarithmically transformed autocorrelation function is a parabola with its maximum at the origin. Each parabola becomes a Gaussian curve when the antilog is taken. Four profiles across the fitted function are shown in Fig. 7.18. The maximum of the fitted function fell at 0.325 indicating that only 32½ % of the variance of the residuals is explained by the signal versus 67½ % by the noise. The function has elliptical contours which are elongated in approximately the NW-SE direction. This directional anisotropy would indicate that the actual topography features (channels?) which constitute the signal are elongated in the NW-SE direction.

The signal was extracted from the residuals by using the theoretical autocorrelation function. The kriging method was used to estimate the values of the signal at the intersection points of a grid (UTM grid) with 1-km spacing. Every kriging value was estimated from all observations falling within the ellipse described by the 0.01 contour of the theoretical Gaussian autocorrelation function. This ellipse is approximately 13 km long and about 7 km wide. The estimated kriging values on the 1-km grid defined a relatively smooth pattern which was contoured yielding the pattern of Fig. 7.15.

The pattern of Fig. 7.15 shows a number of minima and maxima. The amplitudes of these fluctuations are very small in comparison with the variations described by the trend (Fig. 7.16) as can be seen in Fig. 7.17 representing the sum of trend and signal. Ideally, the signal would correspond to small depressions and uplifts in the Archean basement, as it seems to do in most of the southern part of the area. However, other features of the estimated signal can be interpreted as belonging to the later structural deformation pattern. The lack of a clearly developed negative at Quirke Lake in the northern part of the area where significant uranium mineralization occurs may be caused by the fact that the channel direction is approximately parallel to the structural trend in this part of the area. This would make it hard and perhaps impossible to discriminate between the two patterns by means of the present statistical model.

7.2.3 Sulphur in Coal: Lingan Mine Example

The Lingan Mine was located in the Sydney coalfield on Cape Breton Island in Nova Scotia. All seams in this coalfield belong to the Morien Group that is about 2 km thick and Pennsylvanian in age. The strata of the Morien Group dip eastward, usually between 4° and 20°. When production of coal from the Lingan Mine commenced in 1972, there was great interest in prediction of the sulphur content of the 1.5–2 m thick Harbour and Phalen seams from which about two million tons of coal was to be extracted annually. From mined-out areas close to the shoreline, these two seams were known to contain less sulphur than other seams. Figure 7.19 shows the study area. Lingan Mine reserves consisted of (1) a 4-km wide zone between two worked areas (No. 12 and No. 26 Collieries on the Harbour seam); and (2) an east-west zone, more than 16 km wide, extending from the initial area and adjacent worked areas toward the 4,000 ft. depth (below sea level) of the seams. The area that was to be mined could not be sampled beforehand because it was too far offshore.

In general, Caper Breton seams have relatively high sulphur content with pyrite as the main sulphide-bearer. However, a large part of the Harbour seam in the No. 26 Colliery averages about 1 % sulphur, yielding a high-grade metallurgical coal. It was planned that much of the future output of the Lingan Mine would be sold by the Cape Breton Development Corporation (DEVCO) as a metallurgical coal. Coal with a sulphur content of up to 2 % could be desulphurized at the Lingan

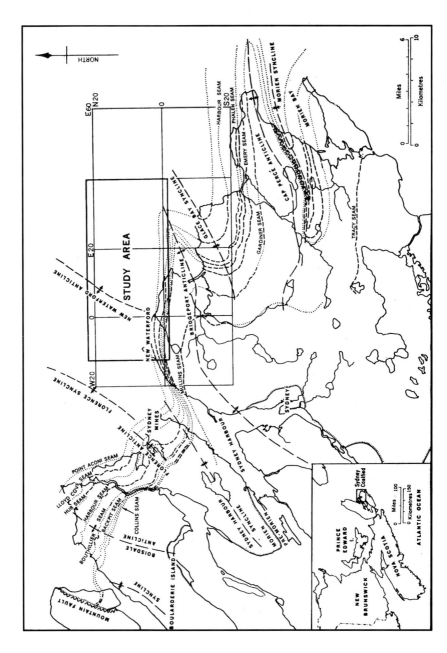

Fig. 7.19 Outcrop pattern of coal seams in the Sydney coalfield, Nova Scotia; DEVCO mining grid (in thousands of feet); outline of study area in Figs. 7.20

7.2 Kriging and Polynomial Trend Surfaces

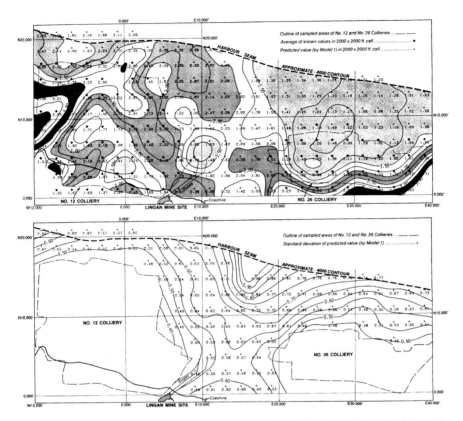

Fig. 7.20 Predicted sulphur variation (in percent) for (2,000 ft. × 2,000 ft.) cells (Model 1), Harbour seam, Lingan Mine area. *Top diagram*: average values in worked areas and predicted values in target areas. *Bottom diagram*: standard deviations on predicted values in target areas. Contours at regular 0.25 % intervals in A are used for pattern enhancement (Source: Agterberg and Chung 1973, Fig. 2)

Mine site by using the compound water cyclone (Walsh et al. 1969). By the end of the last century, nearly all coal mining had been discontinued on Cape Breton Island. This included closure of the Lingan Mine.

Two statistical methods were used for offshore prediction (Agterberg and Chung 1973). Results for (2,000 ft. × 2,000 ft.) blocks in the Harbour Seam are shown in Figs. 7.20 and 7.21. Model 1 is based on kriging and Model 2 on universal kriging. Details of the underlying mathematics and the computer algorithms used are given in Agterberg and Chung (1973). Kriging (Model 1) was not based on the arithmetic mean but on the Best Linear Unbiased Estimator (BLUE) which is efficient and, theoretically, gives slightly better results than the arithmetic mean that is unbiased only. It involves using a Lagrangian multiplier. The autocorrelation function used for Fig. 7.20 is shown in Fig. 7.22. The BLUE for the universal kriging (Model 2) application could not be estimated but was approximated by an asymptotically efficient estimator. Use was made of exponential trends similar to those used in

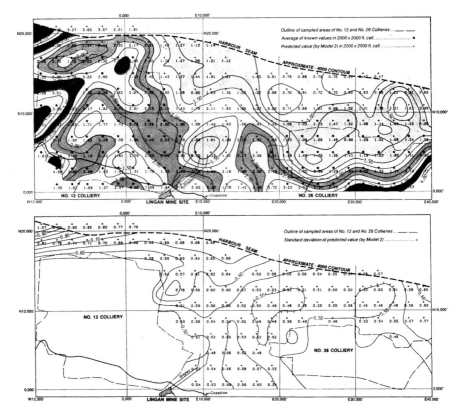

Fig. 7.21 Same as Fig. 7.20 using Model 2. It is assumed that the trend is according to pattern shown in Fig. 7.23 and that residuals from this trend are autocorrelated according to pattern shown in Fig. 7.22 (Source: Agterberg and Chung 1973, Fig. 3)

Sect. 7.1.3. The quadratic exponential trend surface for the Harbour Seam is shown in Fig. 7.23. Next the kriging method of Model 1 was applied to the residuals to complete the computations required for Model 2. For the autocorrelation function for residuals from the surface of Fig. 7.23, see Agterberg and Chung (1973).

The basic concepts behind the two models are illustrated by a hypothetical example in Fig. 7.24. Extrapolations by Model 1 are based on known data at points in the immediate vicinity. Every predicted value (crosses) in Fig. 7.20 is based on all known values (dots) within a circle of 9,000 ft. radius, unless there occur ten or more values within a smaller circle of 7,000 ft. radius; then, the smaller circle was used. The mean of the known data is used for prediction in the unknown area. This is called the (kriging) mean in Fig. 7.24. Separate values for blocks anywhere in the unknown area are estimated next. Uncertainty increases with distance from the worked area, and the predicted values converge to the kriging mean which provides the best estimate when the distance is great. In the more immediate area, close to the shoreline, the estimates are more precise than this mean value.

A basic assumption for Model 1 is that the mean value remains constant. It is known, however, that average sulphur content of Cape Breton coal, even for a

7.2 Kriging and Polynomial Trend Surfaces

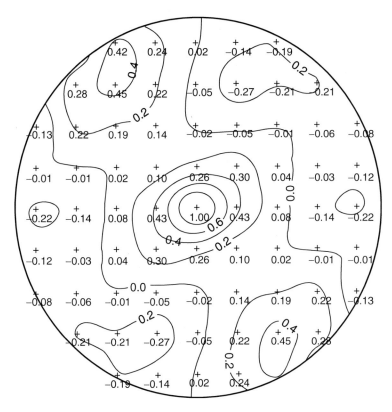

Fig. 7.22 Two-dimensional autocorrelation function estimated from residuals from quadratic exponential trend surface shown in Fig. 7.23 (Source: Agterberg and Chung 1973, Fig. 8)

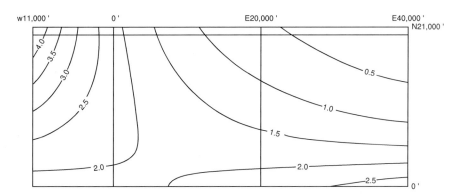

Fig. 7.23 Quadratic exponential trend surface for Harbour seam; based on 442 average sulphur values for overlapping 2,000 × 2,000 ft. cells. This pattern was used as trend component in Model 2 (Source: Agterberg and Chung 1973, Fig. 7)

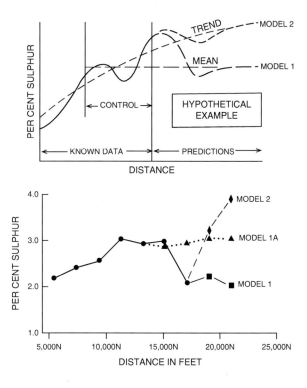

Fig. 7.24 *Top diagram*: Artificial example of sulphur variations along a line to illustrate basic concepts underlying Models 1 and 2. *Bottom diagram*: Comparison of results in westernmost parts of Figs. 7.20 and 7.21; triangles (for Model 1A) represent predictions by Model 1 applied to known data south of 14,000 N only (Source: Agterberg and Chung 1973, Fig. 4)

single seam, changes from colliery to colliery and within the same colliery. Model 2 first attempts to capture this gradual change by means of trend surface analysis. The trend replaces the constant mean (Fig. 7.24, top). Next kriging was applied to the residuals. Again, uncertainty increases with distance into the unknown area but the predicted values converge toward the trend instead of to a constant mean. To some extent, predictions by the two models can be tested against reality. A simple test is illustrated in Fig. 7.24 (bottom). The five westernmost values in the rows of numbers for known sulphur content printed in Figs. 7.20 and 7.21 were averaged to give mean values for rectangular (2,000 ft. × 10,000 ft.) blocks of coal located between the W2,000 ft. and W12,000 ft. grid-lines. They represent averages of known data in the area south of the N18,000 ft. line and predictions to the north of this line. Models 1 and 2 yield different results: the known data in this part of the Harbour seam show a rather abrupt decrease in sulphur content near the northern edge of the worked area (between N15,000 and N17,000 ft.). This relatively low sulphur content area influences the results of Model 1 more strongly than those of Model 2. As an experiment, all known data north of N14,000 ft. were omitted and statistical analysis by Model 1 was applied to the reduced data set. The new (Model 1A) results also are shown in Fig. 7.24 (bottom). The decrease in sulphur content at the northern edge of the No. 12 Colliery could not be predicted in advance.

It was concluded that during coal mining, prospects for predicting sulphur content in coal were excellent over shorter distances (up to 2,000 ft.). This recommendation was accepted by DEVCO for routine planning of actual mining operations. On the other hand, prediction over longer distances from sulphur content of mined-out areas remained speculative. No detailed hindsight studies have been undertaken to evaluate the validity of the patterns based on Models 1 and 2 (Figs. 7.19 and 7.20). However, from generalized production records and relatively few later offshore bore-holes (Hacquebard 2002), it can be concluded that Model 2 worked better than Model 1, because there turned out to exist general decline in sulphur content of the Harbour seam in the northeastern direction.

7.3 Logistic Trend Surface Analysis of Discrete Data

Previous applications of trend analysis can be regarded as applications of the general linear model of least squares. The term "linear" in this context applies to the coefficients b_{pq} which only occur in linear form in the polynomial equation. In several applications (Whalesback and Lingan Mine case history studies), the general linear model was applied to logarithmically transformed element concentration values. Corrections had to be applied to eliminate bias arising when the fitted surfaces were recomputed to apply to original concentration values. Further modifications of the general linear model were used in the universal kriging applications to top of Arbuckle Formation, base of Matinenda Formation, and sulphur in Lingan Mine coal. One advantage of the general linear model is its simplicity in that a final solution is obtained by matrix inversion. It is more cumbersome to obtain solutions for nonlinear functions of the coefficients because then an iterative process with many inversions may have to be used. An example is provided by logistic trend surface analysis.

The logistic model (cf. Chap. 5) is commonly used for the estimation of probabilities of the occurrence of a discrete event. For trend surface analysis, the logistic model can be written in the form $S(u,v) = \frac{1}{1+e^{-T(u,v)}}$ where $T(u,v)$ is as in our previous applications of trend surface analysis. Although $T(u,v)$ can assume any real value depending on location and the values of the coefficients, $S(u,v)$ only can assume values in the interval between 0 and 1. It cannot be negative or greater than 1 and, for this reason, can be used to represent the probability of occurrence of a discrete event.

Figure 7.25 shows the Island of Newfoundland subdivided into square cells measuring 10 km on a side. Cells known to contain one or more massive sulphide deposits are shown in black. Occurrence of massive sulphide deposits in a cell is an event for which the probability of occurrence can be estimated. In Fig. 7.25, the results of two methods of trend surface analysis are shown. In both situations, the column vector **Y** of observed values consisted of elements equal to 1 for the 21 cells with deposits and elements equal to 0 for the 1,388 cells with deposits. First, $T(u,v)$ was fitted directly to the data for the second degree ($r=2$). The result is shown in Fig. 7.25 (left side). Next $S(u,v)$ was fitted with $T(u,v)$ for the second degree (Fig. 7.25, right side). In some situations, the general linear model of least squares

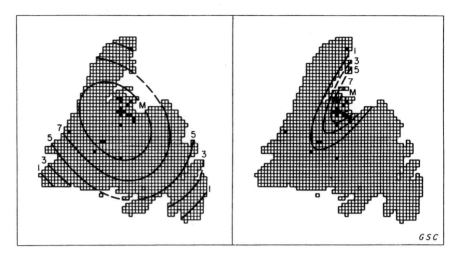

Fig. 7.25 Trend surface analysis of occurrence of massive sulphide deposits in cells measuring 10 km on a side, Island of Newfoundland. Calculated values for cells were subjected to the transformations 9*(Value − MIN)/(MAX − MIN) where MIN and MAX represent smallest and largest calculated value. Transformed values were contoured (M = 8.5); (**a**) Contours of quadratic trend surface; MIN = −0.0733, MAX = 0.0741; (**b**) Contours of logistic quadratic trend surface. Solution obtained after 13 iterations for level of convergence equal to 0.001 (*cf*. Chung 1978); MIN = 0.000; MAX = 0.2188 (Source: Agterberg 1984, Fig. 11)

can be used to estimate probabilities. However, in the example of Fig. 7.25, the pattern for the logistic model is obviously more meaningful than the pattern for the corresponding linear model, if the contours are to be interpreted as indicative of the chances that a cell contains one or more massive sulphide deposits.

The main reason that that $S(u,v)$ is more suitable than $T(u,v)$ in this type of application is related to the fact that $T(u,v)$ in Fig. 7.25 (left side) is an elliptic paraboloid. It means that any vertical intersection of it is a parabola without inflection points at either side of its maximum value. The many 0 values for empty cells then result in small positive values for area where there are known deposits and small negative values for areas without known deposits. The pattern on the right side of Fig. 7.25, on the other hand, is positive everywhere, with small positive values in the areas without known deposits. This is because the logistic trend surface is more flexible than the corresponding ordinary polynomial trend surface.

7.4 Harmonic Trend Surface Analysis

Many features in the Earth's crust tend to be periodic in that they repeat themselves at more or less the same intervals. Examples are anticlines and synclines and fault systems. In these situations harmonic trend analysis may produce better results than ordinary polynomial trend surface analysis. The 2-D power spectrum and autocorrelation function are important tools even when there are no spatial periodicities.

> **Box 7.2: 2-D Harmonic Analysis**
>
> The inverse Fourier transform $A(p,q)$ of a 2-D array of gridded $(m \times n)$ data, $X_A(i,j)$ satisfies: $A(p,q) = \frac{1}{mn} \sum_{i=0}^{m-1} \sum_{j=0}^{n-1} X_A(i,j) \cdot \exp\left[-2\pi I \left(\frac{ip}{m} + \frac{jq}{n}\right)\right]$ where $I = \sqrt{-1}$. Also, $X_A(i,j) = \frac{1}{mn} \sum_{i=0}^{m-1} \sum_{j=0}^{n-1} A(p,q) \cdot \exp\left[-2\pi I \left(\frac{ip}{m} + \frac{jq}{n}\right)\right]$. Without loss of generality, it can be assumed that $\overline{X}_A$ is zero. Any value $A(p,q)$ consists of a real part Re $(A_{p,q})$ and an imaginary part Im $(A_{p,q})$. Any wave can be represented by the continuous function: $X(p,q;u,v) = \text{Re}\left(A_{p,q}\right) \cos\left[2\pi\left(\frac{ip}{m} + \frac{jq}{n}\right)\right] - \text{Im}\left(A_{p,q}\right) \sin\left[2\pi\left(\frac{ip}{m} + \frac{jq}{n}\right)\right]$ where u and v are geographical coordinates as before. The square of amplitude is given by: $P(p,q) = \text{Re}^2(A_{p,q}) + \text{Im}^2(A_{p,q})$. The 2-D autocovariance satisfies: $C(r,s) = \sum_{p=0}^{m-1} \sum_{q=0}^{n-1} P(p,q) \cdot \exp\left[2\pi I \left(\frac{ip}{m} + \frac{jq}{n}\right)\right]$. $X(p,q;u,v)$ functions for a block of values (p,q) form a so-called harmonic trend surface with: $Y(u,v) = \sum_p \sum_p X(p,q;u,v)$.

7.4.1 Virginia Gold Mine Example

A map of average gold values for the southwestern portion of the Virginia Mine is shown in Fig. 7.26. Every contoured value is the average of all individual values in the surrounding 100 ft. × 100 ft. area. Moving average values for a 50-ft. grid were subjected to various types of statistical analysis (Krige 1966; Krige and Ueckermann 1963; Whitten 1966). Agterberg (1974) took (18 × 18) values for a 900 × 900 ft. area in the southeast corner of Fig. 7.26. Some gaps in the array in the array were filled in by using interpolation values. In order to make use of the Fast Fourier Transform method (Cooley and Tukey 1965), the array was enlarged to size (32 × 32). The 2-D spectrum $P(p,q)$ and autocorrelation function are shown in Fig. 7.27. Both maps are symmetrical with respect to the central point (origin) and only part of the lower half is shown.

In the 2-D power spectrum of Fig. 7.27 the relatively large values of $P(p,q)$ are located within a block around the origin. This indicates that the second degree harmonic trend surface provides a reasonable approximation in this application. Values of the autocorrelation function for points less than 100 ft. from the origin are biased because original values for the 50-ft. grid are averages for overlapping cells measuring 100 ft. on a side.

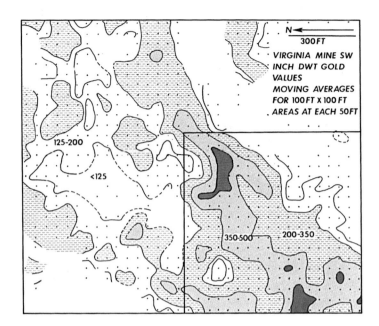

Fig. 7.26 Manually contoured map of moving-average data for gold in inch-dwt (After unpublished map by Prof. D.G. Krige; also Fig. 1 in Whitten 1966). Each value of 50-ft. grid is average for enclosing 100 × 100 ft. cell; outline for subarea selected for harmonic analysis (Source: Agterberg 1974, Fig. 75)

7.4.2 Whalesback Copper Deposit Exploration Example

Figure 7.28 shows a longitudinal cross-section of the Whalesback deposit, Newfoundland. It contains the intersection points of 188 underground drill-holes with the central part of the ore zone. The average copper value was calculated for each hole over the width of the orebody. This value was multiplied by horizontal width yielding so-called percent-foot values contoured by mining staff. The central part of this diagram shows a relatively rich copper zone that dips about 45° downward to the west. The mining grid was previously used for location in Figs. 7.8 and 7.9.

The array of Table 7.4, which was based on Fig. 7.28, was enlarged to size (32 × 32). 2-D autocorrelation function and power spectrum are shown in Figs. 7.29 and 7.30. Both diagrams illustrate the strong zoning in the deposit. The central copper zone is flanked by two elongated minima and at about 460 ft. from it there occur two other relatively copper-rich zones, although their maxima are not as high as that for the central zone. The following experiment indicates that harmonic analysis can be an excellent exploration tool in a situation of this type.

Development of the Whalesback copper deposit was performed in two different stages. Copper percent-foot values from the first surface boreholes are shown in Fig. 7.31b. There are only 20 exploration values, which are irregularly distributed. The question can be asked of how the pattern of Fig. 7.28, which is based on 188 more

7.4 Harmonic Trend Surface Analysis

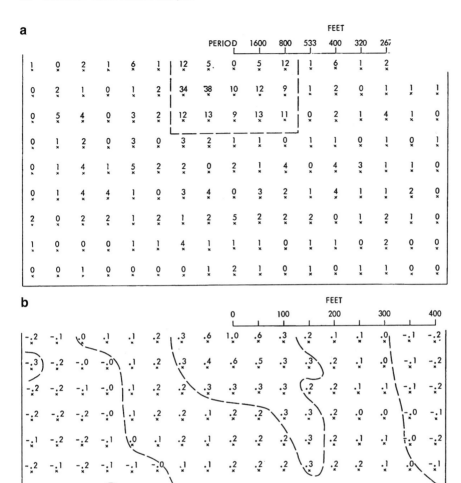

Fig. 7.27 Part of two-dimensional power spectrum (**a**) and autocorrelation (**b**) functions for array depicted in Fig. 7.26. Note that largest values in **a** are concentrated in block around the origin, suggesting a trend which may be described by harmonic trend surface of second degree. Autocorrelation surface is elongated in the direction of the trend (Source: Agterberg 1974, Fig. 76)

precise underground drill-holes that are more evenly spaced and sampled in more detail as well, can be predicted from these 20 values. This problem is analogous to that of constructing a topographical contour map of a mountain chain when only a few elevations would be provided for interpolation. Polynomial trend surface analysis applied to the 20 development values produces patterns that show no resemblance to

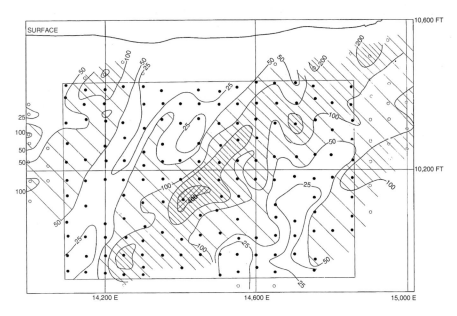

Fig. 7.28 Manually contoured map of percent-foot values for copper in longitudinal section across Whalesback deposit, Newfoundland. Each *dot* represents the approximate intersection point of an underground borehole with the center of the mineralized zone (Source: Agterberg 1974, Fig. 47)

Table 7.4 Matrix of underground drill-hole data for area depicted in Fig. 7.28

29	146	116	76	22	6	(23)	11	18	48	60	94	9	32	137	128
51	47	91	(60)	12	21	43	32	40	52	158	139	94	77	118	77
43	60	94	34	19	67	31	22	65	112	126	78	258	80	(86)	110
91	(70)	60	34	66	29	(41)	32	164	286	160	(126)	(97)	(97)	74	70
91	73	68	21	75	44	35	73	297	89	140	52	66	49	52	43
105	43	36	42	40	77	83	272	209	107	177	40	29	42	52	199
66	32	(47)	58	128	114	354	(194)	284	169	(91)	25	(34)	9	(61)	27
47	(38)	31	86	76	(159)	(150)	113	43	78	(69)	40	21	64	(51)	35
41	8	41	125	196	128	118	74	76	33	70	65	29	64	100	57
19	19	62	260	135	174	135	163	23	18	33	24	36	95	(67)	(78)
20	35	59	168	199	(146)	91	(85)	16	13	58	30	(40)	23	(59)	(68)

Source: Agterberg (1974, Table XXVIII)
Values represent products of horizontal width of orebody (in feet) and average concentration value of copper (in percent)

the pattern of Fig. 7.28. In fact, conventional analysis of variance indicated that the linear ($R^2 = 0.16$) and quadratic ($R^2 = 0.32$) fits were not statistically significant.

Agterberg (1969) fitted the function $X(p, q; u, v)$ (see Box 7.2) by least squares to estimate the coefficients $b_1 = \text{Re}(A_{p,q})$ and $b_2 = \text{Im}(A_{p,q})$ for many possible directions of axes and periods for the waves. Ideally, a least-squares model should have been used by which the four parameters (direction of axis, period, amplitude

7.4 Harmonic Trend Surface Analysis

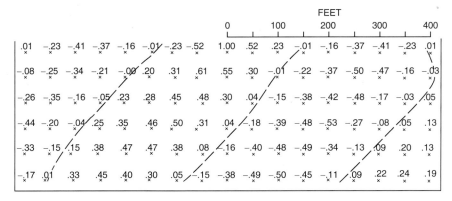

Fig. 7.29 Two-dimensional autocorrelation function for data from Table 7.4, Whalesback copper deposit (Source: Agterberg 1974, Fig. 61)

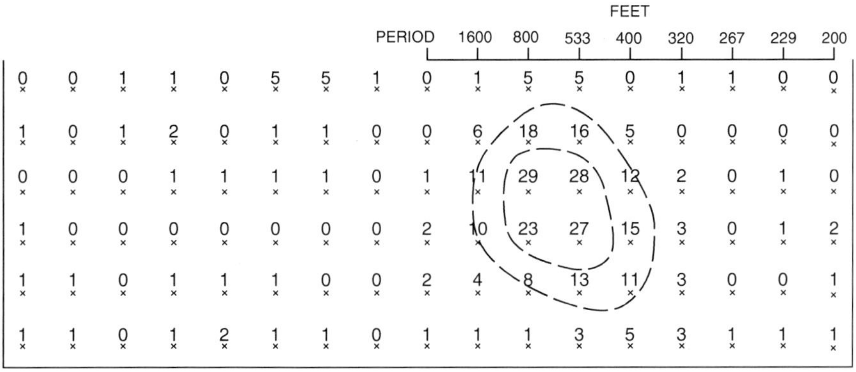

Fig. 7.30 Part of two-dimensional power spectrum for data in Table 7.4; sharply defined peak reflects zoning of copper percent-foot values shown in Fig. 7.28 (Source: Agterberg 1974, Fig. 77)

and phase) are estimated simultaneously but then a nonlinear model should have been developed and applied. The linear model can only be used when p/m and q/n are assumed to be known. For each separate sine wave, R^2 was plotted as a percentage value in Fig. 7.31a. The result can be regarded as a 2-D power spectrum for irregularly spaced data. Since three coefficients were fitted for each sine wave (including the constant term that is related to the mean value), each R^2 value could be converted into an F-value at the upper tail of a theoretical F-distribution with 3 and 17 degrees of freedom. The $F_{0.95}$- and $F_{0.99}$-values correspond to 27-% and 37-% values for R^2 and these were contoured. There are two contoured peaks in Fig. 7.31a and one of these corresponds to the peak in Fig. 7.30. The contour map of the sine-wave whose amplitude falls on the well-developed peak of Fig. 7.31a is shown in Fig. 7.31b, and its profile for a line along the surface ($V = 10{,}600$) in Fig. 7.31c. The 20 original observations were projected onto Fig. 7.31c along lines parallel to the contours. Obviously, the contours in Fig. 7.31b are an

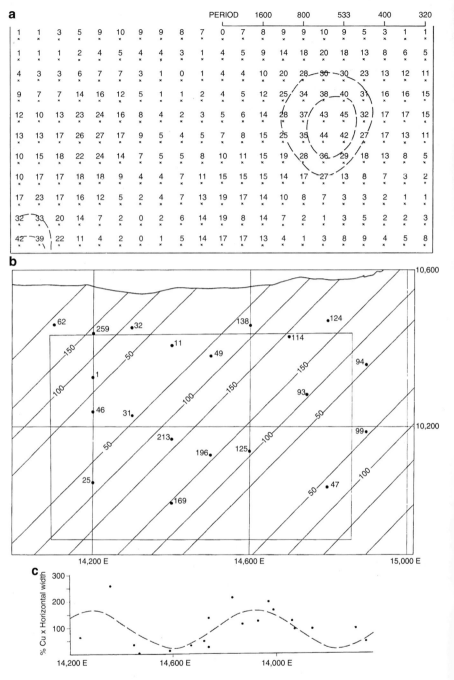

Fig. 7.31 Harmonic analysis of irregularly spaced data (After Agterberg 1969). Intersection points of 20 exploratory bore-holes from topographic surface with ore-zone shown in **b**. Two-dimensional sine-waves were fitted to 20 values by least squares method; percentage explained sum of squares for each fitted wave is shown in **a** with period scale in feet. Wave for value 44 on peak in **a** is shown by straight-line contours in **b**, and also in cross-section for horizontal line (V = 10,600) near topographic surface in **c** (Source: Agterberg 1974, Fig. 78)

7.4 Harmonic Trend Surface Analysis

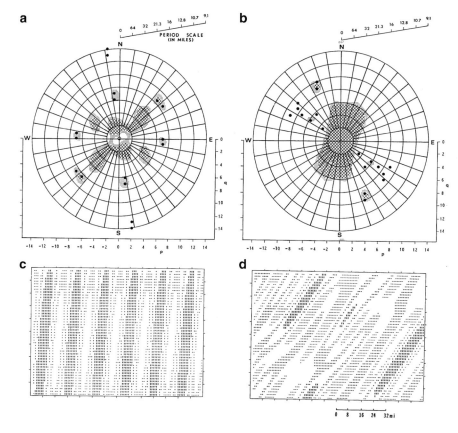

Fig. 7.32 Two-dimensional power spectra (**a** and **b**) and phase maps (**c** and **d**) for distribution of copper (N-S set only) and gold occurrences in east-central Ontario (gridded area in Fig. 3.4). *Dots* in **a** and **b** denote *P*-values selected for constructing phase maps (Source: Agterberg 1974, Fig. 81)

oversimplification of those shown in Fig. 7.28 but existence of zones with alternately high and low copper was correctly predicted.

7.4.3 East-Central Ontario Copper and Gold Occurrence Example

The final example of harmonic trend analysis is concerned with the distribution of copper and gold occurrences in the western Abitibi volcanic belt (Fig. 4.12). The pattern of copper deposits in Fig. 4.13 was coded as an array with 28 columns and 20 rows with numbers of copper deposits per cell. All data were corrected for their mean and the array was augmented to size (32 × 32) by zeros. The inverse Fourier transform was computed with the resulting 2-D power spectrum shown in Fig. 7.32. Individual *P*-values are distributed as chi-square with a single degree of freedom. A. moving average for four *P*-values was contoured of the 95-% and 99-% fractiles

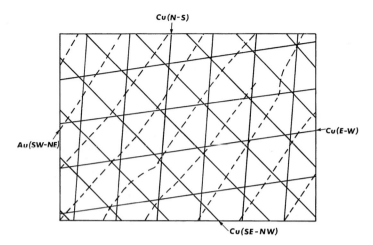

Fig. 7.33 Crest-lines for selected phase maps of copper and gold occurrences (Source: Agterberg 1974, Fig. 82)

of the chi-square distribution with four degrees of freedom (cf. Agterberg and Fabbri 1973). A similar procedure was followed for gold deposits in the same area. Relatively large P-values (with period ≤ 20 miles) at some distance from the origin were selected and marked as dots in Fig. 7.32a, b. Most of these are accompanied by a contoured maximum. Directionally, these dots form small clusters of which there are three for copper and two for gold. The crest-lines of the three (N-S, E-W and SE-NW) copper clusters and the large (SW-NE) gold cluster are shown in Fig. 7.33b. These sets of lines tend to intersect at the same points. Agterberg and Fabbri (1973) speculated that patterns of lines indicating larger concentrations of copper and gold deposits may be useful for extrapolation into parts of the Abitibi area where bedrock is hidden because of relatively thick glacial cover.

References

Agterberg FP (1964) Methods of trend surface analysis. Q Colo Sch Mines 59:111–130
Agterberg FP (1966) Trend surfaces with autocorrelated residuals. In: Proceedings of the symposium on computers and operations research in the mineral industries. Pennsylvania State University, pp 764–785
Agterberg FP (1968) Application of trend analysis in the evaluation of the Whalesback Mine, Newfoundland. Can Inst Min Metall 9:77–88
Agterberg FP (1969) Interpolation of a really distributed data. Q Colo Sch Mines 64(3):217–237
Agterberg FP (1970) Autocorrelation functions in geology. In: Merriam DF (ed) Geostatistics. Plenum, New York, pp 113–142
Agterberg FP (1974) Geomathematics. Elsevier, Amsterdam
Agterberg FP (1984) Trend surface analysis. In: Gaile GI, Willmott CJ (eds) Spatial statistics and models. Reidel, Dordrecht, pp 147–171

References

Agterberg FP, Chung CF (1973) Geomathematical prediction of sulphur in coal, new Lingan Mine area, Sydney coal field. Canadian Mining and Metallurgical Bulletin, October, pp 85–96

Agterberg FP, Fabbri AG (1973) Harmonic analysis of copper and gold occurrences in the Abitibi area of the Canadian Shield. In: Proceedings of the 10th symposium on application of computers in the mineral industries. South African Institute of Mining and Metallurgy, Johannesburg, pp 193–201

Bain GW (1960) Patterns to ores in layered rocks. Econ Geol 55:695–731

Chung CF (1978) Computer program for the logistic model to estimate the probability of occurrence of discrete events. Geological Survey of Canada Paper 78(11)

Cooley JW, Tukey JW (1965) An algorithm for the machine calculation of complex Fourier series. Math Comput 19:297–307

Coons RL, Woolard GP, Hershery G (1967) Structural significance and analysis of mid-continent gravity high. Bull Am Assoc Pet Geol 51:2381–2399

Cressie NAC (1991) Statistics for spatial data. Wiley, New York

Davis JC (1973) Statistics and data analysis in geology. Wiley, New York

Derry DR (1960) Evidence of the origin of the Blind River uranium deposits. Econ Geol 55:906–927

Good GJ (1964) FOTRAN II trend-surface program for the IBM 1620. Kansas Geological Survey, Special District Publication 14

Hacquebard PA (2002) Potential coalbed methane resources of Atlantic Canada. Int J Coal Geol 52:3–28

Haining RP (1987) Trend-surface models with regional and local scales of variation with an application to aerial survey data. Technometrics 29:461–469

Huijbregts C, Matheron G (1971) Universal kriging (an optimal method for estimating and contouring in trend surface analysis). Can Inst Min Metall 12:159–169

Krige DG (1966) Two-dimensional weighted moving average trend surfaces for ore valuation. In: Proceedings of the symposium on mathematical statistics and computer applications in ore valuation, Johannesburg, pp 13–38

Krige DG, Ueckermann HJ (1963) Value contours and improved regression techniques for ore reserve valuations. J S Afr Inst Min Metall 63:429–452

Krumbein WC, Graybill FA (1965) An introduction to statistical models in geology. McGraw-Hill, New York

MacGregor ID, Smith CH (1963) The use of chrome spinels in petrographic studies of ultramafic intrusions. Can Mineral 7:403–412

Matalas NC (1963) Autocorrelation of rainfall and streamflow minimums. US Geological Survey Professional Paper 434-B (and -D)

Matheron G (1967) Kriging or polynomial interpolation procedures? Trans Can Inst Min Metall 70:240–244

Robertson JA (1966) The relationship of mineralization to stratigraphy in the Blind River area, Ontario. Geol Assoc Can Spec Pap 3:121–136

Stanton RL (1972) Ore petrology. McGraw-Hill, New York

Tukey JW (1970) Some further inputs. In: Merriam DF (ed) Geostatistics. Plenum, New York, pp 163–174

Vistelius AB, Hurst VJ (1964) Phosphorus in granitic rocks of America. Geol Soc Am Bull 75:1055–1092

Walsh JH, Visman J, Whalley BJP, Ahmed SM (1969) Removal of pyrite from Cape Breton Coals destined for the use in metallurgical processes. In: Proceedings of the 9th commonwealth mining and metallurgical congress, Institution of Mining and Metallurgy, London, Paper 35

Watson GS (1971) Trend surface analysis. J Int Assoc Math Geol 3:215–226

Whitten EHT (1966) The general linear equation in prediction of gold content in Witwatersrand rocks. In: Proceedings of the symposium on mathematical statistics and computer applications in ore valuation, Johannesburg, pp 124–156

Whitten EHT (1970) Orthogonal polynomial trend surfaces for irregularly spaced data. J Math Geol 2:141–152

Chapter 8
Statistical Analysis of Directional Features

Abstract Many geological features are directed in 2-D or 3-D space, either as undirected or directed lines. Unit vectors are used for their spatial representation. Vectors with magnitudes can be used in some applications. Axes of pebbles in glacial drift provide an example of undirected lines; strengths of magnetization in rock samples exemplify the situation of directed lines to be represented as unit vectors or as vectors with magnitudes. Methods of unit vector field construction can be used to extract regional variation patterns. If there is no significant change of direction within the domain of study, various statistical frequency distribution models can be used for estimating the mean direction or pole and measures of dispersion. Well-known examples are the Fisher distribution for directed lines and the Scheidegger-Watson distribution for undirected lines. In this chapter, unit vector fields are fitted to regional data with variable mean directions using extensions of polynomial trend surface analysis. A relatively simple example consists of determining the preferred paleocurrent directions in sandstones of the Triassic Bjorne Formation on Melville Island, Canadian Arctic Archipelago. Later examples are from structural geology of the Eastern Alps. Directions of the axes of Hercynian minor folds in the crystalline basement of the Dolomites in northern Italy show relatively strong spatial variability, both locally and regionally. Averaging measurements from different outcrops within relatively small sampling domains shows patterns of systematic regional variations that represent Alpine reactivation of Hercynian schistosity planes (*s*-planes) causing rotations of the original minor fold axis directions to the south of the Periadriatic Lineament. Interpretation of seismic data from along the north-south TRANSALP profile that intersects the Periadriatic Lineament near Bruneck (Brunico) in the Pustertal (Pusteria) indicates rotation of Hercynian basement rocks into subvertical positions with subvertical to steeply east-dipping Hercynian minor fold axes. Subsequently, Late Miocene northward and north northeastward movements of the Adria microplate underneath the Eurasian plate resulted in sinistral motion of the crystalline basement rocks in the Bruneck area and strong neo-Alpine compression of basement rocks in the Pustertal to the east. At the same time there was overthrust sheet formation in the Strigno area along the Sugana Fault located to the south of the Italian Dolomites.

Keywords Directed and undirected lines • Unit vector fields • Bjorne formation paleodelta • San Stefano quartzphyllites • TRANSALP profile • Ductile extrusion model • Pustertal tectonites • Italian dolomites • Adria microplate • Defereggen "Schlinge"

8.1 Directed and Undirected Lines

Various geological attributes may be approximated by lines or planes. Measuring them results in "angular data" consisting of azimuths for lines in the horizontal plane or azimuths and dips for lines in 3-D space. Although a plane usually is represented by its strike and dip, it is fully determined by the line perpendicular to it, and statistical analyses of data sets for lines and planes are analogous. Examples of angular data are strike and dip of bedding, banding and planes of schistosity, cleavage, fractures or faults. Azimuth readings with or without dip are widely used for sedimentary features such as axes of elongated pebbles, ripple marks, foresets of cross-bedding and indicators of turbidity flow directions (sole markings). Then there are the B-lineations in tectonites; problems at the microscopic level include that of finding the preferred orientation of crystals in a matrix (e.g., quartz axes in petrofabrics). Another example is the direction of magnetization in rocks. Statistical theory for the treatment of unit vectors in 2-D (Fisher 1993) and 3-D (Fisher et al. 1987) is well developed.

8.1.1 Doubling the Angle

It is useful to plot angular data sets under study in a diagram before statistical analysis is attempted. Azimuth readings may be plotted on various types of rose diagrams (Potter and Pettijohn 1963). If the lines are directed, the azimuths can be plotted from the center of a circle and data within the same class intervals may be aggregated. If the lines are undirected, a rose diagram with axial symmetry may be used such as the one shown in Fig. 8.1. The method of doubling the angle then is used to estimate the mean azimuth (Krumbein 1939). The average is taken after doubling the angles and dividing the resulting mean angle by two. In Agterberg (1974) problems of this type and their solutions are discussed in more detail. Krumbein's solution is identical to fitting a major axis to the points of intersection of the original measurements with a circle. Axial symmetry if preserved when the major axis is constructed. This method can be extended to the situation of undirected lines in 3-D space. Then the first principal component or dominant eigenvector is computed using the coordinates of the intersection points of the lines, which pass through the center of a sphere, and the surface of the sphere (*cf.* Sect. 1.4.1).

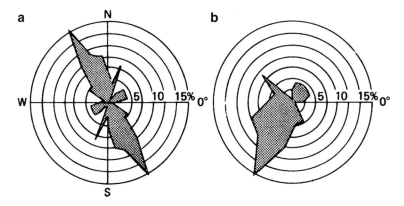

Fig. 8.1 Axial symmetric rose diagram of the direction of pebbles in glacial drift in Sweden according to Köster (1964). Step from (a) to (b) is accomplished by doubling the angles (After Koch and Link 1971, and Batschelet 1965) (Source: Agterberg 1974, Fig. 99)

8.1.2 Bjorne Formation Paleodelta Example

The Bjorne Formation is a predominantly sandy unit of Early Triassic age. It was developed at the margin of the Sverdrup Islands of the Canadian Arctic Archipelago. On northwestern Melville Island, the Bjorne Formation consists of three separate members which can be distinguished, mainly on the basis of clay content which is the lowest in the upper member, C. The total thickness of the Bjorne Formation does not exceed 165 m on Melville Island. The formation forms a prograding fan-shaped delta. The paleocurrent directions are indicated by such features as the dip azimuths of planar foresets and axes of spoon-shaped troughs (Agterberg et al. 1967). The average current direction for 43 localities of Member C is shown in Fig. 8.3a. These localities occur in a narrow belt where the sandstone member is exposed at the surface. The azimuth of the paleocurrents changes along the belt (Fig. 8.2).

The variation pattern shows many local irregularities but is characterized by a linear trend. In the (U,V) plane for the coordinates (see Fig. 8.3), the linear trend surface is: $x = F(u,v) = 292.133 + 27.859u + 19.932v$ (degrees). At each point on the map, x represents the tangent of a curve for the paleocurrent trend that passes through that point, or $\frac{dv}{du} = -\tan x$. It readily is shown that: $du = \frac{dx}{a_1 - a_2 \tan x}$ where $a_1 = 27.859$ and $a_2 = 19.932$. Integration of both sides gives:

$$u = C + \frac{1}{a_1^2 + a_2^2}[a_1 x - a_2 \log_e(a_1 \cos x + y \sin x)]$$

with $y = a_2$ if $\tan x > \frac{a_1}{a_2}$ and $y = -a_2$ if $\tan x < \frac{a_1}{a_2}$. The result is applicable when x is expressed in degrees. When $x \geq 360°$, the quantity $360°$ may be subtracted from

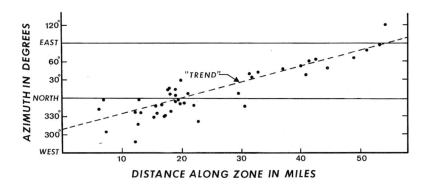

Fig. 8.2 Variation of preferred paleocurrent direction along *straight line* coinciding with surface outcrop of Member C, Bjorne Formation (Triassic, Melville Island, Northwest Territories, Canada); systematic change in azimuth is represented by *trend line* (Source: Agterberg 1974, Fig. 12)

x, because azimuths are periodic with period 360°. The constant C is arbitrary. A value can be assigned to it by inserting specific values for u and v into the equation. For example, if $u = 0$ and $x = 0$, $C = -0.7018$. It can be used to calculate a set of values for u forming a sequence of values for x; the corresponding values for v follow from the original equation for the linear trend surface. The resulting curves (1) and (2) are shown in Fig. 8.3b. If the value of C is changed, the curves (1) and (2) become displaced in the $x = 54°.4$ direction. In this way, a set of curves is created that represents the paleocurrent direction for all points in the area.

Suppose that the paleocurrents were flowing in directions perpendicular to the average topographic contours at the time of sedimentation of the sand. If curve (1) of Fig. 8.3b is moved in the 144°4 direction over a distance that corresponds to 90° in x, the result represents a set of directions which are perpendicular to the paleocurrent trends. Four of these curves which may represent the shape of the paleodelta are shown in Fig. 8.4c. These contours, which are labeled a, b, c, and d satisfy the preceding equation for different values of C and with x replaced by $(x + 90°)$ or $(x - 90°)$. Definite values cannot be assigned to these contours because x represents a direction and not a vector with both direction and magnitude.

In trend surface analysis, the linear trend surface (also see Fig. 8.4a) had explained sum of squares $ESS = 78\%$. The complete quadratic and cubic surfaces has ESS of 80% and 84%, respectively. Analysis of variance for the step from linear to quadratic surface resulted in $\hat{F}(3, 37) = \frac{(80-78)/3}{(100-80)/37} = 1.04$. This would correspond to $F_{0.60}$ (3,37) showing that the improvement in fit is not statistically significant. It is tacitly assumed that the residuals are not autocorrelated as suggested by their scatter around the line of Fig. 8.1. Consequently, the linear trend surface as shown in Fig. 8.4a is acceptable in this situation. The 95% confidence interval for this linear surface is shown in Fig. 8.4b. This is a so-called half-confidence interval with values equal to $[(p+1) \cdot F \cdot s^2(\hat{Y})]$ with $F = F_{0.95}(3,40) = 2.84$ and $s^2(\hat{Y}) = s^2 X'_k (X'X)^{-1} X_k$ with residual variance $s^2 = 380$ square degrees.

8.1 Directed and Undirected Lines 281

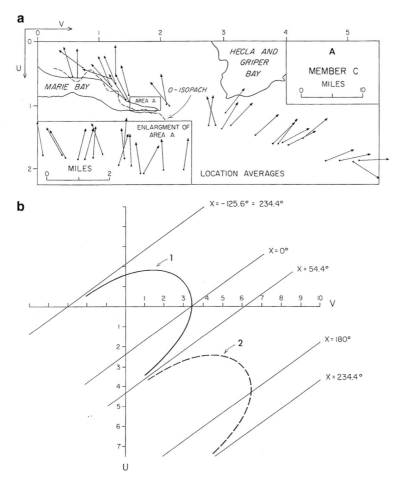

Fig. 8.3 Reconstruction of approximate preferred paleocurrent direction and shape of paleodelta from preferred current directions, Member C, Bjorne Formation, Melville Island (After Agterberg et al. 1967). (a) Preferred directions of paleocurrent indicators at measurement stations; 0-isopach applies to lower part of Member C only. (b) Graphical representation of solution of differential equation for paleocurrent trends. Inferred approximate direction of paleoriver near its mouth is N54.4°E (Source: Agterberg 1974, Fig. 13)

All flow lines in Fig. 8.4a converge to a single line-shaped source. They have the same asymptote in common which suggests the location of a river. An independent method for locating the source of the sand consists of mapping the grade of the largest clasts contained in the sandstone at a given place. Four grades of larges clasts could be mapped in the area of study. They are: (1) pebbles, granule, coarse sand; (2) cobbles, pebbles; (3) boulders, cobbles; and (4) boulders (max. 60 cm). The size of the clasts is larger where the velocity of the currents was higher. Approximate grade contours for classes 1–4 are shown in Fig. 8.4c for comparison. This pattern corresponds to the contours constructed for the delta.

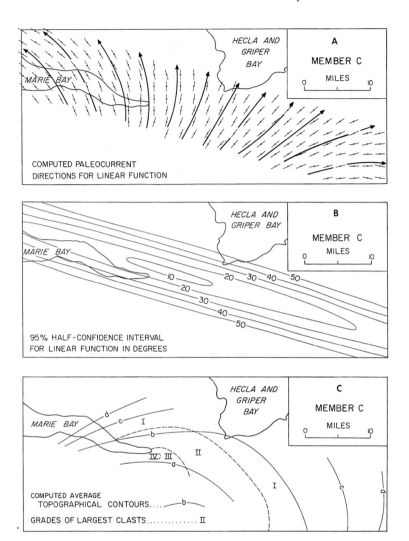

Fig. 8.4 A. Linear trend for data of Fig. 8.3a. (a) Computed azimuths are shown by line segments for points on regular grid, both inside and outside exposed parts of Member C; flow lines are based on curves 1 and 2 of Fig. 8.3b. (b) Contour map of 95 % half-confidence interval for A; confidence is greater for area supported by observation points. (c) Estimated topographic contours for delta obtained by shifting curve 1 of Fig. 8.3b, 90° in the southeastern direction (=perpendicular to isoazimuth lines), and then moving it into four arbitrary positions by changing the constant C. Contoured grades of largest clasts provide independent information on shape of delta (Source: Agterberg 1974, Fig. 41)

8.1 Directed and Undirected Lines

Fig. 8.5 Two examples of minor folds in the Pustertal quartzphyllite belt. *Left side*: East-dipping minor folds near Welsberg (Monguelfo; location near midpoint of western boundary of subarea delineated in Fig. 8.16); *Right side*: ditto along TRANSALP profile near subvertical contact with Permotriassic (grid coordinates 206-767; part of MS112, see Fig. 8.16) (Source: Agterberg 2012, Fig. 1)

8.1.3 Directed and Undirected Unit Vectors

Directional features play an important role in structural geology. Basic principles were originally reviewed and developed by Sander (1948) who associated a 3-dimensional (A, B, C) Cartesian coordinate system with folds and other structures. A-axis and C-axis were defined to be parallel to the directions of compression and expansion, respectively, with the B-axis perpendicular to the AC-plane representing the main plane of motion of the rock particles. Two examples of Hercynian minor folds with clearly developed B-axes are shown in Fig. 8.5 (from Agterberg 1961) for quartzphyllites belonging to the crystalline basement of the Dolomites in northern Italy. Measurements on B-axes from these quartzphyllites were analyzed previously (Agterberg 1959, 1961, 1974, 1985, 2004, 2012) and will again be used here with re-interpretations as needed.

Stereographic projection using either the Wulff net or the equal-area Schmidt net continue to be useful tools for representing sets of directional features from different outcrops within the same neighbourhood or for directions derived from crystals in thin sections of rocks. Contouring on the net is often applied to find maxima representing preferred orientations, which also can be estimated using methods developed by mathematical statisticians, especially for relatively small sample sizes (see, e.g., Fisher et al. 1987).

Reiche (1938) used the vector mean to find the preferred orientation of directional features. Using a paleomagnetic data set, Fisher (1953) further developed this method for estimating the mean orientation from a sample of directed unit vectors. The frequency distribution of unit vectors may satisfy a Fisher distribution, which is the spherical equivalent of a normal distribution. The mean unit vector is estimated as follows. For a sample of measurements, the average direction cosine is determined along each of the three axes of a Cartesian coordinate system. The strength $|R|$ of the vector R described by the three average direction cosines is less than one unless all measured vectors are parallel to one another. The vector R points in the same direction as the vector sum of all observed unit vectors, and the quantity $1-|R|$ provides a measure of the scatter of the measurements around their mean vector.

8.2 Unit Vector Fields

Methods of unit vector field analysis include 2-dimensional polynomial trend-surface fitting applied separately to the three direction cosines of the measurements followed by combining the estimated values to construct the unit vector field. A new mathematical derivation of the underlying theory was presented in Agterberg (2012). Examples of application are to B-axes for Hercynian minor folds and schistosity planes in quartzphyllites belonging to the crystalline basement of the Italian Dolomites. During Alpine orogeny, these s-planes were not only refolded but there were large-scale sliding movements along the Hercynian schistosity resulting in various kinds of rotations of the B-axes. The unit vector fields help to outline these deformation patterns. The unit vector field for the B-axes in quartzphyllites in the San Stefano area east of the Dolomites shows existence of an SW vergent Alpine anticlinal structure. Two examples for quartzphyllites in the Pustertal adjacent to the Periadriatic Lineament can be explained as part of the south vergent upper-crustal response to the north and north-eastward directed subduction of the Adria microplate below the Eastern Alps from the late Miocene onward. Because of this motion the crystalline basement north of the Dolomites was subjected to significant N-S shortening with probably a sinistral component in the vicinity of the TRANSALP profile near Bruneck. Box 8.1 contains a brief review of the mathematics of polynomial unit vector field fitting. It will be followed by applications to B-axes in quartzphyllites south of the Periadriatic Lineament.

Box 8.1: Mathematics of Polynomial Unit Vector Field Fitting

Let U_{oi} represent an observed unit vector with strength $|U_{oi}|=1$ at the i-th observation point ($i = 1, 2, \ldots, n$), and U_{ei} with $|U_{ei}|=1$ the corresponding estimated unit vector. U_{ei} has the same direction as a vector R_i for which the quantity $\Sigma |U_{oi}-R_i|^2$ will be minimized over the n observations. Suppose that the three direction cosines of observed unit vectors U_{oi}, the estimated vectors R_i, and the estimated unit vectors U_{ei} are written as ℓ_{hi}, λ_{hi} and λ_{hi}^* ($h = 1, 2, 3$), respectively. Then $\sum_{h=1}^{3} \ell_{hi}^2 = 1$; $\sum_{h=1}^{3} \lambda_{hi}^2 = |R_i|^2$; and $\sum_{h=1}^{3} \lambda_{hi}^{*2} = 1$. In the Cartesian coordinate system to be used, the x-axis ($h = 1$) points northwards, the y-axis ($h = 2$) eastwards, and the z-axis ($h = 3$) upwards, respectively.

Suppose that trend surface analysis is applied to each direction cosine ℓ_h yielding three polynomial equations of degree p in terms of the geographical coordinates x and y with $\lambda_{hi} = \lambda_h(x_i, y_i) = \sum_{j+k \leq p} b_{hjk} x_i^j y_i^k$ where $j = 0, 1, \ldots, p$;

(continued)

Box 8.1 (continued)

$k = 0, 1, \ldots, p$; and $j + k \leq p$. Because R_i and U_{ei} point in the same direction, the angle θ_i between U_{oi} and U_{ei} satisfies: $\cos \vartheta_i = \sum_{h=1}^{3} \lambda_{hi}^* \ell_{hi} = \frac{\sum_{h=1}^{3} \lambda_{hi} \ell_{hi}}{|R_i|}$; $|R_i| = \left[\sum_{h=1}^{3} \lambda_{hi}^2 \right]^{1/2}$. The difference vector between U_{oi} and R_i, which can be written as $d_i = U_{oi} - R_i$, has strength $|d_i|$ with $|d_i|^2 = 1 + |R_i|^2 - 2|R_i| \cos \vartheta_i = \sum_{h=1}^{3} \ell_{hi}^2 + \sum_{h=1}^{3} \lambda_{hi}^2 - 2 \sum_{h=1}^{3} \lambda_{hi} \ell_{hi} = \sum_{h=1}^{3} (\ell_{hi} - \lambda_{hi})^2$. The sum of squares $T = \Sigma |d_i|^2$ is a minimum when $\frac{\partial T}{\partial b_{mjk}} = 2 \sum_{i=1}^{n} \sum_{h=1}^{3} (\ell_{hi} - \lambda_{hi}) \frac{\partial \lambda_{hi}}{\partial b_{mjk}} = 0$ for all coefficients b_{mjk} ($m = 1, 2, 3$; $j = 1, \ldots, p$; $k = 0, 1, \ldots, p$; $j + k \leq p$). Because the three polynomial functions $\lambda_h (x_i, y_i)$ are linear in b_{mjk}, their partial derivatives satisfy $\frac{\partial \lambda_{hi}}{\partial b_{mjk}} = x_i^k y_i^k$ if $m = h$ and $\frac{\partial \lambda_{hi}}{\partial b_{mjk}} = 0$ if $m \neq h$. This allows us to split the n Gaussian normal equations into 3 groups (for $m = 1, 2$ and 3, respectively) with $\sum_{i=1}^{n} [\ell_{mi} - \sum_{j+k \leq p} b_{hjk} x_i^j y_i^k] x_i^s y_i^t = 0$ where $s = 0, 1, \ldots, p$; $t = 0, 1, \ldots, p$; $s + t \leq p$. The simultaneous linear equations for each group can be solved yielding estimates of the coefficients required to estimate $\lambda_h (x_i, y_i)$. Finally, normalization gives estimates of $\lambda_h^*(x_i, y_i)$ ($h = 1, 2, 3$). For representation of the results as continuous functions on a map, it is convenient to use the azimuth $\alpha (x_i, y_i)$ and dip $\varphi (x_i, y_i)$ of the unit vectors with $\tan \alpha(x, y) = - \lambda_2^*(x, y)/\lambda_1^*(x, y)$ and $\sin \varphi(x, y) = - \lambda_3^*(x, y)$.

8.2.1 San Stefano Quartzphyllites Example

The following example illustrates usefulness of the mean unit vector. Fig. 8.6 (after Agterberg 1961) shows locations and average values of B-axes and schistosity-planes in the San Stefano area east of the Dolomites. Trace of axial plane of neo-Alpine anticline in Fig. 8.6 approximately coincides with the line between E and SE dipping average minor folds, and those dipping W to N. Plots on the Schmidt net for measurements from subareas A and B (outlined on Fig 8.6) are shown in Fig. 8.7. Mean values plotted in Figs. 8.6 and 8.7 were calculated by means of a simple, approximate method of averaging azimuths, strikes and dips. Note that the preferred orientation of the B-axes in subarea A is markedly different from that in Subarea B. Fig. 8.8 (after Agterberg 1985) shows original measurements or average B-axes for (100 m × 100 m) squares in comparison with unit vector means based on larger samples, combining all measurements from within (1 km × 1 km) or (2 km × 2 km) circular neighborhoods. A clear regional pattern confirming the global pattern of Fig. 8.6 emerges in Fig. 8.8b from the mean unit vectors that are based on the larger samples. Structural interpretation of the change in average dip of the B-axes is as follows. The ESE-striking schistosity-planes and the B-axes are Hercynian in age because they occur in large boulders within the

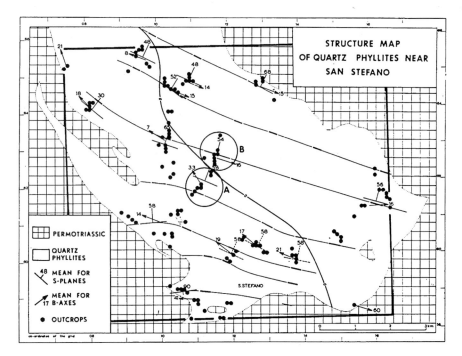

Fig. 8.6 Preferred attitudes of schistosity planes and B-lineations with extrapolated regional pattern for crystalline basement of eastern Dolomites near San Stefano (After Agterberg 1961). Cells of 100 m on a side, where one or more B-lineations were measured, are shown by *dots*. Domains for Fig. 8.7 are indicated by *circles* (a and b); outline of boundary of Fig. 8.9 is also shown (Source: Agterberg 1974, Fig. 110)

Permian Verrucano conglomerate, at the base of the Permotriassic sedimentary succession of the Italian Dolomites. During Alpine orogeny, the San Stefano crystalline formed the core of an anticline in which the Hercynian schistosity planes were re-activated. The trace of its Alpine NW-SE trending axial plane intersects the approximately WNW-ESE striking Hercynian schistosity-planes (Fig. 8.6). Quadratic and cubic unit vector field solutions based on 379 original measurements for the subarea delineated by heavy line in Fig. 8.6 are shown in Fig. 8.9.

A difference between the B-axes of Fig. 8.5 and the paleomagnetic measurements analyzed by Fisher (1953) is that the latter are directed whereas the B-axes of Fig. 8.5 are undirected. Unless the scatter of lines representing individual measurements is very large, the vector mean for undirected lines can be estimated in the same way as that for directed lines. Suppose that θ_i represents the angle of the i-th measurement and the unit vector mean, then the mean vector method is equivalent to maximizing $\Sigma \cos \theta_i$ representing the sum of $\cos \theta_i$ for all measurements. If the scatter is so large that one or more of values of $\cos \theta_i$ for the undirected lines could be negative, $\Sigma \cos^2 \theta_i$ can be maximized instead of $\Sigma \cos \theta_i$. The latter method was first used by Scheidegger (1965) for fault-plane solutions of earthquakes and by Loudon (1964) for orientation data in structural geology. The underlying frequency

8.2 Unit Vector Fields

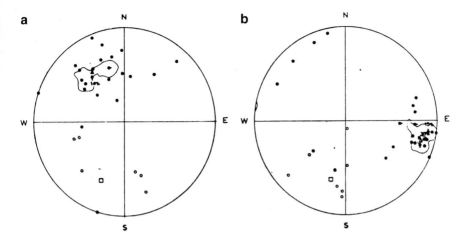

Fig. 8.7 Lineations (*dots*) and *s*-poles (*circles*) plotted on equal-area Schmidt net, lower hemisphere, for domains A and B in central part of Fig. 8.6. Averages plotted in Fig. 8.6 are indicated by *triangles* and *open circles*, respectively (Source: Agterberg 1974, Fig. 111)

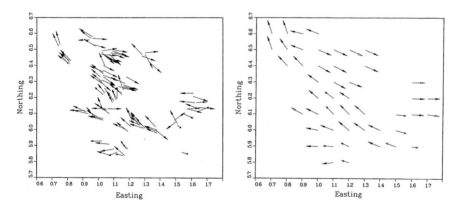

Fig. 8.8 San Stefano area. Unit of distance for coordinates of the grid is 10 km. Length of arrow is proportional to cosine of angle of dip (Maximum *arrow* length is 1 km). *Left side*: B-axes and unit vector means in (100 m × 100 m) cells also shown in Fig. 8.6. *Right side*: Unit vector means of B-axes located within circles with 1 km radius (solid arrow heads) and 2 km radius (open arrow heads) (Source: Agterberg 2012, Fig. 4)

distribution of the measurements, for which maximizing $\Sigma \cos^2 \theta_i$ is optimal, is not the Fisher distribution but the so-called Scheidegger-Watson distribution (Watson 1985). In Sect. 1.4 it was discussed that this method involves extracting the dominant eigenvector from a (3 × 3) matrix with elements determined by the direction cosines of the observed unit vectors. With respect to shapes of frequency distributions for unit vectors, it should be kept in mind that they may not be isotropic. This is illustrated in Fig. 8.7 where it can be seen that the B-axes,

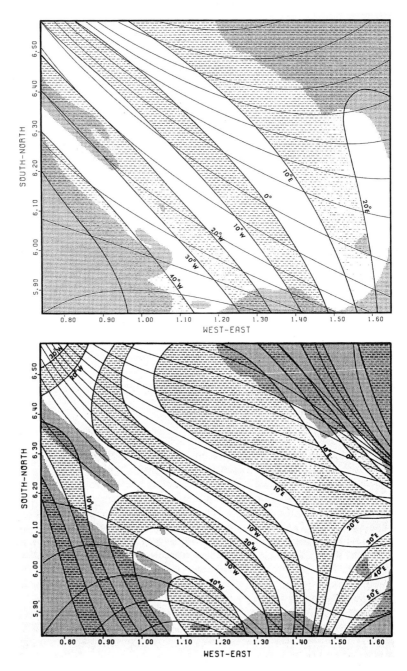

Fig. 8.9 Azimuth (*thin lines*) and dip (*solid lines* marked with amount of dip) of best-fitting unit vector fields on (*top*) quadratic and (*bottom*) cubic solution, for area outlined on Fig. 8.7. Zero-dip line in cubic solution is close to similar line previously drawn by hand in Fig. 8.7 (Note that axes of synclines on limbs of superimposed anticline tend to coincide with isolated segments of Permo-Triassic (*dotted pattern*) enclosed by crystalline rocks in B) (Source: Agterberg 1974, Fig. 112)

which on average are contained within the *s*-planes, tend to scatter more within the average *s*-plane than in the direction perpendicular it.

Agterberg (1974) proposed to subject each of the three direction cosines of B-axes in quartzphyllites belonging to the basement of the Dolomites separately to polynomial trend surface analysis followed by normalization at any point on the map of the study area in order to obtain the unit vector field. Trend surface analysis is based on the assumption that 2-D polynomials can be used for describing trends while the residuals (differences between original and trend values) are randomly distributed. This method does not account for possible spatial autocorrelation of the residuals but the estimated trends would be unbiased if the residuals satisfy a second-order stationary process (*cf.* Watson 1971). The method of separately fitting polynomial functions to direction cosines of unit vectors also was used by Parker and Denham (1979). These authors used cubic smoothing splines instead of ordinary polynomial functions.

An important paper on interpolation and smoothing of directed and undirected linear data was published by Watson (1985). Several new algorithms were proposed by this author using functions according to which the influence of measurements decreases with distance from the points at which interpolation with or without smoothing is to be applied along a line or within a map area. This type of approach also is useful for 3-D applications using drill-hole data to extend the geological map downwards from the topographic surface (Michael Hillier, Geological Survey of Canada, personal information).

8.2.2 Arnisdale Gneiss Example

In the previous section it was shown that Hercynian schistosity in the San Stefano area was reactivated to form the core of an Alpine anticline that can also be seen in the Permotriassic rocks overlying the crystalline basement. This style of folding is analogous to that described by Ramsay (1960, 1967) for gneisses of the Moine Series at Arnisdale, western Highlands of Scotland. An example is shown in Fig. 8.10 where the pattern is for the azimuth of deformed B-lineations. Axial traces (intersections of axial planes with the topographic surface) for the late folds are indicated. A schematic explanation for this style of folding is given in Fig. 8.11. During the late folding, the foliation, which is according to parallel original surfaces r, s, t and u, did not change in attitude. The axial plane of the late folding (ab) makes a small angle with the original surface. The *a*-direction is subvertical and is contained in the original surfaces.

Patterns such the one shown in Fig. 8.10 can become almost completely obscured when the surfaces, on which the lineation was developed were themselves variably oriented or when the amount of compressive strain accompanying the original folding was variable in space. In such situations, one may never be able to reconstruct a pattern of deformation in an accurate manner. However, by developing the vector mean for lineation from observations in different outcrops, trends can be established providing a broad outline of the geometry of the folds and the underlying genetic process.

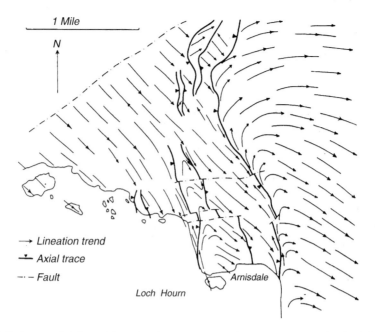

Fig. 8.10 Traces of deformed lineations in gneisses at Arnisdale, western Highlands of Scotland (After Ramsay 1967) (Source: Agterberg 1974, Fig. 108)

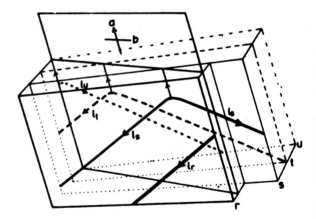

Fig. 8.11 Variation in pattern of deformed lineations on surfaces r, s, t and u resulting from the oblique intersection of the shear plane *ab* with these surfaces (From Ramsay 1967) (Source: Agterberg 1974, Fig. 109)

8.2.3 TRANSALP Profile Example

Figure 8.12 (based on map by Agterberg 1961) shows mean B-axes and schistosity planes in quartzphyllites belonging to the crystalline basement of the Dolomites in an area around the Gaderbach, which was part of the seismic Vibroseis and explosive transects of the TRANSALP Transect. The TRANSALP profile was oriented approximately along a north-south line across the Eastern Alps from

8.2 Unit Vector Fields

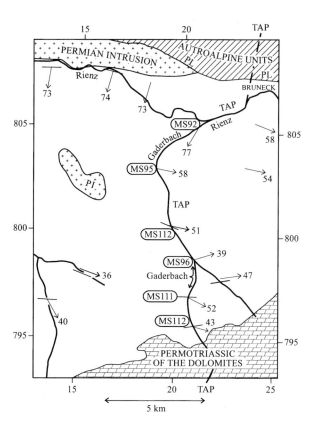

Fig. 8.12 Schematic structure map of surroundings of TRANSALP profile (TAP) in tectonites of crystalline basement of the Dolomites near Bruneck. *Arrows* represent average azimuths of B-axes for measurement samples. Average dips of B-axes are also shown. Average *s*-plane attitudes are shown if they could be determined with sufficient precision. MS: Measurement Sample as in Agterberg (1961, Appendix II). *PI* permian intrusion, *PL* periadriatic lineament. Coordinates of grid are as on 1:25,000 Italian topographic maps (Source: Agterberg 2012, Fig. 5)

Munich to Belluno with locally its position determined by irregular shapes of the topography. This international co-operative effort has led to important new interpretations (see, e.g., Gebrande et al. 2006) to be discussed later in this section. Structurally, the quartzphyllites are of two types: in some outcrops the Hercynian schistosity planes are not strongly folded and their average orientation can be measured. However, elsewhere, like in the vicinity of Bruneck, and in most of the Pustertal to the east of Bruneck, there are relatively many minor folds (as illustrated in Fig. 8.5) of which the orientation can be measured. However, then it may not be possible to measure representative strike and dip of schistosity at outcrop scale. In the parts of the area where the strike and dip of schistosity planes can be measured, there usually are B-lineations for microfolds on the schistosity planes that also can be measured. Consequently, the number of possible measurements on B-axes usually exceeds the number of representative measurements on schistosity planes.

Figure 8.12 shows that the average strike of schistosity is fairly constant in this area but the average azimuth and dip of the B-axes is more variable. There is a trend from intermediate eastward dip of B-axes in the southern parts of the region to subvertical attitude in the north. This systematic change can also be seen on Schmidt net plots for the six measurement samples along the Gaderbach (Agterberg

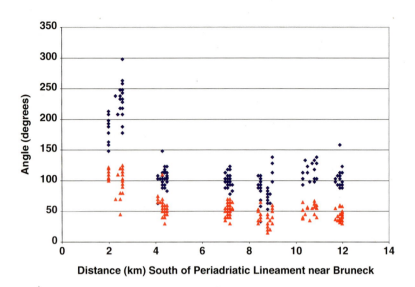

Fig. 8.13 Azimuths (*diamonds*) and dips (*triangles*) of B-axes from six measurement samples along Gaderbach shown in Fig. 8.12 (Note that along most of this section, the B-axes dip nearly 50° East). Closest to the Periadriatic Lineament, their dip becomes nearly 90° SSW (Also see Table 7.1) (Source: Agterberg 2012, Fig. 6)

1961, Appendix III). In Fig. 8.13, the B-axes measured along the Gaderbach are projected on a North-South line approximately parallel to the average orientation of the TRANSALP profile. The method of fitting unit vector fields becomes simplified when the measurements are along a straight line (North-South directed x-axis in Fig. 8.13). Figure 8.14 shows the corresponding direction cosines together with cubic polynomial trend lines fitted by the method of least squares. In Fig. 8.15 the trend lines of Fig. 8.14 are re-plotted as average azimuth and dip curves.

In a general way, the results of the preceding linear unit vector field analysis confirm the earlier conclusions on North-South regional change in average B-lineation attitude. Table 8.1 shows results from another test. Original estimates of mean B-axis orientations for the six measurement samples (Method 1) are compared with ordinary unit vector means (Method 2), and with values on the trend lines of Fig. 8.16 at mean distances south of the (Periadriatic) Pusteria Lineament near Bruneck (Method 3). Differences between the three sets of mean azimuth and dip values are at most a few degrees indicating that regionally, on the average, there is a dip rotation from about 45° dip to the east in the south to about 80° dip in the north accompanied by an azimuth rotation of about 120° from eastward to WNW-ward. It is noted that higher-order polynomials fitted by least squares rapidly become unreliable near the edges of the study area, where there is less data control. Thus, the local maximum at the northern edge (in Fig. 8.15) is not clearly established. Widespread occurrence of steeply dipping B-axes near the Periadriatic Lineament and along the southern border of the Permian (Brixen granodiorite) intrusion, however, is confirmed by results derived from other

8.2 Unit Vector Fields

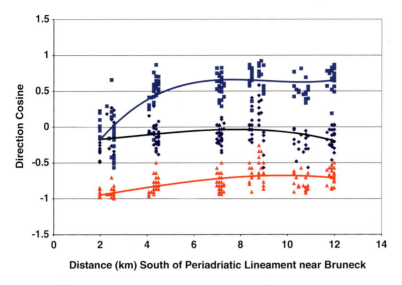

Fig. 8.14 Cubic polynomial curves (*shown as lines*) were fitted to the three direction cosines of the measurements shown in Fig. 8.13. The Cartesian coordinate system used was: x = North; y = East; z = upward. The direction cosines for x, y and z are shown as solid *squares, diamonds and triangles*, respectively (Source: Agterberg 2012, Fig. 7)

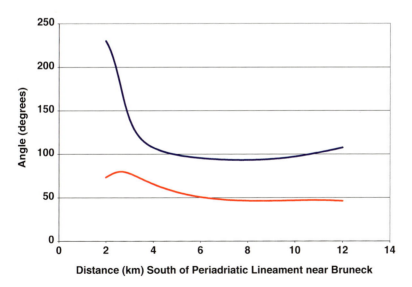

Fig. 8.15 The three cubic polynomials of Fig. 8.14 were combined to show azimuth (solid line) ranging from about 90° to 230° and dip (broken line) ranging from about 45° to 80° of best-fitting unit vector field along the N-S profile (Source: Agterberg 2012, Fig. 8)

Table 8.1 Comparison of average azimuth and dip of B-axes for six measurement samples as obtained by three methods (Source: Agterberg 2012, Table 1)

MS #	Easting	Distance South of PL	Method 1 Az./Dip	Method 2 Az./Dip	Method 3 Az./Dip
92	85.1	2.38	217/77	220/78	218/77
95	83.1	4.38	104/58	103/58	103/62
112	80.4	7.08	101/52	100/52	94/49
96	78.8	8.68	86/39	85/41	93/46
111	76.9	10.58	113/52	113/53	98/46
110	75.6	11.88	104/43	103/44	107/47

measurement samples (Fig. 8.12). The rapid change in average azimuth that accompanies the B-axis steepening is probably real but it should be kept in mind that the azimuth of vertical B-axes becomes indeterminate. As will be discussed in more detail in Sect. 8.3.4, the northward steepening of attitudes of s-planes in the northern quartzphyllite belt to the West of Bruneck from gentle southward dip near Brixen is probably due to neo-Alpine shortening. The northward steepening of the B-axes near Bruneck, which on average are contained within the s-planes, could be due to Neo-Alpine sinistral movements along the Periadriatic Lineament and south of the Permian (Brixen granodiorite) intrusion.

Several authors including Ring and Richter (19dextral94) and Benciolini et al. (2006) have studied the Hercynian (or "Variscan") structural evolution of the quartzphyllites in the Brixen area of which the map in our Fig. 8.12 only covers the easternmost portion. These authors were able to distinguish between successive generations of Hercynian microstructures. Ring and Richter (1994) established that an earlier schistosity (S1) was strongly overprinted by S2 during their D2-deformation stage. A mineral stretching lineation (L2) occurs on S2 with attitudes similar to larger F2-folds (as those shown in our Fig. 8.5). These authors also identified younger F3-folds subparallel to the predominant F2-folds. Benciolini et al. (2006) later identified another earlier generation of minor folds. Their B3-folds are equivalent to Ring and Richter's F2-folds. Maps of L2 lineaments (Ring and Richter 1994, Fig. 4a, 4 L2-lineaments) and B3-folds (Benciolini et al. 2006, Fig. 2, 6 B3 fold axes) show strikingly different patterns because these authors plotted original measurements and not average orientations as are shown in our Fig. 5. As illustrated in our Fig. 8.13, azimuths of minor fold can change by more than 90° over distances less than 100 m. Ring and Richter (1994, Fig. 4b) also constructed regional stereograms of 346 L2-lineations and 826 F2-folds within the larger Brixen area for which Agterberg (1961, Appendix II) had listed 4,018 B-axes subdivided into 52 measurement samples similar to the 13 measurement samples shown in Fig. 5 of this paper. These B-axes are mainly Ring and Richter's L2-lineaments and F2-folds. On the whole, the regional stereograms of Ring and Richter (1994) for the Brixen area, and for the Cima d'Asta and Gosaldo areas to the south of the Italian Dolomites, are in agreement with unit vector field constructions for these areas except that the latter better capture

8.2 Unit Vector Fields

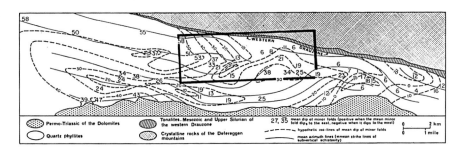

Fig. 8.16 Complex structures of superimposed folding in Pusteria tectonites, northern Italy (After Agterberg 1961). Pattern of mean azimuth and dip of minor fold axes were constructed by averaging for domains and manual contouring. Frame of Fig.8.17 is outlined (Source: Agterberg 1974, Fig. 113)

regional average orientation change. The major D2-deformation occurred in the late Carboniferous and was connected to mainly subhorizontal Hercynian tectonic movements associated with greenschist facies metamorphism (*cf.* Ring and Richter 1994, p. 765). During Alpine orogeny, S2 assumed its current predominantly southward dip In the Brixen area.

8.2.4 Pustertal Tectonites Example

Figure 8.16 (after Agterberg 1961, 1974; also see Whitten 1966) shows mean azimuth lines and average dips of B-axes for measurement samples similar to those shown in Fig. 8.12 in the Pustertal East of Bruneck. As in the immediate vicinity of Bruneck, the quartzphyllites (*cf.* Fig. 8.5) are rather strongly folded in many places. Although B-axis orientations could be measured in about 1,700 outcrops, representative schistosity-plane readings were obtained in 257 outcrops only. In nearly half of these outcrops, *s*-plane attitude is subvertical (dips between 75° northwards and 75° southwards; Agterberg 1961, Fig. 58) although lesser northern dips (<75°) are slightly more prevalent than southern dips (<75°). Thus, the mean azimuth lines in Fig. 8.16 approximately represent mean schistosity-plane strikes. On the whole, the mean azimuth lines converge eastward. Relatively strong local convergence occurs on the Eggerberg East of Welsberg (Fig. 8.16). Measurements taken along the south slope of the East-West elongated Eggerberg show SW-NE strikes and mean azimuths on the average but on its less exposed north slope NW-SE strikes prevail. Figure 8.17 (after Agterberg 1974) shows best-fitting two-dimensional quadratic and cubic polynomial unit vector fields for the Eggerberg and its immediate surroundings. Strong eastward convergence of azimuth lines (Eggerberg structure) in Fig. 8.17 indicates existence of neo-Alpine anticlinal lineament or zone along which material moved upwards with considerable mobility but with preservation of mean attitudes of subvertical *s*-planes and B-axes (after Agterberg 1974, Fig. 115). The cubic solution is probably more

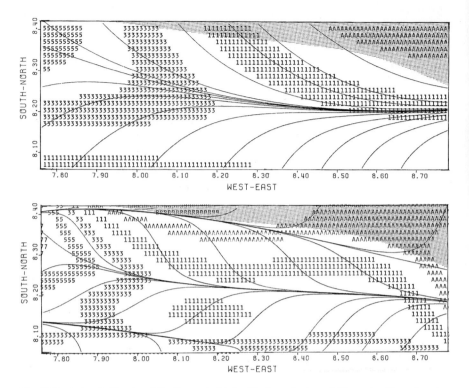

Fig. 8.17 Azimuth (*lines*) and dip (in symbols) for (*Top*) quadratic and (*Bottom*) cubic unit vector field fitted to data within rectangle outlined in Fig. 8.16; printed symbol for dip 1 (10–20° E), 3 (30–40° E), 5 (50–60° E), A (0–10° W). Strong eastward convergence of azimuth lines suggests existence of Alpine lineament or zone trough which material was moved upward with considerable mobility but with preservation of pre-Alpine *s*-planes and *B*-lineations (Source: Agterberg 1974, Fig. 115)

realistic provided that the mean azimuth pattern within the relatively low-dip southwestern part of the area, where there was no data control because of Quaternary cover, as well as the low-dip northeastern pattern north of the Periadriatic Lineament (Western Drauzug), are ignored.

It may be concluded that, in the Pustertal area, the quartzphyllites, which show subvertical *s*-plane strike on the average, contain Alpine anticlinal structures characterized by areas of lower-dipping B-axes that may be accompanied by eastward convergent mean azimuth lines. The most likely explanation (Agterberg 1961) is that these quartzphyllites were squeezed out upwards during the latest phase of Alpine orogeny between the Austroalpine units north of the Periadriatic Lineament and the northern edge of the Permotriassic of the Dolomites (Fig. 8.18). This interpretation would be in accordance with a model advocated by Castellarin et al. (2006, Fig. 9) who assume that the Vibroseis depth-migrated data from the TRANSALP profile show a steeply north-dipping thrust fault that reaches the surface at the Periadriatic Lineament (also see next section).

Fig. 8.18 Uniformly compressed plastic zone narrowing eastward. The replaced Austrian Crystalline (=O) equals the squeezed out material in all places, and
$O = O_0 = O_1 = O_2$ (Source: Agterberg 1961, Fig. 70)

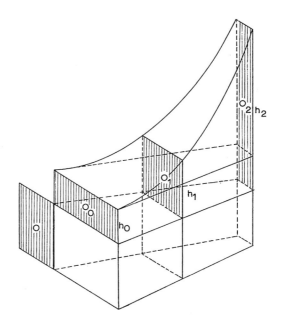

8.2.5 Tectonic Interpretation of Unit Vector Fields Fitted to Quartzphyllites in the Basement of the Italian Dolomites

For the Mediterranean region as a whole, GPS data indicate a roughly NNW-SSE oriented convergence between Africa and Europe of up to 8 mm/year (Picardi et al. 2011). According to Lippitsch et al. (2003) average crustal shortening in the Alps amounted to 5 mm/year during the past 40 million years. Deep seismic lines crossing the Eastern Alps including TRANSALP all show a South-directed European slab below the northward-indenting Adria Microplate lithosphere (Gebrande et al. 2006). As pointed out by Luth et al. (2010), the TRANSALP profile approximately coincides with a major subduction polarity change zone. To the west of this zone, there is southward-directed European subduction below the Western and Central Alps; and to the east of it, there is north-eastward directed subduction of Adria below the Eastern Alps and the Dinarides. This interpretation of neo-Alpine orogeny is primarily based on tele-seismic tomographic data (Lippitsch et al. 2003; Kissling et al. 2006). Luth et al. (2010) and Luth (2011) have developed litho-scale analogue models to investigate the effect of subduction polarity change on the overlying crustal architecture.

Results of seismic Vibroseis and explosive TRANSALP transects across the Alps from Munich to Belluno were described by the TRANSALP Working Group (2002). In this section the unit vector field fitting results for measurements on B-axes and schistosity planes near Bruneck and in the Pustertal east of the TRANSALP Profile are reviewed on the basis of the TRANSALP findings and other newly obtained results on tectonics of the Alpine realm involving movements of the Adria microplate (Picardi et al. 2011) of which the current northern boundary at the surface approximately coincides with the Periadriatic Lineament.

As mentioned in the Introduction, the quartzphyllites of the Dolomites basement were folded during the Hercynian Orogeny. Most B-axes in the Brixen crystalline basement West of the TRANSALP profile and in the Cima d'Asta and Agordo areas south of the Dolomites are relatively low-dipping and have N-S to NNW-SSE orientations. These parts of the Dolomites basement underwent less Alpine deformation than the quartzphyllites in the San Stefano and Pustertal areas (Figs. 8.6 and 8.16). During the early Neogene, the Dolomites and their basement had been subjected to anti-clockwise rotation of about 50° (Channell and Doglioni 1994) indicating that, originally, the Hercynian folds had NE-SW to E-W orientation.

According to Vai (2002) the Italian region contains three different segments of Hercynian mountain chain: (1) Carnian-Dinaric, (2) central-western Southalpine and (3) Apenninic, surrounding a Baikalian or Panafrican consolidated microcraton, roughly corresponding to the core of to-day's Adria microplate. North to northeast directed Hercynian structural axes are almost coaxial with those of Alpine age in the Apennines (3) but cross at high angles in the Southern Alps (2). From the patterns of Fig. 8.6 (San Stefano area) and Fig. 8.16 (Pustertal) it can be inferred that the angle between Hercynian and Alpine structural axes decreases eastward toward the Carnian Alps and southeastward toward the Dinarides (1) where they become subparallel with E-W to SSE-NNW orientations, respectively.

Picardi et al. (2011) provide a recent outline of the Adria microplate, which is a nearly closed basin comprising a Meso-Cenozoic continental block that represents the foreland of the Apennines and Dinarides-Hellenides mountains, two sub-parallel orogenic belts with opposing vergences. Its core is the region currently occupied by the Adrian Sea. The northern boundary of Adria coincides with the Periadriatic Lineament. To the north and south of the Dolomites, the Alpine vergence is southward. The southern boundary of the crystalline basement of the Dolomites consists of the Valsugana Lineament, which is a major thrust fault, in the Strigno area (Fig. 1.1) accompanied by a south vergent overthrust sheet (Agterberg 1961; Castellarin and Vai 1982; Castellarin et al. 2006). The Alpine anticlinal structure in the San Stefano area (Fig. 8.6) has vergence intermediate between the S vergence of the Dolomites and the WSW vergence of the Dinarides. Current deformation in the Adriatic region does not simply reflect the N-S shortening between Africa and Eurasia. Adria's recent motion controls the distribution of earthquakes and tectonic deformation across a broad area of South-central Europe. As in the past, Adria moves in response to the combined motions of the African Plate, Eurasian Plate, Anatolian block, and Hellenic Arc (Picardi et al. 2011). At present it moves north-eastward (Lippitsch et al. 2003).

Stampfli and Hochard (2009) describe the wander path of Adria during the past 180 million years. Initially Adria moved ESE-ward. After about 60 million years, it commenced an anticlockwise rotation moving eastward, northward and NNW-ward. During this motion, the Italian Dolomites and their basement were rotated by about 50° (Channell and Doglioni 1994) and there were very significant Oligocene-Miocene dextral shear movements along the Periadriatic Lineament (Ratschbacher et al. 1991). South vergent faulting and folding within and around the Dolomites commenced in the late Miocene. For example, the overthrust sheet near Strigno associated with the Valsugana Lineament rests partly on Upper Miocene sediments. Its origin may have involved neo-Alpine sliding motion along slightly S dipping quartzphyllites south of the Permian Cima d'Asta intrusion (*cf.* Agterberg 1961, Fig. 83). Laubscher (2010) discusses late Miocene sinistral strike-slip movement totalling about 70 km along the Giudicaria Line, which forms the western boundary of the Dolomites Synclinorium.

Castellarin et al. (2006) have reviewed structure of the lithosphere beneath the Eastern Alps (southern sector of the TRANSALP profile). Participants in the TRANSALP Working Group (2002) originally proposed two different interpretative models of the Vibroseis depth-migrated data called "Crocodile Model" and "Ductile Extrusion Model", respectively. The two interpretations are similar except in the segment of crust below the Periadriatic Lineament. In the second model, a N-dipping break transparent zone in the Vibroseis depth-migrated data was interpreted as the downward extension of the Periadriatic Lineament. In the first model such a thrust fault was not recognized. The deformation patterns of the quartzphyllites indicated by the unit vector field fittings described in the previous two sections support the "Ductile Extrusion Model" in its interpretation that the Periadriatic Lineament in its final stage was a thrust fault that reactivated Hercynian *s*-planes to the south of it by strong neo-Alpine compression between the Austroalpine units to the north and the Permotriassic of the Dolomites to the south. The eastward convergence of the boundaries of these more rigid rocks resulted in upward squeezing out of the more ductile Pustertal quartzphyllites.

The Vibroseis depth-migrated data from the TRANSALP profile across the Periadriatic Lineament also show a number of non-reflective volumes indicating presence of intrusions similar to the Permian Brixen batholith and, to the north of the Periadriatic Lineament, the Oligocene Rieserferner tonalite intrusion. Between these intrusions and the downward extension of the Periadriatic Lineament there are many clearly developed south-dipping high-amplitude reflecting intervals, which are probably equivalent to the south-dipping quartzphyllites of the basement of the Dolomites in most of the Brixen area (to the SW of Fig. 8.12). In accordance with the "Ductile Extrusion Model", these rocks would constitute the northern edge of the Adriatic indenter (Castellarin et al. 2006). Immediately below the TRANSALP profile south of Bruneck, there are no well-developed high-amplitude reflecting intervals. This is probably because of the subvertical attitudes of most *s*-planes in this area. Similar interpretations can be based on the CMP stack section and post-stack depth-migrated section shown and discussed by Lüschen et al. (2006, Figs. 6 and 8).

Castellarin et al. (2006) estimated that upper crustal total shortening in the belt of quartzphyllites between the "Dolomite Synclinorium" and the Periadriatic Lineament amounts to about 15 km and that most of this shortening is due to rotation of the quartzphyllites from low-dipping to high-dipping attitudes. According to the results presented in this section (Figs. 8.12, 8.13, 8.14 and 8.15), this movement probably had a horizontal component as well, as indicated by the subvertical to steeply SSW dipping B-axes In the vicinity of Bruneck and immediately south of the Permian granodiorite intrusion. The shortening probably also involved north vergent thrust faults including the Villnösz Fault south of Brixen. Although it is difficult to estimate the amount of shortening precisely, Castellarin et al.'s estimate is in agreement with the pattern shown in Fig. 8.12, and also with the model of pronounced neo-Alpine shortening in the Pustertal area further eastward. (Fig. 8.18) The main reason that significant extra eastward shortening took place in the Pustertal is probably related to the decrease in width of the quartzphyllite belt in this direction. Agterberg (1961) estimated that the later Pustertal shortening was about 5 km. In his original model, amount of shortening was assumed to decrease with depth. This assumption was probably not correct because extra neo-Alpine shortening in Pustertal was probably between 10 and 15 km. The Hercynian s-planes were reactivated during these strong upward movements. Although there is some evidence of existence of neo-Alpine micro-structures in the area (Agterberg 1961, Appendix I, Figs. 48–51a), the topic of distinguishing between Hercynian and Alpine micro-structures would benefit from further study using methods similar to those applied to microstructures in the Austroalpine units and within the Tauern Window north of the Periadriatic Lineament (Mancktelow et al. 2001).

8.2.6 Summary of Late Alpine Tectonics South of Periadriatic Lineament

The purpose of this section was to revisit the topic of unit vector field fitting in applications to B-axes in the quartzphyllites of the crystalline basement of the Italian Dolomites. A slightly improved mathematical derivation of the use of polynomial trend functions fitted to the three direction cosines of directional features and combining the results was presented. Illustration by practical applications consisted of three parts. First it was shown that in the San Stefano area, to the east of the Dolomites, the relatively simple method of computing vector means for relatively large, overlapping circular areas produces good results. This would imply that neighbourhood-based methods as proposed by Watson (1971) can give good results. During Alpine orogeny the San Stefano quartzphyllites formed the core of a SSW vergent anticline with re-activation of steeply dipping Hercynian schistosity causing opposite rotations of the originally sub-horizontal Hercynian B-axes at the two sides of its axial plane.

Secondly, the polynomial unit vector fitting method was illustrated by means of a relatively simple, linear example. Data from six measurement samples along the Gaderbach south of Bruneck in the Pustertal were projected on a north-south line approximately coinciding with the TRANSALP profile. The quartzphyllite belt is about 12 km wide at this location. Immediately north of the Permotriassic of the

Dolomites, the B-axes, on average, dip about 50° East over a distance of about 8 km. Average dip of the approximately East-West striking s-planes is about 90°. Over the 2 km to the north along the Gaderbach, and probably over the remaining 2 km immediately south of the Periadriatic Lineament, the attitude of B-axes becomes subvertical. The 40° dip steepening is accompanied by an average azimuth rotation from approximately eastward to steeply SSW. This pattern of B-axis and s-plane orientations along the TRANSALP profile south of the Periadriatic Lineament is confirmed by average orientations estimated from other measurement samples to the east and to the west.

The third example was for the Pustertal quartzphyllite belt that becomes narrower to the east reaching its minimum width of about 3 km at a distance of approximately 25 km east of Bruneck. The subvertical Hercynian s-planes contain a number of anticlinal structures characterized by decreases in average dip of minor folds and eastward convergence of strike-lines. The Eggerberg structure is well exposed and the most striking example of these neo-Alpine structures.

Unit vector field results for all three examples can be interpreted in term of the latest (Late Miocene – Pliocene) movements of the Adria microplate. In direction and vergence the San Stefano anticlinal structure fits in with the structures of the Dinarides-Hellenides mountains. As pointed out by Castellarin et al. (2006), the deep seismic TRANSALP data below the Periadriatic Lineament near Bruneck indicate that this major fault dips northwards and was associated with significant neo-Alpine north-south shortening. In the quartzphyllite belt immediately south of the Periadriatic lineament, perhaps 15 km of shortening was mainly achieved by upward rotation of the low-dipping s-planes with attitudes similar to those at greater depths below the Periadriatic lineament, further south in the Brixen area and to the south of the Dolomites in the Cima d'Asta and Agordo areas. In the Pustertal quartzphyllites extra shortening resulted in upward squeezing out along the subvertical Hercynian s-planes because of the significant eastward width decrease of the belt of more ductile quartzphyllites between the Austroalpine units to the north and the Dolomites Synclinorium to the South. The northward rotation of B-axes to subvertical positions immediately south of the Periadriatic Lineament near Bruneck and the Permian Brixen granodiorite intrusion suggests upward sinistral neo-Alpine movements in this vicinity. This displacement would fit in with the late Miocene sinistral strike-slip motion along the Giudicaria Fault to the west (Laubscher 2010). The north-south orientation of the subhorizontal fold axes in the Hercynian basement of the Brixen and Cima d'Asta regions to the north and south of the eastern Dolomites indicates that, originally, the dominant Hercynian strike direction was SE-NW because of the earlier Alpine rotation of the Adria microplate as is well documented by paleomagnetic investigations and evidence of large-scale dextral strike-slip movements along the Periadriatic lineament. The fold patterns in Figs. 8.16 and 8.17 are more complex than those in other parts of the crystalline basement of the Dolomites. It remains possible that parts of the Pustertal quartzphyllites originally were not part of Adria but a westward continuation of the predominantly east-west striking Hercynian crystalline rocks in the Carnian Alps to the east.

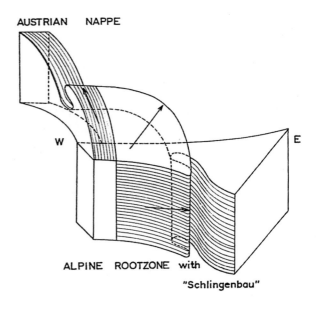

Fig. 8.19 Late Alpine deformation of Austrian crystalline north of Pusteria Lineament. Transition from northward overturned folds approximately along TRANSALP profile into the Defereggen "Schlinge" with sinistral vergency (Source: Agterberg 1961, Fig. 104)

8.2.7 Defereggen Schlinge Example

Significant Alpine compression of the crystalline basement of the Dolomites in northern Italy was shown to have taken place, probably during the Miocene. On a larger scale this involved a 70 km northward displacement of the Dolomites along the Giudicaria Line. It is likely that crystalline rocks north of the Periadriatic Lineament in this area were involved in this late Alpine displacement as well.

Crystalline rocks in the Defereggen and Örtler regions of the Eastern Alps in Austria show "Schlinge"-like structures. Schmidegg (1933, 1936) introduced the term "Schlinge" (=sinuosity) for folds with subvertical axes occurring in the infrastructure of katazonal gneisses. They would have formed by bending of subvertical S-planes in consequence of subhorizontal compression. It is generally assumed that the Schlingen are Variscan (Hercynian) in age (see, e.g., Schulz et al. 2008). However, Agterberg (1961) argued that they originated near the end of the Alpine orogeny (Fig. 8.19) in conjunction with the sinistral movement along the Giudicaria Fault.

References

Agterberg FP (1959) On the measuring of strongly dispersed minor folds. Geol Mijnb 21:133–137
Agterberg FP (1961) Tectonics of the crystalline basement of the Dolomites in North Italy. Kemink, Utrecht
Agterberg FP (1974) Geomathematics. Elsevier, Amsterdam

References

Agterberg FP (1985) Spatial analysis in the earth sciences. In: Glaesner PS (ed) The role of data in scientific progress. Proceedings of 9th international CODATA conference. Elsevier, Amsterdam, pp 9–18

Agterberg FP (2004) Time, concepts and mathematics in geology. Zeitschr D Geol Gesell 155(2–4):211–219

Agterberg FP (2012) Unit vector field fitting in structural geology. Zeitschr D Geol Wis 40(4/6):197–211

Agterberg FP, Hills LV, Trettin HP (1967) Paleocurrent trend analysis of a delta in the Bjorne Formation (Lower Triassic) of northeastern Melville Island, Arctic Archipelago. J Sediment Petrol 37:852–862

Batschelet E (1965) Statistical methods for the analysis of problems in animal orientation and certain biological rhythms. American Institute of Biological Sciences, Washington, DC

Benciolini L, Poli ME, Visonà D, Zanferrari A (2006) Looking inside Late Variscan tectonics: structural and metamorphic heterogeneity of the Eastern Southalpine Basement (NE Italy). Geodin Acta 19(1):17–32

Castellarin A, Vai GB (1982) Introduzione alla geologia strutturale del Sudalpina. In: Castellarin A, Vai GB (eds) Guida alla geologia del Sudalpino centro-orientale. Soc Italiana, Bologna

Castellarin A, Nicolich R, Fantoni R, Cantelli L, Sella M, Selli M (2006) Structure of lithosphere beneath the eastern Alps (southern sector of the TRANSALP transect). Tectonophysics 414:259–282

Channell JET, Doglioni C (1994) Early Triassic paleomagnetic data from the Dolomites (Italy). Tectonics 13(1):157–166

Fisher RA (1953) Dispersion on a sphere. Proc R Soc Lond Ser A 217:295–305

Fisher NI (1993) Statistical analysis of circular data. Cambridge University Press, Cambridge

Fisher NI, Lewis T, Embleton BJJ (1987) Statistical analysis of spherical data. Cambridge University Press, Cambridge

Gebrande H, Castellarin A, Lűschen E, Millahn K, Neubauer F, Nicolich R (2006) TRANSALP – A transect through a young collisional orogen: Introduction. Tectonophysics 414:1–7

Kissling E, Schmid SM, Lippitsch R, Ansorge J, Fűgenschuh B (2006) Lthosphere structure and tectonic evolution of the Alpine arc: new evidence from high-resolution teleseismic tomography. Geol Soc Lond Mem 32(1)

Koch GS Jr, Link RF (1971) Statistical analysis of geological data. Wiley, New York

Köster E (1964) Granulometrische und Morphometrische Messmethoden an Mineralkörnern, Steinen und sonstigen Stoffen. Enke, Stuttgart

Krumbein WC (1939) Preferred orientations of pebbles in sedimentary deposits. J Geol 47:673–706

Laubscher H (2010) Jura, Alps and the boundary of the Adria subplate. Tectonophysics 438(3/4):223–239

Lippitsch R, Kissling E, Ansorge J (2003) Upper mantle structure beneath the Alpine orogen from high-resolution teleseismic tomography. J Geophys Res 108:1–15

Loudon TV (1964) Computer analysis of orientation data in structural geology. Off Nav Res Geogr Br Tech Rep 13, pp 1–130

Lűschen E, Borrini D, Gebrande H, Lammerer B, Millahn K, Neubauer K, Nicolich R (2006) TRANSALP-deep crustal Vibrosis and explosive seismic profiling in the Eastern Alps. Tectonophysics 414:9–38

Luth S (2011) Orogenesis and subduction in analog models. Gildeprint, Enschede

Luth S, Willingshofer E, Sokoutis D, Cloetingh S (2010) Modelling surface expression of subduction polarity change: the Eastern Alps as transition zone? http://geomod2010.fc.ul.pt/abstracts/Luth%20et20al.pdf

Mancktelow NS, Stöckli G, Grollimund B, Műller W, Fűgenschuh B, Viola G, Seward D, Villa IM (2001) The DAV and Periodriatic fault systems in the Eastern Alps South of the Tauern window. Int J Earth Sci 90:593–622

Parker RL, Denham CR (1979) Interpolation of unit vectors. Geophys J R Astron Soc 58(3):685–687

Picardi L, Sani F, Moratti G, Cunningham D, Vittori E (2011) Present-day geodynamics of the circum-Adriatic region: an overview. J Geodyn 51:81–89

Potter PE, Pettijohn FJ (1963) Paleocurrents and basin analysis. Academic, New York

Ramsay JG (1960) The deformation of early linear structures in areas of repeated folding. J Geol 68:75–93

Ramsay JG (1967) Folding and fracturing of rocks. McGraw-Hill, New York

Ratschbacher L, Frisch W, Linzer G, Merle O (1991) Lateral extrusion in the eastern Alps: Part 2. Structural analysis. Tectonics 10:257–271

Reiche P (1938) An analysis of cross-lamination of the Coconino sandstone. J Geol 46:905–932

Ring U, Richter C (1994) The Variscan structural and metamorphic evolution of the eastern Southalpine basement. J Geol Soc Lond 151:755–766

Sander B (1948) Einführung in die Gefügekunde der geologischen Körper. Springer, Vienna

Scheidegger AE (1965) On the statistics of the orientation of bedding planes, grain axes and similar sedimentological data. US Geol Surv Prof Pap 525:C164–C167

Schmidegg O (1933) Neue Ergebnisse in den südlichen Ötztaler Alpen. Verh Geol B Anst 5–6:83–95

Schmidegg O (1936) Steilachsige Tektonik und Schlingenbau auf der Südseite der Tiroler Zentralalpen. Geol B Anst 86(2):115–149

Schulz B, Steenken A, Sigismund S (2008) Geodynamic evolution of an Alpine terrane – the Austroalpine basement to the South of the Tauern Widow as a part of the Adriatic Plate (Eastern Alps). Geol Soc Lond Spec Publ 298:5–44

Stampfli GM, Hochard C (2009) Plate tectonics of the Alpine realm. Geol Soc Lond Spec Publ 327:89–111

TRANSALP WORKING Group (2002) First deep seismic images of the Eastern Alps reveal giant crustal wedges and transcrustal ramps. Geophys Res Lett 29(10):921–924. doi:10.1029/2002GLO14911

Vai GB (2002) Palaeozoic palaeotectonics of the eastern Southern Alps: implication for the TRANSALP Profile. Mem Sci Geol 54:97–100

Watson GS (1971) Trend-surface analysis. Math Geol 3(3):215–226

Watson GS (1985) Interpolation and smoothing of directed and undirected line data. In: Krishnaia PR (ed) Multivariate analysis VI. Academic, New York, pp 613–625

Whitten EHT (1966) Structural geology of folded rocks. Rand McNally, Chicago

Chapter 9
Quantitative Stratigraphy, Splining and Geologic Time Scales

Abstract Quantitative stratigraphy uses logical and mathematical tools to help define the stratigraphic framework of the Earth's crust. Biostratigraphy uses observations on fossil taxa. Biostratigraphic events commonly used for this purpose are the observed first and last occurrence (abbreviated to FO and LO) of each fossil taxon considered. Co-occurrences of fossil taxa in the biostratigraphic record can be used as well. Methods for the integration and long-distance correlation of observed biostratigraphic events include the RASC method for RAnking and SCaling. The main difference between RASC and other methods of regional biostratigraphic correlation is that RASC estimates the relative positions of average fossil events instead of maximal time-stratigraphic ranges, although maximal ranges also can be obtained by using RASC. Different methods of quantitative stratigraphy are briefly reviewed in this chapter. Initially, ranking is illustrated by application to a simple, artificial dataset. Scaling is explained as a refinement of ranking. Implications of techniques of sampling stratigraphic sections are discussed. RASC probable positions with error bars can be determined in different sections for CASC correlation over long distances. This process makes use of spline-curve fitting (splining). For method comparison, several datasets published by others are re-analyzed, not only to establish regional biostratigraphic standards but also to perform correlations between stratigraphic sections. These datasets include FOs and LOs of Eocene nannofossils in wells drilled in California and trilobites from the Cambrian Riley Formation in central Texas. Large-scale RASC/CASC applications involving many thousands of observations include results for well data from the Cenozoic North Sea basin, northwestern Atlantic margin and the Cretaceous seaway between Norway and Greenland. Paleoceanographic interpretations of RASC biozonations supplemented by analysis of variance to study diachronism and correlations between wells are exemplified as well.

The international numerical geologic time scales have been and continue to be partially based on spline-curves fitted to relate age determinations on rock samples to their positions in the relative geologic time scale that is based on classifications of rock units that can be correlated worldwide. Methods of time scale construction are discussed at the end of this chapter.

Keywords Quantitative stratigraphy • Biostratigraphic correlation • RASC • Ranking • Scaling • Smoothing splines • Geologic timescale • Northwestern Atlantic margin • Grand banks • North Sea basin • Californian Eocene nannofossils • Tojeira sections • Cenozoic microfossils • Cambrian trilobites • Riley formation • Cretaceous Greenland-Norway seaway

9.1 Ranking and Scaling

RASC is an acronym for RAnking and SCaling of biostratigraphic events. Code of the RASC computer program was originally published in *Computers & Geosciences*. During the past 30 years this program has been continuously maintained and updated. Its purpose is to combine biostratigraphic data from land-based sections or exploratory wells drilled in sedimentary basins to construct a regional biozonation that can be used for correlation between sections within a study area. The companion program CASC (Correlation And Scaling) makes use of splining which is a powerful method of curve-fitting. It is applied to correlate between stratigraphic sections on the basis of ranking and scaling results for samples taken at irregular sampling intervals. Implications of sampling of stratigraphic sections are discussed in detail. Several examples of past successful large-scale RASC/CASC applications will be given. The original RASC method has been discussed in detail in Gradstein et al. (1985), Agterberg (1990), Agterberg and Gradstein (1999) and Agterberg et al. (2013). CASC was introduced in Agterberg et al. (1985). In this section, calculation of the "optimum" sequence resulting from ranking will be explained using a simple, artificial example. Subsequent scaling of the ranked optimum sequence can be useful for the construction of regional biostratigraphic event zonations.

Regional standard zonations resulting from either ranking or scaling can be used for correlation between stratigraphic sections. Initially, the RASC/CASC programs were written in FORTRAN for mainframe computers. Original code has remained part of all later versions although it was modified and extended repeatedly. An executable file for RASC & CASC Version 20 with manual and documentation can be downloaded from a website maintained by the University of Oslo (http://www.nhm2.uio.no/norlex/rasc). Complete code for Version 20 was made available together with Agterberg et al. (2013). This latest software aims to provide easy access to the RASC outputs in a user-friendly, interactive fashion and incorporates CASC code for correlation between sections. Additionally, it contains code for less-known techniques such as RASC analysis of variance and depth scaling. ActiveX (OCX) Development Environment was chosen to implement all graphic modules on a Windows platform (Liu et al. 2007).

9.1.1 Methods of Quantitative Stratigraphy

Quantitative stratigraphy uses logical and mathematical tools to help define the stratigraphic framework of the Earth's crust. Biostratigraphy uses observations on fossil taxa. Biostratigraphic events commonly used for this purpose are the observed first and last occurrence (abbreviated to FO and LO) of each fossil taxon considered. Co-occurrences of fossil taxa in the biostratigraphic record can be used as well. The RASC method for ranking and scaling of the FOs and LOs (also known as "rascing") was first published 30 years ago (Agterberg and Nel 1982a, b). This method has become well-established. In paleontological textbooks by Benton and Harper (2009), Foote and Miller (2007) and Hammer and Harper (2005) it is listed as one of several methods described to construct regional biostratigraphic zonations. The main difference between RASC and the other methods is that it estimates the relative positions of average fossil events instead of maximal time-stratigraphic ranges, although maximal ranges also can be obtained by RASC, as will be seen in the next two sections. Zhou (2008) has reviewed RASC/CASC in the context of quantitative stratigraphy and presents interesting novel applications.

Quantitative biostratigraphy methods are statistical because of the large uncertainties commonly associated with the positioning of biostratigraphic events. Land-based biostratigraphic sections are more continuous and normally more complete than km-deep exploratory wells drilled in sedimentary basins that usually are sampled by collecting pieces of drill-core at discrete, regular intervals. RASC can be applied to both kinds of data. It is based on a statistical model in which averages are computed from samples of biostratigraphic events supposedly drawn at random from an infinitely large population. Here the term "sample" is used for a set of observations such as biostratigraphic events. Computationally, RASC is very fast and can deal with up to 1,000 biostratigraphic events observed in dozens of wells or land-based sections.

An early statistical method in quantitative stratigraphy was developed by Shaw (1964) for use in hydrocarbon exploration. In this book, Shaw illustrates his technique of "graphic correlation" on first and last occurrences of trilobites in the Cambrian Riley Formation of Texas for which range charts had been published by Palmer (1954). RASC/CASC results for this dataset will be briefly discussed in this section for method comparison. Shaw's method consists of constructing a "line of correlation" on a scatter plot showing the locations of LOs and FOs of taxa in two sections. If quality of information is better in one section, its distance scale is made horizontal. LOs stratigraphically below and FOs above the line of correlation are moved horizontally or vertically toward the line of correlation in such a way that the ranges of the taxa become longer. The reason for this procedure is that it can be assumed that observed highest occurrences of fossil taxa generally occur below truly highest occurrences and the opposite rule applies to lowest occurrences. Consequently, if the range of a taxon is observed to be longer in one section than in the other, the longer observed range is accepted as the better approximation. The objective is to find approximate locations of what are truly the First Appearance Datum (or FAD) and Last Appearance Datum (LAD) for each of the taxa considered. True FADs and LADs probably remain unknown but, by combining FOs and LOs from many stratigraphic sections, approximate FADs and LADs are obtained and the intervals between them can be plotted on a range chart.

Two sections usually produce a crude approximation but a third section can be plotted against the combination of the first two, and new inconsistencies can be eliminated as before. The process is repeated until all sections have been used. The final result or "composite standard" contains extended ranges for all taxa. Software packages in which graphic correlation has been implemented include GraphCor (Hood 1995) and STRATCOR (Gradstein 1996). The method can be adapted for constructing lines of correlation between sections (Shaw 1964). Various modifications of Shaw's graphic correlation technique with applications in hydrocarbon exploration and to land-based sections can be found in Mann and Lane (eds., 1995).

Instead of attempting to statistically maximize the ranges of taxa in relative time by successively adding sections, RASC estimates average positions of biostratigraphic events by simultaneously combining events from all sections that contain them. Often, such average values are more precise than estimates of FADs and LADs that are based on single event occurrences, because these can be anomalous for various reasons including possible local reworking. Because average or "probable" event positions are used, RASC-ranges generally are much shorter than ranges obtained by graphic correlation. In later versions of RASC, approximate LADs (and FADs) are estimated by outward projection from every estimated average event position by adding (or subtracting) the single largest deviation for each event. Consequently, RASC can be used for FAD and LAD estimation as well. CASC correlations, however, remain based on average event occurrences and not on estimates of FADs and LADs. A simple example of construction of RASC lines of correlation followed by FAD/LAD estimation will be provided in the next section.

In their chapter on quantitative biostratigraphy, Hammer and Harper (2005) present separate sections on five methods of quantitative biostratigraphy: (1) graphic correlation, (2) constrained optimization, (3) ranking and scaling, (4) unitary associations and (5) biostratigraphy by ordination. Theory underlying each method is summarized by these authors and worked-out examples are provided. Their book on paleontological data analysis is accompanied by the free software package PAST (available through www.blackwellpublishing.com/hammer) that has been under continuous development since 1998. It contains simplified versions of CONOP for CONstrained OPtimization (Sadler 2004) and RASC, as well as a comprehensive version for Unitary Associations (Guex 1991), a method that puts much weight on observed co-occurrences of fossil taxa in time. For comparison of RASC and Unitary Associations output for a practical example, see Agterberg (1990). Multivariate statistical methods also can make useful contributions to the spatial and temporal analysis of biostratigraphic events. These include principal component analysis (Hohn 1993), correspondence analysis (Agterberg and Gradstein 1999) and archeological seriation (Brower 1985).

CONOP was originally developed as a biostratigraphic adaptation of simulated annealing by Kemple et al. (1989). Like RASC, it works on all sections simultaneously. A single line of correlation is constructed in N-dimensional space where N is the number of sections. Biostratigraphic positions of all events are ranked subject to constraints including preservation of known superpositional relations and co-occurrences of taxa and events. Hammer and Harper (2005, p. 296) conclude that constrained optimization is "an excellent, flexible method for biostratigraphy, based on simple but sound theoretical concepts and potentially providing high-resolution results". For a recent summary of this method with an application, see Cody et al. (2008).

CONOP also has been used for constructing numerical geologic time scales for Paleozoic periods (Gradstein et al. 2004, 2012). In this approach, known age determinations of rock samples for a period are plotted against the positions of these samples along a relative geologic time-scale provided by the constrained optimum sequence for a suitable group of fossils. When the ranges are plotted along the vertical scale, their ordinates can be used as the abscissae in this new application. A smoothing spline (Sect. 9.2) then is fitted with consideration of uncertainties in both the age determinations and the stratigraphic positions of the rock samples. This spline curve is used to estimate the ages of stage boundaries with 95 % confidence intervals (Agterberg 2004; Agterberg et al. 2012).

Other methods of quantitative stratigraphy include Appearance Event Ordination (Alroy 2000) abbreviated to AEO, and graphic biostratigraphic correlation using genetic algorithms (Zhang and Plotnick 2006). Alroy uses super-positional information on taxa as follows: If it can be established that the FO of taxon A occurs below the LO of taxon B in a section, this can be interpreted as a definitive F/L (First/Last) statement because the FO of A can only be extended downwards toward its FAD and the LO of B upwards towards B's LAD. AEO attempts to honor all F/Ls while minimizing the number of F/Ls that would be implied but are not observed. This method is more sensitive to the possible occurrence of local reworking than RASC. For a recent AEO application, see Crampton et al. (2012).

Zhang and Plotnick (2006) proposed to use genetic algorithms, a branch of artificial intelligence that solves complex optimizations problems by imitating neo-Darwinian evolution to solve the "traveling salesman" (TSM) problem. In TSM there are N cities to be visited. In total, there would be $0.5 \times N!$ possible alternative trajectories. This number quickly becomes too large for the trajectories to be investigated separately. A close-to-optimum trajectory is found by imposing constraints so that sets of unsuitable solutions are successively eliminated. In biostratigraphic applications the cities are replaced by biostratigraphic events and the constraints include the super-positional (and co-existential) relations between events. Zhang and Plotnick's method resembles CONOP.

Especially during the first 20 years of its existence, RASC has been widely applied to last occurrences of microfossils observed in cuttings obtained at discrete, regular intervals (e.g., every 10 m) from exploratory wells drilled by oil companies in sedimentary basins. To some extent, this sampling procedure affects the output as will be discussed later (Sect. 9.3.2). Often in a sedimentary basin that is being studied, there occur other types of stratigraphic events such as ash layers, seismic events and gamma ray peaks, which are not subject to biostratigraphic uncertainty. Some of these, especially ash layers, can be used for correlation without uncertainty between stratigraphic sections that may be tens of kilometers apart. Seismic events

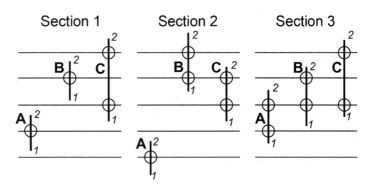

Fig. 9.1 Simple artificial example to illustrate construction of a RASC optimum sequence. Three fossil taxa (**a**, **b** and **c**) have been observed at five equally spaced levels in three sections. *Circles* indicate observed lowest (*1*) and highest (*2*) occurrences of the taxa. Distance between levels could represent a 10-m sampling interval between cuttings taken from three exploratory wells drilled in a sedimentary basin. The *vertical heavy black lines* represent ranges of the three taxa in the hypothetical situation that the sampling interval would be very narrow. Even if the sampling interval is not narrow, it can be assumed that any observed range of a fossil taxon is too short because it is unlikely that the lowest (*1*) and highest (*2*) occurrence would coincide exactly with the truly first and last occurrence of the taxon elsewhere in the sedimentary basin (Source: Agterberg et al. 2013, Fig. 1)

and gamma ray peaks can be diachronous. In RASC "marker horizons" without biostratigraphic uncertainty receive more weight in the calculations. Basic principles of RASC are introduced in the next section. A simple artificial example will be used to illustrate ranking followed by determination of probable positions of stratigraphic events in a RASC biozonation that can be used for CASC correlation between sections. Scaling, equi-distant sampling effects and RASC analysis of variance will be illustrated in Sect. 9.3 on the basis of large exploratory well data sets.

9.1.2 Artificial Example of Ranking

Figure 9.1 shows three artificial stratigraphic sections in which three fossil taxa (A, B and C) were observed to occur. Each taxon has first and last occurrence (FO and LO) labelled 1 and 2, respectively. Consequently, there are six stratigraphic events in total. The equidistant horizontal lines represent five regularly-spaced sampling levels. The discrete sampling procedure changes the positions of the FOs and LOs. As discussed in the previous section, any observed range for a taxon generally is much shorter than its true range of occurrence. In our example, the observed ranges become even shorter because of the discrete sampling. FOs and LOs coincide in three places. In reality, the FO of a taxon always must occur stratigraphically below its LO. It is because of the sampling scheme that these two events can occur at the same level in our example (see circles in Fig. 9.1).

9.1 Ranking and Scaling

Table 9.1 This matrix shows superpositional frequencies f_{ij} ($i \neq j$) for pairs of events defined for each fossil taxon in the artificial example of Fig. 9.1

		C2	C1	B2	B1	A2	A1
1	C2	x	3	2	3	3	3
2	C1	0	x	0	1	2	3
3	B2	1	3	x	3	3	3
4	B1	0	2	0	x	2	3
5	A2	0	1	0	1	x	3
6	A1	0	0	0	0	0	x

Source: Agterberg et al. (2013, Table 1)
For example, in Sections 1 and 3, the heavy line segment range of occurrence for taxon C has its highest point (C2) above the highest point (B2) for taxon B but the reverse hold true in Section 2. Consequently, the superpositional frequency of C2 (column $i = 2$) occurring above B1 (row $j = 1$) is $f_{13} = 2$. The corresponding lower triangle frequency f_{ji} of the i-th event occurring below the j-th event satisfies $f_{ji} = 3 - f_{ij}$; e.g., $f_{31} = 1$

For comparison, coincident FO and LO would occur in land-based studies when only a single fossil for a taxon would be observed in a stratigraphic section. In the artificial example of Fig. 9.1 it is assumed that all three taxa occur in each section. In practical applications, many taxa generally are missing from many sections. This aspect of missing data will be discussed later on the basis of large exploratory well data sets.

Table 9.1 shows the superpositional relationships of the six events defined by the end points of solid lines for the three taxa in Fig. 9.1. Suppose that elements of this matrix (F) are written as f_{ij} ($i \neq j$), where $i = 1$, 2 or 3 represent columns and $j = 1$, 2 or 3 are rows. Elements with $i > j$ in the upper triangle of F have corresponding elements in the lower triangle of F that are equal to $f_{ji} = 3 - f_{ij}$. Suppose further that, on the average, the six events succeed one another according to a true sequence that is the same for all sections, which belong to an infinitely large (statistical) population of sections. Any inconsistency such as the LO of taxon C occurring above the LO of taxon B in Section 1 but below it in the other two sections is assumed to be due to lack of information on the true ranges between FADs and LADs of the taxa in any section. The "optimum sequence" is assumed to provide an approximation of the true sequence of events. It is obtained by re-ordering the events along the rows and columns of the matrix F in such a way that every event occurs more frequently above any other event that occurs stratigraphically below it. In an optimum sequence of biostratigraphic events, the frequencies f_{ij} ($i \neq j$) should satisfy the relationship $f_{ij} > f_{ji}$.

Table 9.2 shows the slightly different superpositional relationships resulting from the sampling at discrete, regular intervals. Because the FO and LO of an event can coincide when rock samples are taken at discrete intervals, any observed co-occurrence is scored as 0.5. If the events would form an optimum sequence, $f_{ij} \geq f_{ji}$ is required instead of $f_{ij} > f_{ji}$. Table 9.3 shows the optimum sequence occurrence matrix for this example. It was obtained simply by moving the event C1 in the stratigraphically downward direction to its new position between events

Table 9.2 Suppose that the fossil taxa of Fig. 9.1 are only sampled at the equally spaced levels

		C2	C1	B2	B1	A2	A1
1	C2	x	3	2	2.5	3	3
2	C1	0	x	0	0.5	2.5	3
3	B2	1	3	x	3	3	3
4	B1	0.5	2.5	0	x	2.5	3
5	A2	0	0.5	0	0.5	x	3
6	A1	0	0	0	0	0	x

Source: Agterberg et al. (2013, Table 2)
Points for pairs of events, which are observed to be coincident, are circled in Fig. 9.1. For example, event A2 coincides with event A1 in Sections 1 and 2. In RASC, coincident events in a section both receive a score of 0.5. Consequently, $f_{56} = 2$ (instead of $= 3$ in Table 8.1), and $f_{65} = 1$

Table 9.3 The so-called "optimum" sequence is obtained by interchanging rows and columns in Table 9.2 so that in the new matrix $f_{ij} \geq f_{ji}$ ($i, j = 1, 2, 3$)

		C2	B2	B1	C1	A2	A1
1	C2	x	2	2.5	3	3	3
2	B2	1	x	2.5	3	3	3
3	B1	0.5	0.5	x	2.5	3	3
4	C1	0	0	0.5	x	2.5	3
5	A2	0	0	0	0.5	x	3
6	A1	0	0	0	0	0	x

Source: Agterberg et al. (2013, Table 3)
In the RASC ranked optimum sequence for the artificial example, C1 was moved downward between A2 and B1. Optimum sequence based on Table 9.1 would be the same. RASC also can be used when sampling is more continuous as in land-based sections

A2 and B1. The first step in the RASC computer program consists of finding the ranked optimum sequence of events by systematically interchanging events until the relation $f_{ij} \geq f_{ji}$ is satisfied for all pairs of events in the new matrix. In practical applications, it may not be possible to find such an order of events that is optimal because cyclical inconsistencies involving three or more events can occur. For example, if there are three events called E1, E2 and E3 with E1 occurring more frequently above E2, E2 above E3, and E3 above E1, an optimum sequence cannot be found by event position interchange. In RASC, various methods are used to cope with inconsistencies of this type; for example, more weight is assigned to frequencies f_{ij} based on larger samples.

In Fig. 9.2, the observed event level locations for the three taxa in the three sections are plotted against their optimum sequence positions. The lines of correlation were fitted by linear least squares. Each "probable" position line (PPL) provides every event with a most likely position along the depth scale for the section in which it occurs. The three PPL plots (Fig. 9.2a–c) are only slightly different. In practical applications, sampling levels normally are defined along a

9.1 Ranking and Scaling

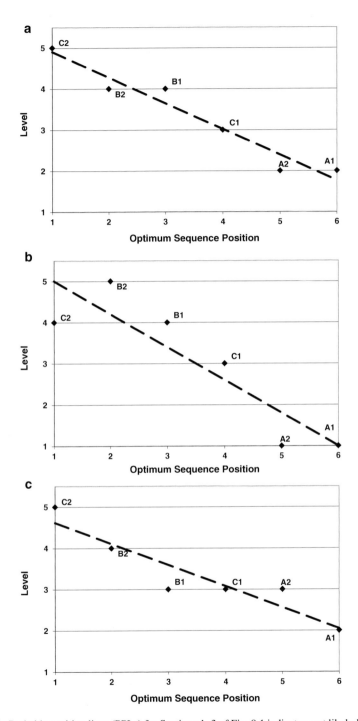

Fig. 9.2 Probable position lines (PPLs) for Sections 1–3 of Fig. 9.1 indicate most likely locations of the six stratigraphic events in the three sections. In the CASC program such "probable" locations, along with their error bars, are connected by *straight lines* between sections. **a**, **b** and **c** are for Sections 1, 2 and 3, respectively (Source: Agterberg et al. 2013, Fig. 2)

Table 9.4 Statistics derived from vertical deviations between the six events and the section PPLs in Fig. 9.2a–c

Taxon	s(1)	s(2)	Min	Max	Range
A	0.61	1.19	−0.05	0.56	0.61
B	2.18	1.71	−0.59	0.80	1.39
C	1.82	1.66	−0.08	0.38	0.46

Source: Agterberg et al. (2013, Table 4)
Each standard deviation is computed from three observations only; $s(1)$ and $s(2)$ are for FOs and LOs, respectively. Standard deviations of this type are used in RASC variance analysis and for estimation of error bars in CASC. Max and Min are for largest LO and lowest FO. The RASC "range" (= Max − Min) is comparable to ranges for taxa derived by graphic correlation methods

depth scale that is different for each section. PPL values are later used in CASC for correlation of average event occurrences between sections.

The residuals, which are the differences between observed event locations and corresponding PPL values in Fig. 9.2a–c, can provide useful information as illustrated in Table 9.4 for the artificial example. The rows in this table are for the three taxa; $s(1)$ and $s(2)$ are standard deviations for FOs and LOs estimated from the three residuals for each taxon. The next two columns in Table 9.4 represent lowest FO and highest LO of each taxon. The "range" in the last column is the difference between maximum LO and minimum FO. It has its largest value for Taxon B, mainly because, according to the optimum sequence, B is out of place with respect to A and C in Section 2. Table 9.4 illustrates how RASC, like other methods of quantitative biostratigraphy, can be used for FAD/LAD approximation.

9.1.3 Scaling

The optimum sequence obtained by ranking (or "ranked optimum sequence") often can be refined by "scaling", which consists of estimating intervals between successive optimum sequence events along a relative time scale. The ranked optimum sequence serves as input for scaling. As the first step in scaling, each frequency f_{ij} is converted into a relative frequency $p_{ij} = f_{ij}/n_{ij}$ where n_{ij} is sample size. Next, the values of p_{ij} are converted into intervals (interevent distances) along a relative time scale. Unlike ranking, scaling is not subject to problems of cyclical inconsistencies involving three or more events. The final order of events after scaling usually differs slightly from the order of events in the ranked optimum sequence.

Basically, scaling was introduced to circumvent the following problem associated with ranking. Events in a ranked optimum sequence can be regarded as equidistant along a linear scale such as the horizontal scales in Fig. 9.2a–c. However, along a relative time scale, the events generally should not be equidistant. For example, if two events are coeval on the average, their interevent distance along a relative time scale should be zero. Events that occurred at approximately the same

9.1 Ranking and Scaling

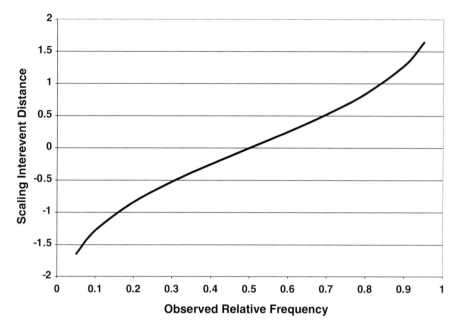

Fig. 9.3 Graphical representation of probit transformation. Every observed relative frequency is transformed into an interevent distance along the relative time scale used in scaling. Because all sample sizes are small, relative frequencies outside the range shown in this figure generally do not occur (Source: Agterberg et al. 2013, Fig. 3)

time should be clustering along this scale. In scaling every relative frequency p_{ij} is changed into a "distance" D_{ij} by means of the following transformation:

$$D_{ij} = \Phi^{-1}(p_{ij})$$

where Φ^{-1} is the probit function representing the inverse of the standard normal distribution. Figure 9.3 illustrates how observed relative frequencies of superpositional relationships between two biostratigraphic events (labeled i and j) are transformed into interevent distances along the RASC scale when the probit transformation is used. This transformation is not very different from a simple linear transformation. In practice, sample sizes n_{ij} generally are rather small. Consequently, the individual relative frequencies p_{ij} are rather imprecise. Their standard deviations can be estimated by means of the binomial frequency distribution model. Re-phrasing the problem in terms of probabilities, it can be said that it is attempted to estimate Δ_{ij} representing the expected value of D_{ij} in an infinitely large population. The concept of direct distance estimation is graphically illustrated in Fig. 9.4. The problem of lack of precision of individual distance estimates D_{ij} can be circumvented by incorporating for estimation all other biostratigraphic events (labeled k) in the vicinity of the pair for which the "true" interevent distance Δ_{ij} is being estimated, by using the theorem

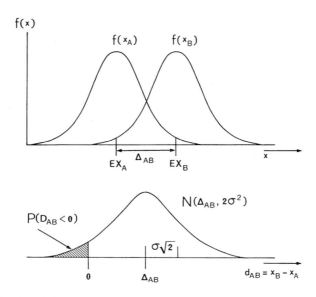

Fig. 9.4 Estimation of distance Δ_{AB} between A and B from two-event inconsistencies frequency $p(D_{AB} < 0) = 1 - P(D_{AB} > 0)$. If X_A and X_B are random variables for location of events A and B along the horizontal distance scale (x-axis), the random variable $D_{AB} = x_A - x_B$ is negative only when order of A and B in section is reverse of order of the expected average locations EX_A and EX_B. The variance of D_{AB} is twice as large as the variance σ^2 of the individual events A and B (Source: Agterberg 1990, Fig. 6.5)

$$P(D_{ij} > 0) = P(D_{ij \cdot k} > 0) = \Phi(\Delta_{ij})$$

where P denotes probability, and both D_{ij} and $D_{ij \cdot k} = D_{ik} - D_{jk}$ are normally distributed with expected values equal to Δ_{ij} (cf. Agterberg 1990, Equation 6.4). In principle, this means that all other events (k) can be used to estimate the interevent distance between any pair of events (i and j). In practice, other events labeled k that are relatively far removed from the events labeled i and j cannot be used because this would result in values of $-\infty$ or $+\infty$. RASC uses a variety of end corrections to prevent this particular problem from significantly affecting the interevent distance estimation.

The statistical model of scaling also can be clarified as follows. Suppose that, along the RASC scale, all biostratigraphic events are normally distributed with different mathematical expectations but with the same variance (σ^2). Interevent distances D_{ij} then also are normally distributed with mean Δ_{ij} but variance $2 \cdot \sigma^2$, and differences between interevent distances $D_{ij \cdot k}$ are normally distributed with mean Δ_{ij} and variance $3 \cdot \sigma^2$. Even if the original events do not have the same variance, different $D_{ij \cdot k}$ variances are not as different as original event variances because these are being averaged. These considerations apply to an infinite statistical population from which small samples are being drawn. For equations of small-sample variances, see Agterberg (1990). Finally, it is noted that setting $D_{ij} = \Phi^{-1}(p_{ij})$ implies $\sigma^2 = 2$. This arbitrary choice of variance controls the unit for plotting events along the RASC interevent distance scale, which is relative.

9.1 Ranking and Scaling

Fig. 9.5 Locations of sections in the Sullivan (1965, Table 6) database for the Eocene used by Hay (1972) for example (see Fig. 9.4) (Source: Agterberg 1990, Fig. 4.1)

9.1.4 Californian Eocene Nannofossils Example

Hay (1972) used stratigraphic information on calcareous nannofossils from sections in the California Coast Ranges for example of application of his original ranking method (see Fig. 9.5 for locations). These sections originally had been studied by Sullivan (1964, 1965) and Bramlette and Sullivan (1961). The distribution of Lower Tertiary nannoplankton described in the latter three papers also was used for example by Davaud and Guex (1978) and Guex (1987) for testing various types of stratigraphic correlation techniques. The original paper by Hay (1972) resulted in extensive discussions (e.g., Edwards 1978; Harper 1981) and applications of other techniques to the Hay example (e.g., Hudson and Agterberg 1982).

Hay (1972) restricted his example to Lower Tertiary nannofossils for samples shown on Sullivan's (1965) correlation chart augmented by stratigraphic information on the Lodo Gulch section from Bramlette and Sullivan (1961). Several of the nannofossil taxa used are known to occur in older Paleocene strata in the Media Ague Creek and Upper Canada de Santa Anita sections (see Sullivan 1964). Incorporation of this other information in the example slightly improves results of the quantitative stratigraphic analysis with respect to lowest occurrences of nannofossils in these two sections (see Agterberg 1990). Tables 9.5 and 9.6 are examples of dictionary input files used this example. Hay (1972) selected lowest occurrences of nine taxa and the highest occurrence of one taxon (*Discoaster tribrachiatus*) shown by symbols in Fig. 9.6 (Table 9.7). Original sample location information is shown in Fig. 9.7 for three sections. Table 9.8 contains original data

Table 9.5 RASC dictionary
(DIC file) for Hay example

1 LO DISCOASTER DISTINCTUS
2 LO COCCOLITHUS CRIBELLUM
3 LO DISCOASTER GERMANICUS
4 LO COCCOLITHUS SOLITUS
5 LO COCCOLITHUS GAMMATION
6 LO RHABDOSPHAERA SCABROSA
7 LO DISCOASTER MINIMUS
8 LO DISCOASTER CRUCIFORMIS
9 HI DISCOASTER TRIBRACHIATUS
10 LO DISCOLITHUS DISTINCTUS

Source: Agterberg (1990, Table 4.1)
LO and HI represent lowest and highest occurrences of nannofossils, respectively

Table 9.6 Nannofossil name file (preliminary DIC file) for Sullivan database as originally coded by Davaud and Guex (1978) and used by Agterberg et al. (1985)

1	CHIPHRAGMALITHUS CRISTATUS	27	DISCOLITHUS FIMBRIATUS	53	RHABDOSPHAERA TRUNCATA	79	DISCOASTER BINODOSUS
2	CHIPHRAEMALITHUS ACANTHODES	28	DISCOLITHUS OCELLATUS	54	RHABDOSPHAERA INFLATA	80	DISCOASTER DEFLANDREI
3	CHIPHRAGMALITHUS CALATUS	29	DISCOLITHUS PANARIUM	55	ZYGODISCUS SIGMOIDES	81	DISCOASTER DELICATUS
4	CHIPHRAGMALITHUS DUBIUS	30	DISCOLITHUS PUNCTOSUS	56	ZYGODISCUS ADAMAS	82	DISCOASTER DIASTYPUS
5	CHIPHRAGMALITHUS PROTENUS	31	DISCOLITHUS SOLIDUS	57	ZYGODISCUS HERLYNI	83	DISCOASTER DISTINCTUS
6	CHIPHRAGMALITHUS QUADRATUS	32	DISCOLITHUS VESCUS	58	ZYGODISCUS PLECTOPONS	84	DISCOASTER FALCATUS
7	COCCOLITHUS BIDENS	33	DISCOLITHUS VERSUS	59	ZYGOLITHUS CONCINNUS	85	DISCOASTER LODOENSIS
8	COCCOLITHUS CALIFORNICUS	34	DISCOLITHUS PERTUSUS	60	ZYGOLITHUS CRUX	86	DISCOASTER MULTIRADIATUS
9	COCCOLITHUS EXPANSUS	35	DISCOLITHUS EXILIS	61	ZYGOLITHUS DISTENTUS	87	DISCOASTER NONARADIATUS
10	COCCOLITHUS GRANDIS	36	DISCOLITHUS EUOCAVUS	62	ZYGOLITHUS JUNCTUS	88	DISCOASTER STRADNERI
11	COCCOLITHUS SOLITUS	37	DISCOLITHUS INCONSPICUUS	63	ZYGRHABLITHUS SIMPLEX	89	DISCOASTER TRIBRACHIATUS
12	COCCOLITHUS STAURION	38	DISCOLITHUS ROBUSTUS	64	ZYGRHABLITHUS BIJUGATUS	90	DISCOASTER CRUCIFORMIS
13	COCCOLITHUS GIGAS	39	ELLIPSOLITHUS MACELLUS	65	BAARUDOSPHAERA BIGELOWI	91	DISCOASTER GERMANICUS
14	COCCOLITHUS DELUS	40	ELLIPSOLITHUS DISTICHUS	66	BAARUDOSPHAERA DISCULA	92	DISCOASTER LENTICULARIS
15	COCCOLITHUS CONSUETUS	41	HELICOSPHAERA SEMILUNUM	67	MICRANTHOLITHUS FLOS	93	DISCOASTER MARTINII
16	COCCOLITHUS CRASSUS	42	HELICOSPHAERA LOPHOTA	68	MICRANTHOLITHUS INAEQUALIS	94	DISCOASTER MINIMUS
17	COCCOLITHUS CRIBELLUM	43	LOPHODOLITHUS NASCENS	69	MICRANTHOLITHUS VESPER	95	DISCOASTER SEPTEMRADIATUS
18	COCCOLITHUS EMINENS	44	LOPHODOLITHUS RENIFORMIS	70	MICRANTHOLITHUS BASQUENSIS	96	DISCOASTER SUBLODOENSIS
19	CYCLOCOCCOLITHUS GAMMATION	45	LOPHODOLITHUS MOCHOLOPHORUS	71	MICRANTHOLITHUS CRENULATUS	97	DISCOASTER HELIANTHUS
20	CYCLOCOCCOLITHUS LUMINIS	46	RHABDOSPHAERA CREBRA	72	MICRANTHOLITHUS AEQUALIS	98	DISCOASTER LIMBATUS
21	DISCOLITHUS PECTINATUS	47	RHABDOSPHAERA MORIONUM	73	CLATHROLITHUS ELLIPTICUS	99	DISCOASTER MEDIOSUS
22	DISCOLITHUS PLANUS	48	RHABDOSPHAERA PERLONGA	74	RHOMBOASTER CUSPIS	100	DISCOASTER PERPOLITUS
23	DISCOLITHUS PULCHER	49	RHABDOSPHAERA RUBIS	75	POLYCLADOLITHUS OPEROSUS	101	DISCOASTEROIDES KUEPPERI
24	DISCOLITHUS PULCHEROIDES	50	RHABDOSPHAERA SCABROSA	76	SPHENOLITHUS RADIANS	102	DISCOASTEROIDES MEGASTYPUS
25	DISCOLITHUS RIMOSUS	51	RHABDOSPHAERA SEMIFORMIS	77	FASCICULOLITHUS INVOLUTUS	103	HELIOLITHUS KLEINPELLI
26	DISCOLITHUS DISTINCTUS	52	RHABDOSPHAERA TENUIS	78	DISCOASTER BARBADIENSIS	104	HELIOLITHUS RIEDELI

Source: Agterberg (1990, Table 4.2)
RASC DIC file of Table 9.5 was obtained automatically from this file

for the Media Agua Creek section that contains nine of the ten stratigraphic events used for example. Guex (1987) used graph theory to construct "unitary associations", which have essentially the same properties as Oppel zones in biostratigraphy. Emphasis in this approach is on coexistences of taxa. Adjoining samples are combined into levels representing "maximal horizons" (*cf.* Guex 1987, p. 20) as illustrated for the Media Agua Creek example in the bottom row of Table 9.8.

The two columns on the right of Fig. 9.6 represent a subjective ranking and Hay's original optimum sequence, respectively. This optimum sequence has the property that every event in it is observed more frequently above all events below it than it is observed below these other events. Every frequency for superpositional relationship of a pair of events can be tested for statistical significance by means of

9.1 Ranking and Scaling

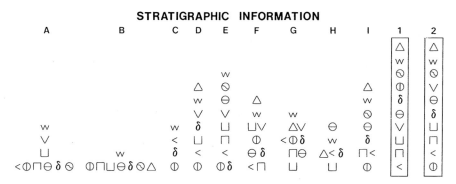

Fig. 9.6 Hay's (1972) example. One last occurrence and nine "first" occurrences of Eocene nannofossils selected by Hay (1972) from the Sullivan Eocene database. Explanation of symbols: δ = LO, *Coccolithus gammation*; Φ = LO *Coccolithus cribellum*; Θ = LO, *Coccolithus solitus*; V = LO, *Discoaster cruciformis*; < = LO, *Discoaster distinctus*; Π = LO, *discoaster germanicus*; U = LO, *discoaster minimus*; W = HI, *Discoaster tribrachiatus*; Δ = LO, *Discolithus distinctus*; □ = LO, *Rhabdosphaera scabrosa*. See Fig. 8.3 for locations of the nine sections (A–I). Some LOs are for nannofossils also found in Paleocene (Sullivan 1964, Table 3). The *columns* on the *right* represent subjective ordering of the events and Hay's original optimum sequence (Source: Agterberg 1990, Fig. 4.2)

Table 9.7 Two possible RASC sequence (SEQ) files for Hay example

A											A									
	9	8	7	6	-5	-4	-3	-2	-1	-999		1	-2	-3	-4	-5	-6	7	8	9-999
B											B									
	9	10	-6	-5	-4	-7	-3	-2	-999			2	-3	-7	-4	-5	-6	-10	9-999	
C											C									
	9	1	5	2-999								2	5	1	9-999					
D											D									
	10	9	8	5	7	1	2-999					2	1	7	5	8	9	10-999		
E											E									
	9	6	4	8	7	3	1	5	-2-999			2	-5	1	3	7	8	4	6	9-999
F											F									
	10	9	8	-7	2	5	-4	3	-1-999			1	-3	4	-5	2	7	-8	9	10-999
G											G									
	9	8	-10	5	-2	-1	4	-3	7-999			7	3	-4	1	-2	-5	10	-8	9-999
H											H									
	4	9	5	-1	-10	7-999						7	10	-1	-5	9	4-999			
I											I									
	10	9	6	4	5	1	-3	2-999				2	3	-1	5	4	6	9	10-999	

Source: Agterberg (1990, Table 4.3)
Minus signs (or hyphens) denote coeval events (*cf.* Fig. 9.2). The last entry is followed by −999 to indicate end of sequence. Left side: SEQ file for stratigraphically downward direction; right side: SEQ file for stratigraphically upward direction

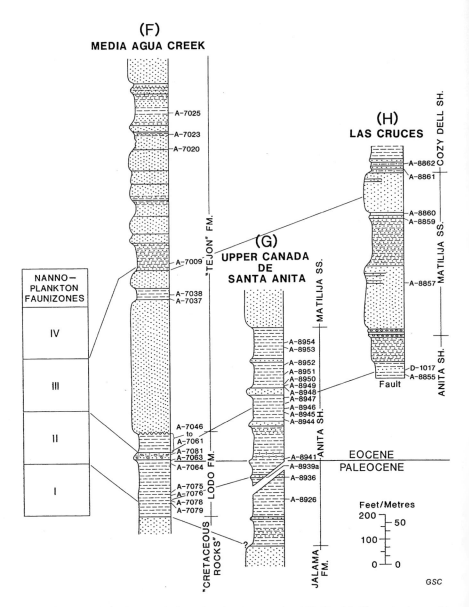

Fig. 9.7 Original stratigraphic information for three sections (F–H) of Sullivan database with stratigraphic correlation based on nannoplankton faunizones according to Sullivan (1965). Table 9.4 contains information on distribution of nine taxa in core samples (Source: Media Agua Creek section (Source: Agterberg 1990, Fig. 4.3)

the binomial frequency distribution model. Figure 9.8 shows the difference between 1 and the cumulative probability $P(k,n)$ that an event occurs k times above another one in a sample of size n. If $1 - P(k,n)$ exceeds 0.95, the fraction k/n is non-random in that it is greater than 0.5 with a probability of at least 95 %. Only six of nine pairs

9.1 Ranking and Scaling

Table 9.8 Stratigraphic distribution of nine taxa of fossil nannoplankton for individual core samples in the Media Agua Creek area, Kern County, California (according to Sullivan 1964, Table 3, and Sullivan 1965, Table 6)

Source: Agterberg (1990, Table 4.4)

Stratigraphic distance (D) in feet was measured upward and downward from the "Tejon" Formation; Paleocene-Eocene boundary occurs between 103 and 118 ft. Fossil (F) numbers in first column are as in Table 9.6. A-abundant, C-common, 0-few, x-rare. Single bar indicates stratigraphic events E1 to E10 used in Table 9.5 and Fig. 9.3 (as defined for core samples extending up to 88 ft below the "Tejon" Formation); relative superpositional relations (as shown in Fig. 9.2) would be changed if lowest occurrences of four taxa in Paleocene shown in lower part of this table were used. Level (L) as in Guex (1987, p. 228)

involving the event W, which occurs at or near the top of all (nine) sections (labeled A to I in Fig. 9.6) are statistically significant for this level of significance. This binomial test has the drawback that it can consider pairs of events only. A multivariate approach as followed in RASC is more appropriate. This is illustrated in Fig. 9.9 for some of the events in the Hay example.

.8750	.8750	.7500	.8750	.8750	.8750	.5000	.0000	.0000	Δ HIGHEST
.9922	.9922	.9688	.9844	.9961	.8750	.9375	.8750	W	
.5000	.5000	.5000	.0000	.5000	.5000	.0000	☾		
.9375	.9375	.8750	.8750	.9375	.3750	V			
.0000	.6250	.7500	.0000	.3750	⊖				
.3750	.0000	.3750	.0000	δ					
.6250	.3125	.3750	⊔						
.0000	.0000	⊓							
.6250	<								
LOWEST Φ									

Fig. 9.8 Values of $(1 - P)$ where P represents the probability that the sequential relation between two events is non-random according to Hay (1972) (Source: Agterberg 1990, Fig. 5.2)

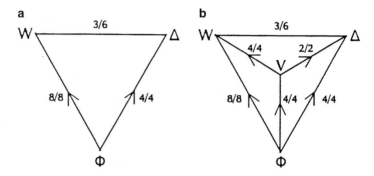

Fig. 9.9 Diagrams to illustrate superpositional relations between (**a**) three events and (**b**) four events in the Hay example. Although Δ and Φ both occur only in four sections, their superpositional relation is probably non-random because of their relations with other events (Source: Agterberg 1990, Fig. 5.3)

As mentioned in Sect. 9.1.3, it usually is not possible to obtain a unique optimum sequence in practice because of cycling events. This problem is illustrated for a three-event cycle in Fig. 9.10. More than three events can be involved in cycling (Agterberg 1990) In the RASC computer program, a ranking solution is obtained by first identifying all subgroups of events involved in cycling and then followed by "breaking" the cycles using sample size considerations. Scaling is not subject to cycling problems and has the additional advantage that it quantifies the strengths of all superpositional relationships. Ranked and scaled optimum sequences for the Hay example are shown in Fig. 9.11. Intermediate steps before the final scaling solution of Fig. 9.11b was obtained are illustrated in Tables 9.9, 9.10, 9.11, and 9.12.

(a) ABCD	(b) CBAD	(c) BCAD	(d) ACBD	(e) CABD	(f) BACD	(g) ABCD
× 232	× 243	× 511	× 322	× 423	× 151	× 232
1 × 51	5 × 11	2 × 43	4 × 23	3 × 22	2 × 32	1 × 51
42 × 3	32 × 2	23 × 2	15 × 1	51 × 1	24× 3	42× 3
074 ×	470 ×	740 ×	047 ×	407 ×	704 ×	074 ×

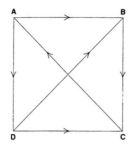

Fig. 9.10 Example of cycling events (based on initial matrix of Worsley and Jorgens 1977). In each matrix, the order of the rows is the same as the order of the columns. Unlike the examples of Figs. 9.1 and 9.3, the algorithm for ordering does not yield an optimum sequence because the initial matrix (a) returns after six iterations as the matrix (g). The directed graph at the *bottom* has the four events at its corner points with *arrows* denoting directions of the super-positional relationships. The three-event cycle is characterized by the closed loop with *arrows* on edges connecting A, B, and C pointed in the same, clockwise direction. The other closed loop (ADC) is subsidiary to ABC because the event D does not participate in the cycling (Source: Agterberg 1990, Table 5.13 and Fig. 5.7)

9.2 Spline-Fitting

Spline-curve fitting or splining is a powerful different approach to finding the functional relationship between two variables. The smoothing spline combines spline interpolation with curve-fitting by least squares if one of the variables is free of uncertainty. It can be made applicable to the situation that both variables are subject to uncertainty. Together with the jackknife method, spline-fitting is helpful in numerical timescale construction and other applications in quantitative stratigraphy. Spline functions have a long history of use for interpolation; e.g. in numerical integration. Their use for smoothing is a later development which commenced with the discovery of smoothing splines by Schoenberg (1964) and Reinsch (1967). Whittaker (1923) already had proposed an early variant. There is a close relationship between smoothing splines and kriging (Watson 1984; Dubrule 1984).

Figure 9.1 illustrates the concepts of interpolation and smoothing spline functions. An excellent introduction to smoothing splines is provided in Eubank (1988). Although splines of higher and lower orders can be constructed, the third-order or cubic smoothing spline seems to be optimum for the treatment of observations at irregularly spaced intervals. During the past 20 years splining has been used to aid in creating the Mesozoic timescale (Agterberg 1988) and the international geologic

OPTIMUM FOSSIL EVENT SEQUENCE

a

SEQUENCE POSITION	FOSSIL NUMBER	RANGE	FOSSIL NAME
1	9	0 – 3	HO DISCOASTER TRIBRACHIATUS
2	10	0 – 3	LO DISCOLITHUS DICTINCTUS
3	8	2 – 5	LO DISCOASTER CRUCIFORMIS
4	6	1 – 5	LO RHABDOSPHAERA SCABROSA
5	4	4 – 6	LO COCCOLITHUS SOLITUS
6	7	5 – 8	LO DISCOASTER MINIMUS
7	5	5 – 8	LO COCCOLITHUS GAMMATION
8	1	7 – 10	LO DISCOASTER DISTINCTUS
9	3	7 – 11	LO DISCOASTER GERMANICUS
10	2	8 – 11	LO COCCOLITHUS CRIBELLUM

b

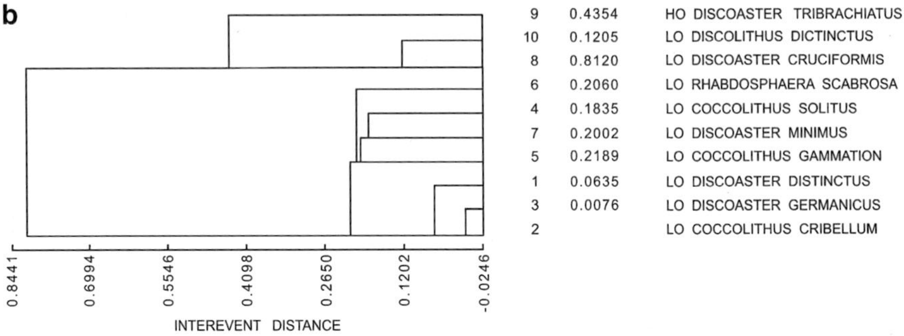

9	0.4354	HO DISCOASTER TRIBRACHIATUS
10	0.1205	LO DISCOLITHUS DICTINCTUS
8	0.8120	LO DISCOASTER CRUCIFORMIS
6	0.2060	LO RHABDOSPHAERA SCABROSA
4	0.1835	LO COCCOLITHUS SOLITUS
7	0.2002	LO DISCOASTER MINIMUS
5	0.2189	LO COCCOLITHUS GAMMATION
1	0.0635	LO DISCOASTER DISTINCTUS
3	0.0076	LO DISCOASTER GERMANICUS
2		LO COCCOLITHUS CRIBELLUM

INTEREVENT DISTANCE: 0.8441, 0.6994, 0.5546, 0.4098, 0.2650, 0.1202, -0.0246

Fig. 9.11 RASC results for Hay example of Fig. 9.6 (After Agterberg and Gradstein 1988); (**a**) ranked optimum sequence; (**b**) scaled optimum sequence. Clustering of events 1 to 7 in the dendrogram (**b**) reflects the relatively large number of two-event inconsistencies and many coincident events near the base of most sections used (Source: Agterberg 1990, Fig. 6.3)

timescale (Agterberg et al. 2013). Accuracy and precision are both important when geologic timescales are constructed. The use of significance tests and confidence intervals can be illustrated by comparing different geologic time scales with one another. As pointed out by Gradstein et al. (2012), the time scale is the tool "par excellence" of the geological trade. Insight into its construction, strengths and limitations greatly enhances its function and its utility. The calibration to linear time of the succession of events recorded in the rocks of the Earth has three components: (1) the international stratigraphic divisions and their correlation in the global rock record; (2) methods of measuring linear time or elapsed durations from the rock record; and (3) methods of joining the stratigraphic scale and the linear scale to assign numerical ages (measured in millions of years ago, or Ma) to the boundaries between the stratigraphic divisions (*cf*. Gradstein et al. 2012, p. 1).

9.2.1 Smoothing Splines

The interpolation spline curve passes through all (*n*) observed values. Along the curve there are a number of knots where various derivatives of the spline function are forced to be continuous. In the example of Fig. 9.12, the knots coincide with the

9.2 Spline-Fitting

Table 9.9 Unweighted distance estimation to obtain intervals between successive events along RASC distance scale for Hay example

A	9	10	8	6	4	7	5	1	3	2
9	x	3.0/6	5.0/5	4.0/4	6.0/7	7.0/7	9.0/9	8.0/8	6.0/6	80/8
10	3.0/6	x	2.5/3	000	35/5	4.5/5	5.0/6	4.5/5	3.5/4	4.5/5
8	00/5	05/3	x	000	3.0/4	4.5/5	5.0/5	5.0/5	4.0/4	5.0/5
6	0.0/4	000	000	x	30/4	1.5/3	3.0/4	2.5/3	3.0/4	3.0/4
4	1.0/7	1.5/5	1.0/4	1.0/4	x	3.5/6	4.5/7	4.5/6	4.5/6	3.0/6
7	0.0/7	05/5	0.5/5	15/3	2.5/6	x	3.5/7	4.0/6	3.5/5	4.5/6
5	0.0/9	10/6	0.0/5	10/4	2.5/7	3.5/7	x	4.5/8	4.0/6	5.0/8
1	0.0/8	05/5	0.0/5	0.5/3	15/6	2.0/6	3.5/8	x	2.5/5	5.0/7
3	0.0/6	05/4	0.0/4	1.0/4	1.5/6	1.5/5	2.0/6	2.5/5	x	3.0/6
2	0.0/8	0.5/5	00/5	1.0/4	3.0/6	1.5/6	3.0/8	2.0/7	3.0/6	x

B	9	10	8	6	4	7	5	1	3	2
9	x	0.000	1.645	1.645	1.068	1.645	1.645	1.645	1.645	1.645
10	0.000	x	0.967	000	0.524	1.282	0.967	1.282	1.150	1.282
8	1.645	0.967	x	000	0.674	1.282	1.645	1.645	1.645	1.645
6	−1.645	000	0.00	x	0.674	0.000	0.674	0.967	0.674	0.674
4	−1.068	−0.524	−0.674	−0.674	x	0.210	0.366	0.674	0.674	0.000
7	−1.645	−1.282	1.282	0.000	−0.210	x	0.000	0.430	0.524	0.674
5	1.645	−0.967	1.645	−0.674	−0.366	0.000	x	0.157	0.430	0.318
1	−1.645	−1.282	1.645	−0.967	−0.674	−0.430	−0.157	x	0.000	0.566
3	−1.645	1.150	1.645	0.674	−0.674	−0.524	−0.430	0.000	x	0.000
2	−1.645	−1.282	−1.645	−0.674	0.000	0.674	−0.318	−0.566	0.000	–

C		10	8	6	4	7	5	1	3	2
9		0.000	1.645	000	−0.577	0.577	000	000	000	000
10		x	0.967	000	000	0.758	0.315	0.315	0.132	0.132
8		0.678	x	000	000	0.608	0.363	0.000	0.000	0.000
6		000	0.000	x	0.674	−0.674	0.674	0.293	0.293	0.000
4		0.544	−0.150	0.000	x	0.210	0.156	0.308	0.000	0.674
7		0.363	0.000	1.282	0.210	x	0.000	0.430	0.094	0.150
5		0.678	−0.678	0.971	0.308	0.366	x	0.157	0.273	0.112
1		0.363	−0.363	0.678	0.293	0.244	0.273	x	0.000	0.566
3		0.495	−0.495	0.971	0.000	0.150	0.094	0.430	x	0.000
2		0.363	0.363	0.971	0.674	−0.674	0.356	−0.248	0.566	x
	Sum/N*	3.98/8	0.56/8	4.87,6	1.16/7	1.56/9	1.60/8	1.69/8	0.51/8	0.06/8

Source: Agterberg (1990, Table 6.2)
A. P-matrix of relative frequencies for the ten events in order of optimum sequence, Values excluded because of threshold $m_{2c} = 3$ are shown as 000. B. Z-values corresponding to P-values. Note that threshold q_c is equal to 1.645. C. Values are differences between values in successive columns of Table 9.2. Zero differences for pairs of q_c-values are shown as 000 and were not used. Bottom row shows sums for columns with number of values (N^*) used for obtaining sum

Table 9.10 Unweighted distance analysis of values shown in Table 9.9 continued to obtain cumulative RASC distances of events

	Events	N*	Sum	Interval	Distance
1	9-10	8	3.98	0.935	0.435
2	10-8	8	0.56	0.070	0.506
3	8-6	6	4.87	0.812	1.318
4	6-4	7	1.16	0.166	1.484
5	4-7	9	1.56	0.174	1.658
6	7-5	8	1.60	0.200	1.858
7	5-1	8	1.69	0.211	2.069
8	1-3	8	0.51	0.064	2.132
9	3-2	8	0.06	0.008	2.140

Source: Agterberg (1990, Table 6.3)
The origin of the scale is set at the first event. Consequently, the distance for event 9 is equal to zero. Event 10 has distance of 0.435; Event 2 has largest cumulative RASC distance (=2.140)

Table 9.11 Weighted distance analysis of values shown in Table 9.10

	Events	W	Sum	Interval	$s(\overline{X})$	Distance
1	9-10	10.3	3.27	0.317	0.100	0.317
2	10-8	7.0	1.24	0.176	0.289	0.493
3	8-6	4.7	3.62	0.770	0.203	1.262
4	6-4	9.2	2.44	0.266	0.163	1.529
5	4-7	15.0	2.34	0.157	0.153	1.686
6	7-5	14.8	2.32	0.157	0.085	1.893
7	5-1	15.2	2.96	0.195	0.082	2.038
8	1-3	12.6	1.47	0.117	0.090	2.155
9	3-2	13.3	−0.08	−0.006	0.124	2.149

Source: Agterberg (1990, Table 6.4)
The Z-values were weighted according to sample size and standard deviations were computed as well. Note that the interval between events 3 and 2 (on *bottom row*) is negative. As a result, event 9 has RASC distance (=2.149) which is less than that of event 8 (=2.155)

Table 9.12 Example of weighted distance analysis after reordering

	Events	Interval	$s(\overline{X})$	Distance
1	9-10	0.317	0.100	0.317
2	10-8	0.176	0.289	0.493
3	8-6	0.770	0.203	1.263
4	6-4	0.266	0.163	1.530
5	4-7	0.157	0.153	1.686
6	7-5	0.157	0.085	1.843
7	5-1	0.195	0.082	2.038
8	1-2	0.118	0.147	2.156
9	2-3	0.006	0.124	2.162

Source: Agterberg (1990, Table 6.5)
The optimum sequence used as input for scaling was not the ranking result used as input for Tables 9.9, 9.10, and 9.11 but the ranking of events in the scaled optimum sequence in the last column of Table 9.11. Differences between Tables 9.11 and 9.12 are restricted to values in two bottom rows only

9.2 Spline-Fitting

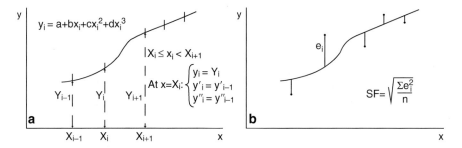

Fig. 9.12 Schematic diagrams of cubic interpolation spline and cubic smoothing spline. The cubic polynomials between successive knots have continuous first and second derivatives at the knots. The smoothing factor (SF) is zero for interpolation splines. In most applications, the abscissae of the knots coincide with those of the data points (Source: Agterberg 1990, Fig. 3.8)

data points. A separate cubic polynomial curve with four coefficients is computed for each interval between two successive data points. These cubics must have continuous first and second derivatives. After setting the second derivative equal to zero at the first and last data points, the continuity constraints yield so many conditions that all $(4n-4)$ coefficients can be computed. Smoothing splines have the same properties as interpolation splines except that they do not pass through the data points. Instead of this, they deviate from the observed values by an amount that can be regulated by means of the smoothing factor (SF) based on the average mean squared deviation. For each specific value of SF, which can be set in advance, or estimated by cross-validation, a single smoothing spline is obtained. Various methods of estimating the smoothing factor were discussed by Wahba (1990).

9.2.2 Irregularly Spaced Data Points

In his book on spline smoothing and non-parametric regression, Eubank (1988, p. 153) discusses that unequally spaced data points may give poor results for smoothing splines. De Boor (1978) pointed this out for interpolation splines. In order to avoid poor results of this type for constructing age-depth curves from biostratigraphic data, Agterberg et al. (1985) proposed the following simple "indirect" method. The following experiment with interpolation splines illustrates how the problem of unrealistic oscillations can be avoided, using this indirect method. It should be kept in mind that the problem of oscillations in data gaps becomes even more serious if the data are subject to much "noise" as in microfossil applications in biostratigraphy.

Figure 9.13 is based on an example of De Boor (1978, Fig. 8.1, p. 224). In total, 49 observations were available for a property of titanium (Y) as a function of temperature (X). These data have regular spacing along the X-axis. Irregular spacing was simulated by De Boor by selecting $n=12$ data points which are closer

Fig. 9.13 De Boor (1978, Fig. 8.1, p. 224) simulated irregular spacing along x-axis by selecting 12 points (*solid circles*) from set of 49 regularly spaced measurements of a variable (y) as a function of another variable (x). The optimum fifth order interpolation spline (with seven knots) provides a poor fit except around the peak (Source: Agterberg 1990, Fig. 3.9)

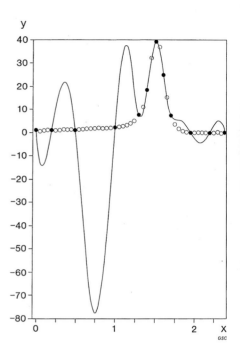

together on the peak than in the valleys. De Boor used this example to illustrate that poor results may be obtained even if use is made of a method of optimum spline interpolation in which best locations are computed for $(n-k)$ knots of a k-th order spline. For the example of Fig. 9.13, $k=5$ so that seven knots are used. Although these seven knots have optimum locations along the X-axis, the result is obviously poor, because the shape of the relatively narrow peak is reflected in non-realistic oscillations in between the more widely spaced data points in the valleys. De Boor (1978, p. 225) pointed out that using a lower-order spline would help to obtain a better approximation. In subsequent applications, use is made of cubic splines only ($k=3$). Figure 9.14a shows the cubic interpolation spline for the 12 regularly spaced points of Fig. 9.2 using knots coinciding with data points. Contrary to the fifth order spline with seven knots, the new result provides a good approximation. However, deletion of three more points from the valleys (Fig. 9.14b) begins to give the relatively poor cubic interpolation spline of Fig. 9.14c which has unrealistic oscillations in the valleys because all intermediate data points were deleted.

The bottom half of Fig. 9.14 shows results obtained by applying the indirect method to the situation that led to the worst cubic-spline result for the previous example (seven data points in Fig. 9.14c). Figure 9.14d is the cubic interpolation spline for seven regularly spaced "levels". Figure 9.14e is a monotonically increasing cubic smoothing spline with a small positive value for SF for the relation between X and level. Figure 9.14f represents the combination of the curves of Fig. 9.14d, e. Its approximation to the original pattern for the 49 values of Fig. 9.13 is only relatively poor in the valleys where no data were used for control. Unrealistic oscillations could be avoided by the use of the three-step indirect method of Fig. 9.14d–f.

9.2 Spline-Fitting

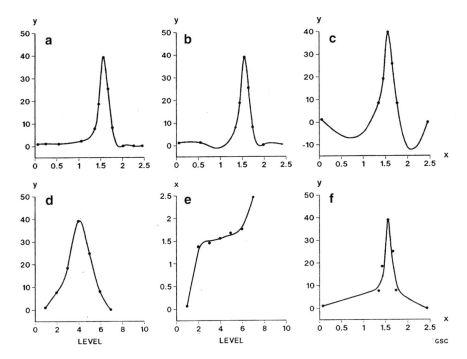

Fig. 9.14 *Top part.* Cubic interpolation splines with knots at data points fitted to irregularly spaced data. (**a**) Use of same 12 data points as in Fig. 9.13 gives good result; (**b**) Deletion of three points in the valleys still gives a fair interpolation spline although local minima at both sides of the peak are not supported by original data set of 49 measurements; (**c**) deletion of two more points in the valleys results in poor cubic interpolation spline. *Bottom part*: Indirect method of cubic spline-fitting. (**d**) The six intervals along the x-axis between data points were made equal before calculation of cubic interpolation spline; (**e**) non-decreasing cubic spline with small positive value of smoothing factor (SF = 0.0038) was fitted to interval as function of "levels"; (**f**) curves of (**d**) and (**e**) were combined with one another and re-expressed as cubic spline function which does not show the unrealistic fluctuations of the cubic interpolation spline of Fig. 9.14c (Source: Agterberg 1990, Fig. 3.10)

9.2.3 Tojeira Sections Correlation Example

A great variety of methods are available for eliminating noise from geoscience data. Several of these methods including (1) curve-fitting and 2-D or 3-D trend analysis using polynomial or Fourier series, (2) geostatistical kriging, (3) signal-extraction (from statistical theory of engineering), (4) inverse distance weighting, and (5) simple moving averaging, are applied in various case histories in this book. Smoothing splines used later in this chapter for estimation of the ages of stage boundaries in the geologic time scale and in long-distance correlation between biostratigraphic sections in wells drilled in sedimentary basins. The following example is a case history study of using smoothing splines between two land-based sections using microfossil abundance data.

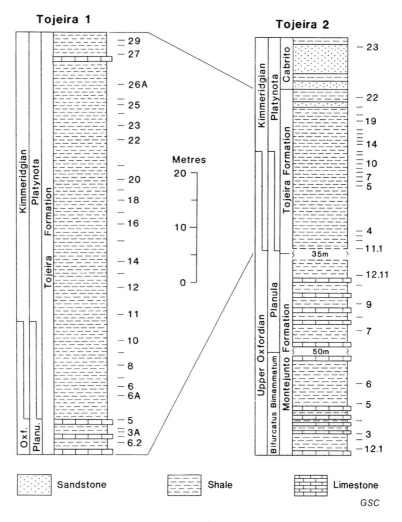

Fig. 9.15 *Left side*. Tojeira 1 section with sample members 6.2–6.29 (After Stam 1987); ammonite zones (*Planula* and *Platynota* Zones) of Mouterde et al. (1973) also are shown. This section is immediately overlain by the poorly exposed sandy Cabrito Formation. *Right side*: Tojeira 2 section with sample numbers 12.1–12.11 and 11.1–11.23 of Stam (1987) (Source: Agterberg 1990, Fig. 3.2)

Figure 9.15 shows two sections in the Montejunto area in Portugal originally sampled by Stam (1987) and later resampled by Agterberg et al. (1990). The purpose of this project was to perform quantitative analysis of occurrence of Middle and Late Jurassic Foraminifera in this area and its implications for the Grand Banks of Newfoundland that can only be sampled by drilling. The Tojeira 1 section with sample numbers 6.2–6.29 (after Stam 1987) is shown on the left side of Fig. 9.15. It is continuously exposed and occurs about 2 km southeast of the Tojeira 2 section

(Fig. 9.15, right side) with Stam's sample numbers 12.1–12.11 and 11.1–11.23. The Tojeira 2 section is not continuously exposed; two missing parts are estimated to be equivalent to 35 and 50 m in the stratigraphic direction, respectively.

Tojeira shale contains a rich and diversified (over 45 taxa) planktonic and benthonic foraminiferal fauna, including *Epistomina mosquensis*. Stam determined from 21 to 43 species per sample in Tojeira 1; between 301 and 916 benthos was counted per sample; proportions were estimated for 14 species. Gradstein and Agterberg (1982) had worked previously with highest occurrences of Foraminifera in offshore wells drilled on the Labrador Shelf and Grand Banks. The samples were small cuttings obtained during exploratory drilling by oil companies. Such samples are small, taken over long intervals and subject to down-hole contamination so that only highest occurrences (not lowest occurrences) of Foraminifera can be determined. Such problems associated with exploratory drilling can be avoided on land if continuous outcrop sampling is possible. According to paleogeographic reconstructions, the Lusitanian and Grand Banks Basins were close to one another during the Jurassic and had comparable sedimentary, tectonic and faunal history (Stam 1987). On land continuous outcrop sampling can be undertaken in the Lusitanian Basin only.

After preliminary statistical analysis of Stam's data, new samples from the two Tojeira sections were collected. Only relatively few samples were taken at exactly the same places where Stam had sampled before. Scattergrams (Agterberg 1990, Fig. 3.3) show typically poor correlations between proportions estimated from counts for species in the same sap-les at the same spots. These scattergrams essentially reflect random counting errors that satisfy the binomial distribution model (Fig. 2.2). Figure 9.15 shows sequences of samples (both Stam's and later data) for the two sections. Distances in the stratigraphic direction are given in meters measuring downward from Stam's stratigraphically highest sample in Tojeira 1 that was taken just below the base of the overlying Cabrito Formation. The stratigraphically highest sample in Tojeira 2 (No. 11.19) occurs about 6 m below this base.

The data for *E. mosquensis* plotted in Fig. 9.16 were tabulated in Agterberg et al. (1990, Table 3). As shown by Nazli (1988), Tojeira microfossil abundances are normalized when the probit transformation is applied. This transformation (*cf*. Sect. 2.3.1) reduces the influence of both relatively high and low proportions. Such normalization is desirable because smoothing splines are fitted by using the method of least squares in which the influence of each deviation from the curve increases according to the square of its magnitude.

Results for the indirect method of Fig. 9.14 applied to *E. mosquensis* in Tojeira 1 and 2 are shown in Fig. 9.16. The two spline-curves were slid with respect to one another until a "best" fit was found (Fig. 9.16, right side). A 10 m downward slid of the Tojeira 2 sequence, which places the base of the overlying Cabrito Formation in nearly the same stratigraphic position in both sections, producers the best correlation. It may be concluded from the pattern of Fig. 9.16 (right side) that both Tojeira 1 and 2 share essentially the same changes in relative abundance of *E. mosquensis* during deposition of the approximately 70 m of Late Jurassic shale in this part of the Lusitanian Basin.

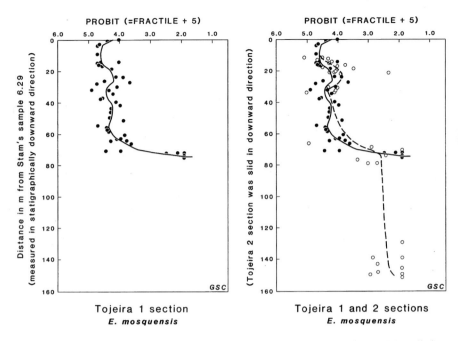

Fig. 9.16 *Left side.* Indirect method of cubic spline fitting illustrated in Fig. 4.9d–f applied to probits of *Epistomina mosquensis* abundance data for Tojeira 1 section. *Right side*: Same with observations and spline curve for Tojeira 2 section superimposed. Patterns were slid with respect to one another until a reasonably good fit was achieved. Zero distance (at sample 6.29 in Tojeira 1) falls just below base of overlying Cabrito Formation (cf. Fig. 9.15). Correlation between the two sections is poorest along the 35 m data gap in Tojeira 2 (Source: Agterberg 1990, Fig. 3.11)

9.3 Large-Scale Applications of Ranking and Scaling

In applications to real datasets, attention has to be paid to the frequency distributions of single events as well as to the two-dimensional frequency distributions of pairs of events included in the dataset. This is because of the prevalence of missing data in practical applications. Biostratigraphic exploratory well data primarily consist of last occurrences (LOs) of taxa, because the sampling procedure normally consists of taking cuttings at regular intervals during the drilling. As mentioned before, lower down a well, material from higher up may contaminate later cuttings so that observed locations of FOs can become biased and, therefore, should not be used. The following real-data examples of cumulative event frequency distributions are for Cenozoic microfossils from (A) 30 North Sea wells, using 1,430 event records of 289 taxa of benthic and planktonic Foraminifera, miscellaneous shelly microfossils, and dinoflagellates (from Kaminski and Gradstein 2005), and (B) 27 Labrador and northern Grand Banks wells, using 960 event records of 178 taxa of benthic and planktonic Foraminifera, and miscellaneous other microfossils (from Gradstein and Agterberg 1982). Figure 9.17 shows cumulative frequency distributions for all events in datasets (A) and (B). These

9.3 Large-Scale Applications of Ranking and Scaling   333

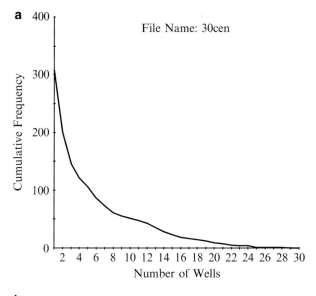

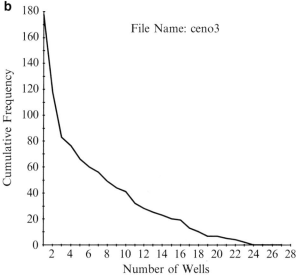

Fig. 9.17 Cumulative fossil event frequency distributions for Cenozoic Foraminifera and other microfossil events within North Sea Basin (**a**) and along northwestern Atlantic margin (**b**). The RASC threshold parameter for Minimum Number of Sections (MNS) was set equal to six and seven for datasets (**a**) and (**b**), respectively (Source: Agterberg et al. 2013, Fig. 4)

curves are "hollow" because relatively many events are observed in one or a few exploratory wells only. By means of the chi-square test for goodness of fit, Agterberg and Liu (2008) have shown that the frequencies for dataset (A) satisfy the logarithmic series distribution model (see Sect. 2.2.4). The frequencies shown in Fig. 9.17 are also the marginal frequencies of the two-dimensional frequency distributions for all possible pairs of events, which are subject to many further occurrences of missing data.

9.3.1 Sample Size Considerations

Because of the prevalence of missing data, an important input parameter in any RASC run is the Minimum Number of Sections (MNS) in which an event should occur (in RASCW program documentation, MNS is called k_c). From a statistical point of view, RASC results rapidly improve when MNS is increased because of greater precision of the relative frequencies (p_{ij}). The downside of large MNS is that low-frequency but biostratigraphically important events could become excluded from the RASC standard zonation. This problem is circumvented as follows: Commonly, the user chooses a value of MNS together with up to 20 "unique events" that occur in fewer than MNS wells. Unique events are biostratigraphically significant (e.g. they may be for index fossils) although they occur rarely, perhaps in a single well. They are added to the ranked optimum sequence and also incorporated in the RASC zonation after scaling. Figure 9.18 shows an output example of scaling for dataset (A) with MNS = 6 and ten unique events. The method by which RASC distances involving unique events are estimated is shown in Fig. 9.19.

Figure 9.18 is an example of a RASC scaling dendrogram in which the estimated interevent distances between successive events in the scaled optimum sequence are plotted to the left in the horizontal direction and connected by lines dropped in the stratigraphically downward direction. In the original RASC dictionary for any dataset, all events have a name and an event number (last two columns of Fig. 9.18). Names of unique events are preceded by double asterisks. The North Sea (NS) log events with single asterisks in Fig. 9.18 are marker events that received more weight during the scaling because they are without biostratigraphic uncertainty. The method by which RASC distances involving marker horizons are estimated is graphically illustrated in Fig. 9.20. Successive clusters in the dendrogram represent RASC zones consisting of events that are relatively close together along the RASC scale. Further biostratigraphic interpretation of RASC clusters will be given in the next section for another practical example.

9.3.2 Cenozoic Microfossils Example

Table 9.13 shows ranked biostratigraphic events observed in part of a single exploratory North Sea well taken from dataset (A). Sequence numbers in columns 1 and 2 are based on depth (column 3) in the well and position of event (columns 4 and 5) in the ranked optimum sequence for MNS = 6. Figure 9.21 is a scattergram showing relative sample position versus ranked optimum sequence number for all events observed in this well together with a best-fitting quadratic PPL curve. Use of

9.3 Large-Scale Applications of Ranking and Scaling

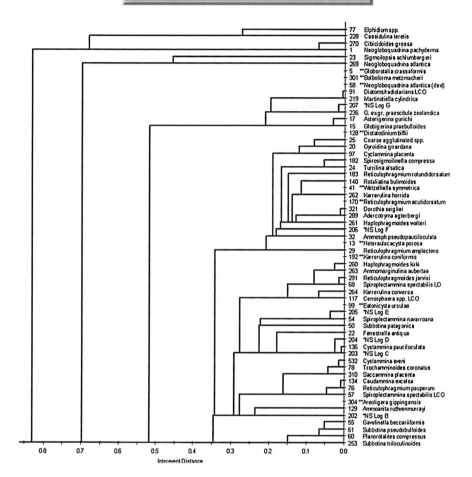

Fig. 9.18 Example of a RASC scaling output obtained for dataset (A) with MNS = 6. Events with *double stars* are unique events occurring in fewer than six wells. *Single star events* are marker events, which are not subject to significant biostratigraphic uncertainty and are weighted more strongly than other events in the scaling algorithm. The order of events is approximately equal to that in the ranked optimum sequence partially listed in Table 9.13. In this diagram, as well as other diagrams in this chapter, the axis of relative geologic time points upwards (Source: Agterberg et al. 2013, Fig. 5)

relative depth instead of observed depth can have the advantage of reducing effects of differences in sedimentation rates between wells. However, real depths are to be used for correlation between wells. The relation between relative and true depth is determined separately for each well. In RASCW Version 20, interpolation spline functions are used for this purpose. For our example this changes the pattern of Fig. 9.21 into that of Fig. 9.22. Every event considered has a "probable" depth in the well on the curve in Fig. 9.22. Probable locations of events in different wells can be connected by lines of correlation in CASC. In Agterberg (1990) other methods are discussed for obtaining the PPL curves to be used in CASC.

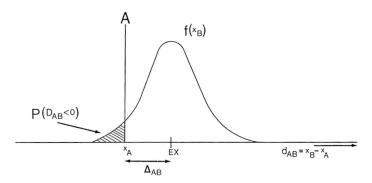

Fig. 9.19 Estimation of distance Δ_{AB} between events A and B from relative frequency of inconsistencies $p(D_{AB} < 0) = 1 - p(D_{AB} > 0)$ when A is a marker horizon with zero variance. The variance of the distance between A and B is equal to the variance of B. Consequently, marker horizons receive more weight than biostratigraphic events when the RASC scaled optimum sequence is calculated (Source: Agterberg 1990, Fig. 6.11)

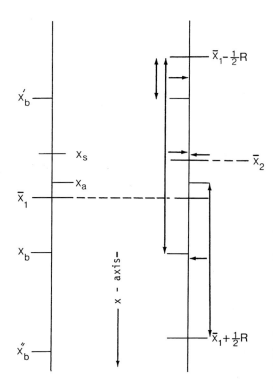

Fig. 9.20 Simple example to illustrate application of unique event option. A unique event was observed in a single section where it coincided with the event S, stratigraphically below A and above the events B, B′ and B″. Cumulative RASC distances of the latter five events are shown along the scale on the *left*. First, the positions of S, A and B were averaged to obtain first approximation for position of the unique event. Second approximation was based on the RASC distances of all events within the range R (Source: Agterberg 1990, Fig. 6.12)

The ranked optimum sequence for the current example can be used to determine first-order depth differences measured by subtracting the depth of an observed event from the depth of the event below it in a well. This option is part of depth-scaling routines later added to the RASC/CASC program (Agterberg et al. 2007). A first-order

9.3 Large-Scale Applications of Ranking and Scaling

Table 9.13 First 28 fossil events in order of ranked optimum sequence for WELL # 18 of dataset (A): Esso (N) 16(1-1); these events are Last Occurrences (LOs) except for a single Last Common Occurrence (LCO) and two log markers

#	Sequence#	Depth (m)	Event#	Event name
1	1	381	77	Elphidium spp.
2	2	381	228	Cassidulina teretis
3	4	637	1	Neogloboquadrina pachyderma
4	7	381	270	Cibicidoides grossa
5	9	637	4	Globorotalia inflata
6	13	637	23	Sigmoilopsis schlumbergeri
7	14	707	269	Neogloboquadrina atlantica
8	15	726	266	Globorotalia puncticulata
9	17	643	219	Martinotiella cylindrica
10	19	908	285	Caucasina eiongats
11	20	822	91	Diatoms/radiolarians LCO
12	21	1,036	282	Uvigerina ex.gr. semlornata
13	23	1,128	207	NS Log G
14	24	899	125	Neogloboquadrina continuosa
15	25	734	71	Epistomina elegans
16	26	899	15	Globigerina prebulloides
17	28	899	236	G, ex.gr. praescitula zealandica
18	29	908	17	Asterigerina gurichi
19	34	1,395	24	Tumilina alsatica
20	36	1,219	25	Coarse agglutinated spp.
21	37	1,395	97	Cyclammina placenta
22	38	1,395	182	Splrosigmollinella compressa
23	39	1,584	262	Karrerulina horrlda
24	42	1,395	140	Rotaliatina bulimoides
25	43	1,554	261	Haplophragmoides walteri
26	46	1,554	321	Dorothia seigliei
27	47	1,554	289	Adercotryma agterbergi
28	49	1,531	206	NS Log F

Source: Agterberg et al. (2013, Table 5)

depth difference is positive when depth of the event below it is greater and negative when it is less. Figure 9.22 can be used to illustrate this procedure. On average, the cluster of points for observed depths dips to the right on this graph. It reflects the fact that events stratigraphically lower in the optimum sequence tend to have greater depths. A negative first order depth arises when the depth of an event is less than that of its neighbor to the right in Fig. 9.22. Histograms of first-order depth differences in all wells for a dataset have average value greater than zero. In RASCW Version 20 their frequency distribution can be modeled in various ways. Figure 9.23 (after Agterberg et al. 2007) shows normal Q-Q plots of square root transformed first-order differences for datasets (A) and (B). The straight lines in Fig. 9.23 represent bilateral gamma distributions that were fitted as straight lines by least squares excluding data points on the commonly occurring anomalous upward bulges near the center of each plot. These anomalies are due to the discrete sampling method used when the wells

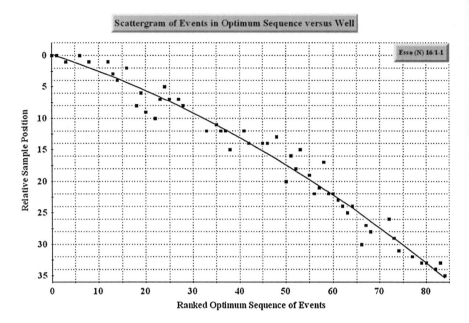

Fig. 9.21 Ranking scattergram with best-fitting (quadratic) PPL for Well # 18 in dataset (A) (Source: Agterberg et al. 2013, Fig. 6)

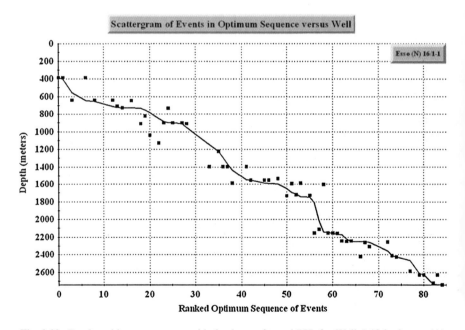

Fig. 9.22 Depth ranking scattergram with depth-transformed PPL for Well # 18 in dataset (A). Relative sample positions along vertical axis in Fig. 8.16 have been replaced by depths of samples. Event positions on this line can be used for CASC correlation between wells (Source: Agterberg et al. 2013, Fig. 7)

were being drilled. This regular sampling effect was previously explained in Sect. 9.1.2. In general, events with equal observed depths in a well record are coeval but their exact superpositional relationship cannot be established. However, it can be seen from Fig. 9.23 that the regular sampling effect is relatively minor in the large-scale RASC applications used for example.

RASC & CASC Version 20 allows application of many other statistical tests. Of special interest are the so-called normality tests by means of which the location of each event in each well can be checked to see that it is not a statistical outlier.

9.4 Automated Stratigraphic Correlation

The first RASC-CASC studies were carried out for wells on the Labrador Shelf, Grand Banks (offshore Newfoundland) and the Scotian Shelf (Gradstein and Agterberg 1982). Some results for these wells were already presented in this chapter (Fig. 9.17, dataset B). Scaling and correlation results for this same dataset and Foraminifera from Jurassic to Lower Cretaceous rocks on the Grand Banks will be reviewed in this section. Shaw's (1964) book uses first and last occurrences of trilobites and some other fauna in the Cambrian Riley Formation of Texas for example (for detailed data, see Shaw's 129-page appendix). His Riley Composite Standard (RST) results will be compared with a subjective zonation originally proposed by Palmer (1954) and RASC-CASC results. Finally, large-scale biostratigraphic zonation and correlations for a Cretaceous Greenland-Norway Seaway microfossil data base (Kaminski and Gradstein 2005) also will be briefly reviewed.

9.4.1 NW Atlantic Margin and Grand Banks Foraminifera Examples

Figure 9.24 (after Williamson 1987) shows a RASC scaling diagram based on 21 wells drilled into the Upper Jurassic and Lower Cretaceous on the northern Grand Banks, offshore Newfoundland. Eleven zones from Kimmeridgian to Cenomanian primarily were based on last occurrences of Foraminifera. Lowest occurrences of events were not used. Index taxa including several with last occurrences included as unique events were used to construct the regional stratigraphic zonation. This was a 4/3 RASC run meaning that events used for statistical calculations were found in at least four wells and pairs of events in at least three wells. Some of the relatively large interfossil (or interevent) distances coincide with major sedimentary changes or breaks that separate the majority of events below from those occurring above it. For example, the large interfossil distance between zones X (Tithonian) and IX (Valanginian) is mainly due to intermediate non-marine or very shallow marine facies probably or Berriasian age, which has a paucity of microfossils. This break may be associated with a condensed limestone sequence believed to cause seismic marker 1 and to be related

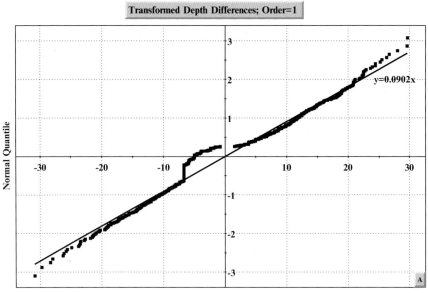

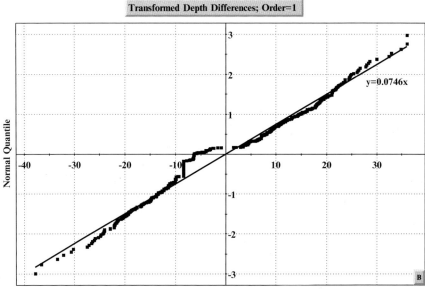

Fig. 9.23 Normal Q-Q plots of first order square root transformed depth differences for datasets (**a**), and (**b**). Approximate normality is demonstrated except for anomalous upward bulge in the center of each plot, which is caused by use of discrete sampling interval (Source: Agterberg et al. 2013, Fig. 8)

9.4 Automated Stratigraphic Correlation

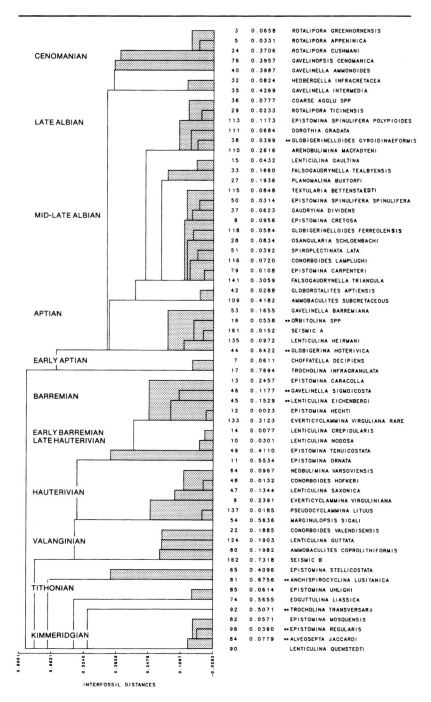

Fig. 9.24 Another example of a RASC scaled optimum sequence. This is an 11-fold interval zonation derived by RASC for Upper Jurassic and Lower Cretaceous foraminiferal record, northern Grand Banks. *Asterisks* indicate unique events included in the scaled optimum sequence, after statistical calculations had been carried out using events occurring in at least four sections (Source: Agterberg 1990, Fig. 9.19)

to changes in seawater level also observed in Portugal now at the other side of the Atlantic Ocean, which originated during the Jurassic.

In order to use RASC ranking or scaling results for correlation between stratigraphic sections in CASC, it is necessary to project their positions in the regional standard, which is an average based on sections, back onto the lines for individual lines. This topic already has been discussed before in this chapter. Figure 9.25 illustrates how the indirect method previously explained in Sect. 9.2.2 (Fig. 9.14) was used in earlier versions of CASC. In post-1998 RASC-CASC computer programs (e.g., the current Version 20) several simplifications were made including that the smoothing factor for the step of event levels to depths (Fig. 9.25b) was set to zero so that the level-depth curve passes through the data points. Because of differences in rates of sedimentation through time, error bars (e.g., 95 % confidence intervals), which are assumed to be symmetrical along the scales constructed for ranking or scaling, can become asymmetrical after projection onto individual wells. The degree of asymmetry depends of the curvature of the final curve for probable positions of the events in each individual well. Anther post-1998 simplification is that lines of correlation in scattergrams, in which observed events in wells are plotted against optimum sequence, are fitted as downward decreasing quadratic curves by least squares. The main reason for these simplifications was that the original method (Fig. 9.25) had to be applied separately to all individual wells. This procedure turned out to be very time-consuming in practice. The use of quadratic curves as shown previously in Fig. 9.6 turned out to be a good and fast substitute.

Figure 9.26 is an early example of a CASC multiwell comparison produced by means of the original CASC program (Agterberg et al. 1985). The underlying scaled optimum sequence was based on 54 last occurrences of Cenozoic Foraminifera in 7 or more wells (out of a set of 21 NW Atlantic Margin wells). The CASC version used had an additional step in that cumulative RASC distances were transformed into millions of years on the basis of a sub-group of 23 Cenozoic foraminiferal events for which literature-based ages were available. Because of significant uncertainties associated with this extra step, long-distance correlations in later versions of CASC are based on the ranked or scaled optimum sequence only. Three types of error bar are shown in Fig. 9.26. A local error bar is estimated separately for each individual well. It is two standard deviations wide and has the probable isochron at its center. Use is made of the assumption that rate of sedimentation is linear in the vicinity of each isochron computed. Consideration of variable sedimentation rates results in the asymmetrical modified local error bar of Fig. 9.26b. Like a local error bar, the global error bars of Fig. 9.26c are symmetric but they incorporate uncertainty in age derived from uncertainties in RASC distances for all 54 foraminiferal events in the scaled optimum sequence based on all (21) wells.

Figure 9.27 shows correlation of ten Cenozoic isochrons between six wells on the Grand Banks and Labrador Shelf including the three wells used for Fig. 9.26. In this study performed by Gradstein and Agterberg (1985) CASC-derived positions are compared with observed depths. Conventional chronostratigraphic correlation only uses observed depths. The CASC-based depths result in slight up or down adjustments of the age boundaries. The data used to obtain Figs. 9.26 and 9.27 was published as the Gradstein-Thomas database in the Appendix of Gradstein et al. (1985).

9.4 Automated Stratigraphic Correlation

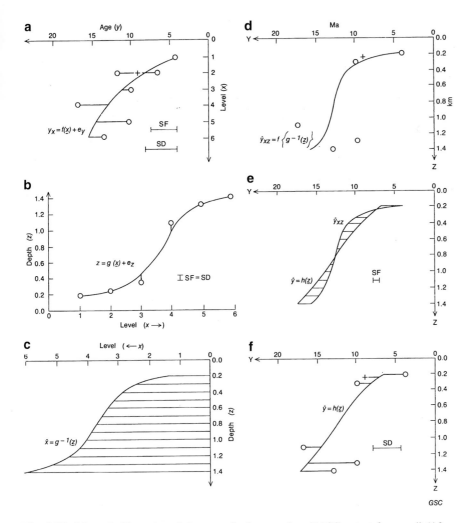

Fig. 9.25 Schematic illustration of the event-depth curve from RASC output for a well (After Agterberg et al. 1985); (**a**) RASC distance has been replaced by age (y) and a cubic smoothing spline is used to express y as a function of x representing relative event level in a well; bar in x denotes use of regular sampling interval for x; smoothing factor (SF), which was selected before the curve-fitting using one age per level, is slightly smaller than standard deviation (SD) for all original values; (**b**) spline curve $g(x)$ is fitted to express depth as a function of event level; SF = SD is some very small value; (**c**) spline curve $g(x)$ recoded as set of values for x at regular interval of z; (**d**) curve passing through set of values of y at regular interval of z obtained by combining spline curve of Fig. 9.25a with that of Fig. 9.25c; (**e**) spline curve fitted to values of Fig. 9.25d using new smoothing factor SF; (**f**) standard deviation SD is computed after curve-fitting using one age per level (Source: Agterberg 1990, Fig. 9.6)

Ordinary scaling in RASC is based on the assumption that all events have normal frequency distributions with equal variance along the interval scale. In general, different events have different frequency distributions. In the analysis of variance option of RASC, approximate estimates of these frequency distributions can be

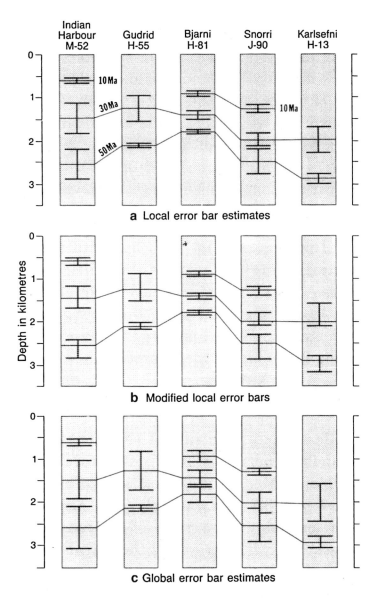

Fig. 9.26 Example of CASC multi-well comparison with three types of error bars. The probable positions of the time-lines were obtained from event-depth curves fitted to the biostratigraphic information of individual wells (Source: Agterberg 1990, Fig. 1.2)

obtained by collecting deviations between observed and probable positions for all events from within the subsets of wells in which the events were observed to occur. The standard deviations of these distributions can be used to distinguish between good and bad marker events. All deviations between observed and estimated positions are small for good markers. Lines of correlation connecting two good markers in different wells generally do not cross each other unless the two events

9.4 Automated Stratigraphic Correlation

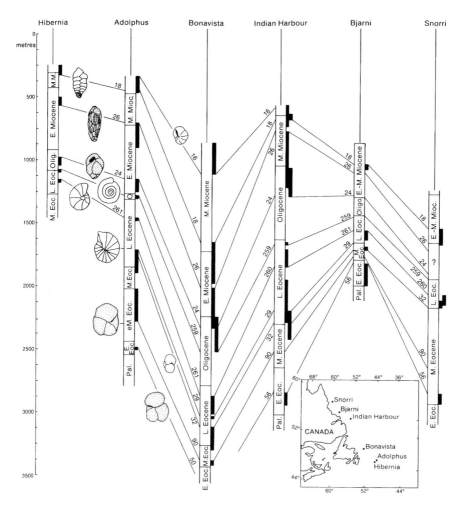

Fig. 9.27 Tracing of ten Foraminiferal events through six wells using the CASC method to calculate the most likely depths. *Black bars* show deviations from the observed depths (Source: Gradstein and Agterberg 1985, Fig. 15)

considered were nearly coeval. Scaling can be modified by assigning different variances to different events. D'Iorio (1988) and D'Iorio and Agterberg (1989) conducted the following study. An iterative procedure was developed in which the spline-based correlation method of Fig. 9.25 was applied alternately with modified scaling with different variances for different events. The procedure was continued until a stable set of variances was reached upon convergence. Results of this modified RASC method as applied to the Gradstein-Thomas database are shown in Fig. 9.28. Mean deviation and maximum observed deviation are plotted in addition to RASC position reached after convergence. During the iterative process there was relatively little change in estimates of the cumulative RASC distances. However, for some events such as highest occurrence of taxon 50 (*Subbotina patagonica*) there was a significant decrease in

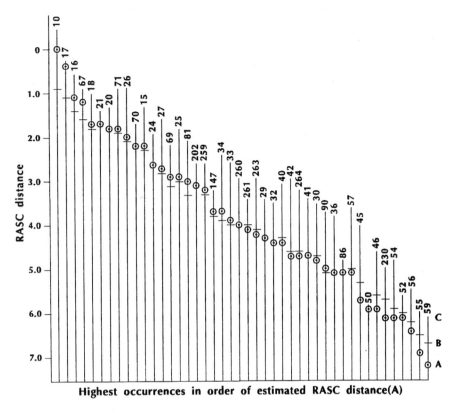

Fig. 9.28 Extended RASC ranges for Cenozoic Foraminifera in Gradstein-Thomas database. Letters for taxon 59 on the right represent (A) estimated RASC distance, (B) mean deviation from spline-curve, and (C) highest occurrence of species (i.e., maximum deviation from spline-curve). B is shown only if it differs from A. Good markers such as highest occurrence of taxon 50 (*Subbotina patagonica*) have approximately coinciding positions for A, B And C. Note that as a first approximation it could be assumed that the highest occurrences (C) have RASC distances which are about 1.16 units less that the average position. Such systematic difference in RASC distance is equivalent to approximately 10 million years (Source: Agterberg 1990, Fig. 8.9)

variance of deviations between observed and estimated mean event positions. It was concluded that taxon 50 is an exceptionally good marker. Maximum deviations between observed and estimated positions tend to be the same (≈1.16) for most other events. This suggests that, on average, the truly last appearance datum (LAD, *cf.* Sect. 9.1.1) occurred at least 10 million year later than average observed last occurrence (LO) for most Cenozoic Foraminiferal taxa.

9.4.2 Central Texas Cambrian Riley Formation Example

As explained in Sect. 9.1.1, the rationale underlying Shaw's method of graphic correlation differs from the one underlying RASC and CASC, which is based on simultaneously averaging biostratigraphic events in all stratigraphic sections for a

9.4 Automated Stratigraphic Correlation

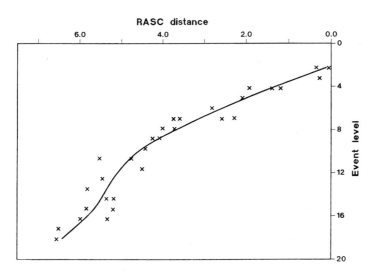

Fig. 9.29 RASC distance-event level plot for Morgan Creek section. Spline-curve is for optimum (cross-validation) smoothing factor SF = 0.382 (Source: Agterberg 1990, Fig. 9.23)

study area. Nevertheless, final results in an application of both approaches to the same dataset can be similar as illustrated in Fig. 9.31 for three of the seven sections used for example in Shaw's (1964) book. These results were obtained as follows. First RASC distances estimated by scaling were plotted against event levels as shown for the Morgan Creek section in Fig. 9.29. Next the event level scales were replaced by distance scales (Fig. 9.30). The spline curves in these diagrams were not only used for positioning the probable positions of cumulative RASC distances in Fig. 9.31 but also to obtain modified error bars that are asymmetrical because of curvatures of the spline curves.

Figure 9.31 shows three different types of correlation. Firstly, there are Palmer's (1955) original zones obtained by conventional subjective paleontological reasoning. Secondly, there is Shaw's set of RTS value correlation lines based on his Riley Composite Standard (RST). Finally, CASC correlation lines are shown for selected values along the relative geologic time axis based on cumulative interevent distances as those previously shown for dataset (A) in Fig. 9.4. Some of these values are shown together with error bars (±one standard deviation transformed to depth scale). The example shows either that sedimentation rate probably decreased from high to low during deposition of the Riley Formation or, less likely, that there was a change in rate of trilobite evolution. More details on this example based on 60 FOs and LOs in at least MNS = 5 of the seven sections can be found in Agterberg (1990). Earlier applications of quantitative biostratigraphic techniques to Palmer's (1954) Riley Formation trilobite data include Edwards and Beaver (1978), Hudson and Agterberg (1982) and Guex (1987); for a more recent application of CONOP to the same data set, see Kemple et al. (1995).

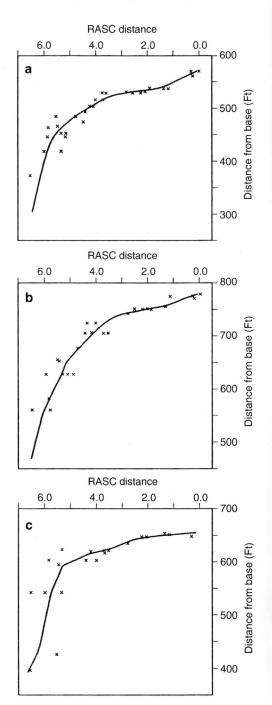

Fig. 9.30 Spline-curves for positions of RASC distance values in three sections obtained by means of indirect method. (**a**) Morgan Creek section. Curve of Fig. 9.29 was combined with curve for positions of event levels according to method of Fig. 9.25. Second (cf. Fig. 9.25b) and third (cf. Fig. 9.25e) smoothing factors were set equal to 0.02 and 0.2, respectively. Final standard deviation of deviations from curve is SD = 0.390; (**b**) White Creek section (SD = 0.357); (**c**) Pontotoc Section (SD = 0.615) (Source: Agterberg 1990, Fig. 9.24)

9.4 Automated Stratigraphic Correlation

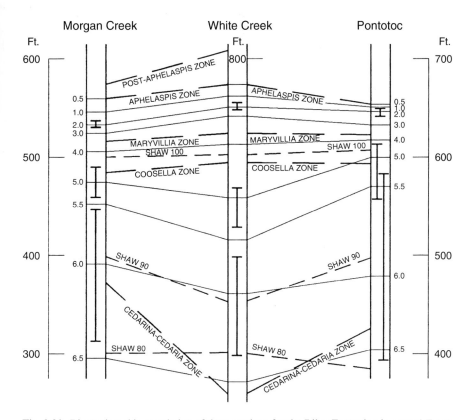

Fig. 9.31 Biostratigraphic correlation of three sections for the Riley Formation in central Texas by means of three methods. Palmer's (1954) original biozones and Shaw's (1964) R.T.S. value correlation lines were superimposed on CASC (Agterberg 1990) results. Error bars are projections of single standard deviation on either side of probable positions for cumulative RASC distance values equal to 2.0, 5.0 and 6.0, respectively. A cumulative RASC distance value is the sum of interevent distances between events at stratigraphically higher levels. Uncertainty in positioning the correlation lines increases rapidly in the stratigraphically downward direction. The example shows that the three different methods used produce similar correlations (Source: Agterberg 1990, Fig. 9.25)

9.4.3 Cretaceous Greenland-Norway Seaway Microfossils Example

The final example of large-scale RASC/CASC applications is concerned with the stratigraphic distribution of 1,755 foraminiferal and dinoflagellates microfossil events in 31 wells from the Cretaceous seaway between Norway and Greenland (after Gradstein et al. 1999). The emphasis in this example is on paleoceanographic interpretation and use of RASC variance analysis, a technique developed in the late 1990s. Setting MNS = 7 left 72 events that were augmented by 9 unique events. Almost all events are last occurrences but several Last Common Occurrences

(LCOs) and a First Common Occurrence (FCO) also could be identified. The scaling result is shown in Fig. 9.32. This dendrogram served as a template to build a Cretaceous zonal model with 16 stratigraphically successive interval zones that are middle Albian through late Maastrichtian in age. Five large breaks (at events 52, 84, 205, 255 and 137 in Fig. 9.32) indicate transitions between natural microfossil sequences; such breaks relate to hiatuses or facies changes, some of which are known from European sequence stratigraphy (see Gradstein et al. 1999 for more details): (1) The *una-schoelbachi* break reflects a latest Albian lithofacies change and hiatus, connected to the Octeville hiatus in NW Europe. (2) The *delrioensis* (LCO)-*brittonensis* break reflects the mid-Cenomanian lithofacies change and hiatus, connected to the mid-Cenomanian non-sequence and Rouen hardground of NW European sequence stratigraphy. (3) The *Marginotruncata-polonica* break, above the level of *Heterosphaeridium difficile* LCO, which represents a maximum flooding surface, may be the turn-around in the middle Coniacian tectono-eustatic phase, near the end of the Lysing sand phase. (4) The *belli-dubia* break, is again (near or) at a maximum flooding event, this time correlated to the LCO of *T. suspectum* in the early middle Campanian, above the change from marly sediments to siliciclasts at the base of the Campanian. (5) The *dubia-szajnochae* break reflects the abrupt change from siliciclasts to marly sediment at the Campanian-Maastrichtian boundary, only noted in the southern part of the study region.

Figure 9.33 is machine output for CASC correlation of nine events in eight wells. The PPLs for these wells were based on the ranked optimum sequence with MNS = 7. In Gradstein et al. (1999) this sequence, which is almost the same as the sequence of events in Fig. 9.32, was used for RASC variance analysis. Individual events deviate from their PPL in each section. These deviations, which are either positive or negative, have standard deviations that differ from event to event. Good markers have small standard deviations. Nine such events are connected by lines in Fig. 9.33, which is CASC output with 95 % confidence limits. The eight wells in Fig. 9.33 are arranged from north to south. CASC has a flattening option according to which the line of correlation for a specific event between sections is made horizontal. Event 16, the last occurrence (LO) of *Hedbergella delrioensis* was used for flattening in Fig. 9.33. These events and other events including several with relatively large standard deviations also are correlated in Fig. 9.34. The poor markers with larger standard deviations show cross-over inconsistencies in this diagram. Large standard deviations can be due to a variety of reasons. For example, Foraminifera that are benthic tend to show more inconsistencies than planktonic forms. Separate RASC plots of deviations for an event in all wells may reveal patterns of diachronism. For example, *L. siphoniphorum*, observed in 19 wells, appears to be time transgressive, ranging into younger strata southward. The same may be true for *E. spinosa*, observed in 13 wells (Gradstein et al. 1999, p. 69).

In Fig. 9.34 variance data on fossil events are used to create a different, but effective type of correlation plot. Cretaceous turbiditic sands (with yellow or gray patterns in Fig. 9.34) occurring offshore mid and southwestern Norway are correlated in five wells. The Lower Cenomanian *Hedbergella delrioensis* FCO and LCO

9.4 Automated Stratigraphic Correlation

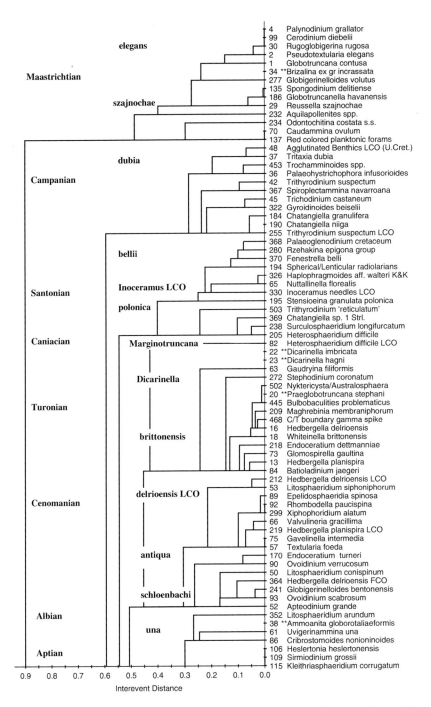

Fig. 9.32 Example of scaling in relative time of RASC applied to 1,755 Cretaceous foraminiferal and dinoflagellates events in 30 exploration wells from the seaway between Greenland and Norway (After Gradstein et al. 1999). The dendrogram with event names and numbers is RASC machine output but cluster and stage names were inserted later (Source: Agterberg et al. 2013, Fig. 10)

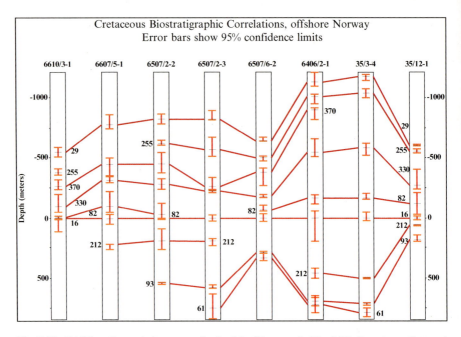

Fig. 9.33 CASC well correlation output for 9 of the 97 events in Fig. 9.32. Output was flattened on event 16 (*Hedbergella delroensis* LO) (Source: Agterberg et al. 2013, Fig. 11)

events and Santonian *Inoceramus* needles LCO event are considered flooding surfaces (FS). These three events represent reliable regional correlation levels, with below-average standard deviations. They reflect considerable marine transgression accompanied by slow sedimentation. In Gradstein and Agterberg (1998) RASC variance analysis was used to demonstrate that the lower Cenomanian Lange sands (between the two *Hedbergella* levels in Fig. 9.34) and the middle Cenomanian to Turonian Lysing sands (above the *Hedbergella* LCO level) are not seismic markers or well log 'sheet sands' in a correlative sense as had been assumed before, but show more complex correlation patterns in line with the sedimentological interpretation that they are debris flow and turbidite sands.

9.5 Construction of Geologic Time Scales

This section contains a review of how geologic timescales were created in the past. During the last 50 years, successive international timescales have been constructed. There has been continuous improvement in geochronological dating methods as well as in chronostratigraphic positioning of the rock samples subjected to age determination (also see Sect. 3.1.6). These improvements have led to changes in statistical methods used to estimate the ages of stage boundaries. The last international geologic time scale (GTS2012) is shown in Fig. 9.35.

9.5 Construction of Geologic Time Scales 353

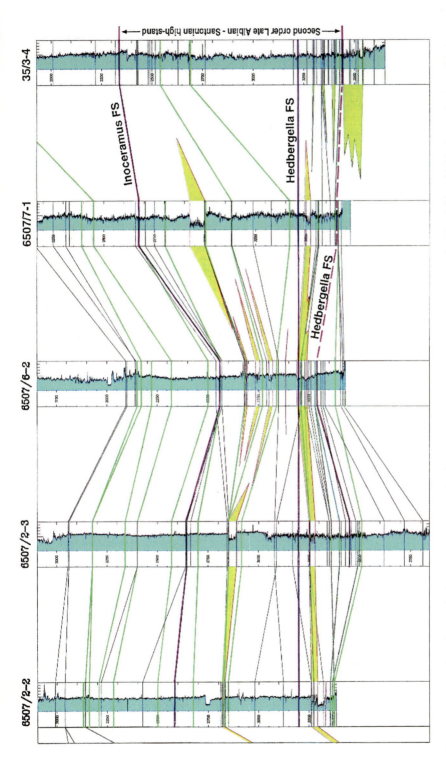

Fig. 9.34 Correlation of events in the optimum sequence for five wells from the Cretaceous dataset also used for Figs. 9.32 and 9.33. Gamma logs are also shown. See text for further explanations (Source: Agterberg et al. 2013, Fig. 12)

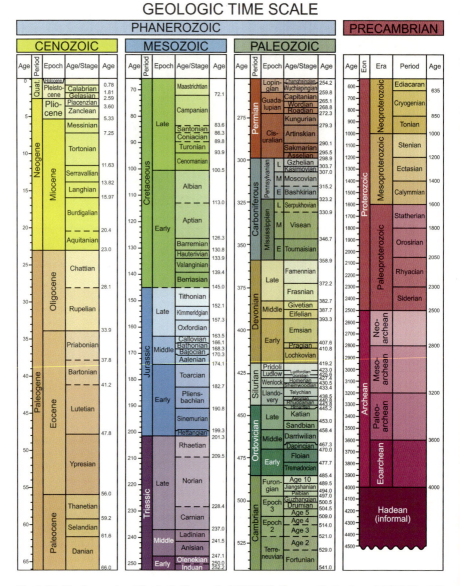

Fig. 9.35 The International Geologic Timescale (GTS2012) (Source: Gradstein et al. 2012, Fig. 1.2)

9.5.1 Timescale History

The very first numerical time scales were constructed between 1911 and 1960 by Arthur Holmes who remarked: "To place all the scattered pages of earth history in their proper chronological order is by no means an easy task" (Holmes 1965, p. 148). In 1911 Holmes examined the decay of uranium to lead in seventeen

9.5 Construction of Geologic Time Scales

Table 9.14 Comparison of successive numerical age estimates of Periods in the International Geologic Time Scale and their 95 %-confidence intervals

Base of period or epoch	Holmes 1960	GTS2004	±2σ(2004)	GTS2012	±2σ(2012)	Difference
Pleistocene	1	1.8		2.59		0.79
Pliocene	1.1	5.3		5.33		0
Miocene	25	23		23.03		0
Oligocene	40	33.9	0.1	33.9		0
Eocene	60	55.8	0.2	56		0.2
Paleocene	70	65.5	0.3	66	0.5	0.5
Cretaceous	135	145.5	4	145	0.8	−0.5
Jurassic	180	199.6	0.6	201.3	0.2	1.7
Triassic	225	251	0.4	252.2	0.5	1.2
Permian	270	299	0.6	298.9	0.2	−0.1
Carboniferous	350	359.2	2.5	358.9	0.4	−0.3
Devonian	400	416	2.8	419.2	3.2	3.2
Silurian	440	443.7	1.5	443.8	1.5	0.1
Ordovician	500	488.3	1.7	485.4	1.9	−2.9
Cambrian	600	542	1	541	1	−1

Source: Agterberg (2013)

radioactive minerals of which he excluded just over half that were considered to be problematic (*cf.* Jackson 2006, p. 245). From the remaining eight analyses he arrived at dates for the bases of the Silurian, Devonian and Carboniferous, setting them at 430, 370 and 340 Ma, respectively. In 1960, after more age determinations of rock samples had become available, Holmes published the ages shown in Table 9.14, where they are compared with ages for the same periods or epochs in the 2004 and 2012 geologic time scales published by the International Commission on Stratigraphy and Geologic TimeScale Foundation, respectively. All numerical time scales involve the conversion of a relative geological time scale into a linear time scale along which rock samples and chronostratigraphic divisions are measured in millions of years. Holmes based his relative time scale on age-thickness interpolations. He was aware of limitations of this method, stating: "I am fully aware that this method of interpolation has obvious weaknesses, but at least it provides an objective standard, and so far as I know, no one has suggested a better one" (Holmes 1960, p. 184). As pointed out by Gradstein et al. (2012, p. 16), Holmes's estimate for the base of the Cambrian is curiously close to modern estimates (see Table 9.14).

Subsequently, Walter Brian Harland spearheaded several broadly based projects between 1960 and 1990 that resulted in two widely used international geologic time scales (Harland et al. 1982, 1990) known as GTS82 and GTS90. Geochronologists commonly report analytical precision of age determinations as a 95 % confidence interval written as $\pm 2\sigma$ because $\Phi(0.975) \approx 2$. Many dates of greatly variable precision were used for constructing GTS82 and GTS90. After 1990, higher precision dates with 2σ-values of 0.5 % or better have become available. Radiometric (e.g., Uranium-Lead, $^{40}Ar/^{39}Ar$ and Rhenium-Osmium) methods have improved significantly and the new dates are at least an order of magnitude more precise than the earlier dates. In time scale calculations the weight of individual age determinations is approximately proportional to the inverse of their measurement variance (σ^2). Consequently, a modern age date, which is ten times more precise than a pre-1990 age determination as obtained by the Rubidium-Strontium or Potassium-Argon method, is weighted 100× stronger in the statistical calculations. This implies that a single modern date roughly receives at least as much weight as 100 earlier dates. New high-precision dates continue to become available regularly.

Since 1990, Felix Gradstein has led several international teams of scientists engaged in constructing new time scales including GTS2004 (Gradstein et al. 2004), which became used by geologists worldwide, and GTS2012 (Gradstein et al. 2012) that has replaced GTS2004. Construction of GTS2012 involved over 65 geoscientists and other experts. Stage boundary age estimates in GTS2004 and GTS2012 are accompanied by approximate 95 % confidence intervals shown as $\pm 2\sigma$ for periods and Cenozoic epochs in Table 9.14. The largest difference between GTS2012 and GTS2004 in this table is 3.2 Ma and occurs at the base of the Devonian.

Constructing a numerical time scale consists of converting a relative geological time scale such as the age thickness scale used by Holmes into a linear time scale along which samples and events are measured in millions of years. Geologists realized early on that for clarity and international communication the rock record was to be subdivided into a chronostratigraphic scale of standardized global stratigraphic units such as "Cambrian" and "Miocene". This became possible because many events in geological history affected the entire surface of the Earth or very large regions. Numerous examples of methods to define the original chronostratigraphic boundaries could be cited. One example is Lyell's early subdivision of the Cenozoic that was partly based on a quantitative model. The early editions of Lyell's (1833) book contain a 60-page appendix with presence-absence information on 7,810 species of recent and fossil shells. By counting for each Series (Epoch) the number of fossil shells of species living today and re-computing the resulting frequencies into percentage values, Lyell established the first subdivision of the Tertiary Period into Pliocene, Miocene and Eocene. Later, Paleocene, Oligocene and Pleistocene were added, thus providing the break-down of the Cenozoic in the geologic time table using names that reflect the magnitudes of these percentage values. Another example is the Maastrichtian-Paleocene boundary marked by an iridium anomaly caused by bolide impact now estimated to have taken place 66.0 million years ago (*cf.* Table 9.14).

The latest international Geologic Time Scale is shown in Fig. 9.35 (from Gradstein et al. 2012). An international initiative that has been helpful to establish GTS2004 and GTS2012 is the definition of GSSPs or "golden spikes". The first GSSP ("Global Boundary Stratotype Section and Point") fixed the lower limit of the Lochkovian Stage (Silurian-Devonian boundary) at the precise level in an outcrop with the name Klonk in the Czech Republic (Martinsson 1977). Each GSSP must meet certain requirements and secondary desirable characteristics (Remane et al. 1996, Table 2.1). Before a GSSP is formally defined, its correlation potential in practice is thoroughly tested. GTS 2004 made use of 28, 8 and 7 GSSP's for the

Paleozoic, Mesozoic and Cenozoic; the corresponding numbers for GTS2012 were increased to 36, 12 and 13 GSSP's, respectively (Gradstein et al. 2012, Table 2.5). In general, precise age dates are determined for relatively many rock samples taken at or close to the boundary on which a GSSP is defined. Especially for the Mesozoic, several stage boundary ages could be based on such local data sets. It is clearly advantageous for all earth scientists to use a single time scale with a common set of stratigraphic subdivisions. However, in some parts of the world it may be difficult to recognize all stages for a period. As a regional companion to GTS 2004, Cooper et al. (2004) produced the "New Zealand Geologic Time Scale" with 72 regional stages. Almost half of these have their own boundary definitions and boundary stratotypes (SSPs), not all of which can be readily correlated to the GSSPs.

Many different geologic time scales have been published after Holmes's first scale of 1911. Twelve different time scale methods and their usage in twenty-eight selected time scales are reviewed in Gradstein et al. (2012, eds., Fig. 1.7). The two methods of dating already mentioned are rate of radioactive decay of elements, and tuning of cyclic sequences to orbital time scale. The other methods listed include Holmes's original maximum thickness of sediments per time period method, and Harland et al.'s equal duration of stages hypothesis (GTS82). Virtually all time scales use stratigraphic reasoning and biostratigraphic/geomagnetic calibration. Approximate constancy of spreading in ocean floor segments is helpful in the Jurassic and Cretaceous. Holmes's constant sedimentation rate hypothesis was refined for calculation of some later time scales to the assumption that zone duration can be proportional to zone thickness. Likewise, Harland et al.'s (1982) equal duration of stages hypothesis was refined in some later time scales to the equal duration of (sub-)zones hypothesis. Trends in the $^{87}Sr/^{86}Sr$ stable isotope scale also have been used in some time scales.

9.5.2 Differences Between GTS2012 and GTS2004

For several chronostratigraphic boundaries in Table 9.14 there is a statistically significant discrepancy between the GTS2004 and GTS2008 estimates. Such differences are mainly due to improvements in accuracy of radiometric methods over the past 10 years. Accuracy and precision can be discussed in the context of the various methods of time scale estimation that were used. Virtual certainty was achieved in GTS2012 for the estimates of the Cenozoic stage boundaries that are based on astrochronology. The relatively large discrepancy in age for the base of the Pleistocene in Table 9.14 is due to redefinition of this boundary.

As pointed out in Sect 1.2, the time scales of the Neogene (23.0 − 2.59 Ma) and Paleogene (66.0 − 23.0 Ma) periods now are entirely based on astronomical calibrations. Other differences between GTS2012 and GTS2004 also are seen in Table 9.14. The differences for several age estimates are significant in that they exceed the widths of the 95 % confidence intervals representing precision. Such lack of accuracy (systematic differences) is mainly due to significant improvements in geochronological practice that took place between 2004 and 2012 (Schmitz 2012). When GTS2004 was constructed, it was already known that there were two main problems: (1) Lack of accuracy of radiometric decay constants, and (2) "External" errors that had not been added to "internal" errors to estimate precision. These two problems that affected GTS2004 have been solved adequately during the past 10 years because geochronologists have succeeded in improving methodologies by re-calibrations.

For example, by calibrating with respect to high-precision reference ages based on the U-Pb system, Kwon et al. (2002) estimated that the decay constant of ^{40}K was $\lambda = 5.476 \pm 0.034 \times 10^{-10}$/year. This estimate fell between $5.543 \pm 0.020 \times 10^{-10}$/year that was used by geochronologists at the time (Steiger and Jäger 1977) and $5.428 \pm 0.068 \times 10^{-10}$/year of nuclear scientists (Endt and van der Leun 1973). The decay constant utilized in GTS2012 is $\lambda = 5.463 \pm 0.107 \times 10^{-10}$/year. With respect to problems of estimating precision when GTS2004 was constructed, Renne et al. (1998) had pointed out that, originally, errors quoted for $^{40}Ar/^{39}Ar$, only included internal inconsistencies related to the measurement of $^{40}Ar/^{39}Ar_K$ and the J-factor (cf. McDougall and Harrison 1999), whereas external errors associated with measurement of the K-Ar age of the fluence monitor and errors related to the determination of the decay constants were not considered. Similar problems, although to a lesser extent, were associated with other published radiometric dates when GTS2004 was constructed.

Schmitz (2012) discusses in detail how radiometric re-calibration was applied to GTS2012. Systematic error propagation was taken into account. Both internal and external errors were incorporated. Legacy data that could not be reproduced from published literature data were rejected. Before using them for GTS2012, ages and their errors were recalculated from published primary isotope ratios. U-Pb and Pb-Pb ages were harmonized using the uranium decay constant ration of Mattinson (2010), in addition to recalculation of all $^{40}Ar/^{39}Ar$ on the basis of the total K decay constant $\lambda_{total} = 0.463 \pm 0.0107 \times 10^{-10}$/year. of Min et al. (2000) that provides best inter-calibration of $^{40}Ar/^{39}Ar$, $^{206}Pb/^{238}U$ with the astronomical clock for the FC sanidine monitor standard age of 28.201 ± 0.046 Ma (cf. Kuiper et al. 2008). These re-calibrations have resulted in significant improvements in both accuracy and precision of GTS2012 with respect to GTS2004.

9.5.3 Splining in GTS2012

One of the geomathematical techniques used for the construction of GTS2004 and GTS2012 is splining (cf. Agterberg et al. 2012). It remains the best method for Paleozoic stage boundary estimation. The smoothing spline is a smooth curve fitted to n data points on a graph in such a way that the sum of squares of deviations between the curve and the points is minimized (Sect. 9.2.1). Suppose that a relative time scale is plotted along the X-axis and age (in Ma) along the Y-axis. If the values x_i ($i = 1, \ldots, n$) along the X-axis are free of error, a cubic smoothing spline $f(x)$ is fully determined by the n pairs of values (x_i, y_i), the standard deviations of the dates

9.5 Construction of Geologic Time Scales

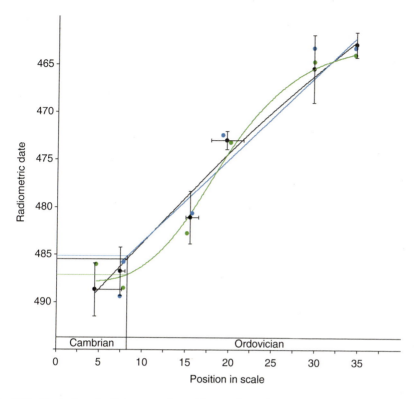

Fig. 9.36 Illustration of splining procedure with a small set of data points around the Cambrian-Ordovician boundary. *Black dots*: Given data with stratigraphic and radiometric error bars (total range and 2-sigma). *Black curve*: Spline of the given data, cross-validated smoothing factor SF = 1.075. *Blue dots*: Random replicate of the input data generated according the given points and their error bars, SF = 2.125. *Blue curve*: Spline of the random replicate. *Green dots and curve*: Another random replicate and its spline, SF = 0.85. The Cambrian-Ordovician boundary age from the spline of the given data is 485.39 Ma. The ages computed from the random replicates are 485.17 and 487.06 Ma (Source: Gradstein et al. 2012, Fig.14.1)

$s(y_i)$, and by a smoothing factor (SF) representing the square root of the average value of the sum of squares of scaled residuals $r_i = \{y_i - f(x_i)/s(y_i)\}$. If all $s(y_i)$ values are unbiased, SF = 1, or SF is a value slightly less than 1 (*cf.* Agterberg 1994, p. 874). If SF significantly exceeds 1, this suggests that some or all of the $s(y_i)$ values used are too small (under-reported). Thus, if SF can be determined independently, the spline-fitting method may provide an independent method of assessing mutual consistency and average precision of published 2σ error bars. The smoothing factor (SF) can be estimated in practice by cross-validation using the "leave-off-one" method that produces an optimum smoothing factor (*cf.* Agterberg 2004).

Figure 9.36 (from Agterberg et al. 2012, Figure 14.1) illustrates splining on a small data set consisting of six Cambrian and Ordovician data points in the vicinity of the Cambrian-Ordovician boundary. In this application, the relative geological scale (X-axis) is part of the Ordovician-Silurian graptolite composite standard obtained by the CONOP method of constrained optimization (Cooper and Saddler 2012, Table 20.1). The CONOP method (also see Sect. 9.1.1), originally developed by Kemple et al. (1995) and Sadler (2001), uses evolutionary programming techniques to find a composite range chart with optimal fit to all the field observations. Four of the six CONOP values have stratigraphic uncertainty that is expressed by the horizontal error bars in Fig. 9.36. The six vertical bars are $\pm 2\sigma$ error bars for the dates that were obtained by the $^{207}Pb/^{208}Pb$ method. These dates and error bars are listed as C11, O1, O2, O3, O4 and O8 in Appendix 2 of GTS2012 (Gradstein et al. 2012). If a point has error bars along one or both (X- and Y-) axes, this means that its "true" position on the graph could be different with probabilities controlled by their supposedly rectangular frequency distribution along the X-axis and Gaussian error distribution along the Y-axis. Monte Carlo simulation can be used to randomly pick points from within their uncertainty intervals in order to create replicates of the given data set. The three splines in Fig. 9.36 for original data set and two random replicates have different SF values obtained by cross-validation and produce slightly different estimates for the age of the Cambrian-Ordovician boundary. Suppose that this procedure of randomly selecting points is repeated not twice but say 10,000 times (bootstrap method). Then the resulting estimates of the chronostratigraphic boundary form a histogram from which a 95 % confidence interval can be derived. This is the splining procedure used for estimating 2σ values on Paleozoic and some other stage boundary ages in GTS2012 (*cf.* Agterberg et al. 2012).

9.5.4 Treatment of Outliers

It often happens that one or a few data points end up relatively far away from the smoothing spline, farther than indicated by their error bars. Such outliers are handled by assuming that their standard deviations must have been underestimated. Underestimation of 2σ can occur if not all so-called "external" sources of uncertainty, e.g. imprecision of decay constants, were considered when a date was published. The procedure used for both GTS2004 and GTS2008 contains a step where outliers were identified and their standard deviations adjusted. The spline was then recomputed.

Individual scaled residuals are either positive or negative and should be approximately distributed as Z-values (from the "normal" Gaussian distribution in standard form). Their squares are chi square distributed with one degree of freedom, and can be converted into probabilities to test the hypothesis that they are not greater than can be expected on the basis of the set of all $s(y)$ values used for scaling the residuals. The sum of squares of several scaled residuals is also approximately distributed as chi-square but with a larger number of degrees of freedom. A statistical test can therefore be used to identify the relatively few outliers exhibiting error bars that are much narrower than expected on the basis of most ages in the same data set. The $s(y)$ values of outliers are revised by replacing probabilities (p) that are too small (e.g., $p < 0.05$) by 0.5. Setting the probability equal to 50 % is equivalent to replacing the chi-square value by 0.4549. This is equivalent to adopting a new Z value of 0.674, because chi-square with a single degree of freedom is Z^2. The new $s(y)$ value is then computed by dividing the old Z value (scaled) residual by 0.674, and multiplying the result by the original $s(y)$.

It is good to keep in mind the following consideration: any potential outlier with a very small probability p resulting from the preceding chi-square test belongs to a data set of n dates that are analyzed simultaneously. For example, the Devonian spline curve in Becker et al. (2012, Figure 22.14) is based on $n = 19$ dates. Applications of the chi-square test with a single degree of freedom, indicates that six dates have $p < 0.05$. However, in terms of probability calculus, the problem can be rephrased as follows. What is the probability that a date will yield a p' value in the chi-square test with one degree of freedom when $n = 19$? The answer to this question is $p' = 1 - (1 - p)^{19}$. If $p' = 0.05$, it follows that $p = 0.0027 < 0.05$. If this revised significance test is used, only one of the six dates (D5) would have a reported 2σ confidence interval that is too narrow. A full statistical analysis of 2σ values of dates used for time scale construction may involve several successive spline-fittings and study of all residuals in comparison with one another. Other statistical procedures to treat outliers are discussed in Agterberg (2004) and Agterberg et al. (2012).

In general, a data set used for splining contains the dates for a limited time interval; e.g. an epoch. This can lead to edge effects in that age boundary estimates and their 95 % confidence intervals, which are based on one data set, may differ from those based on another data set. This problem can be alleviated by use of overlapping data sets: some of the oldest dates in one data set can be made the same as the youngest dates in the other data set. Additionally, differences between estimates of 2σ values at the edges may be smoothed out by "ramping" (Agterberg 2004, p. 113).

9.5.5 Early Geomathematical Procedures

The stratigraphic record that is to be combined with radiometric dates and their standard deviations includes litho-, bio-, chrono-, cyclo- and magnetostratigraphy. For the construction of several previous time scales, ages of stage boundaries were estimated by applying the chronogram method (Harland et al. 1982, 1990), or maximum likelihood method (Gradstein et al. 1994, 1995), to a worldwide database of chronostratigraphically classified dates (*cf.* Sect. 3.1). These methods, which can also be applied to more closely spaced zone boundaries (Pálfy et al. 2000), resulted in age estimates accompanied by approximate 95 % confidence intervals. A final time scale was obtained by calibration using graphical or curve-fitting methods, including cubic smoothing splines. As explained before, these earlier time scale construction methods made use of very large numbers of imprecise age determinations. New radiometric dating methods are much more precise, and GTS2004 and GTS 2012 are based on relatively few high-precision dates. It can be expected, however, that many more high-precision dates will become available in future. Earlier geomathematical procedures that were developed for large data sets then may become relevant again, especially if time scales are to include finer stratigraphic subdivisions.

Odin (1994) discussed three separate approaches to numerical time scale construction: statistical, geochronological and graphical methods. Gradstein et al. (1994, 1995) used all three approaches in a stepwise procedure involving maximum likelihood, use of stratigraphically constrained dates, and recalibration by curve-fitting. The chronogram method used by Harland et al. (1982, 1990) and its maximum likelihood extension are suitable for estimation of the age of chronostratigraphic boundaries from a radiometric database, when most rock samples used for age determination are subject to significant relative uncertainty. Inconsistencies in the vicinity of chronostratigraphic boundaries then can be ascribed to imprecision of the age determination method.

A general disadvantage of the chronogram and maximum likelihood methods is that the relative stratigraphic position of any rock sample is generalized with respect to stage boundaries that are relatively far apart in time. The relative stratigraphic of one sample with respect to others within the same stage is not considered. A better approach is to incorporate precisely known stratigraphic positions for which high-precision age determinations are available.

9.5.6 Re-proportioning the Relative Geologic Time Scale

McKerrow et al. (1980) described an iterative method to construct a numerical time scale for the Ordovician, Silurian and Devonian. A sequence of diagrams was constructed wherein the isotopic age of the sample was plotted along the X-axis and its stratigraphic age along the Y-axis. On each diagram, the samples were plotted as rectangles representing their analytical uncertainty (2σ) as well as their stratigraphic uncertainty. Successive diagrams had slightly differing vertical scales, until a scale was obtained that allowed a straight line to pass through almost all the rectangles. Cooper (1999) adopted a modified version of this method for an Ordovician time scale based on 14 analytically reliable and stratigraphically controlled high-resolution TIMS U-Pb zircon and a single Sm-Nd date. These Ordovician dates were plotted along a relative time scale that was then re-proportioned as necessary to achieve a good fit with a straight line obtained by linear regression. This method of re-proportioning the Ordovician time scale by relative shortening and lengthening of parts of the relative time scale was based on a comparison of sediment accumulation rates in widely different regions and, to some extent, on empirical calibration. Agterberg (2002) subjected Cooper's (1999) data to splining and found that the optimum smoothing factor (SF) corresponds to a straight-line fit. He then used Ripley's MLFR (Maximum Likelihood fitting method for Functional

Relationship) to fit a straight line in which stratigraphic uncertainty was considered in addition to the analytical uncertainty (2σ). The MLFR method (Ripley and Thompson 1987) generalized the major axis method to the situation that the variances of X and Y are not equal and different for every data point (cf. Chap. 4).

For GTS2004, the preceding method was further modified because, in most applications, the optimum spline is not a straight line but significantly curved. The empirical straightening procedure used by McKerrow et al. (1980) was automated by splining after replacing the standard deviation for radiometric uncertainty $s(y_i)$ by $s_t(y_i) = \{s^2(x_i) + s^2(y_i)\}^{0.5}$ to incorporate stratigraphic uncertainty with s $(x_i) = 0.2875 \cdot q$ where q represents length of stratigraphic error bar. This relation between with $s(x_i)$ and q was based on the assumption that stratigraphic uncertainty is according to a rectangular frequency distribution (Agterberg 2002). If the values x_i are not free of error but have standard deviations $s(x_i)$, the ordinary spline-smoothing technique remains valid provided that $s(y_i)$ is replaced by $s_t(y_i)$ and the best-fitting smoothing spine does not deviate strongly from a straight line (cf. Lybanon 1984; Agterberg 2004).

Suppose that the observed dates (y_i) are plotted against the best-fitting spline-values $X_i = f(x_i)$ instead of against x_i. Provided that the relative geologic time scale (X-axis) is approximately linearly related to the age (Y-axis), the plot of the observed dates y_i against x_i is approximately according to a straight line that passes through the origin and dips 45°. Representing this line by the equation $Y = a + b \cdot X$, it follows that, approximately, $a = 0$ and $b = 1$. A plot of this type is equivalent to the final plot obtained by trial and error by McKerrow et al. (1980). The data points scatter around this line. The modified data set can be subjected to Ripley's MLFR method to fit the straight line. The main purpose of this exercise is not to estimate the two coefficients (a and b) that are already known approximately although they can be refined, but their standard deviations $s(a)$ and $s(b)$ and their covariance $s(a,b)$. A 95 % confidence belt around the best-fitting straight line can be constructed by using these supplementary statistics. The widths of this belt at the locations of chronostratigraphic boundaries along the X-axis then provide estimates of their 2σ-values.

References

Agterberg FP (1988) Quality of time scales – a statistical appraisal. In: Merriam DF (ed) Current trends in geomathematics. Plenum, New York
Agterberg FP (1990) Automated stratigraphic correlation. Elsevier, Amsterdam
Agterberg FP (1994) Estimation of the mesozoic geological time scale. Math Geol 26:857–876
Agterberg FP (2002) Construction of numerical geological time scales. Terra Nostra, April 2002, pp 227–232
Agterberg FP (2004) Geomathematics. In: Gradstein JM, Ogg JG, Smith AG (eds) A geologic time scale 2004. Cambridge University Press, Cambridge, pp 106–125 & Appendix 3, pp 485–486
Agterberg FP (2013) Timescale. Earth Syst Environ Sci Refer Mod 1–9
Agterberg FP, Gradstein FM (1988) Recent developments in stratigraphic correlation. Earth Sci Rev 25:1–73

Agterberg FP, Gradstein FM (1999) The RASC method for ranking and scaling of biostratigraphic events. Earth Sci Rev 46:1–25

Agterberg FP, Liu G (2008) Use of quantitative stratigraphic correlation in hydrocarbon exploration. Int J Oil Gas Coal Technol 1(4):357–381

Agterberg FP, Nel LD (1982a) Algorithms for the ranking of stratigraphic events. Comput Geosci 8:69–90

Agterberg FP, Nel LD (1982b) Algorithms for the scaling of stratigraphic events. Comput Geosci 8:163–189

Agterberg FP, Oliver J, Lew SN, Gradstein FM, Williamson MA (1985) CASC FORTRAN IV interactive computer program for correlation and scaling in time of biostratigraphic events, Geological Survey of Canada open file report 1179. Geological Survey of Canada, Ottawa

Agterberg FP, Gradstein FM, Nazli K (1990) Correlation of Jurassic microfossil abundance data from the Tojeira section, Portugal. Geol Surv Can Pap 89–9:467–482

Agterberg FP, Gradstein FM, Liu G (2007) Modeling the frequency distribution of biostratigraphic interval zone thicknesses in sedimentary basins. Nat Resour Res 16(3):219–233

Agterberg FP, Hammer O, Gradstein FM (2012) Statistical procedures, Chap 14. In: Gradstein FM, Ogg JG, Schmitz M, Ogg G (eds) The geologic time scale 2012. Elsevier, Amsterdam, pp 269–274

Agterberg FP, Gradstein FM, Cheng Q, Liu G (2013) The RASC and CASC programs for ranking, scaling and correlation of biostratigraphic events. Comput Geosci 54:279–292

Alroy J (2000) New methods for quantifying macroevolutionary patterns and processes. Paleobiology 26(4):707–733

Becker RT, Gradstein FM, Hammer O (2012) The Devonian period. In: Gradstein FM, Ogg JG, Schmitz M, Ogg G (eds) The geologic time scale 2012. Elsevier, Amsterdam, pp 559–601

Benton MJ, Harper DAT (2009) Introduction to paleobiology and the fossil record. Wiley-Blackwell, Chichester

Bramlette MN, Sullivan FR (1961) Coccolithophorids and related nannoplankton of the Early Tertiary in California. Micropaleontology 7:129–188

Brower JC (1985) Archeological seriation of an original data matrix. In: Gradstein FM, Agterberg FP, Brower JC, Schwarzacher WS (eds) Quantitative stratigraphy. UNESCO, Paris/Reidel/Dordrecht, pp 95–108

Cody RD, Levy RH, Harwood DM, Sadler PM (2008) Thinking outside the zone: high-resolution quantitative diatom biochronology for the arctic Neogene. Palaeogeogr Palaeoclimatol Palaeoecol 260:92–121

Cooper RA (1999) The Ordovician time scale – calibration of graptolite and conodont zones. Acta Univ Carol Geol 43(1/2):1–4

Cooper RA, Sadler PM (2012) The Ordovician period. In: Gradstein FM, Ogg JG, Schmitz M, Ogg G (eds) The geologic time scale 2012. Elsevier, Amsterdam, pp 489–523

Cooper RA, Agterberg FP, Alloway BV, Beu AG, Campbell HJ, Crampton JS, Crouch EM, Crundwell MP, Graham IJ, Hollis CJ, Jones CM, Kamp PJJ, Mildenhall DC, Morgans HEG, Naish TR, Raine JI, Roncaglia I, Sadler PM, Schiøler P, Scot, GH, Strong CP, Wilson GJ, Wilson GS (2004) The New Zealand geological timescale. Institute of Geological & Nuclear Sciences Monograph 22. Institute of Geological & Nuclear Sciences, Lower Hutt

Crampton JS, Foote M, Cooper RA, Beu AG, Peters SA (2012) The fossil record and spatial structuring of environments and biodiversity in the Cenozoic of New Zealand. In: McGowan AJ, Smith AB (eds) Comparing the geological and fossil records: implications for biodiversity studies, Geological Society London special publication 358. Geological Society, London, pp 105–122

D'Iorio MA (1988) Quantitative biostratigraphic analysis of the Cenozoic of the Labrador Shelf and Grand Banks. Unpublished PhD thesis, University of Ottawa, Ottawa

D'Iorio MA, Agterberg FP (1989) Marker event identification technique and correlation of Cenozoic biozones on the Labrador Shelf and Grand Banks. Bull Can Soc Petrol Geol 37:346–357

Davaud E, Guex J (1978) Traitement analytique 'manuel' et algorithmique de problèmes complexes de corrélations biochronologiques. Ecl Gel Helv 71(10):581–610

References

De Boor C (1978) A practical guide to splines. Springer, New York
Dubrule O (1984) Comparing splines and kriging. Comput Geosci 10:327–338
Edwards LE (1978) Range charts and space graphs. Comput Geosci 4:247–258
Edwards LE, Beaver RJ (1978) The used of paired comparison models in ordering stratigraphic events. Math Geol 10:261–272
Endt PM, van der Leun C (1973) Energy levels of A = 21-44 (V) nuclei. Nucl Phys A 214:1–625
Eubank RI (1988) Spline smoothing and nonparametric regression. Dekker, New York
Foote M, Miller AJ (2007) Principles of paleontology, 3rd edn. Freeman, New York
Gradstein FM (1996) STRATCOR – Graphic zonation and correlation software – user's guide, version 4
Gradstein FM, Agterberg FP (1982) Models of Cenozoic foraminiferal stratigraphy – northwestern Atlantic margin. In: Cubitt JM, Reyment RA (eds) Quantitative stratigraphic correlation. Wiley, Chichester, pp 119–173
Gradstein FM, Agterberg FP, Brower JC, Schwarzacher WS (1985) Quantitative stratigraphy. UNESCO, Paris/Reidel, Dordrecht
Gradstein FM, Agterberg FP (1985) Quantitative correlation in exploration micropaleontology. In: Gradstein FM, Agterberg FP, Brower JC, Schwarzacher WS (eds) UNESCO, Paris/Reidel, Dordrecht, pp 309–357
Gradstein FM, Agterberg FP (1998) Uncertainty in stratigraphic correlation. In: Gradstein FM, Sandrik FM, Milton D (eds) Sequence stratigraphy-concepts and applications. Elsevier, Amsterdam, pp 9–29
Gradstein FM, Agterberg FP, Ogg JG, Hardenbol J, van Veen P, Thierry J, Huang Z (1994) A mesozoic time scale. J Geophys Res 99(B12):24,051–24,074
Gradstein FM, Agterberg FP, Ogg JG, Hardenbol J, van Veen P, Thierry J, Huang Z (1995) A Triassic, Jurassic and cretaceous time scale. In: Berggren WA (ed) Geochronology, time scales and global stratigraphic correlation, SEPM (Society for Sedimentary Geology) special publication 54. SEPM, Tulsa, pp 95–126
Gradstein FM, Kaminski MA, Agterberg FP (1999) Biostratigraphy and paleoceanography of the cretaceous seaway between Norway and Greenland. Earth Sci Rev 46:27–98 (erratum 50:135-136)
Gradstein FM, Ogg JG, Smith AG (eds) (2004) A geologic time scale 2004. Cambridge University Press, Cambridge
Gradstein FM, Ogg JG, Schmitz M, Ogg G (eds) (2012) The geologic time scale 2012. Elsevier, Amsterdam
Guex J (1987) Corrélations biochronologiques et associations Unitaires. Press Polytech Rom, Lausanne
Guex J (1991) Biochronological correlations. Springer, Berlin
Hammer O, Harper D (2005) Paleontological data analysis. Blackwell, Ames
Harland WB, Cox AV, Llewellyn PG, Pickton CAG, Smith AG, Walters R (1982) A geologic time scale. Cambridge University Press, Cambridge
Harland WB, Armstrong RI, Cox AV, Craig LA, Smith AG, Smith DG (1990) A geologic time scale 1989. Cambridge University Press, Cambridge
Harper CW Jr (1981) Inferring succession of fossils in time. The need for a quantitative and statistical approach. J Paleontol 55:442–452
Hay WW (1972) Probabilistic stratigraphy. Ecl Helv 75:255–266
Hohn ME (1993) SAS program for quantitative correlation by principal components. Comput Geosci 11:471–477
Holmes A (1960) A revised geological time-scale. Trans Edinb Geol Soc 17:117–152
Holmes A (1965) Principles of physical geology. Nelson Printers, London
Hood KC (1995) GraphCor – interactive graphic correlation software, version 2.2
Hudson CB, Agterberg FP (1982) Paired comparison in biostratigraphy. Math Geol 14:69–90
Jackson PW (2006) The chronologers' quest. Cambridge University Press, Cambridge

Kaminski MA, Gradstein FM (2005) Atlas of Paleogene cosmopolitan deep-water agglutinated Foraminifera, Grzybowski Foundation special publication 10. Grzybowski Foundation, Kraków

Kemple WG, Sadler PM, Strauss DG (1989) A prototype constrained optimization solution to the time correlation problem. In: Agterberg FP, Bonham-Carter GF (eds) Statistical applications in the earth sciences. Geological Survey of Canada Paper 89-9, Ottawa, pp 417–425

Kemple WG, Sadler PM, Strauss DG (1995) Extending graphic correlation to many dimensions: stratigraphic correlation as constrained optimisation. In: Mann KO, Lane HR, Scholle PA (eds) Graphic correlation, vol 53. Society of Economic Paleontologists and Mineralogists, Tulsa, pp 65–82

Kuiper K, Deino A, Hilgen FJ, Krijgsman W, Renne PR, Wijbrans JR (2008) Synchronizing rock clocks of Earth history. Science 320(5875):500–504

Kwon J, Min K, Bickel PJ, Renne PR (2002) Statistical methods for jointly estimating the decay constant of ^{40}K and the age of a dating standard. Math Geol 34(4):457–474

Liu G, Cheng Q, Agterberg FP (2007) Design and application of graphic modules for RASC/CASC quantitative stratigraphic software. In: Zhao P, Agterberg FP, Cheng Q (eds) Proceedings of IAMG annual meeting, Beijing, August 2007, pp 719–722

Lybanon M (1984) A better least-squares method when both variables have uncertainties. Am J Phys 52(1):22–26

Lyell C (1833) Principles of geology. Murray, London

Mann KO, Lane HR (1995) Graphic correlation, SEPM (Society for Sedimentary Geology) special publication 53. SEPM (Society for Sedimentary Geology), Tulsa

Martinsson A (ed) (1977) The Silurian- Devonian boundary: final report of the Committee of the Siluro-Devonian boundary within IUGS Commission on Stratigraphy and a state of the art report for Project Ecostratigraphy. IUGS Series A 5, Beijing

Mattinson JM (2010) Analysis of the relative decay constants of ^{235}U and ^{238}U by multi-step CA-TIMS measurements of closed-system natural zircon samples. Chem Geol 99:1–13

McDougall I, Harrison TM (1999) Geochronology and thermochronology by the ^{40}Ar/^{39}Ar method, 2nd edn. Oxford University Press, New York

McKerrow WS, Lambert BSJ, Chamberlain VE (1980) The Ordovician. Silurian and Devonian time scales. Earth Planet Sci Lett 51:1–8

Min K, Mundil R, Renne PR, Ludwig KR (2000) A test for systematic errors in ^{40}Ar//^{39}Ar geochronology through comparison with U/b analysis of a 1.1-Ga rhyolite. Geoch Cosmoch Acta 64(1):73–98

Mouterde R, Ruget C, Tintant H (1973) Le passage Oxfordien – Kimmeridgien en Portugal (regions de Torres Vedras et du Montegunto). Com Rend Acad Sci Paris 277(D):2645–2648

Nazli K (1988) Geostatistical modelling of microfossil abundance data in upper Jurassic shale, Tojeira sections, central Portugal. Unpublished MSc thesis, University of Ottawa, Ottawa

Odin GS (1994) Geologic time scale (1994). Com Rend Acad Sci Paris s II 318:59–71

Pálfy J, Smith PI, Mortensen JK (2000) A U-Pb and ^{40}Ar/^{39}Ar time scale for the Jurassic. Can J Earth Sci 37:923–944

Palmer AR (1954) The faunas of the Riley formation in central Texas. J Paleontol 28:709–788

Reinsch CH (1967) Soothing by spline functions. Num Math 10:177–183

Remane J, Basset MG, Cowie JW, Gohrband TM, Lane HR, Michelsen O, Wang N (1996) Revised guidelines for the establishment of global chronostratigraphic standards by the international commission on stratigraphy (ICS). Episodes 19(3):77–81

Renne PR, Swisher CC III, Deino AL, Karner DB, Owens TL, DePaolo DJ (1998) Intercalibration of standards, absolute ages and uncertainties in ^{40}Ar/^{39}Ar dating. Chem Geol 145:117–152

Ripley BD, Thompson M (1987) Regression techniques for the detection of analytical bias. Analyst 112:377–383

Sadler PM (2001) Constrained optimisation – approaches to the paleobiologic correlation and seriation problems: a users' guide and reference manual to the CONOP program family. Unpublished, PM Sadler, Riverside

Sadler PM (2004) Quantitative biostratigraphy achieving finer resolution in global correlation. Annu Rev Earth Planet Sci 32:187–231

Schmitz MD (2012) Radiometric isotope chronology. In: Gradstein FM, Ogg JG, Schmitz M, Ogg G (eds) The geologic time scale 2012. Elsevier, Amsterdam, pp 115–126

Schoenberg IJ (1964) Spline functions and the problem of graduation. Proc Natl Acad Sci U S A 52:947–950

Shaw AG (1964) Time in stratigraphy. McGraw-Hill, New York

Stam B (1987) Quantitative analysis of middle and late Jurassic foraminifera from Portugal and its implications for the Grand Banks of Newfoundland, Utrecht micropaleontal bulletins 34. State University of Utrecht, Utrecht

Steiger RH, Jäger E (1977) Subcommission on geochronology: convention on the use of decay constants in geo- and cosmo-chronology. Earth Planet Sci Lett 36:359–362

Sullivan FR (1964) Lower Tertiary nannoplankton from the California coast ranges, 1 Paleocene, University of California publications in geological sciences. University of California Press, Berkeley

Sullivan FR (1965) Lower Tertiary nannoplankton from the California coast ranges. II. Eocene, University of California publications in geological sciences. University of California Press, Berkeley

Wahba G (1990) Spline models for observational data. SIAM, Philadelphia

Watson GS (1984) Smoothing and interpolation by kriging and with splines. J Int Assoc Math Geol 16(6):601–625

Whittaker ET (1923) On a new method of graduation. Proc Edinb Math Soc 41:63–75

Williamson MJ (1987) Quantitative biozonation of the late Jurassic and early cretaceous of the East Newfoundland Basin. Micropaleontology 33:37–65

Worsley TR, Jorgens ML (1977) Automated biostratigraphy. In: Ramsay ATS (ed) Oceanic micropaleontology. Academic, London

Zhang T, Plotnick RE (2006) Graphic biostratigraphic correlation using genetic algorithms. Math Geol 38(7):781–800

Zhou D (2008) RASC/CASC: example of creative application of statistics in geology. In: Bonham-Carter G, Cheng Q (eds) Progress in geomathematics. Springer, Heidelberg, pp 377–390

Chapter 10
Fractals

Abstract As illustrated in previous chapters, many geological features display random characteristics that can be modeled by adopting methods of mathematical statistics. A question to which new answers are being sought is: Where does the randomness in Nature come from? Nonlinear process modeling is providing new clues to answers. Benoit Mandelbrot discovered about 50 years ago that many objects on Earth can be modeled as fractals with non-Euclidean dimensions. Spatial distribution of chemical elements in the Earth's crust and features such as the Earth's topography that traditionally were explained by using deterministic process models, now also are modeled as fractals or multifractals, which are spatially intertwined fractals. Which processes have produced phenomena that are random and often characterized by non-Euclidian dimensions? The physico-chemical processes that have resulted in the Earth's present configuration were essentially deterministic and "linear" but globally as well as locally they may display random features that can only be modeled by adopting a non-linear approach that is increasingly successful in localized prediction. The situation is analogous to the relation between climate and weather. Longer term climate change can be modeled deterministically but short-term weather shows random characteristics that are best modeled by adopting the non-linear approach in addition to the use of conventional deterministic equations. This chapter reviews fractal modeling of solid Earth observations and processes with emphasis on topography, thickness measurements, geochemistry and hydrothermal processes. There is some overlap with multifractals that will be discussed in more detail in the next two chapters. Special attention is paid to improvements in goodness of fit and prediction obtained by non-linear modeling. The spatial distribution of ore deposits within large regions or within worldwide permissive tracts often is fractal. Illustrative examples to be discussed include lode gold deposits on the Canadian Shield and worldwide podiform Cr deposits, volcanogenic massive sulphide and porphyry copper deposits. The Pareto distribution is closely associated with fractal modeling of metal distribution within rocks or surficial cover in large regions. The Concentration-Area (C-A) method is a useful new tool for geochemical prospecting to help delineate subareas with anomalously high element concentration values that can be targets for further exploration with drilling.

Keywords Fractals • Fractal dimension • Cluster density determination • Concentration-area (C-A) method • Cascade models • Earth's topography • Mitchell-Sulphurets mineral district • Chemical element concentrations • Abitibi lode gold deposits • Volcanogenic massive sulphide deposits • Iskut River stream sediment data • Model of de Wijs • Model of Turcotte

10.1 Fractal Dimension Estimation

Fractals are objects or features characterized by their fractal dimension that is either greater than or less than the integer Euclidian dimension of the space in which the fractal is imbedded. The word "fractal" was coined by Mandelbrot (1975). On the one hand, fractals are often closely associated with the random variables studied in mathematical statistics; on the other hand, they are connected with the concept of "chaos" that is an outcome of some types of non-linear processes. Evertsz and Mandelbrot (1992) explain that fractals are phenomena measured in terms of their presence or absence in boxes belonging to arrays superimposed on the domain of study in 1-D, 2-D, or 3-D space, whereas multifractals apply to "measures" representing of how much of a feature is present within the boxes used for measurement. Multifractals are either spatially intertwined fractals (*cf.* Stanley and Meakin 1988) or mixtures of fractals in spatially distinct line segments, areas or volumes that are combined with one another. During the past 40 years, the fractal geometry of many natural features in Nature has become widely recognized (see, e.g., Mandelbrot 1983; Barnsley 1988; Cheng 1994; Turcotte 1997; Raines 2008; Carranza 2008; Ford and Blenkinsop 2009; Agterberg 2012; Cheng 2012). Fractals in geology either represent the end products of numerous, more or less independent processes (e.g., coastlines and topography), or they result from nonlinear processes, many of which took place long ago within the Earth's crust. Although a great variety of fractals can be generated by relatively simple algorithms, theory needed to explain fractals of the second kind generally is not so simple, because previously neglected nonlinear terms have to be inserted into existing linear, deterministic equations.

Both fractals and multifractals are commonly associated with local self-similarity or scale-independence, which generally results in power-law relations that can be represented as straight lines on log-log paper. Frequency distribution models closely associated with fractals and multifractals include the Pareto, lognormal and various extreme-value distributions. Computer simulation experiments can be used to generate artificial fractals. Multiplicative cascade models are a useful tool for generating artificial multifractals in computer simulation experiments. Multifractals often result from non-linear process modeling. Lovejoy et al. (2010) have pointed out that non-linear process modeling has made great strides forward in the geosciences, especially in geophysics, but is not yet as widely accepted as it should be. These authors briefly describe a number of recent "success stories" in which non-linear process modeling led to results that could not have been obtained otherwise. Multifractals will be discussed in more detail in Chap. 11.

10.1 Fractal Dimension Estimation

The concept of self-organized criticality (Bak 1996) has resulted in experimentally produced fractal phenomena. One of Bak's physical experiments consisted of dropping a grain of sand on a pyramid of grains in a "critical" state so that an additional grain creates a sand avalanche. The number of grains per avalanche then satisfies a Pareto frequency distribution with a relatively thick high-value tail. Another application of self-organized criticality is in the study of seismicity: Rundle et al. (2003) showed that the Gutenberg-Richter earthquake frequency-magnitude relation is a combined effect of the geometrical (fractal) structure of fault networks and the non-linear dynamics of seismicity.

Other successful applications of non-linear modeling are the following. Most weather-related processes taking place in the atmosphere including cloud formation and rainfall are multifractal (Lovejoy and Schertzer 2013). Other space-related non-linear processes include "current disruption" and "magnetic reconnection" scenarios (Sharma 1995; Uritsky et al. 2008). Within the solid Earth's crust, processes involving the release of large amounts of energy over very short intervals of time including earthquakes (Turcotte 1997), landslides, flooding (Gupta et al. 2007) and forest fires (Malamud et al. 1998) are non-linear and result in fractals or multifractals.

Increasingly it is realized that various processes that took place millions of year ago during the geologic past also were non-linear. These include, for example, the formation of columnar basalt joints (Goehring et al. 2009) and self-organization in geochemical reaction-diffusion systems resulting in banding within Mississippi Valley-type lead-zinc deposits (Fowler and L'Heureux 1996). Several types of ore deposits resulting from hydrothermal processes display geometric characteristics that have been known to be fractal or multifractal for a long time (Blenkinsop 1995; Cheng 2008). The same consideration applies to oil and gas pools (Mandelbrot 1995; Barton and La Pointe 1995). Non-linear process modeling in these fields has two practical applications. On the one hand it has resulted in new exploration techniques for the discovery of new mineral deposits; on the other hand, it allows statistical modeling of the size-frequency distributions of populations of known deposits.

The relationship between fractal point pattern modeling and statistical methods of parameter estimation in point-process modeling will be reviewed in Sect. 10.2. Statistical estimation of the cluster fractal dimension by using Ripley's (1976) K-function has advantages in comparison with the more commonly used methods of box-counting and cluster fractal dimension estimation because it corrects for edge effects, not only for rectangular study areas but also for study areas with curved boundaries determined by regional geology. Application of box-counting to estimate the fractal dimension of point patterns has the other disadvantage that, in general, it is subject to relatively strong "roll-off" effects for smaller boxes. Point patterns used for example in this section are mainly for gold deposits in the Abitibi Volcanic Belt on the Canadian Shield. Additionally, it will be proposed that, worldwide, the local point patterns of podiform Cr, volcanogenic massive sulphide and porphyry copper deposits, which are spatially distributed within irregularly shaped favorable tracts, satisfy a fractal clustering model with similar fractal

Fig. 10.1 Brownian landscape after Mandelbrot (1977, Plate 211) intersected by horizontal plane showing contours with fractal dimension $D = 1.5$. Landscape below this horizontal plane is not shown (Source: Agterberg 1980, Fig. 6)

dimensions. The problem of deposit size (metal tonnage) also is considered. Several examples are provided of cases in which the Pareto distribution, which is closely connected with fractals, provides good results for the largest deposits in metal size-frequency distribution modeling (also see Sect. 4.4.2).

The dimension of a fractal is either greater than or less than the integer Euclidian dimension of the space in which the fractal is imbedded. On the one hand, fractals are often closely associated with the random variables studied in mathematical statistics; on the other hand, they are connected with the concept of "chaos" that is an outcome of some types of non-linear processes. An excellent introduction to fractals and chaos in the geosciences is provided in Turcotte (1997). Local singularity analysis (Cheng 2005; Cheng and Agterberg 2009) is an example of non-linear modeling of geochemical data in mineral exploration and environmental applications that produce new types of maps, which are significantly different from conventional contour maps that tend to smooth out local neighborhoods with significant enrichment of ore-forming and other minerals (*cf.* Sect. 11.5).

10.1.1 Earth's Topography and Rock Unit Thickness Data

Mandelbrot (1977) developed the novel approach for the modeling of irregular natural phenomena as fractals characterized by their fractal dimension D. His first example consisted of measuring the length of the coastline of Britain ($D \approx 1.3$). Irregular curves usually have a fractal dimension that exceeds the Euclidian dimension ($E = 1$) of a straight line or geometric curve such as a circle satisfying an algebraic equation. Feder (1988) pointed out that different coastlines have different fractal dimensions. For example, the Australian coastline has $D \approx 1.1$ but the Norwegian coastline with its fjords has $D \approx 1.52$. Irregular surfaces have fractal dimensions exceeding $E = 2$ but their contours have $2 > D > 1$. An example is shown in Fig. 10.1 that was generated as follows (Mandelbrot 1977, p. 207). A horizontal plateau was broken along a straight line chosen at random to introduce a kind of vertical fault with a random difference between the levels at the two sides of the fault plane. This process was repeated many times resulting in the "ordinary" Brownian landscape of Fig. 10.1. By "ordinary" Mandelbrot meant that this landscape is closely related to the well-known process of

10.1 Fractal Dimension Estimation

Fig. 10.2 "Perimeter" measured for Fig. 1.10a by method illustrated in bottom diagram of Fig. 1.9 (Case 1). Results for smaller area have smaller total perimeters. The subarea of Fig. 1.9 gave results shown as Case 1a but more detailed contour map of Sparky Sandstone thickness for same subarea gave results shown as Case 2. See text for measurements of fractal dimensions (Source: Agterberg 1980, Fig. 7)

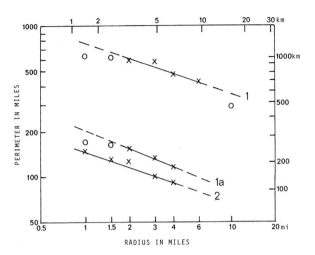

Brownian motion of a particle in the plane. Any cross-section of the landscape of Fig. 10.1, such as the jagged line topping the black stripe at the bottom, is an ordinary Brownian function with the following property: It represents the distance (measured in the vertical direction) in a given direction between (1) a particle subject to Brownian motion in the plane, and (2) an arbitrarily selected starting point as a function of time (plotted in the horizontal direction). Every profile of the ordinary Brownian landscape of Fig. 10.1, and also any contour created by intersecting the landscape with a horizontal plane, has $D = 1.5$. More complicated generating mechanisms result in other types of Brownian landscapes that are either smoother or more regular.

Agterberg (1980) subjected the pattern of Fig. 1.12 to the following measurement procedure. The "perimeter" of the thickness contour was measured by using the method explained in Fig. 1.11 for seven different unit radii. When the unit radius was 1.5 miles (2.4 km) of less, the measured perimeter provides a good approximation of the combined length of the contours in Fig. 1.12. For radius greater than 1.5 miles (2.4 km), a shorter "perimeter" was measured. These measurements are shown in Fig. 10.2 (Case 1) with a logarithmic scale along both axes. Lengths cannot be measured when the radius becomes too large; e.g., >10 miles (16 km), and measurements then should not be considered. The pattern of Fig. 1.12 suggests a curve that approximately coincides with the straight line fitted by least squares to the four points shown as crosses. The fractal dimension is equal to 1 minus the slope of the straight line (see, e.g., Mandelbrot 1983). Hence, $D \approx 1.34$. This result could be corroborated as follows: Burnett and Adams (1977, Fig. 5b on p. 347) also published a more detailed map of the 30 ft (9.14 m) contour thickness contour of the Sparky sandstone for a subarea (not shown here) of study area shown in Fig. 1.12. The results labeled 1a and 2 in Fig. 10.2 are based on the part of Fig. 1.12 for this subarea and the other large-scale map, respectively. Estimates of the fractal dimension based on these other two sets of measurements of the perimeter were $D \approx 1.38$ and $D \approx 1.34$, respectively. The pattern labeled 2 on Fig. 1.12 illustrates that estimated length of the perimeter for unit radius set at

1 mile (1.6 km) fits in with the patterns for larger radii. This illustrates that the so-called roll-off effect had not yet set in on for the shapes of contours on the large-scale map. This example illustrates that measuring fractal dimension can have value in quantitatively assessing degree of smoothing of geological features on maps at different scales.

Lovejoy and Schertzer (2007) have reviewed early history of fractal geometry pointing out that, early on, several mathematicians and geophysicists had pointed out problems of differentiability and integrability in connection with some types of natural phenomena. For example, Perrin (1913) wrote: "Consider the difficulty in finding the tangent to a point of the coast of Brittany ... depending on the resolution of the map the tangent would change. The point is that a map is simply a conventional drawing in which each line has a tangent. On the contrary, an essential feature of the coast is that ... at each scale we guess the details which prohibit us from drawing a tangent". With respect to problems of integrability, Steinhaus (1954) stated: "The left bank of the Vistula when measured with increased precision would furnish lengths ten, hundred, or even 1,000 times as great as the length read off a school map. A statement nearly adequate to reality would be to call most arcs encountered in nature as not rectifiable." Among the early pioneers, Lovejoy and Schertzer (2007) list Vening Meinesz (1951) who argued that the power spectrum $P(k)$ of the Earth's topography has the scaling form $k^{-\beta}$ where k is a wave number ($=2\pi \times$ frequency) and $\beta = 2$, which according to Lovejoy and Schertzer (2007, p. 466) is "close to the modern value $\beta = 2.1$". Vening Meinesz (1951) derived his scaling model of the Earth's topography as follows.

Prey (1922) originally had developed the Earth's topography in terms of spherical harmonics up to the 16th order. For the mathematics of spherical functions; see, e.g., Freeden and Schreiner 2008. Vening Meinesz (1951) normalized Prey's coefficients after making a separation between continents and oceans. He showed that mean square elevation (y) is approximately related to order x of harmonic according to the power law relation $y = C \cdot x^{-\beta}$ where C and β are constants. This kind of relationship is in accordance with the concept that the Earth's topography can be described as a universal multifractal (cf. Sect. 12.7). Orders of spherical harmonics are analogous to wave numbers in a periodogram. On the original graph (Vening Meinesz 1951, Fig. 3), y multiplied by $\{n \cdot (n-1)\}$ was related to n according to a curve that was approximately horizontal for $x > 2$. Heiskanen and Vening Meinesz (1958) argued that multiplication by $\{n \cdot (n-1)\}$ is to be preferred to multiplication by n^2 but this is a minor refinement only.

In 1957, at the request of Vening Meinesz, the Netherlands Geodetic Commission completed a new development in spherical harmonics of the Earth's topography using improved data, especially on the topography of water-covered areas, extending Prey's calculations from the 16th to the 31st order. Figure 10.3 shows the resulting refinement of the 1951 analysis subsequently obtained by Vening Meinesz (1964). The new curve is subhorizontal for $n > 5$, although there may exist an increase in y for higher orders suggesting that β is slightly less than 2, especially for continental topography. Although the precise value of β is not known exactly, it is clear that the Earth's topography on the continents as well as on the ocean floor approximately displays the same type of fractal/multifractal behavior.

10.1 Fractal Dimension Estimation

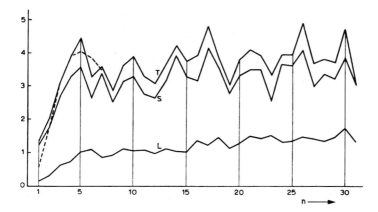

Fig. 10.3 Spherical harmonic development of the Earth's topography (after Vening Meinesz 1964). T = total topography; S = ocean floor topography; L = continental topography. Horizontal scale is for order n (=1 to 31) of spherical harmonics; vertical scale is root mean square of elevation multiplied by $n^{1/2} \cdot (n+1)^{1/2}$. Hatched curve represents inverse values of Rayleigh numbers for hypothetical mantle currents (not discussed here) distributed according the corresponding spherical harmonics (Source: Vening Meinesz 1964, Fig. IV.2)

10.1.2 Chemical Element Concentration Values: Mitchell-Sulphurets Example

Cheng (1994) has studied the spatial distribution of 1,030 concentration values of gold and associated elements in partially altered volcanic rocks in the Mitchell-Sulphurets area, northwestern British Columbia (Fig. 10.4a). Plots of Au content in surface samples (Fig. 10.4b) from altered and unaltered rocks using logarithmic probability paper (=lognormal Q-Q plot) suggest that the gold concentration values satisfy a single, positively skewed frequency distribution that is approximately lognormal (Fig. 10.4c). On the other hand, when geochemical isopleths are constructed (isolines or contours for concentration values in ppb Au) and log-log paper is used to plot the cumulative area of surface rocks with larger concentration values against the contour value, the altered and unaltered rocks show entirely different power-law relations (Fig. 10.4d). This indicates that a power-law-based approach can be useful for the delineation of geo-chemical anomalies in addition to commonly used approaches directly based on element frequency distributions and contour maps (see, e.g., Sinclair 1991).

Figure 10.5 shows Au and Cu maps employing single contours to divide the data set into two parts, above and below the contour's value. It can be seen that the shapes of the areas enclosed by successive contours are changing gradually; total enclosed area decreases as the value of the contour increases. From these contours an optimum threshold for separating anomalies from background areas can be selected by means of a log-log plot for the element concentration-area relation. This threshold coincides with a sudden change in the rate of decrease of the area enclosed by higher value contours on the log-log plot.

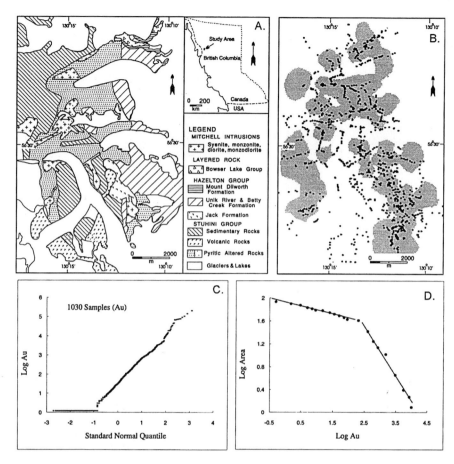

Fig. 10.4 Mitchell-Sulphurets mineral district, northwestern British Columbia. Four plots originally constructed by Cheng (1994): (**a**) Simplified geology after R.V. Kirkham (personal communication, 1993), including outline of alteration zones; (**b**) Sampling sites and (smoothed) isopleth for 200 ppb gold concentration value; (**c**) Lognormal Q-Q plot (Au determinations below 2 ppb detection limit are shown as horizontal line); (**d**) Log-log plot for area enclosed by isopleths (including circumference of pattern for 200 ppb in Fig. 10.4b). Straight lines are least-squares fits (log base 10; gold in ppb) (Source: Agterberg 1995a, Fig. 3)

The Geographic Information System (GIS) used to obtain the contour maps of Fig. 10.5 ("Spatial Analysis System" or SPANS; *cf.* Cheng 1994) allowed several interpolation procedures, including 2-D Kriging and so-called "potential mapping". The latter method consisted of the calculation of simple weighted moving averages with a moving circular window with adjustable parameters to control the weighting of values at neighboring points (*cf.* Sect. 1.4.1). The parameters include radius of the circular window, decay ratio of the weighting function, and maximum number of samples to be included within each window. For the example (*cf.* Cheng et al. 1994) parameters selected were: radius of 0.8 km, decay ratio of 0.5 (corresponding to a linear weighting function with weight 0 for samples located

10.1 Fractal Dimension Estimation

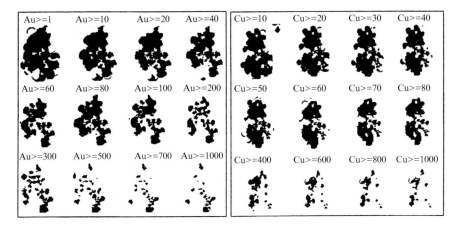

Fig. 10.5 Successive binary patterns for separate contours of Au and Cu, Mitchell-Sulphurets area. (Source: Cheng et al. 1994, Fig. 5)

at the boundary or outside the moving window, and 1 for samples located at the center) and maximum of ten samples per window. (If more than ten samples occur within the window for a given location, only the ten nearest points were used to evaluate the surface at that location.)

Suppose $A(\rho)$ denotes the area with concentration values greater than the contour value ρ. This implies that $A(\rho)$ is a decreasing function of ρ. If ν represents the threshold, the following empirical model generally provides a good fit to the data for different elements in the study area:

$$A(\rho \leq \nu) \propto \rho^{-\alpha_1}; A(\rho > \nu) \propto \rho^{-\alpha_2}$$

where $\propto$ denotes proportionality; α_1 and α_2 are different exponents.

Figure 10.6 shows log-log plots satisfying the preceding two power-law relations for Au, Cu, As, Ag, Sb and Pb. All areas were computed from separate contour maps such as those shown in Fig. 10.5 for Au and Cu. Pairs of estimated exponents and corresponding optimum thresholds for these 6 elements and 22 other elements or oxides were presented by Cheng (1994, Table 1). These thresholds delineate anomalous areas. Comparison of the areas above and below the threshold of 200 ppb Au on the contour map (Fig. 10.5) with the geological map shows significant spatial correlation between the areas with Au concentration above 200 ppb and Au-associated alteration zones (Cheng et al. 1994, Fig. 2). The same type of correlation with alteration zones applies to Cu (400 ppm threshold; see Fig. 10.5). Note, however, that the Au anomalies are more prominent than the Cu anomalies in the southeastern part of the area. It may be concluded that the log-log plots for the element concentration-area relation provide an excellent method for separating anomalies from background in the Mitchell- Sulphurets area. This empirical result will be explained by fractal modeling in Sect. 10.3.1.

Cheng (1994) also subjected element concentration maps as shown for Au and Cu in Fig. 10.5 to perimeter-area analysis as follows. Theoretically, for a group

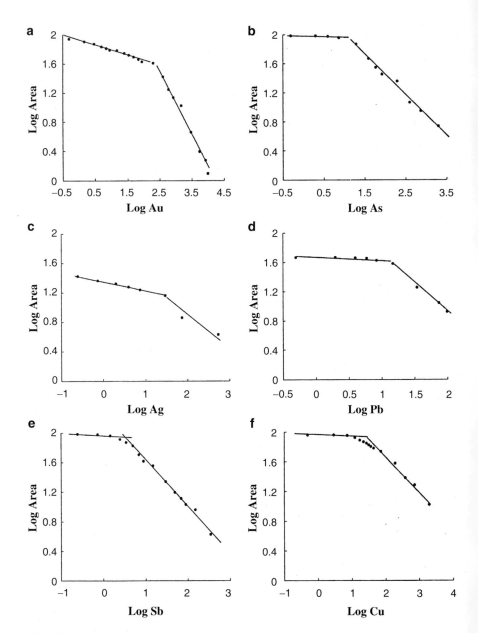

Fig. 10.6 Log-log plots representing the relationships between areas bounded by contours (as shown for Au and Cu in Fig. 10.5), and contour value for four elements, Mitchell-Sulphurets area. Straight line segments were fitted by least squares (Source: Cheng et al. 1994, Fig. 6)

10.1 Fractal Dimension Estimation

of similarly shaped sets, there exist power-law relationships between any two measures of volume or area and perimeters. For example, two areas A_i and A_j enclosed by contours i and j are related to their perimeters L_i and L_j as follows:

$$\frac{L_j(\delta)}{L_i(\delta)} = \left[\frac{A_j(\delta)}{A_i(\delta)}\right]^{D_{AL}/2}$$

where δ is the common yardstick used for measuring both areas and perimeters. The fractal dimension D_{AL} satisfies $D_{AL} = 2D_L/D_A$, where D_L and D_A denote fractal dimensions of perimeter and area, respectively. The contours for Au and Cu in Fig. 10.5 show similar shapes suggesting that the current perimeter-area modeling approach may be valid. Experimental results for Au, Cu and As are shown on log-log plots in Fig. 10.7a, c, and e using perimeters and areas for different contours with concentration values above the thresholds previously defined using the element concentration-area method (Fig. 10.6). Figure 10.7b, d and f show the relationships between estimated lengths of perimeters of contours of concentration values for yardsticks of different lengths. There appear to be two different fractal dimensions in these diagrams on the right side of Fig. 10.7. The first of these (D_1) for yardsticks less than 300 m is equivalent to the so-called "textural" fractal dimension (Kaye 1989, p. 27). It is close to 1 for the elements considered and can be explained as the result of smoothing during interpolation. The other estimate of D_L (D_2) ranges from 1.14 to 1.33 and is probably representative of the true geometry of anomalous areas for the elements considered. It is noted that the least squares estimates of D_1 and D_2 in Fig. 10.7 are subject to considerable uncertainty because they are based on relatively few points.

Box 10.1: Fractal Perimeter-Area Relation

The following perimeter-area relationship was introduced by Mandelbrot (1983) for similarly shaped sets: $L(\delta) = C \cdot \delta^{(1-D)}\sqrt{A\delta^D}$ where C is a constant. A modification of this equation was introduced by Cheng (1994) as follows (cf. Cheng et al. 1994). For yardstick δ, the estimated length and area of a pattern in 2-D can be expressed as: $L(\delta) = L_0\delta^{(1-D)}$; $A(\delta) = A_0\delta^{(2-D)}$. Similarly shaped patterns at different scales can be derived from one another by changing the scale. Suppose that r_i represent the ratio to create the i-th geometrical pattern from the k-th one. Estimates of L and A for these two geometries then can be obtained from: $L_k(\delta) = L_0\delta^{(1-D_L)}$; $L_i(\delta) = L_0(r_i\delta)^{-D_L}\delta$; $A_k(\delta) = A_0\delta^{(2-D_A)}$; $A_i(\delta) = A_0(r_i\delta)^{-D_A}\delta^2$. Consequently, $\frac{L_i(\delta)}{L_k(\delta)} = \left[\frac{A_i(\delta)}{A_k(\delta)}\right]^{D_{AL}/2}$; $L_i(\delta) = \delta\left[\frac{A_i(\delta)}{\delta^2}\right]^{D_{AL}/2} = \delta^{1-D_{AL}}[A_i(\delta)]^{D_{AL}/2}$, and $DAL = 2DL/DA$.

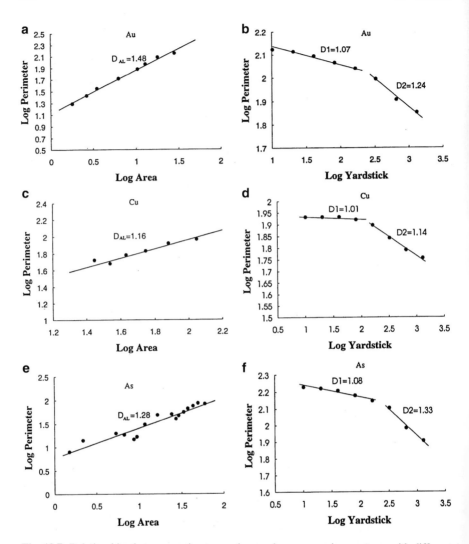

Fig. 10.7 Relationships between perimeters and anomalous areas using contours with different values greater than thresholds for (**a**) Au, (**c**) Cu and (**e**) As. Log-log plots on right side show relationships between estimated lengths of the perimeters of the anomalous areas for (**b**) Au, (**d**) Cu and (**f**) As with variable yardstick. $D1$ and $D2$ are for textural and structural fractal dimensions, respectively. All solid lines were obtained by least squares (Source: Cheng et al. 1994, Fig. 8)

From any two of the three fractal dimensions (D_L for perimeter; D_{AL} for perimeter-area relation; and D_A for area), the third one can be obtained by means of the relation $D_A = 2D_L/D_{AL}$. Theoretically, D_A cannot be greater than 2. In most applications of the perimeter-area method, it is set equal to 2 so that $D_L = D_{AL}$. The latter relation holds approximately true for Cu and As ($D_{AL} = 1.16$ and $D_A = 1.96$

10.1 Fractal Dimension Estimation

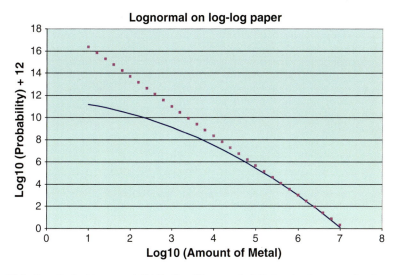

Fig. 10.8 Hypothetical lognormal distribution (*blue curve*) plotted on log-log paper (log base 10). Straight line tangent to lognormal curve represents Pareto distribution (*pink line*) (Source: Agterberg 2007, Fig. 1a)

for Cu and $D_{AL} = 1.28$ and $D_A = 2.07$ for As). For Au, the estimated value of D_{AL} (1.48) is significantly greater than the estimated structural fractal dimension (1.24). From the relation $D_A = 2D_L/D_{AL}$, it follows that $D_A = 1.68$ for gold which is less than 2. These results indicate that the distribution of Au in the alteration zones is more irregular than that of Cu or As.

10.1.3 Total Metal Content of Mineral Deposits: Abitibi Lode Gold Deposit Example

The problem of whether natural resources are best modeled as Pareto- or lognormal-type remains important because both approaches continue to be used extensively for oil and other natural resource size-frequency modeling. As mentioned before, Mandelbrot (1983, p. 262; *cf.* Chap. 4) had challenged geoscientists by asserting that oil and other natural resources have Pareto distributions and this "finding disagrees with the dominant opinion, that the quantities in question are lognormally distributed. The difference is extremely significant, the reserves being much higher under the hyperbolic than under the lognormal law" (*cf.* Sect. 4.4.1). The difference between the two types of frequency distribution is illustrated in Figs. 10.8 and 10.9 using an artificial example for amount of metal contained in all ore deposits from a region. The distribution that plots as a straight line on one kind of diagram plots as a curve on the other diagram. The Pareto not only has a thicker tail, it deviates more strongly from the lognormal at the other (lower value) end,

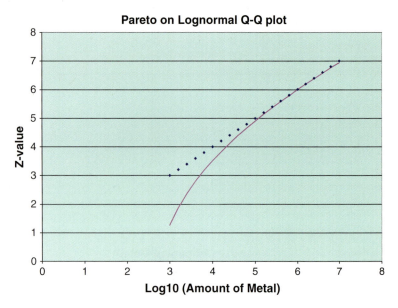

Fig. 10.9 Two distributions of Fig. 10.8 re-plotted as lognormal Q-Q plot (Source Agterberg 2007, Fig. 1b)

because as frequency for the lognormal approaches 0 for decreasing amount of metal, the corresponding Pareto frequency approaches ∞. The number of oil fields containing more than a given amount of oil can be modeled by means of the Pareto distribution, as demonstrated by several authors including Drew et al. (1982) and Crovelli (1995).

Singer and Menzie (2010) used the lognormal distribution as a benchmark for the modeling of worldwide metal resources. Agterberg and Divi (1978) developed a two-dimensional lognormal model for copper, lead and zinc ore deposits in the Canadian Appalachian region. On the other hand, Turcotte (1997) provides various examples of successful fitting of the Pareto to various metal resource data sets. These examples include log-log plots of grade versus cumulative production of mercury, lode gold and copper in the United States originally established by Cargill (1981) and Cargill et al. (1980, 1981). Respective merits of lognormal and Pareto size-modeling already were investigated in Chap. 4 for copper deposits in the Abitibi area on the Canadian Shield. In the Abitibi copper hindsight study, both models were applicable although the Pareto had the advantage in predicting properties of size frequency distributions of amounts of copper ore that was to be discovered later.

Freiling (1966) made a direct comparison of lognormal and Pareto distributions showing that the tails of these distributions are very different. However, extensive testing of the two models on mass-size distributions led him to the conclusion that available data were not sufficient to distinguish between lognormal and power-law distributions in practice.

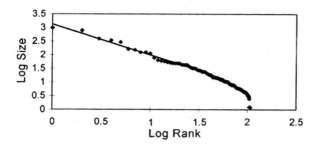

Fig. 10.10 Production and reserves of 107 lode gold deposits (in metric tons Au) in Abitibi volcanic belt, Superior Province, Canadian Shield, ranked according to their sizes (largest deposit first). Pareto distribution fitted as straight line (Source: Agterberg 1995a, Fig. 2)

The simple method of checking whether or not a set of resource estimates satisfies a Pareto distribution introduced in Chap. 4 consists of constructing a log-log plot of amount of metal per deposit and size rank (in descending order). In Fig. 10.10 this method is applied to gold deposits in the Abitibi volcanic belt. If the Pareto model is satisfied, this type of plot produces a straight line, usually with a relatively sharp downward bend at the end because of an economic cut-off effect. Although lognormal and Pareto distributions commonly are fitted to metal from ore deposits in large regions or on a worldwide basis, it should be kept in mind that metal from ore deposits in smaller regions can have different frequency distributions. The Vistelius model that combining lognormal distributions from smaller regions or different rock units can produce a new lognormal distribution with different parameters was discussed in detail in Chap. 3. With respect to the example for the Abitibi volcanic belt, it is shown in Agterberg (1995b) that 53 lode gold deposits from the Superior Province on the Canadian Shield but from outside the Abitibi volcanic belt satisfy a different Pareto distribution.

10.2 2-D Distribution Patterns of Mineral Deposits

Point patterns can be fractal (Feder 1988; Stoyan et al. 1987; Korvin 1992). In general, fractal point patterns are characterized by the fact that average point density (=number of points per unit of area) decreases when size of study area is increased. In this respect, fractal point patterns differ from commonly used statistical point-process models such as the random Poisson and various clustering models in which the mean point density is assumed to be the same within the entire study area regardless of its size. Mathematical statisticians (Ripley 1976, 1981, 1987, 1988; Diggle 1983; Rowlingson and Diggle 1991, 1993; Cressie 2001; Baddeley and Turner 2012; Baddeley et al. 2006; Baddeley et al. 2008) have developed very precise methods to estimate the parameters of constant-density point processes. In Chap. 1, Ripley's method of correcting estimates of point pattern parameters for edge effects was applied to wildcats and gas discoveries in exploration "plays" in Alberta with study area boundaries that were approximated by multi-edge polygons instead of being rectangular in shape. Not only do these "plays" have irregular boundaries, they may contain islands of terrain judged to be unfavorable for drilling but situated within the study area. These edge correction methods can equally well be applied in fractal point process modeling. This section advocates the use of such methods, which account for edge effects, allow irregularly shaped study areas and, generally, are much more precise than various box-counting methods more commonly used for the statistical treatment of fractal point patterns if the domains of study are not rectangular in shape.

A closely related topic not covered in this chapter is that fractal point patterns can be multifractal. The spatial distribution of gold mineral occurrences in the Iskut River map sheet, northwestern British Columbia, is multifractal instead of monofractal (Cheng 1994; Cheng and Agterberg 1995). Although theory of multifractal point processes was developed in these latter two publications, there have been relatively few other applications along these lines, although generalizations to multifractal point pattern modeling can have advantages similar to those documented in the relatively many existing multifractal applications to measures or chemical element concentration methods (Mandelbrot 1999; Cheng 2012). The topic of multifractal point patterns will be discussed in the next chapter. The points in a point pattern can have different values. This has led to the modeling of "marked" point processes (Cressie 2001). This topic also could be of importance in resource potential modeling where the points represent mineral deposits or oil pools with positively skewed size-frequency distributions and possess other features that differ greatly from point to point.

Carlson (1991) applied a fractal cluster model to hydrothermal precious-metal mines in the Basin and Range Province, western United States. The underlying theoretical model is clearly explained in Feder (1988, Chap. 3). For the spatial distribution of points in the plane this model requires that the number of points within distance r from an arbitrary point is proportional to r^{D_c} where D_c represents the so-called cluster fractal dimension which is less than the Euclidian dimension of the embedding space (=2 for two-dimensional space). It implies that the cluster density decreases with increasing r. As mentioned before, a basic difference between this fractal model and commonly used statistical point-process models such as the Poisson and Neyman-Scott models is that the latter are not fractal because their dimension remains equal to the Euclidian dimension (=2) for increasing r.

Two methods (box-counting, and cluster density determination) are commonly used for the fractal modeling of point clusters. The first method consists of superimposing square grids with different spacings (ϵ) on the study region. The number of boxes (N_c) containing one or more points is counted for every square grid. The number N_ϵ decreases with increasing value of ϵ, mainly because there are fewer larger boxes. If the box-counting fractal model is satisfied, this decrease satisfies a power-law relation and shows as a straight line with slope - D_b on a log-log plot where D_b represents the fractal dimension. Cluster density determination can be performed by centering circles with different radii (r) on all points in the study area, counting how many other points occur within these circles, and averaging the results. The relation between cluster densities then is according to a power-law that plots as a straight line with slope $D_c - 2$ on log-log paper if the fractal model with D_c as its fractal dimension is satisfied.

Carlson (1991) obtained the following results: for distances less than 15 km, cluster density determination as applied to 4,775 precious metal deposits gave $D_c = 0.83$ versus $D_b = 0.50$ for box-counting. For greater distances between 15 and 1,000 km, the two methods yielded $D_c = 1.17$ versus $D_b = 1.51$. Note that for short distances, D_b is less than D_c but for larger distances, it is the other way around. For both kinds of fractal cluster, the dimension seemed to be bifractal. Carlson commented that differences between D_b and D_c are "troubling but common in measuring fractal cluster dimensions". It is mentioned here that fractal dimensions for short distances generally are too low because of "roll-off" effects that affect D_b more strongly than D_c (*cf.* Sect. 10.1.1 and next paragraph). Because of possible edge effects, the $D_b = 1.51$ estimate may be most accurate one. Carlson (1991) interpreted his results in terms of a bifractal model assuming that fractal hydrothermal and fracture systems are effective over scales from about 15–1,000 km. Various examples of geological bifractals in other types of applications can be found in Korvin (1992).

Differences in fractal point pattern dimension can be due to different reasons. One explanation is that D_b and D_c are not necessarily the same; e.g., they are expected to be different for a multifractal cluster. With respect to the differences between D_b and D_c: due to limited resolution of maps, the measurements on fractals normally become biased for small distances (ϵ or r) and this can cause measured fractal dimensions to be biased. Theoretically, a fractal cluster model has the property of self-similarity in that point patterns for any enlarged subarea would be similar even if the zooming-in is repeated indefinitely. In practice, graphical representations become increasingly incomplete with increased enlargement. This type of bias does not only apply to fractal point patterns but to other types of fractals as well. This well-known "roll-off" effect has been studied in detail by structural geologists (e.g., Walsh et al. 1991; Pickering et al. 1995; Blenkinsop 1995). It already was discussed earlier in this chapter for contour maps (Sect. 10.1.1). When sample size is large, the "roll-off" effect can be modeled by using a continuous curve that asymptotically approaches the fractal straight line on log-log paper. However, for smaller samples, the observed "roll-off" effect may create an artificial sequence resembling two or more separate straight-line segments. Downward bias in D_b toward the origin is stronger than that in D_c over short distances.

On the other hand, estimates of D_c for large distances may become less precise and biased downward more strongly than those of D_b if edge effects are not taken into account. Unless Ripley's $K(r)$ function is used, it is difficult to account for edge effects in D_c because: (1) much useful information is discarded if all largest circles around points (along with all smaller circles within the largest circles) are required to be fully contained within the study area, and (2) the boundary of the study area may not have a simple (rectangular or circular) shape but can be highly irregular and then would have to be approximated by a polygon representing a curvilinear shape instead of by a rectangle, thus creating additional complications as discussed in Sect. 1.5.3. These two edge effect problems can be avoided by using Ripley's $K(r)$ function. Although estimates of D_b based on the largest boxes are relatively imprecise, because there are relatively few of them, they can have the advantage of remaining unbiased provided that exactly the same study area is covered by all box-counting grids.

The results obtained by Carlson (1991) have been criticized by various authors but Raines (2008) in a comprehensive review concluded (a) in several other applications, differences between D_b and D_c are less than those found by Carlson, and (b) in general, the bifractal cluster model would be satisfied because dimensions for shorter distances are indeed less than those for larger distances. Raines (2008) was able to reduce bias in fractal cluster estimation to some extent by using GIS –based methods. However, it is possible to improve fractal estimation procedures more significantly by adopting the methods developed by mathematical statisticians as will be shown in the next section on the basis of a practical example.

10.2.1 Cluster Density Determination of Gold Deposits in the Kirkland Lake Area on the Canadian Shield

Figure 10.11 (from Agterberg et al. 1993) shows the locations of 295 gold deposits (mines and occurrences) in a rectangular area of 4,185 km^2 in the vicinity of Timmins and Kirkland Lake in the Abitibi area on the Precambrian Canadian Shield. It is part of a much larger study area (=89,600 km^2) containing 1,306 gold deposits with $D_c = 1.514$ to be discussed in more detail in the next section. Cluster density estimation applied to Fig. 10.11 gives $D_c = 1.493$, which is close to the estimate based on the larger area. Figure 10.11 shows large subareas without any gold deposits (lacunarity). This is either because the bedrock containing gold deposits in these subareas is covered by relatively thick Quaternary deposits or it consists of barren rock types such as Archean gneiss belts without gold deposits. The pattern of Fig. 10.11 is anisotropic because of strong east–west structural trends. By means of additional experiments to be summarized later it can be shown that neither lacunarity nor anisotropy significantly alters the estimate of D_c (≈ 1.5).

Most problems of lack of precision and accuracy outlined in the previous section can be avoided by adopting methods of statistical point-process modeling (Diggle 1983; Ripley 1988; Agterberg 1994). In this type of modeling, the first-order and second-order properties of a spatial point process for events are described by its intensity function $\lambda_1(x)$ and second-order intensity function $\lambda_2(x, y)$ where x and y represent the locations of two event points in the plane. For an isotropic, stationary point process, $\lambda_1(x) = \lambda$ and $\lambda_2(x, y) = \lambda_2(r)$ where r is distance between x and y. A powerful way to characterize the second-order properties of a point process is provided by Ripley's $K(r)$ function, which is proportional to expected number of

10.2 2-D Distribution Patterns of Mineral Deposits

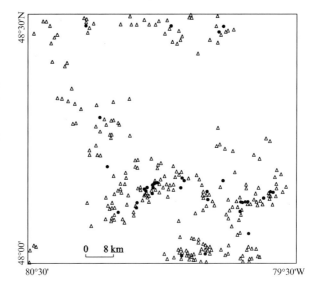

Fig. 10.11 Example of point pattern for 295 gold deposits in the vicinity of Timmins and Kirkland Lake, Abitibi volcanic belt on Canadian Shield. Circles are gold mines (mainly past producers); triangles represent small, unmined deposits (Source: Agterberg 2013, Fig. 2)

further events within distance r from an arbitrary event point. The relation between $\lambda_2(r)$ and $K(r)$ is $\lambda_2(r) \propto K'(r)/r$ where $\propto$ denotes proportionality and $K'(r)$ is the first derivative of $K(r)$ with respect to r. Application of this statistical point process technique to fractal cluster theory yields $\lambda_2(r) \propto r^{D_c-2}$ and $K(r) \propto r^{D_c}$ where $\propto$ again denotes proportionality.

For rectangular areas such as that of Fig. 10.11, the $K(r)$ function can be estimated readily by the method originally developed by Ripley (1976) according to which edge effects due to the boundaries of the study area are avoided. Diggle (1983, p. 72) has published explicit formulae for unbiased estimation of $K(r)$ for event points in a rectangular study area. Application of this method to the pattern of Fig. 10.11 resulted in the pattern of solid circles in Fig. 10.12a. The slope of the straight line fitted to these points gives the estimate $D_c = 1.493$ mentioned before.

Results of two experiments to study the effects of lacunarity and anisotropy are also shown in Fig. 10.12. The largest holes in the point pattern of Fig. 10.11 occur in the northern half of this study area. For this reason, a new study area consisting only of the southern half of Fig. 10.11 was defined. It contains most of the gold deposits in Fig. 10.11 and has three holes (lacunae) that are smaller than the two largest holes in Fig. 10.11. Application of the previous method to the point pattern for the southern half of Fig. 10.11 resulted in the pattern of solid circles in Fig. 10.12b with $D_c = 1.467$ instead of $D_c = 1.493$. The fact that these two estimates are nearly equal to one another illustrates that lacunarity does not significantly affect the estimation of D_c.

The effect of anisotropy was studied as follows. A so-called affine transformation was applied to the pattern of Fig. 10.11 by changing the north–south coordinates of all event points. The north–south coordinate was measured by taking distance from the southern boundary of the study area and these distances were

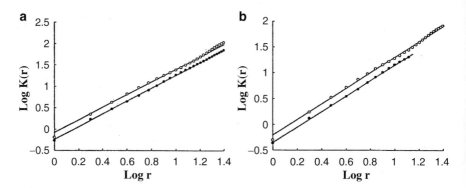

Fig. 10.12 Log-Log plots of $K(r)$ function versus distance r (in km) to estimate cluster fractal dimension D_c (estimated from slopes of best-fitting straight lines). (**a**) Point pattern for gold deposits in Fig. 10.11; (**b**) southern half of point pattern shown in Fig. 10.11. Solid circles show results for original (anisotropic) point patterns; open circles show results for transformed (approximately isotropic) point pattern. Estimates of D_c for solid circles are (**a**) 1.493 and (**b**) 1.498; estimates of D_c for open circles are (**a**) 1.506 and (**b**) 1.524. Solid circles in Fig. 10.12b are based on fewer points because edge correction formula used requires $r < h - 2$ where h is height of map area (measured in north–south direction) in km. (Source: Agterberg 2013, Fig. 3)

multiplied by 2.5 to provide a more isotropic pattern. Results of application of the previous methods are shown as open circles in Fig. 10.12. The resulting estimates are $D_c = 1.498$ (Fig. 10.12a) and $D_c = 1.506$ (Fig. 10.12b). Like the two other estimates these two estimates are approximately equal to 1.5 illustrating that, like lacunarity, anisotropy does not significantly affect the estimation of D_c.

Results of the box-counting fractal method applied to the point pattern of Fig. 10.11 were tabulated in Agterberg et al. (1993) and here are graphically shown in Fig. 10.12. As already mentioned in Sect. 10.1.3, the box-counting fractal method is less accurate and less precise than cluster density estimation for point patterns. Because there are only 295 points in Fig. 10.11, D_b approaches 0 for decreasing box size ($\epsilon \to 0$). For any very small value of ϵ, the number of boxes containing points becomes $N_C = \text{Log}_{10} \, 295 = 2.47$. In Fig. 10.13 the "roll-off" pattern of measurements asymptotically approaches this value that is almost reached for boxes measuring 1 km on a side, because very few of these contain more than a single point. This source of bias can be seen on Fig. 10.13 for values less than N_C. Larger boxes are not subject to this type of bias but are relatively imprecise because their frequency approaches 0 for increasing box size. The lack of precision is shown on Fig. 10.13 as increased scatter of measurements along the straight line pattern (broken line in Fig. 10.13). This line with slope equal to -1.53 was fitted to the values with $^{10}\text{Log} \, \epsilon \geq 0.6$ only. It would result in the estimate $D_b = 1.53$, which is comparable with the estimates of $D_c \approx 1.5$ obtained in the previous section. Although other interpretations remain possible (*cf.* Raines 2008, p. 289), the most reasonable conclusion is that there is no significant difference between D_b and $D_c \approx 1.5$ for the spatial distribution of gold deposits in the Kirkland Lake area.

10.2 2-D Distribution Patterns of Mineral Deposits

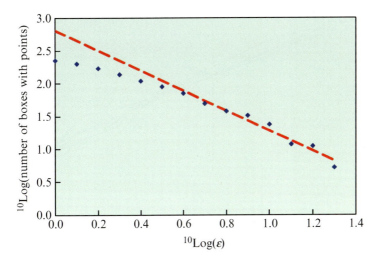

Fig. 10.13 Estimation of box-counting dimension D_b for point pattern of gold deposits shown in Fig. 10.9. Original data were taken from columns 2 and 4 in Table 1 of Agterberg et al. (1993). The straight line with value of $D_b = 1.528$ was fitted to eight largest-box values only, assuming that there is a roll-off effect in values for the smaller boxes. It is noted that Raines (2008) assumed the pattern of Fig. 10.13 to be bifractal (Source: Agterberg 2013, Fig. 4)

10.2.2 Cluster Density Determination of Gold Deposits in the Larger Abitibi Area

The Kirkland Lake area (Fig. 10.9) is part of a larger study area (=89,600 km²) with 1,306 gold deposits (from a compilation by Agterberg et al. 1972). It measures 560 km east–west and 160 km north–south. Subjecting he larger area to fractal cluster modeling by the method described in the previous section gave $D_c = 1.514$ (Agterberg 1993). This is close to $D = 1.493$ and other estimates mentioned previously. The larger study area covers most of the Abitibi Volcanic Belt on the Canadian Shield. It shows many more large holes in the pattern of gold deposits than can be seen in Fig. 10.9. This lacunarity is mainly due to the occurrences of terrains underlain by Archean granite intrusions and gneisses, which are devoid of gold deposits. It was already discussed before that the sizes of the large gold deposits in the Abitibi Volcanic Belt satisfy a fractal Pareto frequency distribution (Fig. 10.8)

In Agterberg (1993), the enlarged study area with 1,306 small and large gold deposits was used to illustrate the calculation of the variance of mean values of numbers of deposits in blocks for regional resource evaluation studies. The study area was partitioned into 3,584 (5 × 5 km) unit cells forming 32 rows and 112 columns. The corresponding estimate of intensity is λ (=1,306/3,584) = 0.3644. In an earlier study, Agterberg (1981) had used the following computer algorithm to estimate the reduced second moment measure $\lambda_2(r)/\lambda$ (cf. Sect. 10.2.1). A grid with (5 × 5 km cells) was successively centered on each of the 1,306 gold deposits.

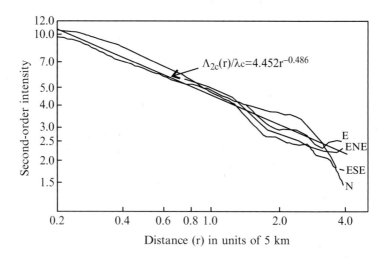

Fig. 10.14 Second-order intensity functions for 1,306 gold deposits in Abitibi volcanic belt. Vertical scale is for $\lambda_{2c}(r)/\lambda_c$ where subscript c denotes relation to cluster fractal dimension. Affine transformation of original point pattern by factor 2.5 in the north–south direction resulted in approximate isotropy as can be seen from observed second-order intensities for four different directions estimated after the transformation. Both scales are logarithmic; the second-order intensity is larger than the covariance density because it has not been centered (Source: Agterberg 1993, Fig. 8)

Frequencies of other deposits occurring within cells of this grid in the vicinity of each deposit were determined and added in order to obtain total frequencies. In each case, only the frequencies for cells falling within the study area were used. This earlier method to correct for edge effects using squares is less precise than when Ripley's $K(r)$ function with circles is used (cf. Sect. 1.5.3). A formal proof that cluster density estimation results in an unbiased estimate of $\lambda_2(r)/\lambda$ in the anisotropic case can be found in Stoyan et al. (1987, p. 125; also see Falconer 2003). The resulting frequency distribution had elliptic contours elongated in the east–west direction. Reduction of the east–west coordinates by the factor 2.5 resulted in approximately circular contours on a compressed map. The (5 × 5 km) unit cells on this transformed map correspond to rectangular cells on the original map measuring 12.5 km in the east–west direction and 5 km in the north–south direction. Estimated intensity for the compressed map is λ_c ($=2.5\lambda$) $=0.9110$. Estimated second-order intensity values for the compressed map are shown in Fig. 10.14. For distances less than 20 km, it is seen that approximately, $\lambda_{2c}(r) \propto r^{-0.486}$ in four different directions. The corresponding cluster dimension is $D_c = 1.514$. Because cluster density estimation is not affected by affine transformation, $D_c = 1.514$ for the original map as well.

The function $\lambda_2(r)$ is not centered with respect to a mean value. The so-called covariance density $C(r) = \lambda_2(r) - \lambda^2$ is the centered form of $\lambda_2(r)$. By means of standard statistical methods, $C(r)$ can be used to estimate the variance of the number of points within a rectangle of any shape (Agterberg 1993, p. 319). For

10.2 2-D Distribution Patterns of Mineral Deposits

Table 10.1 Comparison of fractal cluster model (power-law) variance with sample variances $s^2(x)$ for different (square) cell sizes (Source: Agterberg 1993, Table 1)

Cell size (km)	$S^2(x)$	Power-law estimate
5	1.48	1.450 (0.000 54)
10	13.42	13.257 (0.006 20)
20	128.42	130.903 (0.070 87)
40	111 680	1,296.609 (0.809 56)
Study area		508,777.7 (920.82)

relatively small cells, the variance can also be estimated directly from the observed data (Table 10.1). For large areas, this variance cannot be estimated directly but the power-law model can be used for this purpose. Table 10.1 (from Agterberg 1993) shows variances computed from observed data in comparison with variances based on the fractal power-law model. Standard deviations of the power-law variances are also shown. The model can be used for extrapolation to areas of any size including very large areas. Modeling of uncertainties of this type is useful in regional mineral resource potential studies.

10.2.3 Worldwide Permissive Tract Examples

Singer and Menzie (2010) have developed a three-part method for regional quantitative mineral resource assessments. The three parts are (1) delineation of permissive tracts for selected types of mineral deposits; (2) grade-and-tonnage models for these deposits; and (3) estimating the number of deposits for each type. They have compiled a table of worldwide mineral deposit density control areas for mainly (1) podiform chromite; (2) volcanogenic (Cyprus + Kuroko) massive sulphide and (3) porphyry copper deposits (Singer and Menzie 2010, Table 4.1). For deposit type per permissive area, these authors listed (a) area (km^2), (b) number of deposits, and (c) median and total tons of metal. Figure 10.15 (based on Singer and Menzie 2010, Table 4.1) shows log-log plots for the relations between deposit density and permissive area for the three selected deposit types. In a separate study, Singer and Menzie (2008) have studied the problem of map scale effects on estimating number of undiscovered deposits within permissive tracts.

If the deposits would be randomly distributed across each permissive tract; e.g., according to a Poisson model, one would expect that the three best-fitting straight lines for point density in Fig. 10.15 would be approximately horizontal because number of deposits per tract then would be proportional to tract area for each deposit type. Instead of this, the three best-fitting lines in the log density versus log permissive plots have negative slopes equal to -0.53, -0.62 and -0.61 for chromite, volcanogenic massive sulphide and porphyry copper, respectively. From the relatively large R^2 values of these least squares fits it can be concluded that these three slopes are significantly greater than zero.

Although the permissive tract approach differs from cluster density estimation, it also can be used for fractal cluster estimation. From the slopes of the lines of best fit

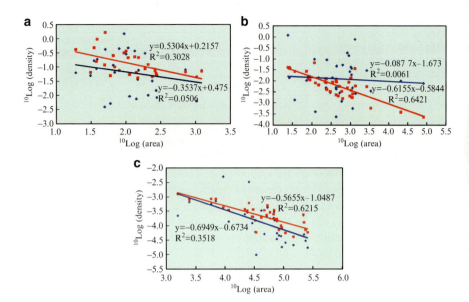

Fig. 10.15 Log-Log plots of original density versus permissive tract area for three types of deposits (original data from Singer and Menzie 2010, Table 4.1). (**a**) Podiform Cr deposits; (**b**) volcanogenic massive sulphide deposits; (**c**) porphyry copper deposits (Source: Agterberg 2013, Fig. 6)

it follows that all three deposit types satisfy a fractal cluster model with D_c equal to 1.47, 1.38 and 1.39, respectively. These results are not only remarkably similar to one another but also fairly close to those obtained previously for gold deposits in the Abitibi Volcanic Belt on the Canadian Shield with $D_c \approx 1.5$. Singer and Menzie (2010) had already established empirically that deposit density for the three types of deposits significantly decreases with size of permissive tract but they did not offer a fractal explanation. A factor probably contributing to the fact that deposit density clearly decreases linearly with tract size in this type of application is that the boundaries of the permissive areas have curved shapes determined by local geology. Roll-off effects should be negligibly small because the tracts are fairly large.

Singer and Menzie (2010, pp. 60–64) also have considered the relation between total tonnage of deposits in their Table 4.1 with area of permissive tract and deposit density simultaneously. In the Fig. 10.15 a second set of points is shown for deposits of each type weighted according to their total metal tonnage. For podiform chromite and volcanogenic sulphide deposits, the slopes of the lines fitted to the second set of points are not statistically significant but for the porphyry coppers there is an indication that size-weighted density decreases with tract area. Figure 10.16 shows that for each of the three deposit types (podiform chromite, volcanogenic massive sulphide and porphyry copper) the largest deposits approximately satisfy a Pareto size frequency distribution. Because the three data sets are not large, Quandt's (1966) method was used for parameter estimation, because it yields unbiased (consistent)

10.3 Geochemical Anomalies Versus Background

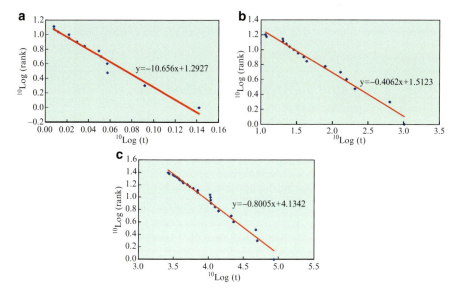

Fig. 10.16 Log-log plots of rank versus metal tonnage for the three types of deposits; the straight lines were fitted to the largest deposits only and represent Pareto distributions (Source: Agterberg 2013, Fig. 7)

estimates. The slopes in the log rank versus log tonnage plots are -10.66, -0.406 and -0.801 for podiform Cr, volcanogenic massive sulphide and porphyry copper deposits, respectively.

10.3 Geochemical Anomalies Versus Background

The main method to be considered in this section is the Concentration-Area (C-A) method, which is based on the fact that log-log plots of element concentration value versus cumulative area often show patterns that can be described by two consecutive straight-line segments (*cf.* Fig. 10.6). These may correspond to anomaly and background, respectively. There are situations, however, that a pattern consists of more than two straight-line segments, or that a curvilinear pattern provides a better fit. The C-A method was originally proposed by Cheng (1994) and also is described with applications in Cheng et al. (1994).

In addition to the C–A method, several other methods for non-linear spatial information extraction have been developed during the past 10 years to aid in the analysis of regional geochemical and geophysical map data (Cheng 1999). These methods, which will not be discussed here, include integrated spatial and spectral analysis for geochemical anomaly separation (Cheng et al. 2001), eigenvalue–eigenvector analysis of multifractal fields (Li and Cheng 2004; Cheng 2005), and use of local singularities for spatial interpolation (Cheng 2006). Most of these techniques were incorporated in the software package GeoDAS (Cheng 2003).

10.3.1 Concentration-Area (C-A) Method

Suppose that a map pattern in 2-D is fractal with dimension $D = f(\alpha)$. The reason for introducing the so-called singularity α at this point is that in multifractal modeling (Chap. 11) a fractal can be considered as a special case of a multifractal in which different element concentration values have different values of α. A multifractal is characterized by its multifractal spectrum which is a plot of $f(\alpha)$ versus α. If areal distribution of element concentration values in a study region is fractal, its multifractal spectrum degenerates into a single spike.

For small cell size $\in$, the estimated area $A(\in)$ and concentration value $\rho(\epsilon)$ can be expressed as $A(\epsilon) \propto \epsilon^{-f(\alpha)+2}; \rho(\epsilon) \propto \epsilon^{\alpha-2}$. Elimination of $\in$ gives: $A(\rho) \propto \rho^{[2-f(\alpha)]/(\alpha-2)}$ where it is assumed that $0 \leq f(\alpha) \leq 2$ and $0 \leq \alpha < 2$. For $\alpha = 2$, ρ becomes a constant that is independent of $A(\rho)$. The power-law relationship between $A(\rho)$ and ρ would plot as a straight line on log-log paper. However, in Fig. 10.6, as in many other applications of this type, a good fit only is obtained when two straight line segments are fitted instead of a single one. In Cheng et al. (1994) it is assumed that, to a first approximation, these two straight line segments represent separate populations representing "background" and "anomaly", respectively. In Sect. 10.1.2 it was discussed that the cumulative frequency distribution of gold does not provide an indication that there would be two separate populations. Thus the C-A method can provide a new tool for geochemists to identify anomalies of element concentration contour maps. For the Mitchell-Sulphurets area, the thresholds for Au and Cu fall at 400 ppb and 200 ppm, respectively (Sect. 10.1.2). In the next section, stream sediment data from the Iskut River area, northwestern British Columbia, will be analyzed using the C-A method with results that are similar to those obtained for the bedrock samples from the Mitchell Sulphurets area.

In general, fractal and multifractal modeling can produce useful information on separating anomalies from geochemical background (fractal C-A method, Cheng et al. 1994; and S-A method, Cheng 2001). At the end of this chapter (Sect. 10.4), it will be discussed that patterns such as those shown in Fig. 10.6 probably are not bifractal but can be explained as a combination of a fractal anomaly pattern that is superimposed on multifractal background with approximately lognormal element concentration frequency distribution.

10.3.2 Iskut River Area Stream Sediments Example

The Iskut River area is located in northwestern British Columbia (Fig. 10.17). In this relatively isolated area, 6 Au deposits and 177 Au mineral occurrences of

10.3 Geochemical Anomalies Versus Background

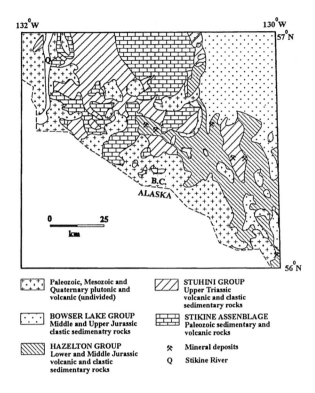

Fig. 10.17 Regional geology map of the Iskut River area according to Anderson (1993) (Source: Cheng et al. 1996, Fig. 7)

predominantly hydrothermal types have been documented (B.C. Minfile Map 104B, 1989; Anderson 1993). The area is underlain by Paleozoic and Mesozoic sedimentary, volcanic and plutonic rocks and has been subjected to low-grade regional metamorphism, heterogeneous penetrative deformation and complex fault history. Paleozoic sedimentary and volcanic rocks are mainly exposed in the central and western parts of the NTS 104B map sheet. Major rock types include greenstones, limestones, shales and clastic sedimentary rocks. Mesozoic assemblages are divided into three major groups (Anderson 1993): (1) Upper Triassic Stuhini group (volcanic and clastic sedimentary sequences); (2) Lower and Middle Jurassic Hazelton Group (volcanic and clastic sedimentary sequences); (3) Middle and Upper Jurassic Bowser Lake group (clastic sedimentary sequences) which outcrop mainly in the northeastern parts of the map sheet. Paleozoic and early Mesozoic rocks (up to Middle Jurassic) were intruded during two episodes of magmatism. Late Triassic plutonic rocks consist of I-type hornblende-biotite metadiorite, quartz monzonite and monzodiorite. Early Jurassic plutonic activity was characterized in the southwestern parts of NTS 104B by biotite-hornblende granodiorite and quartz monzodiorite intrusions. In the northeastern area alkali-feldspar-rich, biotite- or hornblende-rich syenite, quartz monzonite and alkali-feldspar porphyry intrusions predominate. Stockwork vein-type epithermal precious-metal mineralization and mesothermal base- and

Fig. 10.18 Locations of 698 stream sediment samples in the Iskut map sheet, northwestern British Columbia, part of National Geochemical Reconnaissance survey conducted by Geological Survey of Canada (Source: Cheng et al. 1996, Fig. 8)

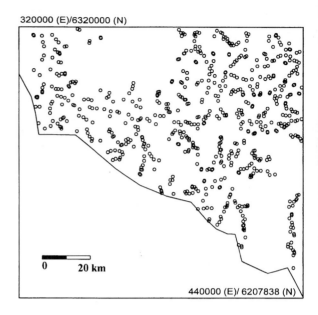

precious-metal mineralization are commonly spatially related and may be genetically associated with alkali-feldspar porphyry intrusions (Anderson 1993; Cheng, 1 994). Sedimentary and volcanic clastic rocks of both the Stuhini and Hazelton groups are favourable for Au mineralization in the area (Alldrick 1987; Anderson 1989, 1993; Cheng 1994).

In total, 698 stream sediment samples were collected from the area (Geological Survey of Canada, GSC Open file 1645), providing coverage at a reasonably uniform density, but irregularly distributed (Fig. 10.18). Samples were analyzed for a suite of trace elements including Au, Ag, Cu, Mo, As, Sb and Pb. The data were used to characterize some of the geochemically distinct rock units. Cheng (1994) showed that stream sediment elements Au, Ag, Cu, As and Sb are spatially associated with Au mineralization. Gold content of the 698 samples is shown i n a Q-Q plot (Fig. 10.19). In the present context, the goal is to delineate areas that are anomalous with respect to stream sediment Au. Results of application of the C-A method are shown in Fig. 10.20. As in the previous example of gold in the Mitchell-Sulphurets area, a model of two distinctly different straight-line segments provided a good fit suggesting that gold anomalies are superimposed on background. The study area is underlain by different rock types but this heterogeneity is not visible in the Q-Q plot for gold (Fig. 10.19). Additional results for other elements and anomaly maps are given in Cheng et al. (1996).

More recent publications in which the C-A method is applied for anomaly detection include Lima et al. (2008), Carranza (2008), Park and Chi (2008) and Zuo et al. (2009). The approach also has been extended from 2-D to 3-D (Afzal et al. 2011; Wang et al. 2013).

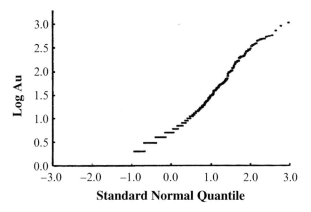

Fig. 10.19 Lognormal Q-Q plot of gold values (in ppb) of stream sediment samples at locations shown in Fig. 10.18. Secondary peak at zero in (a) is for gold values below detection limit (Source: Cheng et al. 1996, Fig. 9b)

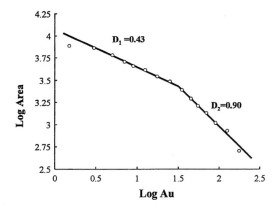

Fig. 10.20 Concentration-Area (C-A) model for Au. Areas were measured on contour map of gold values generated by Cheng et al. (1996). Notice that values fall on two distinct straight-line segments, and are interpreted as background and anomaly populations (Source: Cheng et al. 1996, Fig. 11)

10.4 Cascade Models

In a paper concerned with multifractal measures for the geoscientist, assuming that ore tonnage is equivalent to volume, Mandelbrot (1989) uses the model of de Wijs (1951) as a starting point for spatial distribution of metals in the Earth and concludes that the zinc ore "curdled" into a multifractal. The model of de Wijs will be taken as a starting point in this section. In general, it leads to lognormal distribution of metal concentrations without Pareto tails and modifications of the approach will be needed to explain separate straight-line segments commonly resulting from applications of the C-A method.

The model of de Wijs is a simple example of a binomial multiplicative cascade (*cf.* Mandelbrot 1989). The theory of cascades has been developed extensively over the past 25 years by Lovejoy and Schertzer (1991) and Schertzer et al. (1997) and other geophysicists, particularly in connection with cloud formation and rainfall (Schertzer and Lovejoy 1987; Over and Gupta 1996; Veneziano and Furcolo 2003; Veneziano and Langousis 2005). More recently, multifractal modeling of solid-Earth processes has been advanced by Qiuming Cheng and colleagues (Cheng 1994; Cheng and Agterberg 1996; Cheng 1999; Cheng 2005). In applications concerned with turbulence, the original binomial model of de Wijs has become known as the binomial p-model (Schertzer et al. 1997). It is noted several advanced cascade models in meteorology (Schertzer and Lovejoy 1991; Veneziano and Langousis 2005) result in frequency distributions that resemble the lognormal but have Pareto tails. An example of a cascade model resulting in a frequency distribution with Pareto tail will be given in Sect. 12.7.

The Pareto is characterized by a single parameter that can be related to a fractal dimension. The lognormal has two parameters (*cf.* Sect. 4.4). A basic difference between the two models is that Pareto frequency density approaches infinity in its low-value tail whereas lognormal frequency density at zero-value is zero. During the last 25 years it became increasingly clear that the Pareto often performs better than the lognormal in modeling the upper-value tails of frequency distributions in geochemistry and resource analysis. Turcotte (1997, 2002) has developed a variant of the model of de Wijs that results in a Pareto distribution, which is truncated in its lower tail. In practical applications, frequency density at zero-value always is observed to be zero. Thus the lognormal generally provides a more realistic model for modeling low-value tails. This is probably one reason why the lognormal has been preferred to the Pareto in the past. The other reason was that, in practical applications, the largest values in the upper tails of frequency distributions become increasingly rare when value goes to infinity. In goodness-of-fit tests the largest values generally are combined into a single class so that it is not possible to distinguish between lognormal and Pareto (Agterberg 1995a). In Sect. 4.4 a new method of intercept analysis was applied according to which the Pareto performs better than the lognormal for large copper deposits in the Abitibi area on the Canadian Shield.

This section is concerned with combining data from large regions. The problem of interest is how to explain and model a regional frequency distribution of element concentration values that resembles the lognormal but displays a power-law tail. Two ways to solve this problem are (1) to adopt a single-process model with an end product that is lognormal except in its upper tail that is Pareto, and (2) to consider the end product to be a mixture of two or more separate processes resulting in lognormal and Pareto distributions, respectively. Single-process models include the previously mentioned meteorological models (e.g., the beta-lognormal cascades of Veneziano and Langousis 2005). Already in the 1980s, Schertzer and Lovejoy (1985) had pointed out that the binomial p-model can be regarded as a "micro-canonical" version of their α-model in which the strict condition of local preservation of mass is replaced by the more general condition of preservation of mass within larger neighborhoods (preservation of ensemble averages). Cascades of this type can result in pure lognormals or in lognormals with Pareto tails. The applicability of such single cascade approaches to geological processes that took place within the Earth's crust remains to be investigated. With respect to mixtures of two separate cascades, a promising approach to be discussed at the end of this section consists of superimposing Turcotte's Pareto-type models on a lognormal background.

10.4 Cascade Models

One of the basic assumptions in geochemical abundance models (Brinck 1974; Harris 1984; Garrett 1986) is that trace elements are lognormally distributed. Originally, Ahrens (1953) postulated lognormality as the first law of geochemistry. In general, it cannot be assumed that the element concentration values for very small blocks of rock collected from a very large environment satisfy a single lognormal frequency distribution model. However, the lognormal model often provides a valid first approximation especially for trace elements. Reasons why the lognormal model may not be applicable include the following: Concentration values for all constituents form a closed number system and this prevents major constituents from being lognormally distributed. Also, discrete boundaries (contacts) between different rock types commonly exist in regional applications and mixtures of two or more lognormals would occur if rock types have lognormals with different parameters. Three types of generating mechanisms or explanations have been suggested to explain lognormality. The first two were previously discussed in Chap. 3. Aitchison and Brown (1957) already had reviewed processes in which random increases of value are proportional to value do result in lognormal distributions, in the same way that processes subject to conditions underlying the central limit theorem of mathematical statistics lead to normal distributions. The second type of explanation was advocated by Vistelius (1960): mixtures of populations with mean values that are proportional to standard deviations tend to result in positively skewed distributions that resemble lognormal distributions even if the original populations are normal. Thirdly, multiplicative cascade models such as the model of de Wijs can help to explain lognormality (*cf.* Agterberg 2001, 2007; Sect. 12.5.2).

Allègre and Lewin (1995) provided an overview of geochemical distributions that are either lognormal or Pareto. A relatively simple generalization of the lognormal model is to assume that the concentration values for a chemical element in a large region or 3-D environment originate from two different populations representing background and anomalies, respectively. The largest concentration values then primarily represent anomalies. This type of modeling either uses lognormal Q-Q plots (Sinclair 1991), or use is made of concentration-area (C-A) log-log plots (Sect. 10.3.1) to distinguish between two or more separate populations. Often it can be assumed (Agterberg 2007) that (a) the relatively small concentration values (background) represent a mixture of different populations, and (b) the largest values satisfy a Pareto distribution with a tail that is thicker than lognormal.

10.4.1 The Model of de Wijs

The simplest multiplicative cascade model in 1-D, 2-D or 3-D is the model of de Wijs (1951) . This model is graphically illustrated in Figs. 10.21 and 10.22. In the original model of de Wijs, any block of rock is divided into two equal parts (*cf.* Sect. 6.2). The concentration value (ξ) of a chemical element in the block then

Fig. 10.21 First stages of two-dimensional cascade model of de Wijs. Overall mean concentration value was set equal to one; d = dispersion index. Non-Random index matrix corresponds to (4 × 4) squares distribution of concentration values (Source: Agterberg 2007, Fig. 2a)

$$(1+d) \quad (1-d)$$

$$(1+d)^2 \quad (1+d)(1-d)$$
$$(1+d)(1-d) \quad (1-d)^2$$

$$\begin{array}{llll}(1+d)^4 & (1+d)^3(1-d) & (1+d)^3(1-d) & (1+d)^2(1-d)^2 \\ (1+d)^3(1-d) & (1+d)^2(1-d)^2 & (1+d)^2(1-d)^2 & (1+d)(1-d)^3 \\ (1+d)^3(1-d) & (1+d)^2(1-d)^2 & (1+d)^2(1-d)^2 & (1+d)(1-d)^3 \\ (1+d)^2(1-d)^2 & (1+d)(1-d)^3 & (1+d)(1-d)^3 & (1-d)^4\end{array}$$

40	31	31	22		04	13	22	31
31	22	22	13		13	22	13	22
31	22	22	13		13	22	40	31
22	13	13	04		22	31	31	22

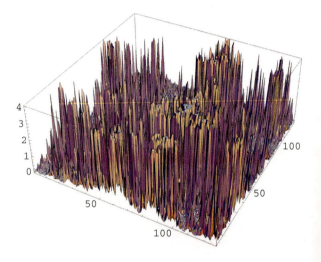

Fig. 10.22 Realization of model of de Wijs (see Fig. 10.21 in 2-D for $d = 0.4$ and $N = 14$). Overall average value is equal to 1. Values greater than 4 were truncated (Source: Agterberg 2007, Fig. 3)

can be written as $(1 + d) \times \xi$ for one half and $(1 - d) \times \xi$ for the other half so that total mass is preserved. The coefficient of dispersion d is assumed to be independent of block size. This approach can be modified by replacing d by a random variable (random-cut model; Sect. 12.3.1). Figure 10.21 illustrates the original model of de Wijs: any cell containing a chemical element in 1-, 2-, or 3- dimensional space is divided into two halves with element concentration values $(1+d) \times \xi$ and $(1 - d) \times \xi$. For the first cell at the beginning of the process, ξ can be set equal to unity. This implies that all concentration values are divided by their overall regional average concentration value (μ). The index of dispersion (d) is independent of cell-size. In 2-D space, two successive subdivisions into quarters result in

10.4 Cascade Models

4 and 16 cells with concentration values as shown in Fig. 10.21. The maximum element concentration value after k subdivisions is $(1+d)^k$ and the minimum value is $(1-d)^k$; k is kept even in 2-D applications in order to preserve mass but the frequency distribution of all concentration cannot be distinguished from that arising in 1-D or 3-D applications of this multiplicative cascade model.

In a random cascade, larger and smaller values are assigned to cells using a discrete random variable. Multifractal patterns generated by a random cascade have more than a single maximum. The frequency distribution of the element concentrations at any stage of this process is called "logbinomial" because logarithmically transformed concentration values satisfy a binomial distribution. The logbinomial converges to a lognormal distribution although its upper and lower value tails do remain weaker than those of the lognormal (Agterberg 2007). Notation can be simplified by using indices that are powers of $(1+d)$ and $(1-d)$, respectively; for the example of Fig. 10.21, $(1+d)^3 (1-d)$ is written as 31 in the 16-cell matrix on the left in the next row. If at each stage of subdivision, the location of higher and lower concentration cells is determined by a Bernoulli-type random variable, the arrangement of cells may become as shown in the 16-cell matrix on the right. Because of its property of self-similarity, the model of de Wijs was recognized to be a multifractal by Mandelbrot (1983, 1989) who adopted this approach for applications to the Earth's crust.

Figure 10.22 shows a 2-D logbinomial pattern for $d=0.4$ and $k=14$. Increasing the number of subdivisions for the model of de Wijs (as in Fig. 10.21) to 14 resulted in the 128×128 pattern shown in Fig. 10.22 in which values greater than 4 were truncated to more clearly display most of the spatial variability. The frequency distribution of all 214 values is logbinomial and approximately lognormal except in the highest-value and lowest-value tails that are thinner than lognormal (also see Sect. 12.4). When the number of subdivisions becomes large, the end product cannot be distinguished from that of multiplicative cascade models in which the dispersion index D is modeled as a continuous random variable with mathematical expectation equal to 1 instead of the Bernoulli variable allowing the values $+d$ and $-d$ only (Sect. 12.5). The lognormal model often provides good first approximations for regional background distributions of trace elements.

10.4.2   The Model of Turcotte

Figure 10.23 shows Turcotte's variant of the model of de Wijs: After each subdivision, only the half with larger concentration in further subdivided into halves with concentration values equal to $(1+d) \times \xi$ and $(1-d) \times \xi$. This simplifies the process as illustrated for 16 cells in 2-D space. At each stage of this process the concentration values have a Pareto-type frequency distribution. In analogy with Turcotte's (1997) derivation for blocks in 3-D space, it can be shown that a fractal dimension equal to $D = 2 \times \log_2 (1+d)$ can be defined for this process. The final element concentration map has only one maximum value contrary to patterns generated by

Fig. 10.23 Turcotte's variant of model of de Wijs as shown in Fig. 10.21 (Source: Agterberg 2007, Fig. 2b)

$(1+d)$ $(1-d)$

$(1+d)^2$ $(1-d)$

$(1+d)(1-d)$ $(1-d)$

$(1+d)^4$ $(1+d)^2(1-d)$ $(1-d)$ $(1-d)$
$(1+d)^3(1-d)$ $(1+d)^2(1-d)$ $(1-d)$ $(1-d)$
$(1+d)(1-d)$ $(1+d)(1-d)$ $(1-d)$ $(1-d)$
$(1+d)(1-d)$ $(1+d)(1-d)$ $(1-d)$ $(1-d)$

40	21	01	01		11	11	01	01
31	21	01	01		11	11	01	01
11	11	01	01		01	01	40	21
11	11	01	01		01	01	31	21

the model of de Wijs that have many local maxima. Figure 10.23 is a mosaic of four patterns resulting from Turcotte "fractal" cascades with $d=0.4$ and $k=12$; vertical scale is logarithmic (base 10). Contrary to the multimodal logbinomial patterns, the Turcotte fractal cascade develops a single peak. However, the same Turcotte cascade could have been operative in different parts of a study area. If the index of dispersion (d) remained the same for all separate cascades, the combined frequency distribution after many subdivisions for each cascade would satisfy a single Pareto distribution plotting as a straight-line with slope determined by d.

Turcotte's cascade model is a modification of the original multifractal-generating cascade (Sect. 10.4.1). Only cells with largest concentration value during a previous subdivision are further subdivided into parts with different element concentration values. It is assumed that the same type of cascade was operational at n different random locations generating a pattern with n maxima. The frequency distribution of the concentration values then would remain the same except for enlargement of all frequencies by the factor n. Element concentrations generated by Turcotte's cascade satisfy a Pareto distribution which is associated with a fractal instead of a multifractal. The slope β of the straight-line representing this Pareto distribution on log-log paper satisfies $\beta = -1/\log_2(1+d)$. Figure 10.25 is a C-A diagram for the Turcotte model with $d=0.4$ and $k=14$. Consequently, $\beta = -2.060$.

10.4.3 Computer Simulation Experiments

The following computer simulation experiments (from Agterberg 2007) illustrate that an unbiased estimate of the Pareto parameter can be obtained in the hypothetical situation of a study area where background satisfies the model of de Wijs with overall average concentration value set equal to 0.1 and $d=0.3$. Suppose that one or

10.4 Cascade Models

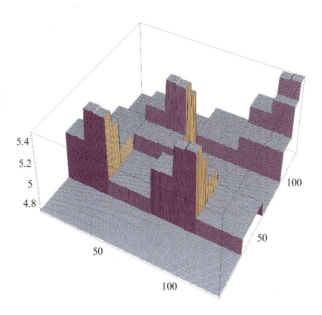

Fig. 10.24 Four 2-D realizations of Turcotte's fractal model (Fig. 10.23) for $d=0.4$ and $N=12$. Total area was subdivided into four quadrants. Vertical scale is for $5+\log_{10}$ (Value) (Source: Agterberg 2007, Fig. 4)

more Turcotte cascades with overall average value equal to 1.0 and $d=0.4$ (cf. Fig. 10.24) are superimposed on this background. Figure 10.26 is a C-A diagram for this hypothetical situation. The pattern in this diagram is approximately linear with slope of approximately -2. It was obtained by random sampling of concentration values resulting from de Wijs and Turcotte cascades combining the resulting values with one another.

Suppose now that the Turcotte cascades were operational in only 25 % of the area. Use of Cheng's (2003) method of piecemeal fitting of successive straight-line segments in a C-A diagram indicates that the largest concentration values approximately fall on a straight line with slope equal to -2 as in Figs. 10.25 and 10.26. Consequently, the estimated value of d is 0.4. These experiments (Figs. 10.27 and 10.28) illustrate that unbiased estimates of this type can be obtained irrespective of how many Turcotte cascades were operative in the area or how much of the study area consists of approximately lognormal background without anomalies. Figure 10.28 is a plot of the frequency density values for the second experiment. The smaller peak on the right corresponds to the line-segment for largest concentration values in Fig. 10.27.

A practical example of dispersion index (d) estimation for Turcotte's model is as follows. Suppose a measure μ of the amount of a chemical element in a square cell measuring ϵ km on a side satisfies $\mu = c \cdot \epsilon^\alpha$ where c is a constant, and α is the singularity also known as Hölder exponent (Mandelbrot 1989; Evertsz and Mandelbrot 1992); then α can be estimated by measuring the slope of the straight line in a log-log plot of μ against $\in$. In this 2-D application, $\mu = \xi \cdot \epsilon^2$ where $\xi = c \cdot \epsilon^{\alpha-2}$ represents average element concentration value in the cell. If element concentration values for samples taken at the surface of a study area are realizations of a stationary random variable with constant population mean, then $\alpha=2$ represents non-singularity. "Singular" locations (where $\alpha < 2$) may indicate anomalous

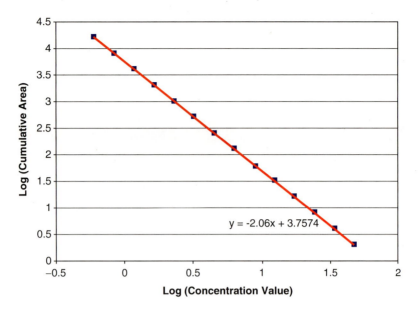

Fig. 10.25 C-A plot for Turcotte model with $d = 0.4$ and $N = 14$ (log base 10) (Source: Agterberg 2007, Fig. 7)

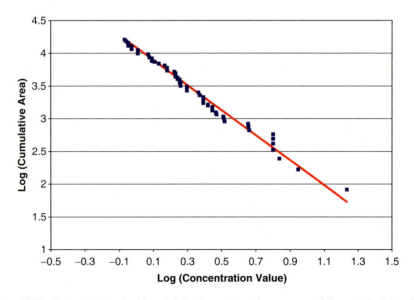

Fig. 10.26 C-A plot for logbinomial background with mean $\mu = 0.1$ and $d = 0.5$ with superimposed Turcotte anomalies with $\mu = 1.0$ and $d = 0.4$ (log base 10) (Source: Agterberg 2007, Fig. 8)

10.4 Cascade Models

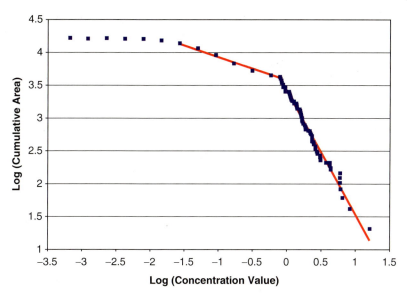

Fig. 10.27 C-A plot for lognormal background with superimposed Turcotte anomalies restricted to 25 % of total study area (Source: Agterberg 2007, Fig. 9)

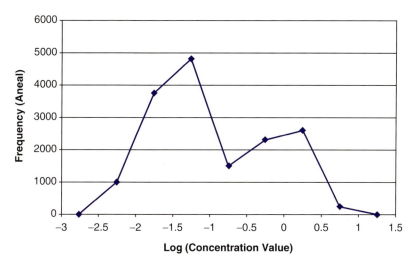

Fig. 10.28 Frequencies of lognormal-Pareto mixture of Fig. 10.27 (Source: Agterberg 2007, Fig. 10).

enrichment of the chemical element. Examples of local singularity mapping were given in Cheng (2007) and Cheng and Agterberg (2009) for various elements in stream sediments from the Gejiu area, Yunnan Province, China. The topic of singularity analysis will be discussed in more detail in the next chapter.

Like several other chemical elements in the Gejiu area, arsenic in the surficial deposits that were sampled shows two types of anomalies. The arsenic local singularity map for arsenic shows many relatively small anomalies (where $\alpha < 2$) across the entire Gejiu area (Fig. 11.25, also to be discussed later). A significant fraction of these anomalies is spatially correlated with occurrences of (mined and unmined) mineral deposits. The arsenic concentration values are highest in the eastern part of the area. Together the highest values describe a large irregularly shaped anomaly that is probably caused by mining activities restricted to this part of the area.

Figure 11.26g, h will show frequencies of As local singularities and As concentration values, respectively. Suppose that parameters describing the two preceding anomaly types are identified by the subscripts 1 (for local singularity anomalies) and 2 (for the high-concentration anomaly). From $\beta_2 = -3.0178$ (estimated slope of best-fitting line in Fig. 11.26h) it follows immediately that $d_2(As) = 0.258$. From $\xi = c \cdot \epsilon^{\alpha-2}$ with $\epsilon = 2$ km in this application, it follows that estimated slope of straight-line in Fig. 11.26g ($= -2.6474$) provides an estimate of $\beta_1 = -8.7945$. Consequently, $d_1(As) = 0.082$. Suppose that β is a parameter estimated by the slope of the best-fitting straight-line on a C-A plot. Then this estimate can be converted into either the fractal dimension D ($= -2/\beta$) or into the index of dispersion d ($= 2^{-D/2} - 1$) characterizing the non-linear process. In terms of Fig. 10.21: if a block with high-concentration value (ξ) is divided into two halves, the concentration values of the halves are, on average, equal to $(1+d) \times \xi$ and $(1-d) \times \xi$, respectively. Thus a higher index of dispersion means stronger spatial variability. The small anomalies (where $\alpha < 2$) with $d_1(As) = 0.082$ have lower dispersion index than the broad regional anomaly restricted to the eastern part of the area with $d_2(As) = 0.258$. Several other elements (tin, copper, silver, gold, cadmium, cobalt, iron, nickel lead, and zinc) show anomalies similar to those for arsenic (Cheng and Agterberg 2009). The first type (local singularities) is useful for exploration because it provides indicators for buried ore-bodies. The second type helps to describe regional pollution due to mining activities. The shapes of two kinds of anomalies (1 and 2) are markedly different, and this probably is the main reason that a clear distinction could be made between the two underlying enrichment processes (proximity to buried mineral deposits and pollution due to mining activities) in this example of application.

References

Afzal P, Alghalandis YF, Khakzad A, Moarefvand P, Omran NR (2011) Delineation of mineralization zones in porphyry Cu deposits by fractal concentration-volume modeling. J Geochem Explor 108:220–232

Agterberg FP (1980) Mineral resource estimation and statistical exploration. In: Miall AD (ed) Facts and principles of world oil occurrence, Canadian Society of Petroleum Geologists Memoir 6. Canadian Society of Petroleum Geologists, Calgary, pp 301–318

Agterberg FP (1981) Geochemical crustal abundance models. Trans Soc Min Eng AIME 268:1823–1830

Agterberg FP (1993) Calculation of the variance of mean values for blocks in mineral potential mapping. Nonrenew Resour 2:312–324

Agterberg FP (1994) FORTRAN program for the analysis of point patterns with correction for edge effects. Comput Geosci 20(2):229–245

Agterberg FP (1995a) Multifractal modeling of the sizes and grades of giant and supergiant deposits. Int Geol Rev 37:1–8

Agterberg FP (1995b) Power laws versus lognormal models in mineral exploration. In: Proceedings of CAMI'95, 3rd Canadian conference on computer applications in the mineral industries, Montreal, pp 17–26

Agterberg FP (2001) Multifractal simulation of geochemical map patterns. In: Merriam D, Davis JC (eds) Geologic modeling and simulation: sedimentary systems. Kluwer, New York, pp 327–346

Agterberg FP (2007) Mixtures of multiplicative cascade models in geochemistry. Nonlinear Process Geophys 14:201–209

Agterberg FP (2012) Sampling and analysis of element concentration distribution in rock units and orebodies. Nonlinear Process Geophys 19:23–44

Agterberg FP (2013) Fractals and spatial statistics of point patterns. J Earth Sci 24(1):1–11

Agterberg FP, Divi SR (1978) A statistical model for the distribution of copper, lead and zinc in the Canadian Appalachian region. Econ Geol 73:230–245

Agterberg FP, Chung CF, Fabbri AG, Kelly AM, Springer JS (1972) Geomathematical evaluation of copper and zinc potential of the Abitibi area, Ontario and Quebec, Geological Survey of Canada Paper 71-41. Department of Energy, Mines and Resources, Ottawa

Agterberg FP, Cheng Q, Wright DF (1993) Fractal modelling of mineral deposits. In: Proceedings of APCOM XXIV, international symposium on the applications of computers and operations research in the mineral industries, Canadian Institute of Mining, Metallurgy and Petroleum, Montreal, vol 1, pp 3–53

Ahrens LH (1953) A fundamental law of geochemistry. Nature 712:1148

Aitchison J, Brown JAC (1957) The lognormal distribution. Cambridge University Press, Cambridge

Alldrick DJ (1987) Geology and mineral deposits of the Salmon River Valley, Stewart area, NTS 104A and 104B. British Columbia Ministry of Energy, Mines and Petroleum Resources, Geological Survey Branch, Open File Map 1987-22

Allègre CJU, Lewin E (1995) Scaling laws and geochemical distributions. Earth Planet Sci Lett 132:1–3

Anderson RG (1989) A stratigraphic, plutonic, and structural framework for the Iskut River map area, northwestern British Columbia. Geol Surv Can Pap 89-1E:145–154

Anderson RG (1993) Mesozoic stratigraphic and plutonic framework for northwestern Stikinia (Iskut River area), northwestern British Columbia, Canada. In: Dunne C, McDougall KA (eds) Mesozoic paleogeography of the western United States, II, Pacific Section 71. Society of Economic Paleontologists and Mineralogists, Los Angeles, pp 477–494

Baddeley A, Turner R (2012) Package 'Spatstat', manual, version 1.30.0 (Released 2012-12-23) PDF. http://140.247.115.171

Baddeley A, Bárány I, Scheicher R (2006) Spatial point processes and their applications. Lect Notes Math 192:1–75

Baddeley A, Møller J, Poles AG (2008) Properties of residuals for spatial point processes. Ann Inst Stat Math 60(3):627–649

Bak P (1996) The science of self-organised criticality. Copernicus Press, New York

Barnsley M (1988) Fractals everywhere. Academic Press, Boston

Barton CC, La Pointe PR (1995) Fractals in the earth sciences. Plenum, New York

Blenkinsop TG (1995) Fractal measures for size and spatial distributions of gold mines: economic implications. In: Blenkinsop TG, Tromp PL (eds) Sub-Saharan economic geology, Special Publication Geological Society of Zimbabwe, 3. A.A. Balkema, Rotterdam, pp 177–186

Brinck JW (1974) The geochemical distribution of uranium as a primary criterion for the formation of ore deposits. In: Chemical and physical mechanisms in the formation of uranium mineralization, geochronology, isotope geology and mineralization. International Atomic Energy Agency Proceedings Series STI/PUB/374, International Atomic Energy Agency, Vienna, pp 21–32

Burnett AI, Adams, Vienna, KC (1977) A geological, engineering and economic study of a portion of the Lloydminster Sparky Pool, Lloydminster, Alberta. Bull Can Petr Geol 25(2):341–366

Cargill SM (1981) United States gold resource profile. Econ Geol 76:937–943

Cargill SM, Root DH, Bailey EH (1980) Resources estimation from historical data: mercury, a test case. J Int Assoc Math Geol 12:489–522

Cargill SM, Root DH, Bailey EH (1981) Estimating unstable resources from historical industrial data. Econ Geol 76:1081–1095

Carlson A (1991) Spatial distribution of ore deposits. Geology 19(2):111–114

Carranza EJM (2008) Geochemical anomaly and mineral prospectivity mapping in GIS, Handbook of Exploration and Environment Geochemistry 11. Elsevier, Amsterdam

Cheng Q (1994) Multifractal modeling and spatial analysis with GIS: gold mineral potential estimation in the Mitchell-Sulphurets area, northwestern British Columbia. Unpublished doctoral dissertation, University of Ottawa

Cheng Q (1999) Multifractality and spatial statistics. Comput Geosci 25:949–961

Cheng Q (2001) Selection of multifractal scaling breaks and separation of geochemical and geophysical anomaly. J China Univ Geosci 12(1):54–59

Cheng Q (2003) GeoData Analysis System (GeoDAS) for mineral exploration and environmental assessment, user's guide. York University, Toronto

Cheng Q (2005) A new model for incorporating spatial association and singularity in interpolation of exploratory data. In: Leuangthong D, Deutsch CV (eds) Geostatistics Banff 2004. Springer, Dordrecht, pp 1017–1025

Cheng Q (2006) GIS based fractal and multifractal methods for mineral deposit prediction. In: Leuangthong D, Deutsch CV (eds) Geostatistics Banff 2004. Springer, New York, pp 1017–1026

Cheng Q (2007) Mapping singularities with stream sediment geochemical data for prediction of undiscovered mineral deposits in Gejiu, Yunnan Province, China. Ore Geol Rev 32:314–324

Cheng Q (2008) Non-linear theory and power-law models for information integration and mineral resources quantitative assessments. In: Bonham-Carter G, Cheng Q (eds) Progress in geomathematics. Springer, Heidelberg, pp 195–225

Cheng Q (2012) Multiplicative cascade processes and information integration for predictive mapping. Nonlinear Process Geophys 19:1–12

Cheng Q, Agterberg FP (1995) Multifactal modeling and spatial point processes. Math Geol 27(7):831–845

Cheng Q, Agterberg FP (1996) Multifractal modelling and spatial statistics. Math Geol 28(1):1–16

Cheng Q, Agterberg FP (2009) Singularity analysis of ore-mineral and toxic trace elements in stream sediments. Comput Geosci 35:234–244

Cheng Q, Agterberg FP, Ballantyne SB (1994) The separation of geochemical anomalies from background by fractal methods. J Geochem Explor 51(2):109–130

Cheng Q, Agterberg FP, Bonham-Carter GF (1996) A spatial analysis method for geochemical anomaly separation. J Geochem Explor 56:183–195

Cheng Q, Xu Y, Grunsky EC (2001) Integrated spatial and spectrum analysis for geochemical anomaly separation. Nat Resour Res 9:43–51

Cressie NAC (2001) Statistics for spatial data. Wiley, New York

Crovelli RA (1995) The generalized 20/80 law using probabilistic fractals applied to petroleum field size. Nonrenew Resour 4(3):233–241

De Wijs HJ (1951) Statistics of ore distribution I. Geol Mijnbouw 13:365–375

Diggle PJ (1983) Statistical analysis of spatial point patterns. Academic, London

Drew LJ, Schuenemeyer JH, Bawiee WJ (1982) Estimation of the future rates of oil and gas discoveries in the western Gulf of Mexico, US Geological Survey Professional Paper, 1252. U.S. GPO, Washington, DC

References

Evertsz CJG, Mandelbrot BB (1992) Multifractal measures. In: Peitgen H-O, Jurgens H, Saupe D (eds) Chaos and fractals. Springer, New York

Falconer K (2003) Fractal geometry, 2nd edn. Wiley, Chichester

Feder J (1988) Fractals. Plenum, New York

Ford A, Blenkinsop TG (2009) An expanded de Wijs model for multifractal analysis of mineral production data. Miner Deposita 44(2):233–240

Fowler A, L'Heureux I (1996) Self-organized banded sphalerite and banding galena. Can Mineral 34(6):1211–1222

Freeden W, Schreiner M (2008) Spherical functions of mathematical geosciences. Springer, Heidelberg

Freiling EC (1966) A comparison of the fallout mass-size distributions calculated by lognormal and power-law models, U.S. Naval Radiology Defense Laboratory Rep TR-1105. The Laboratory, San Francisco

Garrett RG (1986) Geochemical abundance models: an update, 1975 to 1987, U.S. Geol Surv Circ 980:207–220

Goehring L, Mahadevan L, Morris SW (2009) Nonequilibrium scale selection mechanism for columnar jointing. Proc Natl Acad Sci U S A 106(2):387–392

Gupta VK, Troutman B, Dawdy D (2007) Towards a nonlinear geophysical theory of floods in river networks: an overview of 20 years of progress. In: Tsonis AA, Elsner JB (eds) Nonlinear dynamics in geosciences. Springer, New York, pp 121–150

Harris DP (1984) Mineral resources appraisal. Clarendon, Oxford

Heiskanen WA, Vening Meinesz FA (1958) The Earth and its gravity field. McGraw-Hill, New York

Kaye BH (1989) A random walk through fractal dimensions. VCH Publishers, New York

Korvin G (1992) Fractal models in the earth sciences. Elsevier, Amsterdam

Li Q, Cheng Q (2004) Fractal singular value decomposition and anomaly reconstruction. Earth Sci 29(1):109–118 (in Chinese with English abstract)

Lima A, Plant JA, De Vivo B, Tarvainen T, Albanese S, Cicchella D (2008) Interpolation methods for geochemical maps: a comparative study using arsenic data from European stream waters. Geochem Explor Environ Anal 8:41–48

Lovejoy S, Schertzer D (1991) Multifractal analysis techniques and the rain and cloud fields from 10^{-3} to 10^6 m. In: Schertzer D, Lovejoy S (eds) Non-linear variability in geophysics. Kluwer, Dordrecht, pp 111–144

Lovejoy S, Schertzer D (2007) Scaling and multifractal fields in the solid earth and topography. Nonlinear Process Geophys 14:465–502

Lovejoy S, Schertzer D (2013) The weather and climate. Cambridge University Press, Cambridge

Lovejoy S, Agterberg F, Carsteanu A, Cheng Q, Davidsen J, Gaonac'h H, Gupta V, L'Heureux I, Liu W, Morris SW, Sharma S, Shcherbakov R, Tarquis A, Turcotte D, Uritsky V (2010) Nonlinear geophysics: why we need it. EOS 90(48):455–456

Malamud BD, Morein G, Turcotte DL (1998) Forest fires: an example of self-organized critical behavior. Science 281:1840–1842

Mandelbrot BB (1975) Les objects fractals: forme, hazard et dimension. Flammarion, Paris

Mandelbrot BB (1977) Fractals, form, chance and dimension. Freeman, San Francisco

Mandelbrot BB (1983) The fractal geometry of nature. Freeman, San Francisco

Mandelbrot BB (1989) Multifractal measures, especially for the geophysicist. Pure Appl Geophys 131:5–42

Mandelbrot BB (1995) The statistics of natural resources and the law of Pareto. In: Barton CC, La Pointe PR (eds) Fractals in petroleum geology and the earth sciences. Plenum, New York, pp 1–12

Mandelbrot BB (1999) Multifractals and $1/f$ noise. Springer, New York

Over TM, Gupta VK (1996) A space-time theory of mesoscale rainfall using random cascades. J Geophys Res 101:26319–26331

Park N-W, Chi K-H (2008) Quantitative assessment of landslide susceptibility using high-resolution remote sensing data and a generalized additive model. Int J Remote Sens 29(1):247–264

Perrin J (1913) Les atomes. NRF-Gallimard, Paris

Pickering G, Bul JM, Sanderson DJ (1995) Sampling power-law distributions. Tectonophysics 248:1–20

Prey A (1922) Darstellung der Höhen- und Tiefen-Verhältnisse der Erde durch eine Entwicklung nach Kugelfunctionen bis zur 16. Ordnung. Nachr Ak Wiss Göttingen Math Phys Kl 11(1):1–29

Quandt RE (1966) Old and new methods of estimation and the pareto distribution. Metrika 10:55–58

Raines GL (2008) Are fractal dimensions of the spatial distribution of mineral deposits meaningful? Nat Resour Res 17:87–97

Ripley BD (1976) The second-order analysis of stationary point processes. J Appl Probab 13:255–266

Ripley BD (1981) Spatial statistics. Wiley-Interscience, New York

Ripley BD (1987) Point processes for the earth sciences. In: Chung CF, Fabbri AG, Sinding-Larsen R (eds) Quantitative analysis of mineral and energy resources. Reidel, Dordrecht, pp 301–322

Ripley BD (1988) Statistical inference for spatial processes. Cambridge University Press, Cambridge

Rowlingson BS, Diggle PJ (1991) Estimating the K-function for a univariate point process on an arbitrary polygon, Lancaster University Math Dep Tech Rep MA91/58. Lancaster University, Lancaster, pp 1–15

Rowlingson BS, Diggle PJ (1993) SPLANCS: spatial point pattern analysis code in S-Plus. Comput Geosci 19(5):627–655

Rundle JB, Turcotte DL, Shcherbakov R, Klein W, Sammis C (2003) Statistical physics approach to understanding the multiscale dynamics of earthquake fault systems. Rev Geophys 41:1019

Schertzer D, Lovejoy S (1985) The dimension and intermittency of atmospheric dynamics multifractal cascade dynamics and turbulent intermittency. In: Launder B (ed) Turbulent shear flow 4. Springer, New York, pp 7–33

Schertzer D, Lovejoy S (1987) Physical modeling and analysis of rain and clouds by anisotropic scaling of multiplicative processes. J Geophys Res 92:9693–9714

Schertzer D, Lovejoy S (1991) Non-linear geodynamical variability: multiple singularities, universality and observables. In: Schertzer D, Lovejoy S (eds) Non-linear variability in geophysics. Kluwer, Dordrecht, pp 41–82

Schertzer D, Lovejoy S, Schmitt F, Chigirinskaya Y, Marsan D (1997) Multifractal cascade dynamics and turbulent intermittency. Fractals 5:427–471

Sharma AS (1995) Assessing the magnetosphere's nonlinear behavior: its dimension is low, its predictability high. Rev Geophys 33:645

Sinclair AJ (1991) A fundamental approach to threshold estimation in exploration geochemistry: probability plots revisited. J Geochem Explor 41(1):1–22

Singer DA, Menzie WD (2008) Map scale effects on estimating the number of undiscovered mineral deposits. In: Bonham-Carter G, Cheng Q (eds) Progress in geomathematics. Springer, Heidelberg, pp 271–283

Singer DA, Menzie WD (2010) Quantitative mineral resource assessments. Oxford University Press, New York

Stanley H, Meakin P (1988) Multifractal phenomena in physics and chemistry. Nature 335:405–409

Steinhaus H (1954) Length, shape and area. Colloq Math 3:1–13

Stoyan D, Kendall WS, Mecke J (1987) Stochastic geometry and its applications. Wiley, Chichester

Turcotte DL (1997) Fractals and chaos in geology and geophysics, 2nd edn. Cambridge University Press, Cambridge

Turcotte DL (2002) Fractals in petrology. Lithos 65(3–4):261–271

References

Uritsky VM, Donovan E, Klimas AJ (2008) Scale-free and scale-dependent modes of energy release dynamics in the night time magnetosphere. Geophys Res Lett 35(21):L21101, 1–5

Veneziano D, Furcolo P (2003) Marginal distribution of stationary multifractal measures and their Haar wavelet coefficients. Fractals 11(3):253–270

Veneziano D, Langousis A (2005) The maximum of multifractal cascades: exact distribution and approximations. Fractals 13(4):311–324

Vening Meinesz FA (1951) A remarkable feature of the earth's topography. Proc K Ned Akad Wet Ser B Phys Sci 54:212–228

Vening Meinesz FA (1964) The earth's crust and mantle. Elsevier, Amsterdam

Vistelius AB (1960) The skew frequency distribution and the fundamental law of the geochemical processes. J Geol 68:1–22

Walsh J, Watterson J, Yielding G (1991) The importance of small-scale faulting in regional extension. Nature 351:391–393

Wang Q, Deng J, Zhao J, Li N, Wan L (2013) The fractal relationship between orebody tonnage and thickness. J Geochem Explor 122:4–8

Zuo RG, Cheng QM, Xia QL, Agterberg FP (2009) Application of fractal models to distinguish between different mineral phases. Math Geosci 41:71–80

Chapter 11
Multifractals and Local Singularity Analysis

Abstract Multifractals are spatially intertwined fractals. For example, a chemical concentration value obtained from rock samples in a study area may be a fractal with fractal dimension different from those of other concentration values for the same element, but together the fractal dimensions may form a multifractal spectrum $f(\alpha)$ that is a continuous function of the singularity α, which depends on the concentration value. Self-similar patterns produce multifractals of this type. If a block of rock with chemical concentration value X is divided into equal parts, the halves have concentration values $(1+d) \cdot X$ and $(1-d) \cdot X$ where d is a constant. The model of de Wijs assumes that the dispersion index d is independent of block size. This cascade process produces a multifractal. The properties of a multifractal can be estimated by the method of moments or by the histogram method. The four-step method of moments has the advantage that the assumption of multifractality is being tested during its first step because this should produce an array of straight lines on a log-log plot of the spatial mass-partition function $\chi\left(\epsilon,q\right)$ against measure of block size (ϵ) used. It should be kept in mind, however, that the multifractal spectrum estimated by the method of moments is primarily determined by the majority of measurements that are clustered around the mean or median. Very large or very small observed values are rare; because of this, the low-singularity and high-singularity tails of a multifractal spectrum generally cannot be estimated with sufficient precision. Because of strong local autocorrelations effects, singularity analysis can provide better estimates of singularities associated with the very large or very small observed values. Practical examples of multifractal modeling include the distribution of gold in the Mitchell-Sulphurets area, northwestern British Columbia, uranium resources in the U.S. and worldwide, lengths of surface fractures in the Lac du Bonnet Batholith, eastern Manitoba, geographic distribution of gold deposits in the Iskut map area, British Columbia. Local singularity mapping is useful for the detection of geochemical anomalies characterized by local enrichment even if contour maps for representing average variability are not constructed. Examples include singularity maps based on various element concentration values from stream sediment samples and their relation to tin deposits in the Gejiu area, Yunnan Province, as well as Ag and Pb-Zn deposits in northwestern Zhejiang

Province, China. The iterative Chen algorithm for space series of element concentration values offers a new way to separate local singularities from regional trends. This technique is applied to the Pulacayo zinc and KTB copper values.

Keywords Multifractals • Self-similarity • Method of moments • Multifractal spatial correlation • Line segments • Point patterns • Local singularity analysis • Chen algorithm • Worldwide uranium deposits • Mitchell-Sulphurets mineral district • Lac du Bonnet Batholith • Iskut River gold deposits • Gejiu mineral district • Zhejiang Pg-Zn deposits • KTB copper variability

11.1 Self-Similarity

Both fractals and multifractals often can be explained as a consequence of self-similarity. This simply means that the physical or chemical laws that controlled patterns and spatial variability in rocks at one scale also controlled patterns and spatial variability at other scales. Thus self-similarity implies scale-independence. Of course, the spatial extent of self-similarity generally is limited at both longer and shorter sampling intervals because other processes with different laws become pre-dominant at different scales.

In Sect. 10.4.1 it was discussed that, if ξ represents average concentration value of a chemical element in a very large block of rock (e.g. the upper part of the Earth's crust), division of this block into halves results in two smaller blocks with element concentration values equal to $(1+d)\,\xi$ and $(1-d)\,\xi$, respectively. The ratio of these two different values is $\eta = (1+d)/(1-d)$. After n successive subdivisions (for large n), there are 2^n small blocks with concentration values ranging from $(1-d)^n\xi$ to $(1+d)^n\xi$. The difference between logarithmically (base e) transformed higher and lower values resulting from subdividing any block into halves is $\log_e \eta$. The frequency distribution of the final set of 2^n concentration values is logbinomial. When d is relatively large, the logbinomial becomes positively skewed with a thin, long high-value tail. The lower and higher values generated at successive iterations can have random spatial locations with respect to one another.

The model of de Wijs provides a simple starting point for parametric spatial modeling. De Wijs (1951) already derived the following equation for the standard deviation of logarithmically transformed element concentration values:

$$\sigma^2 = \frac{n}{4}(\ln \eta)^2$$

where σ^2 is the logarithmic variance (base e) of the concentration values. This equation (to be referred to later as the variance formula of de Wijs) follows directly from the well-known equation for the variance of a binomial distribution with $p = \frac{1}{2}$ that is equal to $n/4$, taking account of the fact that the spacing between

11.1 Self-Similarity

ordered values along the logarithmic scale is equal to $\ln \eta$. It can be rewritten in the form (cf. Matheron 1962, p. 309):

$$\sigma^2 = \alpha \cdot \ln \frac{V}{v}$$

where v represents the volume for which the concentration value is determined; V is a larger volume in which v occupies a random position. In one-dimensional applications, the volumes v and V are reduced to line segments, and $V/v = 2^n$. The constant α satisfies:

$$\alpha = \frac{1}{4\ln 2}[\ln \eta]^2.$$

According to the De Moivre-Laplace theorem (Bickel and Doksum 2001, p. 470), the frequency distribution of $\ln \eta$ converges to normal form when n increases. Frequency density values in the upper tail of the logbinomial are less than those of the lognormal. The logbinomial would become lognormal when n representing the number of subdivisions of blocks is increased indefinitely. Paradoxically, its variance then also would become infinitely large. In practical applications, it is often seen that the upper tail of a frequency density function of element concentration values is not thinner but thicker and extending further than a lognormal tail. Several cascade models (see, e.g., Schertzer and Lovejoy 1991; Veneziano and Furcolo 2003; Veneziano and Langousis 2005) result in frequency distributions that resemble the lognormal but have Pareto tails.

Box 11.1: Self-Similarity and Power Laws

Korvin (1992) offers a heuristic exposition of the idea that self-similarity necessarily leads to power-law type relations. A more rigorous approach can be found in Aczél and Dhombres (1989). Arguing along the same lines as Korvin, suppose that μ_1, μ_2, and μ_3 are the measures of a fractal set with "singularity" α in three cells with different sizes labeled ϵ_1, ϵ_2 and ϵ_3, respectively. Self-similarity would imply that the ratio of the measures for two cells depends on the ratio of their sizes only, or: $\frac{\mu_1}{\mu_2} = f\left(\frac{\epsilon_1}{\epsilon_2}\right); \frac{\mu_2}{\mu_3} = f\left(\frac{\epsilon_2}{\epsilon_3}\right);$ $\frac{\mu_1}{\mu_3} = f\left(\frac{\epsilon_1}{\epsilon_3}\right)$. Hence: $f\left(\frac{\epsilon_1}{\epsilon_3}\right) = f\left(\frac{\epsilon_1}{\epsilon_2}\right) \cdot f\left(\frac{\epsilon_2}{\epsilon_3}\right)$. The function f is such that it satisfies a relation of the type $f(ab) = f(a) \cdot f(b)$ where a and b are constants. Almost 200 years ago, the French mathematician A.L. Cauchy had shown that this implies that f must be power-law type with $f(x) = x^p$ where p is a constant. Other types of functions cannot be of the type $f(ab) = f(a) \cdot f(b)$. It follows that for the measure on the fractal set: $\mu_\alpha = c \cdot \epsilon^\alpha$ where c and the singularity α are constants.

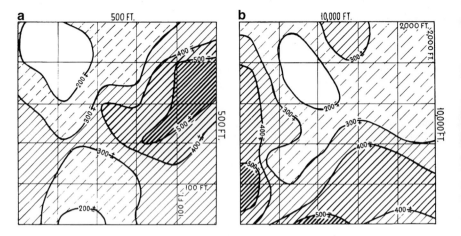

Fig. 11.1 Typical gold inch-dwt trend surfaces in the Klerksdorp goldfield on the basis of two-dimensional moving averages for two areas with similar average grades. (**a**) Moving averages of 100 × 100 ft. areas within a mined-out section of 500 × 500 ft. (**b**) Moving averages of 2,000 × 2,000 ft. areas within mined-out section of 10,000 × 10,000 ft (Source: Krige 1966, Fig. 3)

11.1.1 Witwatersrand Goldfields Example

As shown in Fig. 11.1, Krige (1966) illustrated the concept of self-similarity by means of gold value contour maps for square areas measuring 500 ft and 10,000 ft on a side. The same moving averaging method was used to construct these two gold contour maps at different scales, which exhibit similar patterns. Krige's example of regional self-similarity is based on a very large number of mining assays. Other studies commonly are based on much smaller datasets.

Starting from approximately 75,000 gold determinations for very small blocks (equivalent to channel samples, *cf.* Fig. 2.9), Krige (1966) found an approximate straight line relationship between logarithmic variance and logarithmically transformed size of reef area for gold values in the Klerksdorp goldfield, South Africa (Fig. 11.2). A linear relationship of this type is to be expected when the model of de Wijs is valid. Size of reef area ranged from 0.1 to 1,000 million square feet. Consequently, in Krige's application, scale-independence applies to square cells with side lengths extending over five orders of magnitude. The gold occurs in relatively thin sedimentary layers called "reefs". Average gold concentration value per sample is multiplied by length of sample cut across the reef and unit of gold assay values is expressed as inch-pennyweight in Fig. 11.2 (1 in.-pennyweight = 3.95 cm-g). Three straight-line relationships for smaller blocks within larger blocks are indicated. There are two relatively minor departures from the simple linear model derived at the beginning of this section (variance formula of de Wijs). The first of these two departures is that a small constant term (+20 in.-pennyweights) was added to all gold values. This reflects the fact that, in general, the three-parameter lognormal provides a better fit than the two-parameter lognormal as discussed in Sect. 3.2.3. The second departure consists of the fact that

11.1 Self-Similarity

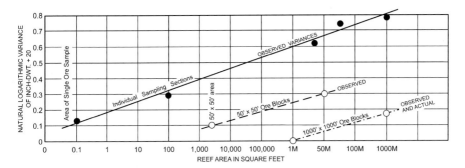

Fig. 11.2 Logarithmic variance of gold values as a function of reef area sampled. The variance increases linearly with log-area if the area for gold values is kept constant. The relationship satisfies the model of de Wijs (Source: Agterberg 2012b, Fig. 1)

constant terms are contained in the observed logarithmic variances plotted in Fig. 11.2. These additive terms are related to differences in the shapes of blocks. For example, channel samples are approximately linear but gold fields are plate-shaped. This constant term (called "sampling error" by Krige (1966), which equals 0.10 units along the vertical scale in Fig. 11.2) is independent of size of area. These two relatively small refinements were discussed in more detail in Agterberg (2012a).

11.1.2 Worldwide Uranium Resources

The logbinomial model of de Wijs was used in mineral resource evaluation studies by Brinck (1971, 1974). A comprehensive review of Brinck's approach can be found in Harris (1984). The original discrete model of de Wijs is assumed to apply to a chemical element in a large block consisting of the upper part of the Earth's crust with known average concentration value ξ commonly set equal to the element's crustal abundance.

According to Brinck (1974), chemical analysis is applied to blocks of rock that are very small in comparison to the large block targeted for study. Let $n=N$ represent the maximum number of subdivisions of the large block. Suppose that the element concentration values available for study: (1) constitute a random sample from the population of 2^N very small blocks within the large block, and (2) show an approximate straight line pattern on their lognormal Q-Q plot. The slope of this line then provides an estimate of σ from which η (and d) can be derived by means of the variance formula of de Wijs. Brinck (1974) set 2^N equal to the average weight of the very small block used for chemical analysis divided by total weight of the environment targeted for study. This model constitutes one of the earliest applications of the model of de Wijs. It is likely that Brinck's approach remains applicable. Estimation of the parameters of the model of de Wijs including d could be improved by adopting the multifractal modeling approach to be discussed in Sect. 11.2.

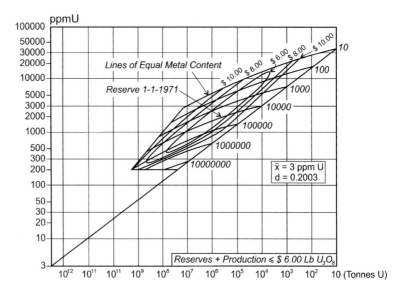

Fig. 11.3 Uranium resources and inferred potential reserves as estimated by the program MIMIC (Brinck 1974). Numbers along diagonal refer to largest possible deposits for given values of $\bar{x}$ and d. Dollar values refer to estimated average exploitation costs

Figure 11.3 (modified from Brinck 1974) is a worldwide synthetic diagram for uranium with average crustal abundance value set equal to 3 ppm and dispersion index $d = 0.2003$. This diagram is equivalent to a cumulative frequency distribution plot with two logarithmic scales. Value (ppm U) is plotted in the vertical direction and weight (tonnes U) is plotted in the horizontal direction. All weight values are based on cumulative frequencies calculated for the logbinomial distribution and are fully determined by the mean and coefficient of dispersion. The diagram shows curved lines of equal metal content. In 1971 it was, on the whole, profitable to mine uranium if the cost of exploitation was less than $6.00 US per pound U_3O_8. Individual orebodies can be plotted as points in Fig. 11.3. In 1971 such deposits would be mineable if their point would fall within the elliptical contour labeled $6.00. The other elliptical contours are for uranium deposits that would have been more expensive to mine.

Later applications of Brinck's geochemical crustal abundance approach include Ruzicka (1976) and Garrett (1986). These authors used other methods for estimating the maximum number of subdivisions of the environment (N). In his study of sedimentary uranium deposits, Ruzicka (1976) based the estimate of N on sedimentological considerations assuming that the environment had been sorted naturally into small homogeneous domains. Garrett (1986) estimated N from maximum "barrier" concentration values chosen for chemical elements in particular deposit types or regions. Both authors acknowledged that estimating the number of subdivisions (N) constitutes a major problem in applications of Brinck's approach (also see Harris 1984, Chap. 8, pp. 184–222).

The Brinck approach to modeling metal occurrence in the upper part of the Earth's crust has not been adopted widely. Economic geologists know that orebodies for uranium and other metals are of many different types with characteristic features that differ from type to type. Also, the genetic processes resulting in different types of orebodies were very different. These facts and concepts do not seem to fit in with the simplicity of Brinck's approach. Nevertheless, comparison of sizes and grades of ore deposits for uranium and other metals (cf. Brinck 1971) fit in remarkably well with diagrams such as Fig. 11.3. Approximate multifractal distribution of some metals in the Earth's crust or within smaller blocks provides an interesting alternative approach for regional mineral resource evaluation studies. This different kind of approach is equivalent to what was discussed in the previous chapter: the Earth's topography can be modeled as a fractal although, obviously, the genetic processes that have caused differences in elevation on Earth were very different. It is also equivalent to the remarkable fact that fractal modeling can be used to model the geographical distribution of mineral deposits within worldwide permissive tracts (Sect. 10.2.3).

11.2 The Multifractal Spectrum

Multifractals arose in physics and chemistry as a generalization of (mono-)fractals (Meneveau and Sreenivasan 1987; Feder 1988; Lovejoy and Schertzer 1991). Multifractals can be regarded as spatially intertwined monofractals (Stanley and Meakin 1988). Mandelbrot (see, e.g., Evertsz and Mandelbrot 1992) has emphasized that multifractals apply to continuous spatial variability patterns, whereas monofractals are for binary Yes-No type patterns. The relation between monofractals and multifractals also was considered by Herzfeld et al. (1995) who showed that the ocean floor could not be modeled as a monofractal. Better results were obtained by a multifractal model after incorporation of a non-stationary component (Herzfeld and Overbeck 1999). The multifractal spectrum is the tool *par excellence* in the study of multifractals. In this spectrum a monofractal plots as a single spike because it has only one singularity associated with it. Examples of multifractal spectra for geoscience data include Gonçalves et al. (2001), Pahani and Cheng (2004) and Arias et al. (2011, 2012).

A 1-D example of a multifractal is as follows. Suppose that $\mu = x \cdot \epsilon$ represents total amount of a metal for a line segment of length ϵ and x is the metal's concentration value. In the multifractal model it is assumed that (1) $\mu \propto \epsilon^\alpha$ where $\propto$ denotes proportionality, and α is the singularity exponent corresponding to concentration value x; and (2) $N_\alpha(\epsilon) \propto \epsilon^{-f(\alpha)}$ represents total number of line segments of length ϵ with concentration value x, and $f(\alpha)$ is the fractal dimension of these line segments.

There are three different ways in which a multifractal spectrum can be computed: histogram method, method of moments and direct determination. The first two methods are described in Evertsz and Mandelbrot (1992). The histogram method is intuitionally more appealing because it is easy to understand. However, in practice it is better to use the method of moments with Legendre transform. This method produces better results faster than the histogram method although the latter can be useful in the study of frequency distributions of multifractals. The third method (direct determination) developed by Chhabra and Jensen (1989) is useful but will not be applied in this chapter.

Box 11.2: Derivation of the Multifractal Spectrum

Evertsz and Mandelbrot (1992) make a clear distinction between (1) Hölder exponent or singularity $\alpha(x) = \lim_{\epsilon \to 0} \frac{\log_e \mu\{B_x(\epsilon)\}}{\log_e \epsilon}$ at a point x, and (2) "coarse" Hölder exponent $\alpha = \frac{\log_e \mu\{B_x(\epsilon)\}}{\log_e \epsilon}$ measured for a volume $B_x(\epsilon)$ around the point x. The mass-partition function is: $\chi_q(\epsilon) = \sum_{N(\epsilon)} \mu_i^q = \int N_\epsilon(\alpha)(\epsilon^\alpha)^q d\alpha$ with $N_\epsilon(\alpha) = \epsilon^{-f(\alpha)}$ where $f(\alpha)$ represents fractal dimension. It follows that $\lim_{\epsilon \to 0} \chi_q(\epsilon) = \int \epsilon^{q\alpha - f(\alpha)} d\alpha$ keeping in mind that α ($=\alpha_q$) also is a function of q. At the extremum: $\frac{\partial\{q\alpha - f(\alpha)\}}{\partial \alpha} = 0$ and $\frac{\delta f(\alpha)}{\delta q} = q$ for any moment q. Writing $\tau(q) = q\alpha - f(\alpha)$, it follows that $\frac{\delta \tau(q)}{\delta \alpha} = q$. The multifractal spectrum satisfies: $f(\alpha) = q\alpha - \tau(q)$. If the multifractal spectrum $f(\alpha)$ exists, the so-called codimension satisfies $C_1 = f(\alpha) - E$ where E is the Euclidian dimension (cf. Evertsz and Mandelbrot 1992).

11.2.1 Method of Moments

In practice, a feature such as element concentration in rock samples is measured in blocks of different sizes. The mass-partition function $\chi(\epsilon,q)$ then is the sum of all measurements raised to the power q for blocks with cell edge ϵ. Terms such as "mass-partition function" were borrowed from physical chemistry (see, e.g., Evertsz and Mandelbrot 1992). The slopes of the lines on a log-log plot of partition function against size measure are estimates of the mass exponent $\tau(q)$. The singularity $\alpha(q)$ is the first derivative of $\tau(q)$ with respect to q, and the fractal dimension satisfies $f(\alpha) = q \cdot \alpha(q) - \tau(q)$. The model of de Wijs results in a log-binomial frequency distribution that converges to a lognormal distribution. The model of de Wijs can be generalized in various ways. It is useful for simulating spatial multifractal patterns. The tails of the negative binomial are thinner than those of the lognormal. The high-value tail of the lognormal, in turn, is thinner than the tail of the Pareto distribution (cf. Figs. 10.8 and 10.9).

11.2 The Multifractal Spectrum

> **Box 11.3: The Binomial Measure**
>
> The Besicovitch process (Mandelbrot 1983, p. 377) divides the unit interval $S = [0, 1]$ into two intervals of equal length $\delta = 2^{-1}$ that are assigned the measures $\mu_0 = p$ and $\mu_1 = 1 - p$. Repeatedly, each part can be subdivided in the same way. After the n-th iteration there is one interval with length $\delta_n = 2^{-n}$ and measure p^n. In general, with $\xi = \frac{k}{n}$ and $k = 0, 1, \ldots, n$, there are $N_n(\xi) =$
>
> $$\binom{n}{\xi n} = \frac{n!}{(\xi n)!\{(1-\xi)n\}!} \approx \frac{1}{\sqrt{2\pi n\xi(1-\xi)}} \exp[-n\{\xi \cdot \log_e \xi + (1-\xi) \cdot \log_e(1-\xi)\}]$$
>
> intervals with measure $\mu_\xi = \Delta^n(\xi)$, with $\Delta(\xi) = \mu_0^\xi \mu_1^{(1-\xi)}$. The $N_n(\xi)$ line segments with length $\delta_n = 2^{-n}$ and measure μ_ξ form a fractal set with dimension $f(\xi) = -\dfrac{\xi \cdot \log_e \xi + (1-\xi) \cdot \log_e(1-\xi)}{\log_e 2}$ (*cf.* Feder 1988).
>
> The singularity α is defined as $(\xi) = \dfrac{\log_e \mu_\xi}{\log_e 2} = -\dfrac{\xi \cdot \log_e p + (1-\xi) \cdot \log_e(1-p)}{\log_e 2}$.
>
> Because of one-to-one correspondence between the parameters ξ and α, $f(\xi)$ can be replaced by $f(\alpha)$, which only depends on α. The multifractal spectrum is a plot of the fractal dimension $f(\alpha)$ against the singularity α.

Figures 11.4 and 11.5 provide a step by step example of a multifractal spectrum for the previously used 2-D example (Fig. 10.22) that was constructed artificially and is known to be a multifractal instead of a monofractal or a mixture of multifractals and monofractals. The multifractal spectrum is constructed in four steps (*cf.* Box 11.2):

1. First, a grid of equally shaped cells (squares in 2-D) is superimposed on the entity to be studied. The mass-partition function $\chi(\epsilon, q)$ of order q and linear block size ϵ is plotted on log-log paper against ϵ. Making q a subscript, it satisfies the equation: $\chi_q(\epsilon) = \sum_{N(\epsilon)} \mu_i^q$ where $\mu_i = x_i \cdot \epsilon^2$ is the measure for the i-th cell. In geochemical applications x_i represents element concentration value. Summation is for measures of all cells after raising these measures to the power q that is also called the "moment". Straight lines are fitted for different values of q that can be any real number with $-\infty < q < \infty$. If the lines are not straight, the object of study is not a multifractal. The array of straight lines for the example of Fig. 10.22 is shown in Fig. 11.4.

2. The slopes $\tau(q)$ of the lines in Fig. 11.4 depend on q only. They satisfy $\tau(q) = \dfrac{\partial \chi_q(\epsilon)}{\partial \epsilon}$. In Fig. 11.5a the mass exponents $\tau(q)$ are plotted against q. This plot shows a curve. If the object of study would be a monofractal, $\tau(q)$ would be linearly related to q.

3. It can be shown that $\dfrac{\partial \tau(q)}{\partial q} = \alpha(q)$. This results in the third plot (Fig. 11.5b) with $\alpha(q)$ as a function of q. The first derivative of the mass exponent was

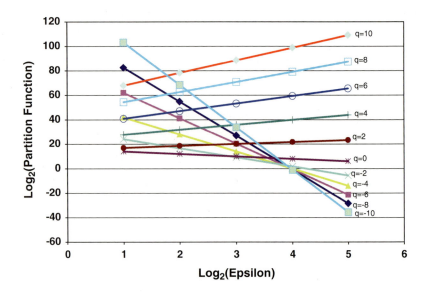

Fig. 11.4 Log-log plot (base 2) of mass-partition function versus length of cell edge for pattern of 128 × 128 concentration values of Fig. 10.22. Smallest values have cell edge with $\log_2(\epsilon) = 1$. Series shown are for integer values of q only. Slopes of straight lines are plotted in Fig. 11.3a (Source: Agterberg 2001, Fig. 2)

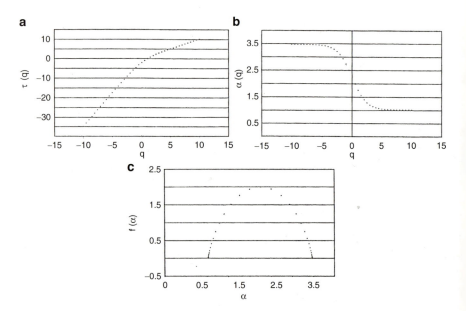

Fig. 11.5 Method of moments continued from Fig. 11.4. (**a**) Relation between mass exponent $\tau(q)$ and q; (**b**) Relation between singularity exponent $\alpha(q)$ and q; (**c**) Relation between multifractal spectrum value $f(\alpha)$ and singularity exponent α. Multifractal spectrum is symmetrical and has maximum value $f(\alpha) = 2$; the endpoints where $f(\alpha) = 0$ also are special in that their position does not depend on number of iterations (n) used for simulation experiment (Source: Agterberg 2001, Fig. 3)

11.2 The Multifractal Spectrum

approximated by the second order difference method. The mass exponent was determined for pairs of closely spaced values $q \pm 0.001$ and successive differences between two of these values divided by 0.002. A monofractal would show as a horizontal sequence of points on this diagram.

4. Finally, because it can also be shown that $f(\alpha) = q \cdot \alpha(q) - \tau(q)$, the multifractal spectrum is obtained with $f(\alpha)$ as a function of α (Fig. 11.5c). The points plotted in this diagram coincide with a theoretical limiting curve $f(\alpha)$ for $n \to \infty$. The maximum value of this curve is equal to 2. A monofractal would show as a spike in the multifractal spectrum. If the object would show a narrow peak, it would still be multifractal but close to monofractal.

A multifractal case history study in 2-D was provided by Cheng (1994) for gold in the Mitchell-Sulphurets area. The example of Fig. 11.6a shows how the multifractal spectrum was obtained for 100 ppb Au cutoff. Different grids with square cells measuring on a side were superimposed on the study area. The average gold value was determined for each grid cell containing one or more samples with Au > 100 ppb. Each average value was raised to the power q. The sum of the powered values satisfies a straight-line relationship for any q in a multifractal model. This aspect is verified in Fig. 11.6a. The slopes of many best-fitting straight lines (including those in Fig. 11.6a) are shown as the function $\tau(q)$ in Fig. 11.6b. The first derivative of $\tau(q)$ with respect to q gives $\alpha(q)$, and the multifractal spectrum $f(\alpha)$ follows from the relation $f(\alpha) = q \cdot \alpha(q) - \tau(q)$ (solid line in Fig. 11.6c).

The multifractal nature of the gold deposits in the Mitchell-Sulphurets area is shown in the spectra for different cutoff values in Fig. 11.6c, d. A fractal model would have resulted in a spectrum consisting of a single spike characterized by two constants: the fractal dimension $f(\alpha)$ and the singularity α. The four spectra of Fig. 11.6c, d are approximately equal on the left side, which is representative of the largest concentration values. The point where the multifractal spectrum reaches the α-axis and the slope of the curve at this point together determine the approximate area-concentration power-law relation on the right side in Fig. 10.4d (Agterberg et al. 1993; Cheng et al. 1994a). The maximum value of $f(\alpha)$ in Fig. 11.6c, d decreases with increasing cutoff value. In general, if $f(\alpha) < 2$ in 2-D space, it can be assumed that the multifractal measure is defined on fractal support (cf. Feder 1988).

The preceding four-step method also can be applied to 1-D or 3-D objects. In Fig. 11.7 the method is applied to the 118 channel samples from the Pulacayo Mine (original example from Cheng 1994). A measure $\mu_i(\epsilon) = \epsilon \cdot x_i(\epsilon)$ where $x_i(\epsilon)$, $i = 1, \ldots, 118$, for cell sizes (ϵ) ranging from 2 to 30 m (total length is 234 m). Some results of estimating the mass-partition function with q ranging from -34 to 34 are shown in Fig. 11.7a. The straight lines were fitted by ordinary least squares. The slopes $\tau(q)$ of all fitted straight lines with q ranging from -35 to 50 are shown in Fig. 11.7b. These results include $\tau(0) = -0.976 \pm 0.011$ where the uncertainty is expressed using the standard deviation ($\pm s$). $\tau(0)$ represents the box-counting dimension. It is noted that the slope for $q = 1$ is approximately equal to 0. This is as it should be because $\tau(0) = 0$ represents the principle of conservation of total mass. The second-order mass exponent $\tau(2) = 0.979 \pm 0.019$ is important for geostatistical modeling and will be used extensively in Sect. 11.4.

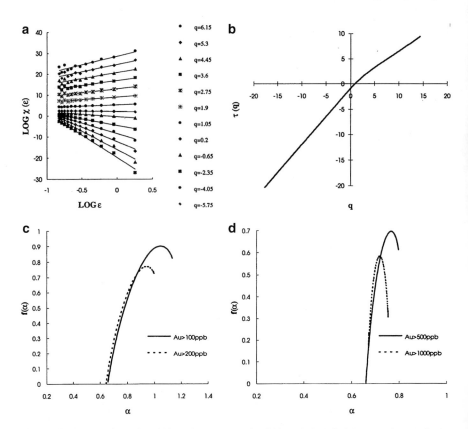

Fig. 11.6 Construction of multifractal spectrum of gold in Mitchell-Sulphurets mineral district (see Fig. 10.4). (**a**) $x(f)$ represent s sum of gold values greater than 100 ppb averaged for cell s and raised to the power q; cell edge ϵ i n km; log base 10; lowest sets of solid triangles and squares are for q equal to -7.45 and -9.15, respectively. (**b**) Slopes $\tau(q)$ of best-fitting straight lines including those shown in Fig. 1.6a as function of q. (**c** and **d**) Multifractal spectra are for four different cut-off values. It is noted that, contrary to a multifractal, a fractal is characterized by a single dimension that would have a spectrum consisting of a spike (Source: Agterberg 1995, Fig. 4)

Successive estimates of $\tau(q)$ were connected by straight-line segments (Fig. 11.7b) which, together, form an approximately differentiable curve. Values of the singularity α (Fig. 11.7c) were estimated by applying the central difference technique to successive sets of three consecutive values $\tau(q)$. The multifractal spectrum $f(\alpha)$ (Fig. 11.7d) was derived from the values shown in Fig. 11.7b, c.

The results of Fig. 11.7 show that the Pulacayo zinc concentration values are multifractal instead of monofractal because an ordinary fractal would have resulted in a single straight line in Fig. 11.7b, a horizontal line in Fig. 11.7c, and a vertical spike in Fig. 11.7d. The straight lines in Fig. 11.7a then would have had interrelated slopes $\tau(q)/(q-1) = \tau(p)/(p-1)$ for any pair of values $q \neq p$ in a monofractal model; for example, $\tau(12) = -\tau(-10) - 10$ when $q = 12$ and $p = 10$. Linear regression for these values gave the estimates $\tau(12) = 9.8709 \pm 0.2192$ and

11.2 The Multifractal Spectrum

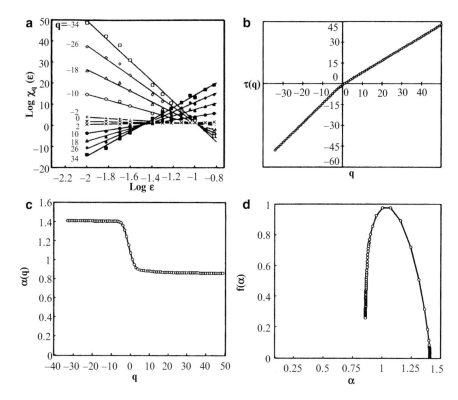

Fig. 11.7 Results of method of moments applied to Pulacayo zinc concentration values. (**a**) Log-log plot for relationship between χ (ϵ,q) and ϵ; straight lines obtained by method of least squares (LS). (**b**) Estimates of $\tau(q)$ include slopes of straight lines in **a**. (**c**) Singularity α estimated from **b** by central difference method. (**d**) Multifractal spectrum $f(\alpha)$. Smallest cell size $\epsilon = 1/100$ corresponds to sampling interval (=2 m); logarithmic scales have base 10 (Source: Cheng and Agterberg 1996, Fig. 2)

$\tau(-10) = -14.1234 \pm 0.4267$. The absolute value of the difference between these slopes is 4.2534 ± 0.4797 which is significantly different from 0. This clearly shows that the underlying model is multifractal instead of monofractal.

11.2.2 Histogram Method

Evertsz and Mandelbrot (1992) discuss that a histogram can be constructed for the singularities α_i associated with measures such as $\mu_i = x_i \cdot \epsilon^2$ in 2-D applications. The binomial frequencies of the concentration values x_i generated by the model of de Wijs (Fig. 10.22) depend on number of iterations n and value of the dispersion index d. Setting $n = 14$ and $d = 0.4$ as for Fig. 10.22 gave the results shown in

Fig. 11.8 (a) Histogram method applied to pattern of Fig. 10.22 with $d = 0.4$ and $n = 14$; (b) ditto with $d = 0.4$ and $n = 30$. In both diagrams (**a** and **b**), the larger values belong to limiting form of $f(\alpha)$ that provides an upper bound and would be reached exactly for $n = \infty$. Difference with limiting form decreases slowly for increasing n. Note that method of moments provides estimates of $f(\alpha)$ for $n \to \infty$ (Source: Agterberg 2001, Fig. 4)

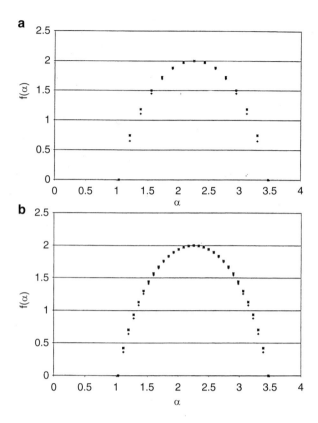

Fig. 11.8a. For infinitely large value n, a limiting multifractal spectrum $f(\alpha)$ (for $n = \infty$) can be derived in analytical form. It is noted that it is paradoxical that the corresponding frequency distribution does not exist because it has infinite variance when the original variance formula of de Wijs is applied. However, values of the limiting form of $f(\alpha)$ (for $n = \infty$) are shown in Fig. 11.8a for comparison with the $f(\alpha)$ values for $n = 14$. There are systematic discrepancies between the two spectra of Fig.11.8a. However, at the extremes, $\alpha_{\min} = \log_2\{2/(1+d)\}^2 = 1.03$ and $\alpha_{\max} = \log_2\{2/(1-d)\}^2 = 3.47$ with $f(\alpha) = 0$, the two spectra coincide. The other point of equality occurs at the center where $f(\alpha) = 2$.

Figure 11.8b shows similar results for $n = 30$. The histogram values are closer to the theoretical limit values in Fig. 11.8b, but it is obvious that convergence is exceedingly slow. On the other hand, the 4-step method immediately resulted in $f(\alpha)$ values (Fig. 11.8c) approximately coinciding with the limit values of Fig. 11.8a, b, illustrating that the method of moments is to be preferred for derivation of the limiting form of the multifractal spectrum $f(\alpha)$ (for $n = \infty$).

Because the realization of the model of de Wijs for a specific value of n is discrete, the multifractal spectrum can be readily interpreted. Each $f(\alpha)$ value represents the fractal dimension of a subset of cells with the same element

11.3 Multifractal Spatial Correlation

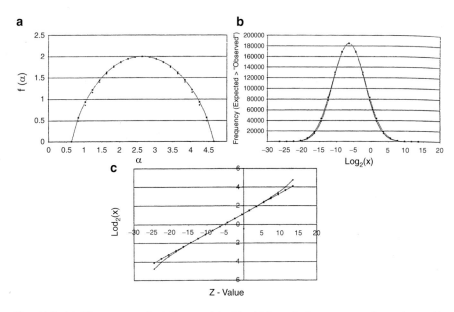

Fig. 11.9 (a) Histogram method illustrated in Fig. 11.7 applied to concentration values with $d = 0.6$ and $n = 20$; (b) Frequency distribution curves corresponding to two multivariate spectra shown in **a**; frequencies of limiting form slightly exceed logbinomial frequencies but difference is zero at center and endpoints; (c) Lognormal Q-Q plot of upper bound frequency distribution shown in **b**; near center, frequency distribution resulting from model of de Wijs is lognormal, and in the tails it is weaker than lognormal (Source: Agterberg 2001, Fig. 9)

concentration value. Thus box-counting of a monofractal binary pattern for a narrow neighborhood centered on any specific value $x \propto \epsilon^{\alpha-2}$ created from a realization such as the one shown in Fig. 10.22 would give a fractal dimension $f(\alpha)$ provided that the number of cells used for box-counting is sufficiently large.

The theoretical multivariate spectrum $f(\alpha)$ (for $n = \infty$) can be used to compute theoretical frequencies of the concentration values x for specific values of n and d. These frequencies are not independent of n, and the frequency distribution curve continues to change when n is increased. Various scaling and rescaling procedures have to be applied in order to derive the results shown in Fig. 11.7 and the frequencies, subsequently, to be derived from the limiting form of $f(\alpha)$. The required calculations were given in detail in a FORTRAN program in Agterberg (2001) (Fig. 11.9).

11.3 Multifractal Spatial Correlation

Cheng (1994; also see Cheng and Agterberg 1996) derived general equations for the semivariogram, spatial covariance and correlogram of any scale-independent multifractal including the model of de Wijs. Their model is for sequences of

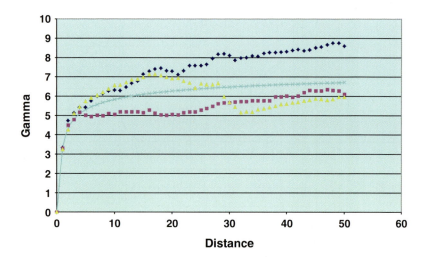

Fig. 11.10 Theoretical form of multifractal semivariogram in comparison with three experimental semivariograms based on 128 rows in patterns similar to the one shown in Fig. 10.22. Deviations between experimental semivariograms and continuous curve are relatively large, but there is probably no significant bias (Source: Agterberg 2001, Fig. 5)

adjoining blocks along a line. Experimentally, their semivariogram resembles Matheron's semivariogram for infinitesimally small blocks. Extrapolation of their spatial covariance function to infinitesimally small blocks would yield infinitely large variance when h approaches zero.

Figure 11.17 shows three experimental semivariograms each based on 128 rows of 128 numbers in patterns similar to the 2-D multifractal previously shown in Fig. 10.22. The theoretical semivariogram satisfies:

$$\gamma_k(\epsilon) = \xi_2(\epsilon) \left[1 - \frac{1}{2} \left\{ (k+1)^{\tau(2)+1} - 2k^{\tau(2)+1} + (k-1)^{\tau(2)+1} \right\} \right]$$

where $k = 1, 2, \ldots$ represents distance between successive cells measured in multiples of ϵ, $\xi_2(\epsilon)$ is the non-centered second-order moment obtained by dividing the mass-partition function for $q = 2$ by number of cells, and $\tau(2)$ is the second-order mass exponent. The two parameters used for the theoretical curve in Fig. 11.10 were in accordance with the results for $q = 2$ shown in Figs. 11.4 and 11.5. The experimental semivariograms are each based on 128^2 individual values. They deviate markedly from the theoretical curve although on the average they are closer to it. Problems associated with lack of precision of experimental semivariograms are well-known in spatial statistics (*cf.* Cressie 1991).

Fig. 11.11 Log-log plot for relationship between $\chi_2(\epsilon)$ and ϵ (Source: Cheng and Agterberg 1996, Fig. 3)

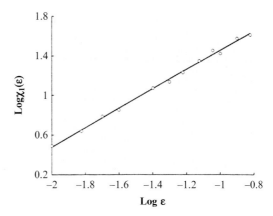

11.3.1 Pulacayo Mine Example

The Pulacayo orebody provides another example of application of multifractal spatial correlation. The most important parameter on which this approach is based is the second-order mass exponent $\tau(2)$ (Fig. 11.11). In this application it is estimated by the slope of the straight line fitted by least squares in Fig. 10.11 with $\tau(2) = 0.979 \pm 0.019$ (*cf.* Sect. 11.2.1). The semi-exponential autocorrelation previously used for smoothing the zinc values of Fig. 2.10 is shown as a broken line in Fig. 10.12 together with the autocorrelation coefficients (for lag distance $h > 0$) to which it was fitted. Its nugget effect would explain about half of total variability of the zinc values. On the other hand, use of the multifractal correlogram (solid line in Fig. 11.12) shows a continuous increase of spatial correlation towards the origin. The method used for fitting the multifractal correlogram will be explained in more detail later in this section. It does not apply when the lag distance becomes very small ($h < 0.07$ in Fig. 10.12) so that the true white noise at the origin cannot be estimated by this method. By means of local singularity mapping (Sect. 11.6.1), white noise will be estimated to represent only about 2 % of total variability of the zinc values. It represents measurement error and strong decorrelation at microscopic scale.

Figures 11.13a, b show the multivariate semivariogram previously shown for the model of de Wijs (Fig. 11.10). It satisfies: $\gamma_k(\epsilon) = \xi_2(\epsilon)$ $\left[1 - \frac{1}{2} \left\{ (k+1)^{\tau(2)+1} - 2k^{\tau(2)+1} + (k-1)^{\tau(2)+1} \right\} \right]$ with $\tau(2) = 0.979$ and $\xi^2(\epsilon) = 391.49$ in comparison with experimental semivariogram values estimated from the 118 Pulacayo zinc values using arithmetic and log-log scales. If $\tau(2)$ is only slightly less than 1, the preceding theoretical equation can be approximated by $\gamma_k(\epsilon) = -\xi_2(\epsilon) \cdot \epsilon^{\tau(2)-1} \log_e \left[\frac{1}{2} \{ \tau(2) + 1 \} \tau(2) k^{\tau(2)-1} \right]$ as shown by Cheng and Agterberg (1996, Eq. 23). Figure 11.13c shows the experimental semivariogram values on a graph with logarithmic scale in the horizontal direction only. The straight line in Fig. 11.13c was fitted by least squares. The approximate

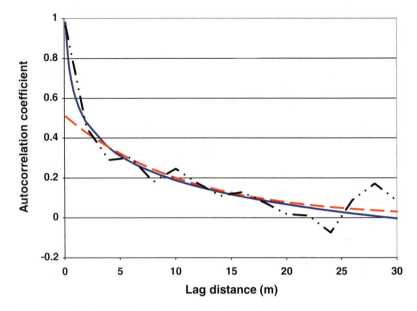

Fig. 11.12 Estimated autocorrelation coefficients (partly broken line) for 118 zinc concentration values of Fig. 2.10. Broken line represents best-fitting semi-exponential function used to extract "signal" in Fig. 2.10. Solid line is based on multifractal model that assumes continuance of self-similarity over distance less than the sampling interval (Source: Agterberg 2012b, Fig. 3)

semivariogram model provides a good fit. It is equivalent to the logarithmic semivariogram model introduced by Matheron (1962, p. 180; also see Table 6.1) and also used by Agterberg (1994a, p. 226).

Cheng and Agterberg (1996) derived the following expression for the autocorrelation function of a multifractal:

$$\rho_k(\in) = \frac{C\varepsilon^{\tau(2)-2}}{2\sigma^2(\varepsilon)} \left[(k+1)^{\tau(2)+1} - 2k^{\tau(2)+1} + (k-1)^{\tau(2)+1} \right] - \frac{\xi^2}{\sigma^2(\varepsilon)}$$

where C is a constant, ϵ represents length of line segment for which an average zinc concentration value is assumed to be representative, $\tau(2)$ is the second-order mass exponent, ξ represents overall mean concentration value, and $\sigma^2(\epsilon)$ is the variance of the zinc concentration values. The unit interval ε is measured in the same direction as the lag distance h. The index k is an integer value that later in this section will be transformed into a measure of distance by means of $k = \frac{1}{2}h$. Estimation for the 118 Pulacayo zinc values using an ordinary least squares model with $\tau(2) = 0.979$ gave:

$$\hat{\rho}_k = 4.37 \left[(k+1)^{1.979} - 2k^{1.979} + (k-1)^{1.979} \right] - 8.00$$

11.3 Multifractal Spatial Correlation

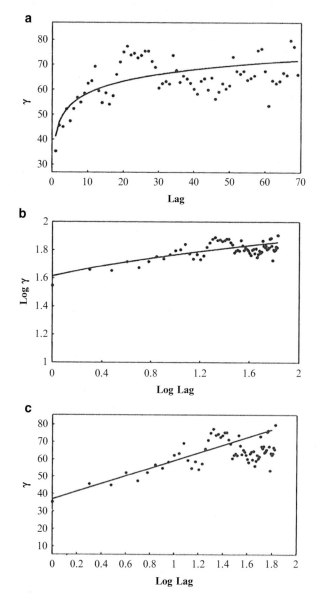

Fig. 11.13 Estimates of semivariogram γ_k. (**a**) Solid line obtained by linear regression after setting $\tau(2) = 0.979$. (**b**) Log-log plot of A. (**c**) Logarithmic approximation (Source: Cheng and Agterberg 1996, Fig. 5)

The first 15 values ($k \geq 1$) resulting from this equation are nearly the same as the best-fitting semi-exponential (broken line in Fig. 10.12). It seems reasonable to use the model for extrapolation toward the origin by replacing the second-order difference on the right side of this expression by the second derivative (*cf.* Agterberg 2012a, Eq. 9):

$$\left[(k+1)^{\tau(2)+1} - 2k^{\tau(2)+1} + (k-1)^{\tau(2)+1}\right] \cong \{\tau(2)+1\}\tau(2)k^{\tau(2)-1}$$

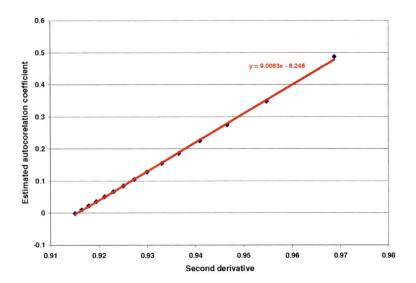

Fig. 11.14 Relation between estimated autocorrelation coefficients (*blue diamonds*) and second derivative of corresponding continuous function. Best fitting straight line (*red*) is used for extrapolation to the origin in Fig. 11.12 (Source: Agterberg 2012a, Fig. 13)

Linear regression of the second derivative for $\tau(2) = 0.979$ on estimated values resulted in the straight-line approximation shown in Fig. 11.14. Although the largest estimated autocorrelation coefficient that could be obtained by this method is only 0.487 (for $k=1$), it now becomes possible to extrapolate toward much smaller values of $k = \frac{1}{2} h$, so that larger autocorrelation coefficients are obtained, by using the second derivative on the right side of the preceding equation instead of the second-order difference. The theoretical autocovariance function shown in Fig. 11.12 was derived by transformation of the straight line of Fig. 11.14 for lag distances with $h \geq 0.014$ m. For integer values ($1 \leq k \leq 15$), the curve of Fig. 11.12 (solid line) reproduces the estimated autocorrelation coefficients obtained by the original multifractal model using second-order differences. Extrapolating toward the origin by means of the second-order derivative results in an overall pattern that closely resembles the hypothetical pattern of Fig. 6.19a consisting of the nested design of two superimposed negative semi-exponentials with a small white noise component. Consequently, the multifractal autocorrelation model of Cheng and Agterberg (1996), which is based on the assumption of scale-independence, confirms the existence of strong autocorrelation over short distances ($h < 2$ m).

11.4 Multifractal Patterns of Line Segments and Points

Multifractal case history studies presented earlier in this chapter were concerned with chemical element concentration values in rocks and orebodies. In this section the method will be applied to objects that are spatially distributed through a 2-D space. In the first example, these objects are fractures in granite measured at the surface. The second example deals with the geographical distribution of orebodies in a study area. Two different kinds of edge effects will be considered in these examples. Boundaries of study area have to be considered in both examples. In the first case, the fractures could only be observed in areas where the granite is exposed at the surface. The irregularities in pattern of rock exposures will be considered to reduce bias in the mass-partition function to be estimated for the multifractal modeling.

11.4.1 Lac du Bonnet Batholith Fractures Example

The Lac du Bonnet Batholith in the Winnipeg River Subprovince of the Archean Superior Province, Canadian Shield, mainly consists of pink porphyritic and gray, more equigranular granite; both contain layers of schlieric and xenolithic granite (Brown et al. 1989). Total area of granite exposed at the surface measures about 1,500 km^2. In 1980, Atomic Energy of Canada Limited (AECL) acquired a 21-year lease on a 3.8 km^2 area of the batholith for construction of the Underground Research Laboratory (URL) as part of geoscience research into the disposal of nuclear fuel-waste in crystalline rocks. Detailed mapping of both lithology and fracturing has been performed at surface, and extensive subsurface information was made available for the URL excavations. About 130 boreholes were drilled, to depths up to 1 100 m; these are mainly cored boreholes, logged in detail for lithological and fracture information. In 1992 a project was commenced to analyze these surface and borehole data from a fractal/multifractal point of view, as it was of interest to AECL to estimate three-dimensionally the relative frequencies of large blocks of sparsely fractured granite. Most faults and mesoscopic fractures are either subvertical or dip 10°–0°. Many subvertical joints die out about 100 m below the surface. Low-intermediate-dipping (10°–30°) fractures are associated with relatively few well-defined fault zones in the subsurface extending to at least 800 m depth (Agterberg et al. 1996a).

Fractal modeling of fractures had been the subject of a number of studies (Korvin 1992; Turcotte 1997; Ghosh and Daemen 1993). The primary purpose of the 1992 study was to show that surface fractures can be modeled as multifractals. Natural fault populations had been shown to possess multifractal scaling properties by Cowie et al. (1995). Multifractal modeling provides a link between different types of fractal measurements and the geostatistical approach using spatial covariance functions or semivariograms. A geographic information system (SPANS, *cf.* Chap. 5) was used to perform the measurements required for multifractal modeling (Agterberg et al. 1996a). Because most surface fractures die out with depth, it is not possible to use the results of this study for downward extrapolation. Figure 11.15 shows a relatively well-exposed, triangular test area (−0.11 km^2) at the URL site where the surface fractures have been mapped in detail. The effect of limited exposure at the surface on the statistical measures required special

Fig. 11.15 Test area near AECL Underground Research Laboratory in the Lac du Bonnet Batholith near Pinawa, Manitoba. (**a**) Surface fractures; (**b**) Outcrop pattern (in black) (Source: Agterberg et al. 1996a, Fig. 1)

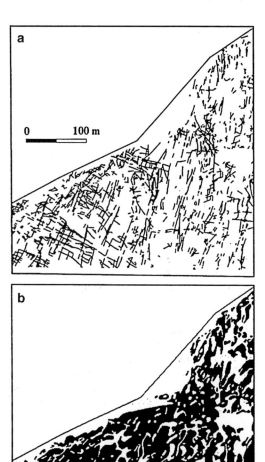

consideration. Several shallow and deep holes had been drilled in the immediate vicinity of this area selected for the pilot study described in this section. A coupled surface-borehole study to relate the surface results from the approximately horizontal plane (Fig. 11.15) to subvertical, linear borehole results was performed later in a separate study (Agterberg et al. 1996b; Agterberg 1997).

Tools developed for numerical treatment of the surface and borehole data for the Lac du Bonnet Batholith and similar crystalline rocks include: (1) determination and analysis of fracture intensity measures for boxes of different sizes; (2) analysis of the precision of the resulting statistical moments; (3) estimation of multifractal spectra; (4) study of spatial covariances and semivariograms; and (5) spatial frequency distribution analysis. This section is concerned with the first four of these topics with applications to surface fractures. Emphasis is on (a) comparison with results obtained by the simpler fractal approach, and (b) implications of the multifractal approach for spatial statistical analysis.

In a paper on fractal measurements, Roach and Fowler (1993) presented computer programs to determine the box-counting dimension and other fractal dimensions from patterns. Another method for dealing with this problem when measuring the so-called mass fractal dimension had been proposed by Lerche (1993), with an application to self-similar fault patterns. An important problem considered in these papers was to avoid bias related to measurements for small cells at one end, and measurements for large cells (close to total size of study region) at the other. For a pattern consisting of line-segments in the plane, the small and large cells yield biased estimates which are approximately equal to 1 and 2, respectively. The latter are estimates of Euclidian dimensions instead of fractal dimensions. These arise from the fact that it is not possible to derive meaningful results from measurements on cells that are nearly as large as the entire study area. Neither is it possible to estimate fractal dimensions from measurements on small cells for which the number of cells with nonzero measurements becomes approximately inversely proportional to cell area.

A fractal dimension must be obtained as the slope of a straight line on a log-log plot representing a fractal (for examples, see later); it should not be estimated from a curve that is gradually changing its slope from 1 to 2. The same type of bias arises in multifractal modeling and should be avoided when the values of $\chi_2(E)$ are estimated. A second problem of bias in fractal measurements occurs when the cells or boxes used for counting or measuring the mass exponents $\tau(q)$ are not restricted to the study area. Use of cells that contain parts of the boundary of the study region results in undesirable edge effects. For patterns of fractures, the importance of this second type of bias was clearly demonstrated by Walsh and Watterson (1993) and Gillespie et al. (1993). In the current application, there is the additional problem that bedrock is not fully exposed (Fig. 11.15). A procedure which can be used to avoid simultaneous bias because of lack of exposure and edge effects is as follows. As illustrated in Fig. 11.16, fractures can be measured only in cells or portions of cells underlain by exposed bedrock within the boundaries of the study region. Suppose that for a square cell with area (a) equal to ϵ^2, the following two measurements are obtained: area of exposed bedrock per cell (s_i), and total length of all fractures per cell (μ_i). The following weighted form then can be used to correct for bias in the multifractal situation:

$$\chi_q(\epsilon) = \sum_{n(\epsilon)} w_i [\mu_i(\epsilon)/w_i]^q$$

where $w_i = s_i/a$ for $i = 1, 2, \ldots, n(\epsilon)$. If $w_i = 1$ for all cells, this expression reduces to: $\chi_q(\epsilon) = \sum_{n(\epsilon)} \mu_i^q$ as was used before. Otherwise, when bedrock is not fully exposed in the study region, it reduces to the original form only if $q = 1$. The modified equation implies adherence to the principle of conservation of total mass within the study region. Note that $s_i \neq a$ for a cell represents either area of covered

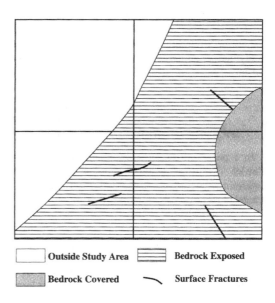

Fig. 11.16 Schematic example of four square cells near boundary of artificial study area illustrating bias prevention method based on considering for each cell: area of exposed bedrock, area of covered bedrock, area outside study region, and combined length of fractures. Measurements of fracture intensity are divided by $w_i = a_i/a$ to correct for bias with a_i representing area of exposed bedrock per cell, and a (constant) total cell area, respectively (Source: Agterberg et al. 1996a, Fig. 2)

bedrock or part of the cell outside the study region (see Fig. 11.16). The modified equation also implies that frequency of fractures in the exposed part of a cell is considered to be representative for the entire cell. This linear extrapolation may result in the assignment of too much weight to fractures in cells with relatively small values of w_i.

In order to assess this source of uncertainty, it is desirable to carry out the following type of sensitivity analysis. The entire procedure can be repeated a number of times, for each ϵ using only those cells with w_i greater than a threshold value denoted as Δ. Estimates then can be regarded as robust if they remain approximately the same for a range of successive Δ-values. A disadvantage of introducing a threshold value for w_i is that the sample size $n(\epsilon)$ is reduced for each ϵ. The sample size decreases when Δ is increased. This, in turn, would result in loss of precision, especially for relatively small values of ϵ. Five different cell sizes (ϵ close to 10, 20, 40, 80, and 160 m) were used for this type of sensitivity analysis. The first three sets of cells are shown in Fig. 11.17. The measurements obtained for each cell (i) were cell area (a_i), exposed area (s_i) and total length of all fractures per cell (μ_i). The cells were constructed by using the Voronoi tessellation procedure in a GIS for points on a regular grid with spacing equal to ϵ. Cells near the edges of the study area are larger or smaller than $a = \epsilon^2$ representing the standard cell size. The variable cell area was used only to determine relative exposed cell area (s_i/a_i) for comparison with successive Δ-values.

Bias correction was applied after setting $w_i = s_i/a$ with $a = \epsilon^2$. Thus, measurements for cells with areas greater than standard cell size were reduced (relatively few of such cells occur at the contact between granite and other rocks; cf. Fig. 11.14a), and those for smaller cells were increased, before raising them to successive powers (q). The analysis was repeated several times, for each ϵ,

11.4 Multifractal Patterns of Line Segments and Points 437

Fig. 11.17 Cells of three different sizes for test area obtained by Voronoi tessellation. (**a**) $\epsilon = 37.5$ m; (**b**) $\epsilon = 18.75$ m; (**c**) $\epsilon = 9.375$ m (see Fig. 11.15 for scale) (Source: Agterberg et al. 1996a, Fig. 3)

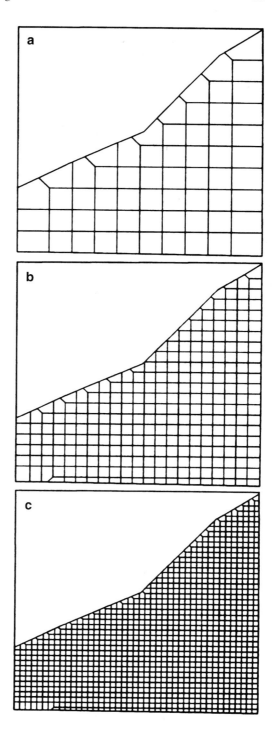

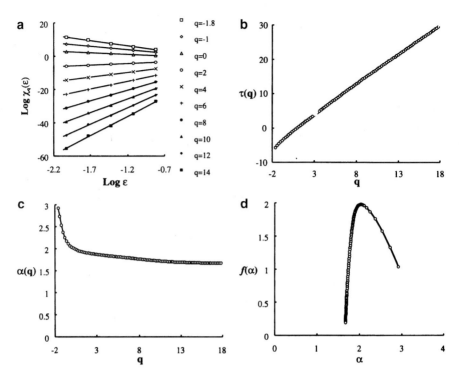

Fig. 11.18 Derivation of multifractal spectrum for cells with more than 20 % exposed surface area. (**a**) Mass-partition function estimated for different values of q; (**b**) least squares estimates (LS) of $\tau(q)$ (=slopes of best-fitting straight lines in Fig. 11.18a) vs. q; (**c**) central difference method applied to values of Fig. 11.18b yields $\alpha(q)$ as first derivative of $\tau(q)$ with respect to q; (**d**) final result: $f(\alpha) = \alpha q - \tau(q)$ (Source: Agterberg et al. 1996a, Fig. 6)

using multiples of 0.1 as threshold values. Results for thresholds $\Delta_0 = 0.0$, $\Delta_1 = 0.1, \Delta_2 = 0.2$, and $\Delta_4 = 0.4$ were Δ_2 given in Agterberg et al. (1996a). The threshold Δ_0 referred to use of all cells containing one or more fractures and this is equivalent to application of $\chi_q(\epsilon) = \sum_{n(\epsilon)} \mu_i^q$ without corrections. Estimates of $\tau(q)$ for the four different threshold values were compared with one another on the basis of estimated standard errors of $\tau(q)$ with q ranging from -0.8 to 2.2, and the estimate for cells with more than 20 % exposure (Δ_2) was selected as the best estimate on this basis. Further results for Δ_2 are shown in Fig. 11.18.

The results displayed in Fig. 11.18 are for q ranging from -1.8 to 18. The reason that values of q outside this range were not used is that the uncertainty of these estimates was considered to be too large. It should be kept in mind that for increased positive q, the estimates of $\tau(q)$ are increasingly determined by the relatively few largest values for amount of fractures per cell. Even small random fluctuations in these largest values then begin to significantly influence the results. On the other hand, the cells with the smallest values primarily determine the estimation of $\tau(q)$ for negative values of q. In general, better estimates are obtained for positive

11.4 Multifractal Patterns of Line Segments and Points

instead of negative values of q. It is not possible to raise the smallest possible observed value for a cell (=0 representing absence of fractures) to a negative power, and large fluctuations are likely to occur when this limit is approached. Figure 11.18 shows how the stepwise derivation of the difference method for the step from Fig. 11.18b, c was used as follows. Suppose that $\tau(q_{k-1})$, $\tau(q_k)$ and $\tau(q_{k+1})$ are three successive estimates of $\tau(q)$ as shown in Fig. 11.18b. Then $\alpha_k(q)$ as it is shown in Fig. 11.18c satisfies $\alpha_k(q) = \dfrac{\tau(q_{k+1}) - \tau(q_{k-1})}{q_{k+1} - q_{k-1}}$. The multifractal spectrum (Fig. 11.18d) deviates significantly from a marrow spike showing that the 2-D distribution of the fractures is multifractal instead of fractal. In Agterberg et al. (1996a), the multifractal semivariogram corresponding to Fig. 11.18 is used for further statistical analysis of fractures at the surface of the Lac du Bonnet Batholith. The frequency of fractures per block of granite decreases rapidly in the downward direction. In Agterberg et al. (1996b) and Agterberg (1997) multifractal analysis was applied to fractures observed along boreholes drilled in a larger area including the area shown in Fig. 11.15. The results of this 3-D analysis were used to estimate probabilities that blocks of granite below the surface are entirely or relatively free of fractures. The practical significance of this study is to help decide on possible underground sites for storage of nuclear waste.

11.4.2 Iskut River Map Gold Occurrences

The area used for this example is the Iskut River map sheet (British Columbia Minfile Map 104B). In this area there are 183 Au mineral deposits and occurrences (Fig. 11.19). It was previously studied using box-counting and number-size methods (Cheng et al. 1994c), fractal pattern integration in Au potential mapping (Cheng et al. 1994b) and geochemical anomaly separation (Cheng et al. 1996; cf. Sect. 10.3.2). Spatial clustering of the Au occurrences is obvious (Fig. 11.19). Different cell sizes (ϵ) ranging from 3 to 10 km were used. Number of Au occurrences per cell was counted on eight grid maps. The multifractal results are shown in Fig. 11.20 where q ranges from 0 to 4. This relatively narrow range was used because the mass-partition function does not exhibit clearly defined power-law relations unless $0 \leq q \leq 4$. Results for $\tau(q)$ include $\tau(0) = -1.335 \pm 0.077$, $\tau(2) = 1.219 \pm 0.037$ and $\tau(4) = 3.070 \pm 0.266$. A fractal model would have resulted in straight lines with slopes interrelated according to $\tau(q)/(q-1) = \tau(p)/(p-1)$. For example, for the fractal model: $\tau(4)/3 = -\tau(0)$. However, $\tau(4)/3 + \tau(0) = 0.312 \pm 0.098$ and this is significantly greater than zero, indicating that the underlying model is multifractal instead of monofractal. The slope of the straight line for $q = 1$ in Fig. 11.20a satisfies $\tau(1) = 0$ representing constant first-order intensity. The maximum value of: $f(\alpha)$ in Fig. 11.20c occurs at $-\tau(0) = 1.335 \pm 0.077$ representing the box-counting dimension of the fractal support.

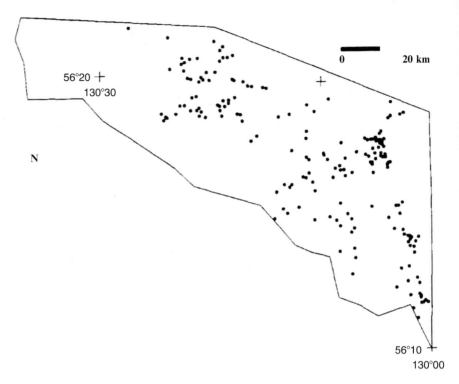

Fig. 11.19 Gold mineral occurrences in Iskut River map sheet, northwestern British Columbia (from B.C. Minfile Map 104B, 1989) (Source: Cheng and Agterberg 1995, Fig. 4)

For spatial analysis, values of $K(r)$ shown as circles in Fig. 11.21 were estimated by using Ripley's edge effect correction that satisfies

$$\hat{K}(r) = n^{-2}|A|\sum_{i\neq j} w_{ij}^{-1} I_i(r_{ij})$$

as discussed in Sect. 1.5.3 and also used in Sect. 10.2.2. The values shown in Fig. 11.21 were estimated using the program CLUST (Agterberg 1994b) to correct for edge effects in a study area bounded by a polygon. Figure 11.21 is a special kind of diagram with the property that a Poisson model for complete spatial randomness (CSR) of points would plot as a straight line through the origin that dips 45°. Ripley's (1981, p. 316) approximate 95 % confidence interval for the largest difference between $[\hat{K}(r)/\pi]^{0.5}$ and r (CSR null hypothesis) is shown as well in Fig. 11.21. Clearly, the CSR model should be rejected.

Cheng (1994) and Cheng and Agterberg (1995) have developed a theoretical multifractal model for 2-D point processes that is equivalent to the multifractal

11.4 Multifractal Patterns of Line Segments and Points

Fig. 11.20 Multifractal analysis results for Au mineral occurrences of Fig. 11.19. (**a**) Log-log plot of mass-partition function for selected values of q, dots represent estimated values, solid lines obtained by LS fitting. (**b**) $\tau(q)$ vs. q. (**c**) Multifractal spectrum showing $f(\alpha)$ vs. α (Source: Cheng and Agterberg 1995, Fig. 5)

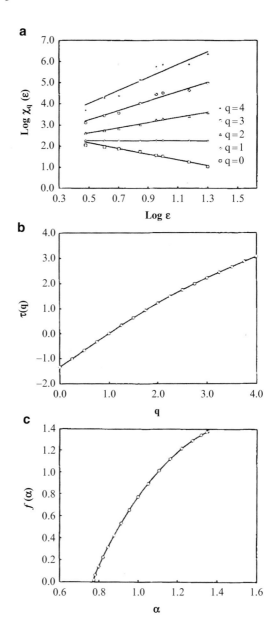

autocorrelation model used in Sect. 11.3. When ϵ is not too large, this model predicts the following approximate relationship:

$$\lambda_2\left(k\epsilon\sqrt{2}\right) \cong \frac{c}{8}\left[\frac{\tau(2)}{2}+1\right]^2 \epsilon^{\tau(2)-2}\left\{(k+1)^{\tau(2)} - 2k^{\tau(2)} + (k-1)^{\tau(2)}\right\}$$

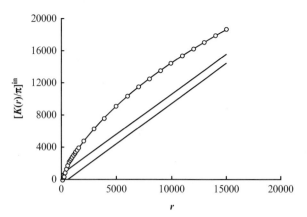

Fig. 11.21 Estimated values of square root of $K(r)/\pi$ vs. r using Ripley's edge effect correction and showing departure from complete spatial randomness (CSR) model (Source: Cheng and Agterberg 1995, Fig. 6)

where c is a constant and $k = 1, 2, \ldots, n$. Replacement of the second-order difference by the second derivative (cf. Cheng and Agterberg 1995, Eq. 20) results in the simple approximate equation:

$$K(r) \cong D_2 r^{\tau(2)}$$

where D_2 is another constant. Estimates of $K(r)$ obtained by this approximate equation were obtained by least squares after substituting $\tau(2) = 1.219$ and are shown in Fig. 11.22a. In general, the preceding relationship in which λ_2 is related to the second-order difference provides a better approximation. Application of this method resulted in the solid line on the log-log plot of Fig. 11.22b. Clearly, both methods provide satisfactory fits in this application. In a separate study Cheng (1994) had shown that this multifractal approach also could be applied to the spatial distribution of trees on a 19.6-acre square plot in Lansing Woods, Clinton, Michigan, used by Diggle (1983) for testing other (non-fractal) statistical models for point processes.

11.5 Local Singularity Analysis

In several recent studies, 2-dimensional applications of local singularity analysis including regional studies based on stream sediment data show local minima that are spatially correlated with known mineral deposits. These minimal singularities, which may provide targets for further mineral exploration, generally are smoothed out when traditional geostatistical contouring methods are used. Multifractal analysis based on the assumption of self-similarity predicts strong local continuity of element concentration values that cannot be readily determined by conventional semivariogram or correlogram analysis. This section is primarily concerned with

11.5 Local Singularity Analysis

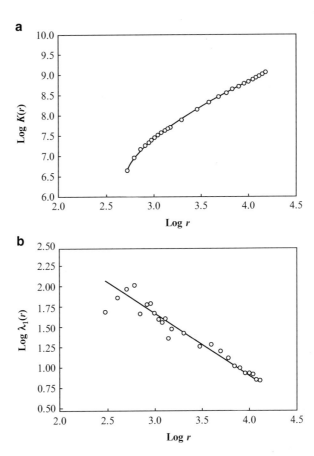

Fig. 11.22 Spatial statistics for Au mineral occurrences. (**a**) Log-log plot of relationship between $K(r)$ and distance r; *solid lines* are LS fits for $\tau(2) = 1.219 \pm 0.037$. (**b**) Log-log plot for relationship between second-order intensity λ_2 and r with theoretical line for $\tau(2) = 1.219$ (Source: Cheng and Agterberg 1995, Fig. 7)

multifractal and geostatistical modeling of the largest and smallest geochemical element concentration values in rocks and orebodies. These extreme values correspond to local singularities with near-zero fractal dimensions that occur close to the minimum and maximum singularity in the multifractal spectrum. These extremities normally cannot be estimated by means of the method of moments because of small-sample size problems that arise when the largest and smallest concentration values are raised to very high powers q. This problem also will be investigated by means of a computer simulation experiment and application to thousands of copper determinations from along the 7-km deep KTB borehole in southeastern Germany, for which local singularity analysis can be used to determine all singularities including the extreme values. The singularities estimated by this method are linearly related to logarithmically transformed element concentration values. This simple relation also can be useful to measure the small-scale nugget effect, which may be related to measurement error and microscopic randomness associated with ore grain boundaries.

Cheng (1999, 2005, 2008) proposed a model for incorporating local spatial association and singularities for interpolation of exploratory data. Recent 2-D applications using stream sediment geochemical data, including Cheng and Agterberg (2009), Zuo et al. (2009) and Xiao et al. (2012), results in relatively small anomalies consisting of minimal singularities on maps. Some of these anomalies are spatially correlated with known mineral deposits and others present new targets for further mineral exploration. On conventional contouring maps such anomalies are smoothed out. Although local singularity mapping uses the same data as those on which element concentration contour maps are based, the new local singularity patterns are strikingly different in appearance. The reason is that only data from points within very small neighborhoods are used for estimating the singularities and data are weighted in a manner that differs from weighting in most contouring methods. Understanding the behaviour of the semivariogram or spatial autocorrelation function over very short distances is important for the study of localized maxima and minima in the variability patterns of element concentration values. Local singularity analysis, which has become important for helping to predict occurrences of hidden ore deposits, is based on the equation $x = c \cdot \epsilon^{\alpha - E}$ where x, as before, represents element concentration value, c is a constant, α is the singularity, ϵ is a normalized distance measure such as block cell edge, and E is the Euclidian dimension. The singularity is estimated from observed element concentration within small neighborhoods in which the spatial autocorrelation is largely determined by functional behavior near the origin that is difficult to establish by earlier geostatistical techniques.

Iterative algorithms proposed by Chen et al. (2007) and Agterberg (2012a, b) for local singularity mapping when $E = 1$ will be discussed in Sect. 11.6. Local singularities obtained by application of these algorithms provide new information on the nature of random nugget effects versus local continuity due to clustering of ore crystals as occurs in many types of orebodies and rock units. Other applications of local singularity analysis include Xie et al. (2007).

In Cheng's (1999, 2005) original approach, geochemical or other data collected at sampling points within a study area are subjected to two treatments. The first of these is to construct a contour map by any of the methods such as kriging or inverse distance weighting techniques generally used for this purpose. Secondly, the same data are subjected to local singularity mapping. The local singularity α then is used to enhance the contour map by multiplication of the contour value by the factor $\epsilon^{\alpha - 2}$ where $\epsilon < 1$ represents a length measure. In Cheng's (2005) application to predictive mapping, the factor $\epsilon^{\alpha - 2}$ is greater than 2 in places where there had been local element enrichment or by a factor less than 2 where there had been local depletion. Local singularity mapping can be useful for the detection of geochemical anomalies characterized by local enrichment even if contour maps for representing average variability are not constructed (*cf.* Cheng and Agterberg 2009; Zuo et al. 2009).

11.5.1 Gejiu Mineral District Example

Where the landscape permits this, stream sediments are the preferred sampling medium for reconnaissance geochemical surveys concerned with mineral exploration (Plant and Hale 1994). During the1980s and1990s, government-sponsored reconnaissance surveys covering large parts of Austria, the Canadian Cordillera, China, Germany, South Africa, UK, and USA were based on stream sediments (Darnley et al. 1995). These large-scale national projects, which were part of an international geochemical mapping project (Darnley 1995), generated vast amounts of data and continue to be a rich source of information. Cheng and Agterberg (2009) applied singularity analysis to data from about 7,800 stream sediment samples collected as part of the Chinese regional geochemistry reconnaissance project (Xie et al. 1997). For illustration, about 1,000 stream sediment tin concentration values from the Gejiu area in Yunnan Province were used. This area of about 4,000 km^2 contains 11 large tin deposits. Several of these, including the Laochang and Kafang deposits, are tin-producing mines with copper extracted as a by-product. These hydrothermal mineral deposits also are enriched in other chemical elements including silver, arsenic, gold, cadmium, cobalt, iron, nickel, lead, and zinc. Applications to be described here are restricted to tin, arsenic, and copper. Tin and copper are the ore elements of most interest for mineral prospecting whereas arsenic is a toxic element. Water pollution due to high arsenic, lead, and cadmium concentration values is considered to present one of the most serious health problems especially in underdeveloped areas where mining is the primary industry such as in the Gejiu area. Knowledge of the characteristics of spatial distribution of ore elements and associated toxic elements in surface media therefore is helpful for the planning of mineral exploration as well as environmental protection strategies.

The Gejiu mineral district (Fig. 11.23a) is located along the suture zone of the Indian Plate and Euro-Asian plates on the southwestern edge of the China sub-plate, approximately 200 km south of Kunming, capital of Yunnan Province, China. The Gejiu Batholith with outcrop area of about 450 km^2 is believed to have played an important role in the genesis of the tin deposits (Yu 2002). The ore deposits are concentrated along intersections of NNE–SSW and E–W trending faults. Stream sediment sample locations in the Gejiu area are equally spaced at approximately 2 km in the north–south and east–west directions. Every sample represents a composite of materials from the drainage basin upstream of the collection site (Plant and Hale 1994). Regional trends are captured in a moving average map of tin concentration values from within square cells measuring 26 km on aside (Fig. 11.23c). Several parameters had to be set for this use of the inverse distance weighted moving average method (Cheng 2003). In this application, each square represents the moving average for a square window measuring 26 km on a side with influence of samples decreasing with distance according to a power-law function with exponent set equal to −2. Original sample locations were 2 km apart both in the north–south and east–west directions. It shows a large anomaly in the eastern part of the Gejiu area surrounding most large tin deposits including the mines. The three large tin deposits in the central part of the Gejiu area are recent discoveries. To illustrate in detail how singularities were estimated 12 different locations were arbitrarily selected on the map as shown on Fig. 11.23b. In total, 1,056 local tin singularities were estimated (Fig. 11.23d) by assembling the tin concentration values from within the same 26×26 km^2 cells used for the moving average map (Fig. 11.23c). For example, Fig. 11.24 illustrates in detail how singularities were estimated at two of the 12 different locations. The singularities were estimated by fitting straight lines on log-log plots of either the concentration value (x) versus ϵ using $x = c \cdot \epsilon^{\alpha-2}$, or amount of metal $\mu = x \cdot \epsilon^2$ versus ϵ using $\mu = c \cdot \epsilon^{\alpha}$. In Fig. 11.23d, both methods are used yielding similar estimates of the singularity α.

The main difference between the patterns of Fig. 11.23c, d is that lower and higher local tin singularity values are more evenly distributed across the Gejiu area than the tin concentration values themselves. The preceding analysis was repeated for arsenic, which is a toxic element. Its moving average and local singularity patterns (Fig. 11.25) are similar to those obtained for tin (Fig. 11.23).

The histogram of all local tin singularities is unimodal (Fig. 11.26a). Strength of spatial correlation between a point pattern and a contour map can be estimated by means of the weights of evidence method (Sect. 5.1.4). The Student's t-value diagram (Fig. 11.26b) was used to express statistical significance of strength of spatial correlation between (a) point pattern of the 11 tin deposits (Fig. 11.23a), and (b) the local tin singularity map (Fig. 11.23d). Although there are relatively few tin deposits, t-values near the peak (where $t = 4.84$ at $\alpha = 1.925$) exceed $t_{0.05} = 2.0$ representing 95 % confidence level for statistical significance indicating positive correlation between the two patterns. In total, 93 local tin singularities have $\alpha < 1.925$. Their combined area measures only 8.8 % of total study area, but 61.3 % of the tin deposits occur within this relatively small low-singularity sub-area. The logarithm (base e) of number of local tin singularities exceeding α is linearly related to α (Fig. 11.23c) with slope of approximately 2.3. Likewise, there is an approximate straight-line relationship (slope $= -3.2$) between logarithmically transformed cumulative area and largest tin concentration values (Fig. 11.23d). This is an example of a concentration–area (C–A) plot (Sect. 10.3.1). The pattern on a C–A plot was automatically broken into three successive straight-line segments. For tin (Fig. 11.26d), $\log_e \alpha$ of the third segment extends from 7.414 to 8.637 with best-fitting straight-line: $y = 26.9230 - 3.1576x$. It represents a Pareto frequency distribution for the highest concentration values. Other chemical elements enriched in the hydrothermal tin deposits show patterns similar to those obtained for tin (Fig. 11.26a–d) as illustrated for arsenic (Fig. 11.26e–h). For arsenic, the second straight-line segment in the C–A plot (Fig. 11.26h; also see Fig. 10.30) extends from 7.147 to 7.955 with best-fitting straight-line: $y = 25.6069 - 3.0178x$ as previously discussed in Sect. 10.4.

11.5 Local Singularity Analysis

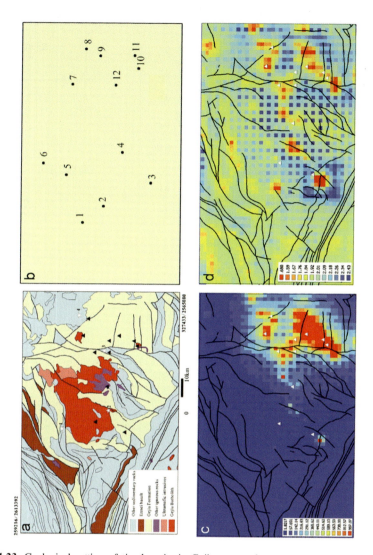

Fig. 11.23 Geological setting of tin deposits in Gejiu area and map patterns derived from tin concentration values in stream sediment samples. (**a**) Simplified geology after Yu (2002). Solid lines indicate faults; triangles represent tin deposits; UTM coordinates apply to map corner points. (**b**) Locations examples of estimation of local tin singularity in Fig. 11.24. (**c**) Distribution of tin concentration values in 1,000 stream sediment samples using inverse distance weighted moving average. Eight of 11 large tin deposits fall on large anomaly in eastern part of area that is due to environmental pollution related to mining (**d**) Distribution of local tin singularities. Tin deposits tend to occur in places where local singularity is less than 2 (Source: Cheng and Agterberg 2009, Fig. 1)

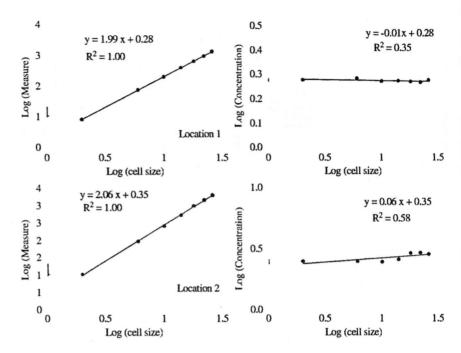

Fig. 11.24 Examples of estimation of local singularity at locations shown in Fig. 11.23b. Cell size is length of side of square cells with half-widths set equal to 1, 3, 5, ..., 13 km. Plots on left and right show relations between values of μ and ξ versus cell size, respectively. Value of μ is sum of tin concentration values in all 2×2 km cells contained within the larger cells centered on a location and value of $\xi = \mu/\epsilon^2$; base of logarithms is 10. Straight lines were fitted by ordinary least squares method (log of value regressed on log of cell size); R^2 = multiple correlation coefficient squared. Typically all seven points fall on the best-fitting straight line although small deviations do occur (e.g., second point at Location 2). Slopes of straight-lines on left provide estimates of local singularity α and those on the right are for $\alpha - 2$ (Source: Cheng and Agterberg 2009, Fig. 2)

11.5.2 Zhejiang Province Pb-Zn Example

The second example shown in Figs. 11.27 and 11.28 (after Xiao et al. 2012) is for lead in northwestern Zhejiang Province, China, that in recent years has become recognized as an important polymetallic mineralization area after discovery of several moderate to large Ag and Pg-Zn deposits mainly concentrated along the northwestern edge of the study area (Fig. 11.27), which has been relatively well explored. The pattern of local singularities based on Pb in stream sediment samples is spatially correlated with the known mineral deposits. It can be assumed that similar anomalies in parts of the less explored parts of the area provide new targets for further mineral exploration. By combining singularities for different chemical elements with one another, spatial correlation between anomalies and mineral deposits can be further increased (Xiao et al. 2012).

11.6 Chen Algorithm 449

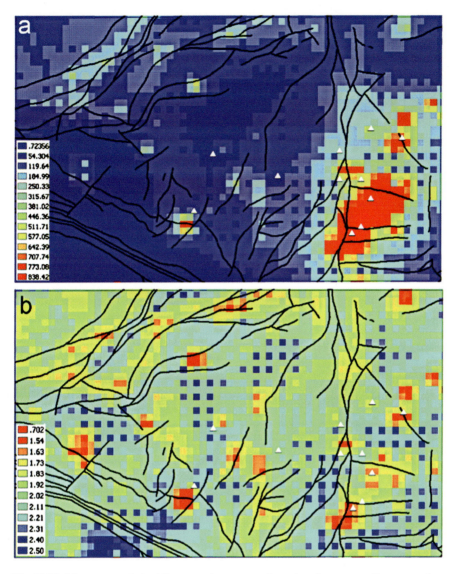

Fig. 11.25 Map patterns derived from arsenic concentration values in stream sediment samples. (**a**) and (**b**) show patterns similar to those for Sn in Fig. 11.23c, d (Source: Cheng and Agterberg 2009, Fig. 3)

11.6 Chen Algorithm

Theory of the model of de Wijs (cf. Sect. 10.4.1) also known as the binomial/p model is presented in great detail in textbooks including Feder (1988), Evertsz and Mandelbrot (1992), Mandelbrot (1989) and Falconer (2003). There have been

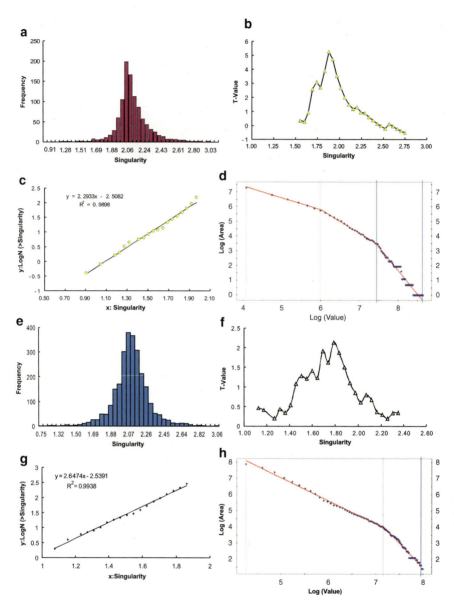

Fig. 11.26 Frequency distributions of local tin and arsenic singularities and concentration values. (**a**) Histogram of tin singularities shown in Fig. 11.23c. (**b**) Student's *t*-value for contrast (*C*) expressing degree of spatial correlation between binary pattern and point pattern (consisting of 11 tin deposits shown in Fig. 11.23a). (**c**) Cumulative frequency plot corresponding to histogram (Fig. 11.26a); $N(>\alpha)$ is number of units with local singularities exceeding α; base of logarithms is 10. Straight-line was fitted by ordinary least squares method ($\text{Log}_{10} N$ regressed on α); R^2 = multiple correlation coefficient squared. (**d**) C-A (concentration-area) plot showing cumulative area versus tin concentration value; base of logarithms is e. (**e**) Histogram of As singularities (shown in Fig. 11.23b). (**f**) Student's *t*-value for arsenic. The peak near $\alpha = 1.7$ indicates positive spatial correlation between arsenic singularities and tin deposits. (**g**) Pattern for arsenic corresponding to pattern for tin in Fig. 11.23c. (**h**) Arsenic C-A plot (Source: Cheng and Agterberg 2009, Fig. 4)

11.6 Chen Algorithm

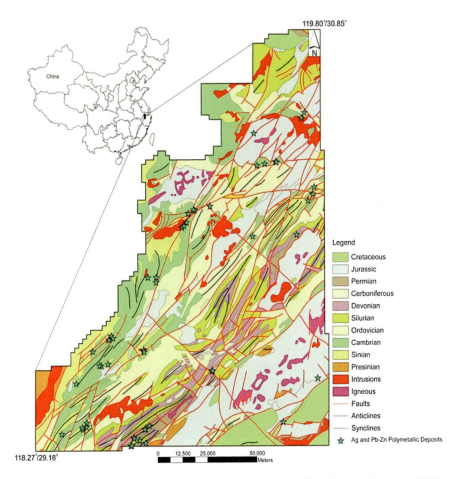

Fig. 11.27 Geological map of northwestern Zhejiang Province, China (Source: Xiao et al. 2012, Fig. 1)

numerous successful applications of this relatively simple model including many to solve solid Earth problems (also see, e.g., Cheng 1994; Cheng and Agterberg 1996; Agterberg 2007a, b; Cheng 2008). Although various departures from the model have been described in these papers and elsewhere, the binomial/p model basically is characterized by a single parameter. In the original model of de Wijs (1951), this parameter is the dispersion index d. In Sect. 6.2 it was discussed that the absolute dispersion index of Matheron satisfies $A = (\log_e \eta)^2 / \log_e 16$, and $\eta = (1+d)/(1-d)$. When the parameter p is used as in Evertsz and Mandelbrot (1992): $p = 0.5(1-d)$. The multifractal spectrum of this model, which shows the fractal dimensions $f(\alpha)$ as a function of the singularities, has its maximum $f(\alpha) = 1$ (for $E = 1$) at $\alpha = 1$, and $f(\alpha) = 0$ at two points with: $\alpha_{min} = -\log_2(1-p)$; $\alpha_{max} = -\log_2 p$. Another parameter that can be used to

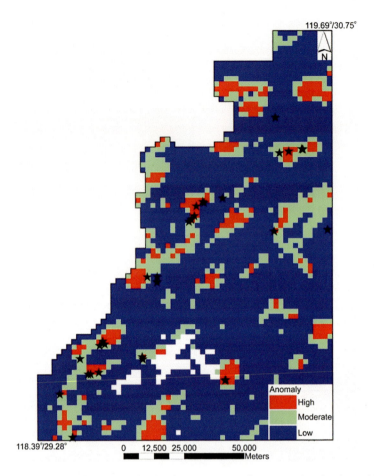

Fig. 11.28 Raster map for Fig. 11.27 showing target areas for prospecting for Ag and Pb-Zn deposits delineated by comprehensive singularity anomaly method (cell size is 2 × 2 km). Known Ag and Pb-Zn deposits are indicated by stars (Source: Xiao et al. 2012, Fig. 13)

characterize the binomial/p model is the second order mass exponent $\tau(2) = -\log_2\{p^2 + (1-p)^2\}$. If the binomial/p model is satisfied, anyone of the parameters p, d, α, $\tau(2)$, α_{min}, α_{max}, or $\sigma^2(\log_e x)$ can be used for characterization. Using different parameters can be helpful in finding significant departures from model validity if these exist.

According to Chen et al. (2007) local scaling behaviour follows the following power-law relationship:

$$\rho\{B_x(\epsilon)\} = \frac{\mu\{B_x(\epsilon)\}}{\epsilon^E} = c(x)\epsilon^{\alpha(x)-E}$$

11.6 Chen Algorithm

where $\rho\{B_x(\epsilon)\}$ represents element concentration value determined on a neighbourhood size measure B_x at point x, $\mu\{B_x(\epsilon)\}$ represents amount of metal, and E is the Euclidean dimension of the sampling space. In general, $\rho\{B_x(\epsilon)\}$ is an average value of element concentration values for smaller B's at points near x with different local singularities. Consequently, use of the power-law relationship as it stands would produce biased estimates of $c(x)$ and $\alpha(x)$. How could we obtain estimates of $c(x)$ that are non-singular in that they are not affected by the differences between local singularities within B_x? Chen et al. (2007) proposed to replace the original model by:

$$\rho(x) = c^*(x)\epsilon^{\alpha^*(x)-E}$$

where $\alpha^*(x)$ and $c^*(x)$ are the optimum singularity index and local coefficient, respectively. In the Chen algorithm the initial crude estimate $c(x)$ is considered to be obtained at step $k=1$ of an iterative procedure. It is refined repeatedly by using:

$$c_{k-1} = c_k(x)\,\epsilon^{\alpha_k(x)-E}$$

This procedure will be explained by application to the 118 zinc concentration values of the Pulacayo Mine example.

11.6.1 Pulacayo Mine Example

In the 1-dimensional Pulacayo example, $E=1$; and, for $\epsilon=1$, B_x extends $\epsilon/2=1$ m in two directions from each of the 118 points along the line parallel to the mining drift. Suppose that average concentration values $\rho\{B_x(\epsilon)\}$ also are obtained for $\epsilon=3, 5, 7$ and 9, by enlarging B_x on both sides. The yardsticks can be normalized by dividing the average concentration values by their largest length ($=9$). Reflection of the series of 118 points around its first and last points can be performed to acquire approximate average values of $\rho\{B_x(\epsilon)\}$ at the first and last 4 points of the series. A straight line can be fitted by least squares to the five values of $\log_e \mu\{B_x(\epsilon)\}$ against $\alpha(x)\cdot\log_e \epsilon$ then provides estimates of both ln $c(x)$ and $\alpha(x)$ at each of the 118 points. Estimates of $c(x)$ and $\alpha(x)$ are shown in Figs. 12.29 (red line) and Fig. 12.31 (Series 1), respectively. These results of ordinary local singularity mapping duplicate estimates previously obtained by Chen et al. (2007) who proposed an iterative algorithm to obtain improved estimates.

Employing the previous least squares fitting procedure at each step resulted in the values of $c_k(x)$ shown in Fig. 12.29 for the first and fourth step of the iterative process, and for $k=1{,}000$ after convergence has been reached. Values for the first

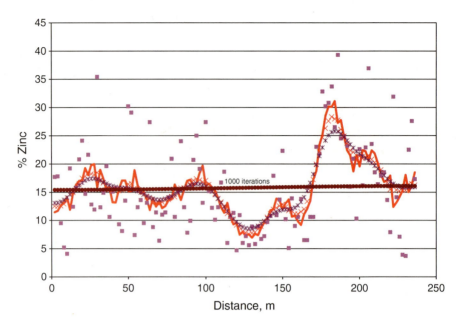

Fig. 11.29 Results of applying iterative method of Chen et al. (2007) continuing the iterations until full convergence is reached. Original zinc values are being smoothed during successive iterations. Second series (*solid line*) obtained after first iteration resembles "signal" in Fig. 2.10. Values after four iterations are shown as *diamonds*. At the end of the process, after 1,000 iterations, when convergence has been reached, the result is approximately a straight line with average value slightly below average zinc content (=15.6 % Zn) (Source: Agterberg 2012a, Fig. 18)

four steps of the iterative process exactly duplicated values plotted in Chen et al. (2007)'s Fig. 1 and partially listed in their Table 11.1 except for the first and last 4 values in all successive series because a slightly different end correction was employed (see before). For $k=1$, the pattern of $c_k(x)$ resembles the signal previously obtained in Fig. 2.10 by eliminating the noise component from the 118 zinc values. Chen et al. (2007) selected $\alpha^*(x) = \alpha_4(x)$ because at this point rate of convergence has slowed down considerably (Figs. 11.29 and 11.30).

In Agterberg (2012a) the iterative process was continued until full convergence was reached in order to obtain more complete information on the nugget effect. In the limit, after about 1,000 iterations, the final pattern is as shown in Fig. 12.29 with an average zinc concentration value that is slightly less than 15.61 % Zn representing the average of the 118 input values. This bias is due to the fact that, at each step of the iterative process, straight-line fitting is being applied to logarithmically transformed variables followed by the results being converted back to original data scale. The small bias in this application can be avoided by forcing the mean to remain equal to 15.61 % during all steps of the iterative process. In other applications, this type of bias can be much larger and preservation of the mean then is more important. End product and some intermediate steps this new run are shown in Fig. 12.30. In comparison with Fig. 12.29, the outputs obtained after the first and

11.6 Chen Algorithm

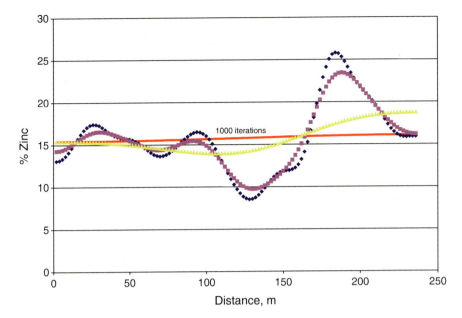

Fig. 11.30 Same as Fig. 11.29 but iterative process was constrained to preserve average zinc value of 15.6 % Zn. Results after 4 iterations ($k = 5$) is same as in Fig. 11.29 but result after 1,000 iterations is slightly different. Intermediate steps for $k = 10$ and 100 are also shown (Source: Agterberg 2012a, Fig. 19)

fourth step of the iterative process remain unchanged. There is only a small difference in results for $k = 1,000$. This confirms that in local singularity analysis it usually is permitted to neglect bias introduced by logarithmic transformation of variables. In Sect. 6.2.5 it was mentioned that the variance of values used in least squares straight line fitting ranges from $\sigma^2_E = 0.0622$ to $\sigma^2_E/9 = 0.0069$ at the beginning of the iterative process, and these variances remain very small at later steps. Estimated singularities for $k = 1$, 54 and 1,000 are shown in Fig. 11.31. The results for $k = 1$ and $k = 5$ duplicate the results previously obtained by Chen et al. (2007).

Full convergence singularities ($k = 1,000$) are significantly different from local singularities and results for $k = 5$ differ in two neighbourhoods along the Tajo vein (approximately from sampling point positions 60 to 75, and 90 to 100, respectively). These two systematic discrepancies merely indicate that full convergence had not yet been reached at $k = 5$. In Fig. 11.32, final singularities are plotted against original zinc concentration values showing a logarithmic curve pattern. In Fig. 11.33 a straight line of least squares was fitted for final singularity versus $\log_{10}$ (%Zn) with the residuals (deviations from this best fitting line) shown separately in Fig. 11.34. The residuals exhibit a white noise pattern with variance equal to 0.001178. Using original zinc values, the variance of residuals is estimated to be 1.3837. Because %Zn variance is 64.13, it follows that the white noise

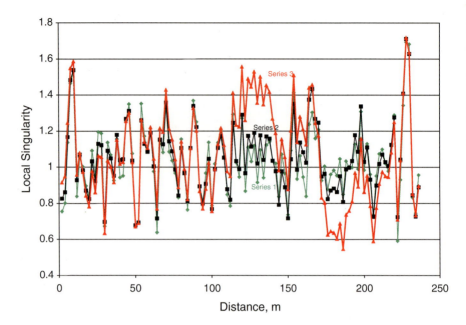

Fig. 11.31 Estimated singularities for iterative process with $k=1$ and $k=1,000$. Full convergence singularities ($k=1,000$) differ significantly from results for $k=1$ (and $k=5$) in two neighborhoods along the Tajo vein (approximately from sampling point locations 60–75 and 90–100) (Source: Agterberg 2012a, Fig. 20)

component is 0.02079. This is only about 4 % of the variance of the noise component previously used to construct the signal of Fig. 2.10. The new sampling error is probably a measurement error associated with the original chemical determinations for zinc and incorporates the crystal boundary effect (Sect. 2.5.1). Incorporation of the nugget effect to estimate zinc content (e.g., by using the theoretical values on the curve previously fitted in Fig. 2.10), approximately reproduces the observed values. This in itself is a trivial result. However, the importance of this application is that it illustrates that the Chen algorithm provides a new method to estimate the nugget effect.

For the Pulacayo Mine example, the $c_k(x)$ values in the Chen algorithm converge to a straight line with very small dip. After 1,000 steps they are all approximately equal to the average Zn value of 15.61 % (Fig. 12.30). This explains why there is an approximate logarithmic relationship between the singularities and the original zinc concentration values (Fig. 11.32).

As explained in Chap. 6, local singularity and this new type of nugget effect are associated with variability over very short distances. Singularities less than 1 signify local Zn enrichment whereas singularities greater than 1 indicate depletion. Minimum and maximum singularities are $\alpha_{min} = 0.547$ and $\alpha_{max} = 1.719$, respectively. These values are only slightly different from $\alpha_{min} = 0.591$ and $\alpha_{max} = 1.693$ obtained by means of the original local singularity approach, without application of the Chen

11.6 Chen Algorithm

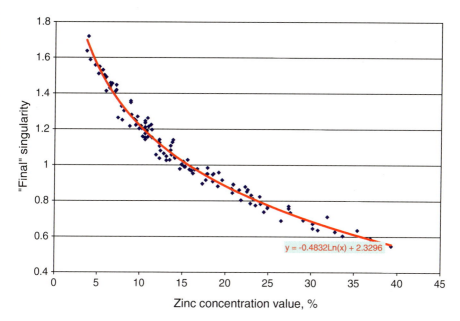

Fig. 11.32 Relationship between final singularity and zinc concentration value is logarithmic. Plotted values (*diamonds*) are same as those shown for $k = 1,000$ in Fig. 11.31. Logarithmic curve was fitted by least squares (Source: Agterberg 2012a, Fig. 21)

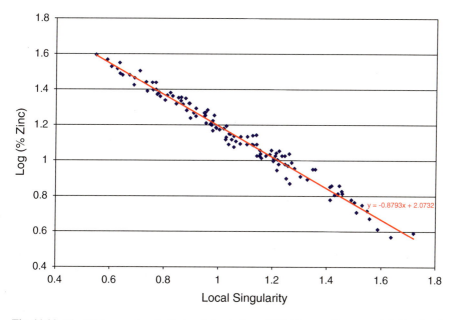

Fig. 11.33 Final Pulacayo singularity in relation to $\log_{10}$ (%Zn) is according to straight line fitted by least squares. Residuals from this line are also shown in Fig. 11.34 (Source: Agterberg 2012a, Fig. 22)

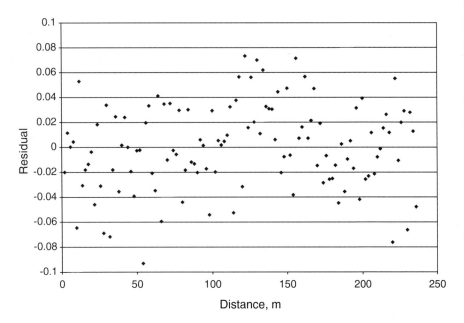

Fig. 11.34 Residuals from straight line in Fig. 11.33 display white noise pattern with variance equal to 0.00118 (Source: Agterberg 2012a, Fig. 23)

algorithm (also see Chen et al. 2007, Table 1). However, the minimum and maximum singularity estimates obtained by singularity analysis differ significantly from estimates based on the multifractal spectrum previously obtained by the methods of moments (see Figs. 11.7 and 11.11) as will be explained in the next paragraph.

Evertsz and Mandelbrot (1992, p. 941) show that the mass exponents satisfy: $\tau(q) = -\log_2[p^q + (1-p)^q]$. Cheng and Agterberg (1996) illustrated that, for q ranging from -35 to 50, the method of moments provides the multifractal spectrum of Fig. 11.7. For small values of q, mass exponent estimates are excellent; e.g., $\tau(2) = 0.979 \pm 0.038$ (Fig. 11.11). From $\tau(2) = 0.979$ it would follow that $d = 0.121$, $\alpha_{min} = 0.835$, and $\alpha_{max} = 1.186$. The latter estimate differs not only from $\alpha_{max} = 1.719$ derived in this section, it also is less than the estimate $\alpha_{max} = 1.402$ on the right side of the multifractal spectrum shown in Fig. 11.7.

The estimate $d = 0.121$ obviously is much too small. Use of the relations $p = 0.5(1-d)$, $\alpha_{min} = -\log_2(1-p) = 0.369$ and $\alpha_{max} = -\log_2 p = 1.719$, produces the much larger estimates $d = 0.369$ and $d = 0.392$, respectively. Clearly, the binomial/p model has limited range of applicability although $\alpha_{max} = 1.402$ on the right side of the multifractal spectrum would yield $d = 0.243$, which is closer to 0.392 than 0.121. The preceding inconsistencies suggest that a more flexible model with additional parameters should be used. Two possible explanations are as follows. A "universal multifractal model" with three parameters was initially developed during the late 1990s by Schertzer and Lovejoy (1991). Lovejoy and Schertzer (2007) have successfully applied this model to the 118 Pulacayo zinc values as will be discussed Sect. 12.6. The other possible explanation is that, in general, α_{min} and

α_{max} cannot be determined by the method of moments because they are would be almost exclusively determined by the largest and smallest observed values. If this second explanation is valid, mass exponents and multifractal spectrum derived by the method of moments such as those shown in Fig. 11.7 (and Fig. 11.11) are valid only for the smaller values of q. For very large values ($q \to \infty$) the multifractal spectrum develops long narrow tails with $f(\alpha)$ intersection the α-axis at α_{min} and α_{max} as estimated by singularity analysis.

11.6.2 KTB Copper Example

The second example is for the long series consisting of 2,796 copper (X-Ray Fluorescence Spectroscopy) concentration values for cutting samples taken at 2 m intervals along the Main KTB borehole, on the Bohemian Massif in southeastern Germany. These data (Fig. 6.12) were previously analysed in Chap. 6. It was assumed that 101-point average copper values representing consecutive 202-m long segments of drill-core captured deterministic trends in this series reflecting systematic changes in rock compositions. The data set was divided into three series (1, 2 and 3) with 1,000, 1,000 and 796 values, respectively. Mean copper values for these three series are 37.8, 33.7 and 39.9 ppm Cu, and corresponding standard deviations are 20.3, 11.0 and 20.6 ppm Cu, respectively. Correlograms for the three series are significantly different (Fig. 6.13), although the patterns for series 2 and 3 resemble one another. Figure 6.14 showed correlograms of the three series using the differences between the copper values and their moving average. Each correlogram is approximately semi-exponential with equation $\rho_h = c \bullet \exp(-ah)$. The slope coefficients (a) of the three curves are nearly equal to one another (0.40, 0.38 and 0.41 for series 1, 2 and 3, respectively). Series 2 and 3 also have similar nugget effect at the origin and were combined to form a single series to be used here for local singularity analysis with white noise variance equal to 37.7 %. Series 1 was not included in this new data set because its nugget effect is markedly different. In Agterberg (2012a) it was pointed out that quantitative modeling of the nugget effect in KTB copper determinations yielded better results than could be obtained for several examples from mineral deposits including the Pulacayo zinc orebody. This is not only because the series of KTB chemical determinations is much longer but also because the nugget effect remains clearly visible over lag distances between 2 m (=original sampling interval) and 10 m.

Mean and variance of the new series are 36 ppm Cu and 268.4, respectively. The variance of logarithmically (base 10) copper values is 0.024447. The coefficient of variation of original data amounts to 0.45. Because this is less than 0.5, the white noise component probably also provides an approximation for the logarithmically transformed copper values. Approximate white noise variance then would be $(0.377 \times 0.024447 =) 0.0092$. This estimate is probably slightly too large because the logarithmic variance of 0.24447 does not account for the deterministic trends. However, it provides a crude estimate that can be used for comparison in the following application of the Chen algorithm.

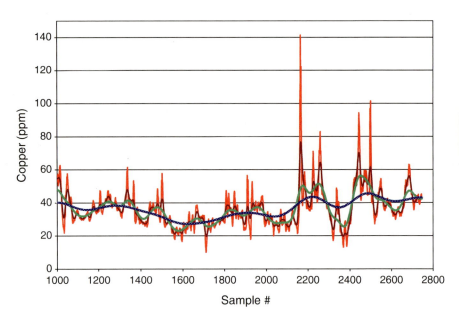

Fig. 11.35 Successive patterns of c_k obtained after 1 (*most scatter*), 5, 100 (*solid line*) and 1,000 (*smoothest curve*) iterations (Source: Agterberg 2012b, Fig. 12)

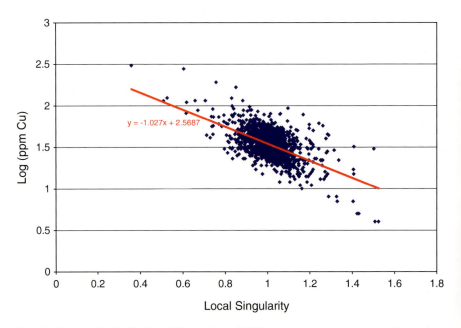

Fig. 11.36 Logarithmically (base 10) transformed KTB copper concentration versus local similarity after single iteration ($k=1$) with regression line (singularity assumed to be error-free). Variance of logarithmically transformed copper concentration value is 0.01636 (Source: Agterberg 2012b, Fig. 13)

11.6 Chen Algorithm

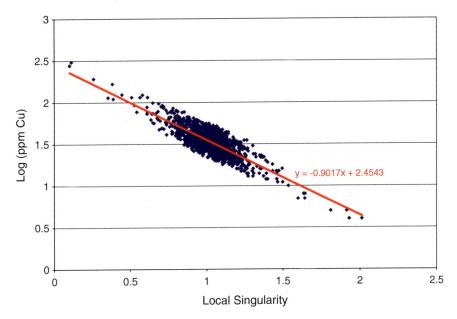

Fig. 11.37 Logarithmically (base 10) transformed KTB copper concentration versus local similarity after $k = 100$ iterations with regression line (singularity assumed to be error-free). Variance of logarithmically transformed copper concentration value is 0.00823 (Source: Agterberg 2012b, Fig. 14)

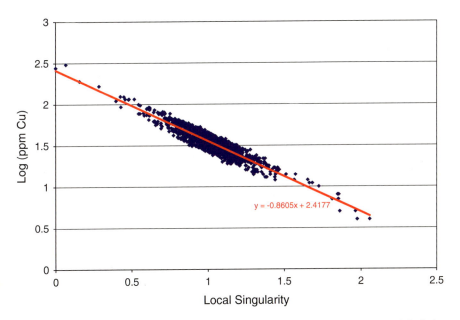

Fig. 11.38 Logarithmically (base 10) transformed KTB copper concentration versus local similarity after $k = 1,000$ iterations with regression line (singularity assumed to be error-free). Variance of logarithmically transformed copper concentration value is 0.00451 (Source: Agterberg 2012b, Fig. 15)

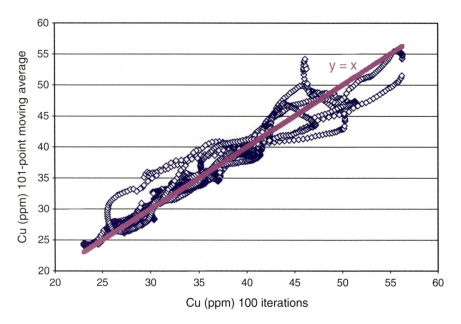

Fig. 11.39 Moving average shown in Fig. 6.22 (Location value >1,000) plotted against c_k obtained after 100 iterations. Because both variables are nearly continuous, their relation plots as a nearly continuous line as well. The $y = x$ line shown for comparison suggests that both variables are equal on the average (Source: Agterberg 2012b, Fig. 16)

Figure 11.35 shows the coefficients c_k for $k = 1, 5, 100$ and 1,000, respectively. A procedure to be explained in more detail in Sect. 12.6 can be used to help decide which pattern is best. Figures 11.36, 11.37 and 11.38 are plots of $\log_{10}$ Cu versus estimated singularities for $k = 1, 100$ and 1,000, respectively. The regression line in Fig. 11.35 provides a poor fit because estimates of the singularities for $k = 1$ cannot be considered to be free of error. Residual variance with respect to this line is 0.01636. The regression lines in Figs. 11.37 and 11.38 for $k = 100$ and $k = 1,000$ provide good fits. However, the result in Fig. 11.36 is to be preferred because it's residual variance (=0.00823) is close to the white noise variance (=0.0092) estimated in the preceding paragraph. Residual variance for the line in Fig. 11.37 amounts to 0.00451 and is about two times too small.

The pattern of c_k for $k = 100$ closely resembles the 101-point moving average (for sample numbers >1,000) shown in Fig. 6.12. This relationship is illustrated in Fig. 11.38. The relatively good fit of the $y = x$ line in this diagram suggests that both variables are subject to approximately the same uncertainty. Figure 11.39 shows the corresponding "final" relation between $\log_{10}$ Cu and singularity (as obtained for $k = 100$). This result in compared with singularities estimated for $k = 1$ in Fig. 11.39 (for 500 sampling points only). The minimum and maximum singularity for $k = 100$ are only slightly less and greater than the result obtained after a single iteration.

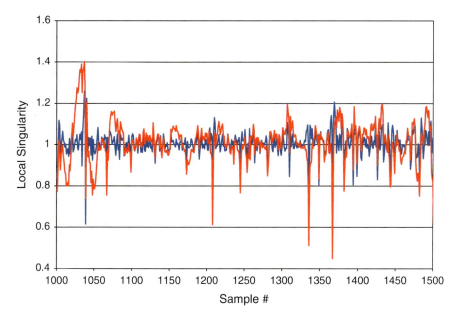

Fig. 11.40 "Final" singularities (*heavy line*) of Fig. 11.37 for KTB copper concentration values at 500 sampling points compared with singularities (thinner line of Fig. 11.39 obtained after single iteration (Source: Agterberg 2012b, Fig. 17)

However, there are several new relatively low values not detected for $k=1$ (Fig. 11.40).

The application of the Chen algorithm described in this section differs from the application to the Pulacayo zinc values described in the previous section. Full convergence for $k \rightarrow \infty$ generally results in a subhorizontal straight line. For the KTB copper values it can be assumed that there exist systematic changes in average copper content along the borehole and convergence was stopped at a point that the c_k closely approximated the deterministic copper trend pattern. The proof that the singularities after approximately 100 iterations are realistic is provided by the fact that the residuals from the trend pattern have nearly the same semi-exponential autocorrelation function that dies out rapidly with distance. This semi-exponential curve is equivalent to the semi-exponential close to the origin in Fig. 6.19.

References

Aczél J, Dhombres J (1989) Functional equations in several variables. Cambridge University Press, Cambridge

Agterberg FP (1994a) Fractals, multifractals, and change of support. In: Dimitrakopoulos R (ed) Geostatistics for the next century. Kluwer, Dordrecht, pp 223–234

Agterberg FP (1994b) FORTRAN program for the analysis of point patterns with correction for edge effects. Comput Geosci 20:229–245

Agterberg FP (1995) Multifractal modeling of the sizes and grades of giant and supergiant deposits. Int Geol Rev 37:1–8

Agterberg FP (1997) Multifractal modelling and geostatistics applied to bore-hole fractures in the Lac du Bonnet Batholith, Manitoba, Canada. In: Proceedings of the 6th international geostatistics congress Wollongong, Australia. Kluwer, Dordrecht, Sept 1996, pp 1163–1172

Agterberg FP (2001) Multifractal simulation of geochemical map patterns. In: Merriam DF, Davis JC (eds) Geologic modeling and simulation: sedimentary systems. Kluwer, New York, pp 327–346

Agterberg FP (2007a) New applications of the model of de Wijs in regional geochemistry. Math Geol 39(1):1–26

Agterberg FP (2007b) Mixtures of multiplicative cascade models in geochemistry. Nonlinear Process Geophys 14:201–209

Agterberg FP (2012a) Sampling and analysis of element concentration distribution in rock units and orebodies. Nonlinear Process Geophys 19:23–44

Agterberg FP (2012b) Multifractals and geostatistics. J Geochem Explor 122:113–122

Agterberg FP, Cheng Q, Wright DF (1993) Fractal modeling of mineral deposits. In: Elbrond J, Tang X (eds) Application of computers and operations research in the mineral industry. In: Proceedings of the 24th APCOM symposium 1, Canadian Institute of Mining, Metallurgy and Petroleum Engineering, Montreal, pp 43–53

Agterberg FP, Cheng Q, Brown A, Good D (1996a) Multifractal modeling of fractures in the Lac du Bonnet Batholith, Manitoba. Comput Geosci 22:497–507

Agterberg FP, Brown A, Sikorsky RI (1996b) Multifractal modeling of subsurface fractures from bore-holes near the Underground Research Laboratory, Pinawa, Manitoba. In: Deep geological disposal of radioactive waste, Proceedings of the international conference on Canadian Nuclear Society, Winnipeg, vol 3, pp 1–12

Arias M, Gumiel P, Sanderson DJ, Martin-Izard A (2011) A multifractal simulation model for the distribution of VMS deposits in the Spanish segment of the Iberian Pyrite Belt. Comput Geosci 37:1917–1927

Arias M, Gumiel P, Martin-Izard A (2012) Multifractal analysis of geochemical anomalies: a tool for assessing prospectivity at the SE border of the Ossa Morena Zone, Variscan Massif (Spain). J Geochem Explor 122:101–112

Bickel PJ, Doksum KA (2001) Mathematical statistics, vol 1, 2nd edn. Prentice-Hall, Upper Saddle River

Brinck JW (1971) MIMIC, the prediction of mineral resources and long-term price trends in the non-ferrous metal mining industry is no longer Utopian. Eurospectra 10:46–56

Brinck JW (1974) The geochemical distribution of uranium as a primary criterion for the formation of ore deposits. In: Chemical and physical mechanisms in the formation of uranium mineralization, geochronology, isotope geology and mineralization. In: IAEA Proceedings of Series STI/PUB/374, International Atomic Energy commission, Vienna, pp 21–32

Brown A, Soonawala NM, Everitt RA, Kamineni DC (1989) Geology and geophysics of the underground research laboratory site, Lac du Bonnet Batholith. Manitoba. Can J Earth Sci 26:404–425

Chen Z, Cheng Q, Chen J, Xie S (2007) A novel iterative approach for mapping local singularities from geochemical data. Nonlinear Process Geophys 14:317–324

Cheng Q (1994) Multifractal modeling and spatial analysis with GIS: gold mineral potential estimation in the Mitchell-Sulphurets area, northwestern British Columbia. Unpublished Doctoral Dissertation, University of Ottawa

Cheng Q (1999) Multifractal interpolation. In: Lippard SJ, Naess A, Sinding-Larsen R (eds) Proceedings of Fifth Annual Conference of the International Association for Mathematical Geology, Tapir, Trondheim, pp 245–250

Cheng Q (2003) GeoData Analysis System (GeoDAS) for mineral exploration and environmental assessment, user's guide. York University, Toronto

Cheng Q (2005) A new model for incorporating spatial association and singularity in interpolation of exploratory data. In: Leuangthong D, Deutsch CV (eds) Geostatistics Banff 2004. Springer, Dordrecht, pp 1017–1025

Cheng Q (2008) Non-linear theory and power-law models for information integration and mineral resources quantitative assessments. In: Bonham-Carter GF, Cheng Q (eds) Progress in geomathematics. Springer, Heidelberg, pp 195–225

Cheng Q, Agterberg FP (1995) Multifractal modeling and spatial point processes. Math Geol 27(7):831–845

Cheng Q, Agterberg FP (1996) Multifractal modeling and spatial statistics. Math Geol 28(1):1–16

Cheng Q, Agterberg FP (2009) Singularity analysis of ore-mineral and toxic trace elements in stream sediments. Comput Geosci 35:234–244

Cheng Q, Agterberg FP, Ballantyne SB (1994a) The separation of geochemical anomalies from background by fractal methods. J Geochem Explor 51(2):109–130

Cheng Q, Agterberg FP, Bonham-Carter GF (1994b) Fractal pattern integration for mineral potential mapping. In: Proceedings of IAMG'94, Mont Tremblant, pp 74–80

Cheng Q, Bonham-Carter GF, Agterberg FP, Wright DF (1994c) Fractal modelling in the geosciences and implementation with GIS. In: Proceedings of 6th Canadian Conference Institute Survey Map, Ottawa, vol 1, pp 565–577

Cheng Q, Agterberg FP, Bonham-Carter GF (1996) A spatial analysis method for geochemical anomaly separation. J Geochem Explor 56:183–195

Chhabra A, Jensen RV (1989) Direct determination of the $f(\alpha)$ singularity spectrum. Phys Rev Lett 62(12):1327–1330

Cowie PA, Sornette D, Vanneste C (1995) Multifractal scaling properties of a growing fault population. Geophys J Int 122(2):457–469

Cressie NAC (1991) Statistics for spatial data. Wiley, New York

Darnley AG (1995) International geochemical mapping – a review. J Geochem Explor 55:5–10

Darnley AG, Garrett RG, Hall GEM (1995) A global geochemical database for environmental and resource management: recommendations for international geochemical mapping, Final report of IGCP project 259. UNESCO, Paris

De Wijs HJ (1951) Statistics of ore distribution I. Geol Mijnbouw 13:365–375

Diggle PJ (1983) Statistical analysis of spatial point patterns. Academic, London

Evertsz CJG, Mandelbrot BB (1992) Multifractal measures. In: Peitgen H-O, Jurgens H, Saupe D (eds) Chaos and fractals. Springer, New York

Falconer K (2003) Fractal geometry, 2nd edn. Wiley, Chichester

Feder J (1988) Fractals. Plenum, New York

Garrett RG (1986) Geochemical abundance models: an update, 1975 to 1987. US Geol Surv Cir 980:207–220

Ghosh A, Daemen JJK (1993) Fractal characteristics of rock discontinuities. Eng Geol 34(1):1–9

Gillespie PA, Howard CB, Walsh JJ, Watterson J (1993) Measurement and characterisation of spatial distributions of fractures. Tectonophysics 226(1):113–141

Gonçalves MA, Mateus A, Oliveira V (2001) Geochemical anomaly separation by multifractal modeling. J Geochem Explor 72:91–114

Harris DP (1984) Mineral resources appraisal. Clarendon, Oxford

Herzfeld UC, Overbeck C (1999) Analysis and simulation of scale-dependent fractal surfaces with application to seafloor morphology. Comput Geosci 25(9):979–1007

Herzfeld UC, Kim II, Orcutt JA (1995) Is the ocean floor a fractal? Math Geol 27(3):421–442

Korvin G (1992) Fractal models in the earth sciences. Elsevier, Amsterdam

Krige DG (1966) A study of gold and uranium distribution pattern in the Klerksdorp goldfield. Geoexploration 4:43–53

Lerche I (1993) A probabilistic procedure to assess the uncertainty of fractal dimensions from measurements. Pure Appl Geophys 140(3):503–517

Lovejoy S, Schertzer D (1991) Multifractal analysis techniques and the rain and cloud fields from 10^{-3} to 106 m. In: Schertzer D, Lovejoy S (eds) Non-linear variability in geophysics. Kluwer, Dordrecht, pp 111–144

Lovejoy S, Schertzer D (2007) Scaling and multifractal fields in the solid earth and topography. Nonlinear Process Geophys 14:465–502

Mandelbrot BB (1983) The fractal geometry of nature. Freeman, San Francisco

Mandelbrot BB (1989) Multifractal measures, especially for the geophysicist. Pure Appl Geophys 131:5–42

Matheron G (1962) Traité de géostatistique appliquée, Mém BRGM 14. Technip, Paris

Meneveau C, Sreenivasan KR (1987) Simple multifractal cascade model for fully developed turbulence. Phys Rev Lett 59(7):1424–1427

Pahani A, Cheng Q (2004) Multifractality as a measure of spatial distribution of geochemical patterns. Math Geol 36:827–846

Plant J, Hale M (eds) (1994) Handbook of exploration geochemistry 6. Elsevier, Amsterdam

Ripley BD (1981) Spatial statistics. Wiley-Interscience, New York

Roach DE, Fowler AD (1993) Dimensionality analysis of patterns: fractal measurements. Comput Geosci 19(6):843–869

Ruzicka V (1976) Uranium resources evaluation model as an exploration tool – exploration for uranium deposits. IAEA Proc Ser 434:673–682

Schertzer D, Lovejoy S (1991) Non-linear geodynamical variability: multiple singularities, universality and observables. In: Schertzer D, Lovejoy S (eds) Non-linear variability in geophysics. Kluwer, Dordrecht, pp 41–82

Stanley H, Meakin P (1988) Multifractal phenomena in physics and chemistry. Nature 335:405–409

Turcotte DL (1997) Fractals and chaos in geology and geophysics, 2nd edn. Cambridge University Press, Cambridge

Veneziano D, Furcolo P (2003) Marginal distribution of stationary multifractal measures and their Haar wavelet coefficients. Fractals 11(3):253–270

Veneziano D, Langousis A (2005) The maximum of multifractal cascades: exact distribution and approximations. Fractals 13(4):311–324

Walsh JJ, Watterson J (1993) Fractal analysis of fracture patterns using the standard boxcounting technique: valid and invalid methodologies. J Struct Geol 15(12):1509–1512

Xiao F, Chen J, Zhang Z, Wang C, Wu G, Agterberg FP (2012) Singularity mapping and spatially weighted principal component analysis to identify geochemical anomalies associated with Ag and Pb-Zn polymetallic mineralization in Northwest Zhejiang, China. J Geochem Explor 122:90–100

Xie X, Mu X, Ren T (1997) Geochemical mapping in China. J Geochem Explor 60:99–113

Xie S, Cheng Q, Chen Z, Bao Z (2007) Application of local singularity in prospecting potential oil/gas targets. Nonlinear Process Geophys 14:185–292

Yu C (2002) Complexity of earth systems – fundamental issues of earth sciences. J China Univ Geosci 27:509–519 (in Chinese; English abstract)

Zuo R, Cheng Q, Agterberg FP, Xia Q (2009) Application of singularity mapping technique to identify local anomalies using stream sediment geochemical data, a case study from Gangdese, Tibet, western China. J Geochem Explor 101:225–235

Chapter 12
Selected Topics for Further Research

Abstract The case history studies described in the preceding eleven chapters leave some questions that could not be answered in full. New theoretical approaches in mathematical statistics and nonlinear physics provide new perspectives for the analysis of geoscience data. For example, bias due to incomplete information continues to be one of the most serious problems in 3-D mapping. How methods such as the jackknife and bootstrap can help to reduce this type of bias is briefly investigated and illustrated using volcanogenic massive copper deposits in the Abitibi area on the Canadian Shield. Compositional data analysis offers new ways to analyze multivariate data sets. Geochemical data from Fort à la Corne kimberlites in central Saskatchewan are used to illustrate the isometric logratio transformation for chemical data that form a closed number system. Three generalizations of the model of de Wijs are: (1) the 3-parameter model with finite number of iterations; (2) the random cut model in which the dispersion index d is replaced by a random variable D; and (3) the accelerated dispersion model in which d depends on concentration value during the cascade process. Universal multifractals constitute a useful generalization of multifractal modeling. As illustrated on the basis of the Pulacayo zinc values, new tools such as use of the first order structure function and double trace analysis generalize conventional variogram-autocorrelation fitting. Measurements on compositions of blocks of rocks generally depend on block size. For example, at microscopic scale chemical elements depend on frequencies of abundance of different minerals. On a regional basis, rock type composition depends on spatial distribution of contacts between different rock types. Frequency distribution modeling of compositional data can be useful in ore reserve estimation as well as regional mineral potential studies. During the 1970, Georges Matheron proposed the theory of permanent frequency distributions with shapes that are independent of block size. The lognormal is a well-known geostatistical example. The probnormal distribution is useful for the analysis of relative amounts of different rock types contained in cells of variable size. It arises when probits of percentage values are normally distributed. Its Q-Q plot has scales derived from the normal distribution along both axes. Parameters (mean and variance) of the probnormal distribution are related to the geometrical covariances of the objects of study. Practical examples are spatial distribution of acidic and mafic volcanics in the Bathurst area, New Brunswick, and in the Abitibi volcanic belt on the Canadian Shield in east-central Ontario and western Quebec.

Keywords Incomplete information • Jackknife method • Compositional data analysis • Non-linear process modeling • Random cut model • Accelerated dispersion • Universal multifractals • Cell composition modeling • Permanent frequency distributions • Probnormal model • Star Kimberlite • South Saskatchewan till • Bathurst acidic volcanics • Abitibi acidic volcanics

12.1 Bias and Grouped Jackknife

In Sect. 7.2.1 elevations of the top of the Arbuckle Formation in Kansas were analyzed as in Agterberg (1970) in order to compare various trend surface and kriging applications with one another. The data set was randomly divided into three subsets: two of these subsets were used for control and results derived for the two control sets were applied to the third "blind" subset in order to see how well results for the control subsets could predict the values in the third subset (*cf.* Sect. 7.2.1). In his comments on this approach Tukey (1970) stated that this form of cross-validation could be used but a better technique would be the Jackknife proposed by Mosteller and Tukey (1968).

Efron (1982) describes cross-validation in the following terms. In original applications of cross-validation, a data set was randomly divided into two halves and one of these halves used for model testing. For example, a regression model fitted to the first half is used to predict the second half. Normally, the first half predictions do less well in predicting the second half than they did for the first half. After computers became widely available, cross-validation could be generalized. It is more common now to leave one data point out at a time and fit the model to the remaining points to see how well it does on the excluded point. The average of the prediction errors, each point being left out once, then is the cross-validated measure of prediction error. This "leave-off-one" cross-validation technique was used in Sect. 9.5.3 to find the best smoothing factor (SF) used for cubic spline-curve fitting of age determinations plotted along a relative geologic timescale. Cross-validation, the jackknife and bootstrap are three techniques that are closely related. Efron (1982, Chap. 7) discusses their relationships in a regression context pointing out that, although the three methods are close in theory, they can yield different results in practical applications.

In Chap. 2 it was pointed out that for a set of n independent and identically distributed (iid) data the standard deviation of the sample mean $(\bar{x})$ satisfies $\hat{\sigma}(\bar{x}) = \sqrt{\frac{\sum_{i=1}^{n}(x_i-\bar{x})^2}{n(n-1)}}$. Although this is a good result it cannot be extended to other estimators such as the median. However, the jackknife and bootstrap can be used to make this type of extension. Suppose $\bar{x}_i = \frac{n\bar{x}-x_i}{n-1}$ represents the sample average of the same data set but with the data point x_i deleted. Let $\bar{x}_{JK}$ represent the mean of the n new values $\bar{x}_i$. The jackknife estimate of the standard deviation is

12.1 Bias and Grouped Jackknife

$$\hat{\sigma}_{JK} = \sqrt{\frac{(n-1)\sum_{i=1}^{n}(\bar{x}_i - \bar{x}_{JK})^2}{n}}.$$ It is easy to show that $\bar{x}_{JK} = \bar{x}$ and $\sigma_{JK} = \sigma(\bar{x})$. The advantage of the jackknife estimate for the standard deviation is that it can be generalized to other types of estimators such as the median.

The bootstrap generalizes the mean and its standard deviation in a different way. Suppose that n samples X_i, all of size n, are drawn with replacement from an empirical probability distribution. The average value of the sample means can be written as $\bar{X}_{BS}$. Then the bootstrap estimate of the standard deviation of $\bar{X}_{BS}$ is

$$\hat{\sigma}_{BS} = \sqrt{\frac{\sum_{i=1}^{n}(\bar{X}_i - \bar{X}_{BS})^2}{n^2}}$$ (cf. Efron 1982, p. 2).

The rationale of bias reduction by using the jackknife is as follows. The jackknife was originally invented by Quenouille (1949) under another name and with the purpose to obtain a nonparametric estimate of bias associated with some types of estimators. Bias can be formally defined as $BIAS \equiv E_F\{\vartheta(\hat{F})\} - \vartheta(F)$ where E_F denotes expectation under the assumption that n iid quantities were drawn from an unknown probability distribution F, $\hat{\vartheta} = \vartheta(\hat{F})$ is the estimate of a parameter of interest with $\hat{F}$ representing the empirical probability distribution. Quenouille's bias estimate (cf. Efron 1982, p. 5) is based on sequentially deleting values x_i from a sample of n values to generate different empirical probability distributions $\hat{F}_i$ based on $(n-1)$ values and obtaining the estimates $\hat{\vartheta}_i = \vartheta(\hat{F}_i)$. Suppose $\overline{\vartheta} = \frac{\sum_{i=1}^{n}\hat{\vartheta}_i}{n}$, then Quenouille's bias estimate becomes $(n-1) \cdot (\overline{\vartheta} - \hat{\vartheta})$ and the bias-corrected "jackknifed estimate" of ϑ becomes $\tilde{\vartheta} = n\hat{\vartheta} - (n-1)\overline{\vartheta}$. This estimate is either unbiased or less biased than $\hat{\vartheta}$.

Box 12.1: Expectation E_n as a Function of $1/n$

Suppose E_n denotes the expectation of $\hat{\vartheta}$ for a sample of size n. Then, $E_n = \vartheta + \frac{a_1(F)}{n} + \frac{a_2(F)}{n^2} + \frac{a_3(F)}{n^3} + \ldots$ for most statistics including all maximum likelihood estimates. The functions $a_1(F)$, $a_2(F)$, ... are independent of n. Also, $E_{n-1} = \vartheta + \frac{a_1(F)}{n-1} + \frac{a_2(F)}{(n-1)^2} + \frac{a_3(F)}{(n-1)^3} + \ldots$ (Schucany et al. 1971). Consequently, $E_F(\tilde{\vartheta}) = nE_n - (n-1)E_{n-1} = \vartheta - \frac{a_2(F)}{n(n-1)} + a_3(F)\left\{\frac{1}{n^2} + \frac{1}{n(n-1)^2}\right\} + \ldots$. Contrary to $\hat{\vartheta}$, which is biased in $O(1/n)$, $\tilde{\vartheta}$ is biased in $O(1/n^2)$. For example, if the maximum likelihood estimate $\hat{\vartheta} = \frac{\sum_{i=1}^{n}(x_i - \bar{x})^2}{n}$ of variance of the mean $\bar{x} = \frac{\sum_{i=1}^{n} x_i}{n}$ is used, $\tilde{\vartheta} = \frac{\sum_{i=1}^{n}(x_i - \bar{x})^2}{n-1}$. Suppose that E_n is plotted against $1/n$ (see Fig. 12.1). Then the unbiased estimator $\vartheta = E_\infty$, and $\frac{E_n - E_\infty}{E_{n-1} - E_n} \approx \frac{1/n}{1/(n-1) - 1/n}$, or $BIAS = E_n - E_\infty = (n-1)(E_{n-1} - E_n)$ and $\vartheta = nE_n - (n-1)E_{n-1}$. Consequently, the jackknife simply replaces E_n by E_{n-1} (cf. Efron 1982, p. 8).

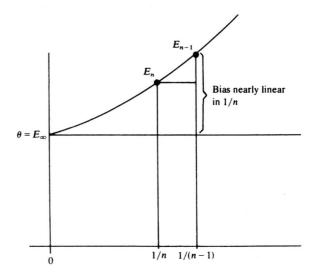

Fig. 12.1 Bias reduction in grouped jackknife; the expectation E_n as a function of n (Source: Efron 1982, Fig. 2.1)

12.1.1 Abitibi Volcanogenic Massive Sulphides Example

Can the bias reducing capability of the jackknife be used in spatial situations in which there is complete information on a set of explanatory variables but incomplete information on the dependent variable or probability of occurrence that is to be estimated? The following example produced results that are encouraging but further research would have to be undertaken to establish more widespread applicability. Other publications reporting tentative results along the same lines include Chung and Agterberg (1980) and Fabbri and Chung (2008).

Agterberg (1973) investigated the possibility of using the jackknife for eliminating bias in mineral potential mapping where bias is introduced by likely existence of undiscovered mineral deposits in a study area. Large deposits for several metals in the Abitibi area on the Canadian Shield were taken for example. Applications of stepwise regression to relate occurrences of large copper deposits to regional geological and geophysical variables were previously discussed in Sect. 4.3.2. Earlier, contours of expected number of cells with large copper deposits contained in (40 × 40 km) cells had been corrected for bias because of undiscovered copper deposits by assuming that all large deposits near the surface of bedrock had been discovered in a control area constructed around the relatively well explored mining camps in the Abitibi area (cf. Sect. 4.3). The jackknife can be used to eliminate this type of bias instead of making the assumption involving a control area.

In the application to copper deposits in the Abitibi area (Agterberg 1973) a grouped jackknife was used. In general, a set of N data can be randomly divided it k subgroups each consisting of n data so that $k \cdot n = N$. Suppose a statistic Y is estimated from all data yielding $\hat{Y}_{all}$; and also k times from sets of $(N-n)$ data, which gives k estimates $\hat{Y}_j$ ($j = 1, \ldots, k$). The so-called pseudo-values $\hat{Y}_j^* = k \cdot \hat{Y}_{all} - (k-1) \cdot \hat{Y}_j$ are averaged to obtain the jackknife estimate $\hat{Y}^*$ and its variance

12.1 Bias and Grouped Jackknife

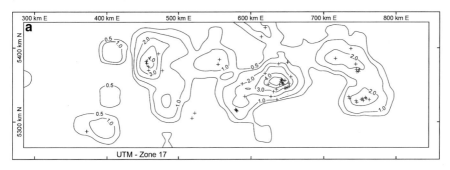

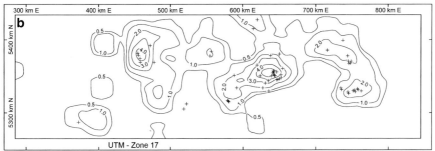

Fig. 12.2 Abitibi area, Canadian Shield; study area as outlined in Fig. 4.12. (**a**) Linear model used to correlate volcanogenic massive sulphide deposits with same lithological variables as used for Fig. 4.16. Contours for (40 × 40 km) cells are based on sum of estimated posterior probabilities summed for the 16 (10 × 10 km) cells contained within these larger unit cells multiplied by the constant F assuming zero mineral potential within "control" area consisting of all (10 × 10 km) "control" cells containing one or more deposits plus all cells that have one or more sides in common with a control cell. (**b**) Jackknife estimate as obtained without control area (modified from Agterberg 1973, Fig. 4)

$s^2(\hat{Y}^*) = s^2(\hat{Y}_j^*)/k$. In this example, there were $N = 35$ control cells each containing one or more volcanogenic massive sulphide deposits. These were randomly divided into $k = 7$ groups of $n = 5$ control cells. A stepwise solution similar to the one previously shown in Table 4.4 was obtained for the entire data set (35 control cells) after setting observed values of Y equal to 1 in control cells (Fig. 12.2a). The contours in this diagram are for expected number of (10 × 10 km) cells per (30 × 30 km) unit cell. For this experiment, cell composition data for 8 different rock types in 1,086 (10 × 10 km) cells were used as explanatory variables augmented by the 28 cross-product variables for these eight rock types. A control area surrounding the relatively well explored mining districts was defined to correct for bias due to undiscovered deposits. Next, similar stepwise solutions were obtained for 7 additional runs, each based on 30 control cells. A control area surrounding the relatively well explored mining districts was not defined for these seven new experiments.

An example of results obtained by this jackknife experiment on Abitibi volcanogenic massive sulphides is as follows (*cf.* Agterberg 1973, p. 8). The explanatory variable first selected when all 35 control cells are used is the

Table 12.1 Comparison of probabilities estimated for ten cells by means of original copper potential determination method and the jackknife

Location	p (original)	p (jackknife)	SD
32/62	0.45	0.44	0.08
16/58	0.33	0.32	0.04
17/58	0.39	0.38	0.05
18/58	0.01	0.00	0.06
16/59	0.35	0.36	0.04
17/59	0.33	0.33	0.14
18/59	0.37	0.40	0.13
16/60	0.43	0.47	0.04
17/60	0.06	0.00	0.06
18/60	0.03	−0.06	0.08

Location is given by UTM grid cell numbers; SD = standard deviation of jackknife estimate (Source: Agterberg 1973, Table 1)

cross-product of acidic volcanics and mafic intrusives. Its coefficient was $B(2,6) = 0.50954 \cdot 10^{-4}$. Dropping the factor 10^{-4}, the seven new estimates were 0.49583, 0.37170, 0.51601, 0.52055, 0.20898, 0.54010 and 0.41653. The average of these 7 new estimates for 30 control cells is 0.43853, which is less than 0.50954 for all 35 control cells. A result of this type is not surprising because bias should increase when fewer control cells are used. The randomly selected five control cells can be regarded as containing volcanogenic massive sulphide deposits that had not been discovered. The seven pseudo-values obtained for $B(2,6)$ are 0.592, 1.337, 0.471, 0.443, 2.313, 0.326 and 1.068, respectively. The jackknife coefficient becomes 0.9356 with variance 0.07198. Similar results were obtained for the other explanatory variables.

The eight stepwise regressions resulted in sets of eight estimated values for all (10 × 10 km) cells in the Abitibi area. Jackknife estimates were obtained for all individual cells and these were combined for overlapping sets of cells to obtain estimated numbers of (10 × 10 km) cells with one or more volcanogenic massive sulphide deposits per larger (30 × 30 km) unit cell (*cf.* Sect. 4.3.2 in which larger unit cells measuring 40 km on a side were used). The result is shown in Fig. 12.2b. There is hardly any difference between the patterns of Fig. 12.2a, b indicating that both methods of bias reduction (use of control area supposedly without undiscovered deposits in Fig. 12.2a and the jackknife in Fig. 12.2b) are about equally good. In Chap. 4, it was discussed in detail that multivariate regression of occurrences of mineral deposits on rock types and other variables measured for the surface of bedrock are only valid for undiscovered deposits occurring relatively close to bedrock surface. Later discovered copper ore was mostly found deeper down but within the same favorable areas within the Abitibi Volcanic Belt as the large copper deposits that already were known in the late 1960s.

Table 12.1 shows a comparison of results for ten selected (10 × 10 km) cells obtained by the original application of the general linear model of least squares and the new jackknife results shown in Fig. 12.2a, b, respectively. These cells constitute a subsample of the 1,086 cells used for this example. Cell 32/62 contains the Magusi River deposit that was discovered in 1972 after publication of the original

Abitibi copper potential map (see Sect. 4.4.1) and cell 17/59 contains the Kidd Creek Mine with the largest volcanogenic massive sulphide deposit in the Abitibi area. All 1,086 probabilities for (10 × 10 km) cells obtained by the original application were multiplied by the factor $F = 1.828$ representing the number of control cells known to contain one or more massive sulphide deposits in the control area constructed around the mining districts divided by the sum of originally estimated probabilities in this control area. For the ten cells in Table 12.1 the jackknife probabilities are all close to the original probabilities multiplied by 1.828. It is noted that one of the estimated probabilities in Table 12.1 is negative. This is because the general linear model was used that does not constrain probabilities to the [0, 1] interval like logistic regression (cf. Sect. 5.2.2). Negative probabilities could be replaced by zeros but in the applications to Abitibi copper deposits a correction of this type would not affect final results.

The jackknife method has two significant advantages with respect to the original solution: (1) it provides estimates of the standard deviations of cell probabilities (shown for ten cells in Table 12.1) which are larger than estimates based on ordinary regression (not shown), which are probably significantly biased; and (2) it does not require the construction of a control area; which in the original applications was a somewhat arbitrary, subjective undertaking.

12.2 Compositional Data Analysis

Geological variables, especially for rock composition data often are subject to various constraints that affect the shapes of their frequency distributions and their relations with other variables. Closed-number systems such as the major oxides measured on rock samples provide a primary example (Chayes 1971). Some very simple examples are as follows: (1) in Sect. 2.5.1 it was pointed out that, if a Pulacayo ore sample would consist of pure sphalerite, its maximum possible zinc value would be about 66 %. This upper limit constrains possible shapes of zinc frequency distributions; (2) olivine (cf. Sect. 4.2.1) is a magnesium-iron silicate in which Mg and Fe can interchange positions in the silicate lattice. It means that if different olivine crystals are analyzed for Mg and Fe, these two elements would show perfect negative correlation ($r = -1.00$) because increase in content of one element implies decrease in content of the other.

Chayes (1971) pointed out that silica (SiO_2) content generally is negatively correlated with other major oxides in collections of rock samples, simply because it usually is the major rock constituent, and increases in other major oxides must be accompanied by decreases in silica. Pearson (1897) already had pointed out that spurious correlations between variables can arise if they are functions of random variables. His reasoning was as follows: "If $x_1 = f_1(w_1, w_2)$ and $x_2 = f_2(w_2, w_3)$ be two functions of the three variables w_1, w_2, w_3, and these variables be selected at random so that there exists no correlation between w_1, w_2 or w_1, w_3 or w_2, w_3, there will still found to exist correlation between x_1, and x_2."

Obviously, application of ordinary statistical methods to data that are subject to spurious correlations can lead to erroneous results. Methods of multivariate statistical analysis often commence by calculating the variance-covariance matrix or the correlation matrix if variances are replaced by ones and covariances by correlation coefficients. The next step of multivariate analysis often consists of calculating the principal components of the variance-covariance or correlation matrix (see, e.g., Agterberg 1974; or Davis 2002). The total variance (sum of all component variances) is decomposed into a number of principal components, which are linear combinations of all variables, accounting for increasingly smaller proportions of total variance. Factor analysis can result in improved results by considering random variability of the variables (Jöreskog et al. 1976). The higher variance linear combinations or component scores can be used in various ways; e.g., on maps where they may provide more information than individual variables. A useful multivariate technique along these lines is correspondence analysis (Greenacre 2009) which results in bi-plots for pairs of components that are shown as vectors together with their scores on the original variables for easy comparison.

Aitchison (1986) has developed the powerful new methodology of compositional data analysis for variables in closed-number systems. His methods avoid misleading results based on spurious correlations. Multivariate analysis is applied to log-ratios instead of to original data. Standard statistical techniques normally apply to the log-ratios. If there are p variables in the closed number-system, the log-ratios approach can be considered to be operational in $(p-1)$-dimensional space allowing unconstrained multivariate analysis. Theory of compositional data analysis is clearly explained in Pawlowsky-Glahn and Buccianti (2011). This book also contains many recent examples of application. The approach also can be less sensitive to outliers, particularly if robust estimators are used (Filzmoser and Hron 2011). The topic of using geostatistics with compositional data is discussed by Pawlowsky-Glahn and Olea (2004). Reyment et al. (2008) applied compositional data analysis to study seasonal variation in radiolarian abundance.

There are three options for log-ratio transformation: (1) additive log-ratio (*alr*), (2) centred log-ratio (*clr*), and (3) isometric log-ratio (*ilr*). The first two techniques (*alr* and *clr*) were developed by Aitchison (1986) and the third one (*ilr*) by Egozcue et al. (2003). The *alr* method works with log-ratios in which the ratios are for original variables divided by one of the variables. This might be problematic because distances between points in the transformed space differ from divisor to divisor. The *clr* method is not subject to this particular drawback because there is a single divisor set equal to the geometric mean of all variables. However, some standard multivariate techniques are difficult to apply under *clr*. The *ilr* method was developed to overcome these problems but it also can be difficult to use for some multivariate statistical techniques. On the whole, *ilr* currently is the preferred method because it often leads to a better understanding of the situation that is being studied. The following example (after Aitchison 1986) is an application of the *clr* method. The next section will contain an *ilr* example.

The ternary diagram for three rock components that form a closed system is a commonly used tool in mathematical petrology and geochemistry. Aitchison's Data

12.2 Compositional Data Analysis

Fig. 12.3 Ternary diagram and log-contrast principal axes for 23 aphyric Skye lava compositions.
A: Alkali or Na_2O+K_2O;
F: Fe_2O_3; M: MgO (Source: Aitchison 1986, Fig. 8.5)

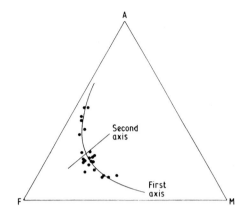

6 example (Fig. 12.3) shows an example with AFM compositions of 23 aphyric Skye lavas (original data from Thompson et al. 1972). Aitchison (1986) first applied ordinary principal component analysis using the variance-covariance matrix. This results in three principal components (PCs). The third PC has zero eigenvalue and zero variance because it simply represents the closure of the system (A + F + M = 1). PC1 and PC2 would plot as a cross in Fig. 12.3 represented by two straight lines which are perpendicular to each other and pass through the center of gravity of the cluster of input data. On the other hand, if the data are divided by their geometric mean before logs are taken, the first two principal components become as shown in Fig. 12.3. Obviously PC1 (first axis) is meaningful in Fig. 12.3. It represents the magma composition spectrum. The second axis represents a noise component that probably does not have any meaning genetically. Potentially this result is of interest from a petrological point of view: from the curved pattern of points in Fig. 12.3, Thompson et al. (1972, p. 235) had concluded that there were two separate trends among the lavas (from relatively sodic, iron-rich hawaiite and mugearite to benmoreite, and from hyperstene-normative basalt via relatively potassic, iron-poor intermediate lavas to trachyte), whereas the *clr* method suggests there is one trend only.

12.2.1 Star Kimberlite Example

Grunsky and Kjarsgaard (2008) have classified discrete eruptive events of the Star kimberlite, Saskatchewan, into five distinctive clusters using statistical methods including compositional data analysis applied to whole rock geochemical data (Fig. 12.4). The data set consisted of 270 kimberlite samples from 38 drill holes that were analyzed for major oxides and trace –element geochemistry. Multivariate techniques after a log-ratio transformation included principal component analysis and discriminant analysis. Five distinct geochemical classes could be distinguished. These are called the Cantiar, Pense, early Joli Fou (eJF), mid Joli Fou (mJF) and late

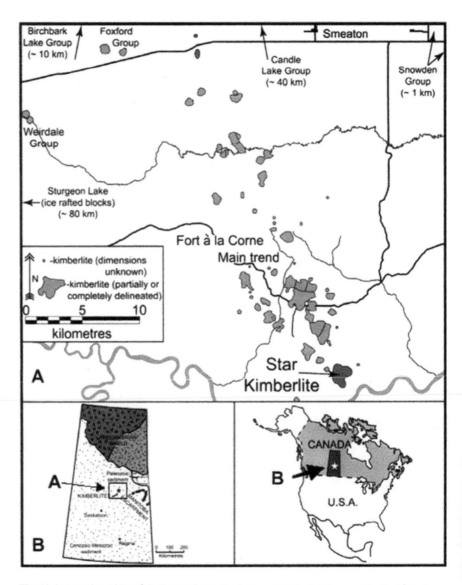

Fig. 12.4 Location of Fort à la Corne kimberlite field, central Saskatchewan. The Fort à la Corne 'main trend' cluster is the largest cluster in the field. The Star Kimberlite occurs at the southeastern end of the field (bottom right on main map) (Source: Grunsky and Kjarsgaard 2008, Fig. 1)

Joli Fou (lJF) equivalent age eruptive phases of the Star kimberlite. Their study area is shown in Fig. 12.5. The Star kimberlite is located in the extreme SE part of the main kimberlite cluster of the Fort à la Corne field situated within the Paleoproterozoic Glennie Domain that overlies the Archean craton.

12.2 Compositional Data Analysis

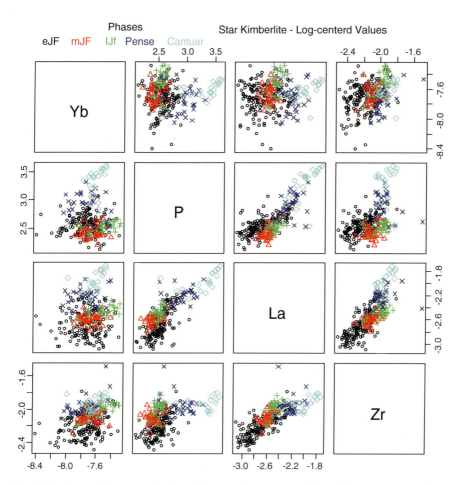

Fig. 12.5 Scatter plot matrix of selected elements that represent lithospheric mantle contamination from the five eruptive phases of the Star kimberlites. The data have been log-centred transformed. Early Joli Fou observations are shown as solid circles, mid Joli Fou as triangles, late Joli Fou as vertical crosses, Pense as diagonal crosses and Cantuar as diamonds (Source: Grunsky and Kjarsgaard 2008, Fig. 3)

Figure 12.6 shows pair plots of four elements (Yb, P, La and Zr) selected to represent kimberlite magma composition variability. The distinctive trends of relative enrichment of these elements indicate progressive fractionation of olivine plus spinel and decreased lithospheric mantle contamination. The evaluation of lithogeochemical data requires transformation from the constant-sum compositional data space to real number space through the application of a log-centered transformation. Principal component analysis was applied to the log-centred data. Figure 12.7 shows the resulting screeplot of ordered eigenvalues and their corresponding significance for the Star dataset. The first two eigenvalues contribute 34.6 and 32.0 % of total variance of the data. Scores of observations and variables

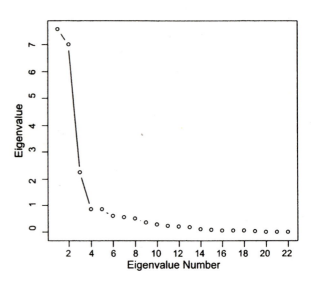

Fig. 12.6 Screeplot of ordered eigenvalues derived from principal component analysis applied to the log-centred Star kimberlite dataset (Source: Grunsky and Kjarsgaard 2008, Fig. 6)

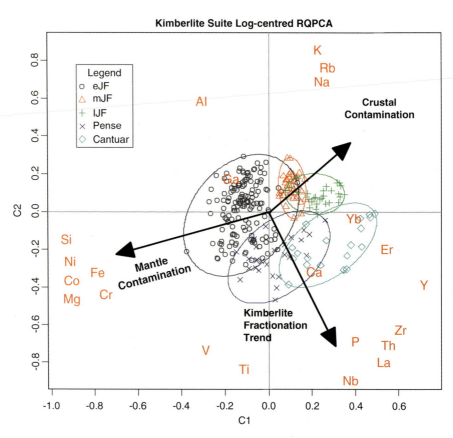

Fig. 12.7 Plot of PC1 versus PC2 scores and loadings of the log-centred Star kimberlite data set. The loadings of the variables are plotted as the element symbols (Source: Grunsky and Kjarsgaard 2008, Fig. 7)

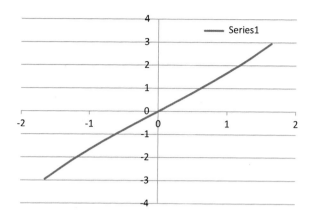

Fig. 12.8 Relationship between logits and probits is approximately linear for logits within the [−3, 3] interval

are shown in Fig. 12.8. More results including scores for other principal components are provided by Grunsky and Kjarsgaard (2008). More than 90 % of the variation of Si is accounted for by the first principal component, whereas 64.0 % of the variability of Ti is accounted for in the second component. The actual contributions provide a measure of how much each element contributes to each component. In the first component, Si, Fe, Co and Ni contributions all exceed more than 10 %. Approximately 45 % of variability of the first component is accounted for by these four elements.

Figure 12.8 is a plot of observations and element scores projected onto the first two principal component axes. The variation of all the data in this diagram describes 68.6 % of the overall variation in the data set. The ellipse boundaries were constructed to encompass all observations for each eruptive and the size of an ellipse is related to the degree of dispersion associated for that phase. Samples that plot close to an element or group of elements are enriched in those elements relative to other samples that plot farther away. It is evident that samples from the eJF show relative enrichment in Si, Cr, Ni, Mg, Fe and Co as would be expected for olivine-rich rocks such as kimberlite or kimberlite contaminated by lithospheric peridotite (which is dominated by olivine). Average composition of two Canadian cratonic mantle peridotite suites have extreme compositions relative to the kimberlite scores and are not plotted in Fig. 12.8, but the mantle contamination vector in this diagram is defined by the scores of these two average compositions. Samples tending toward the negative part of the C1 axis are certainly all mantle peridotite contaminated. Further interpretations are given by Grunsky and Kjarsgaard (2008). The purpose of this example was to show how compositional data analysis combined with classical statistics can result in novel mineralogical and geochemical interpretations.

A final remark on compositional data analysis is as follows. In an interesting paper, Filzmoser et al. (2009) argue that, for statistical analysis, logits (*cf.* Sect. 5.2) of measurements should be used instead original values if the variable of interest is part of a closed number system. They advocate use of logits even when histograms for individual variables are constructed or when standard deviations are being calculated. Suppose such measurements are written as P_i ($i = 1, 2, \ldots, n$). For trace elements the logit transformation $\log_e\{P_i/(1-P_i)\}$ is equivalent to a logarithmic transformation $\log_e P_i$ because P_i then is very small. It is well-known that logits are nearly equal to probits (cf. Sect. 2.3.1) except for the smallest and largest values in a sample. The relationship between logits and probits is approximately linear for the range $0.05 \leq P_i \leq 0.95$ as shown in Fig. 12.8. In Sect. 12.8, measurements will be replaced by their probits (instead of logits) because this transformation is helpful in 2-D cell composition modeling, especially for extrapolation of statistics from smaller to larger cells and vice versa.

12.3 Non-linear Process Modeling

A question to which new answers are being sought is: Where does the randomness in Nature come from? Nonlinear process modeling is providing new clues to answers. Deterministic models can produce chaotic results. A famous early example was described in Poincaré's (1899) study of the three-body problem in mechanics. This author observed that in some systems it may happen that small differences in the initial conditions produce very great ones in the final phenomena, and "... prediction becomes impossible." Nevertheless, the randomness created by non-linear processes obeys its own specific laws. Many scientists including Turcotte (1997) have pointed out that, in chaos theory, otherwise deterministic Earth process models can contain terms that generate purely random responses. Examples include the logistic equation and the van der Pol equation with solutions that contain unstable fixed points or bifurcations. Multifractals, which are spatially intertwined fractals and were anticipated by Hans de Wijs (in 1948), provide a novel way of approach to problem-solving in situations where the attributes display strongly positively skewed frequency distributions. The standard geostatistical model used in ordinary kriging assumes a semivariogram with both range and nugget effect. The range extends to distances at which results from other deterministic or random processes begin to overshadow local variability. The nugget effect often is the result of relatively wide sampling between points that hides short-distance variability. The multifractal semivariogram shows sharp decrease toward zero near the origin. Local singularity mapping uses this short-distance variability to delineate places with relatively strong enrichment or depletion of element concentration on geochemical maps and in other applications (Cheng and Agterberg 2009). This method provides a new approach for mineral exploration and regional environmental assessment (cf. Sect. 11.5). These nonlinear developments are closely related to statistical theory of extreme events (Beirlant et al. 2005).

12.3.1 The Lorentz Attractor

It can be argued that non-linear process modeling originated in 1963 when the meteorologist Edward Lorenz (1963) published the paper entitled "Deterministic nonperiodic flow" (*cf.* Motter and Campbell 2013). Earlier, he had accidentally discovered how miniscule changes in the initial state of a time series computer simulation experiment grew exponentially with time to yield results that did not resemble one another although the equations controlling the process were the same in all experiments (Lorenz, 1963). It turned out that the notion of absolute predictability, although it had been assumed intuitively to hold true in classical physics, is in practice false for many systems. In the title of a talk Lorenz gave in 1972, he aptly referred to this chaotic behaviour as the "butterfly effect" asking if the flap of a butterfly in Brazil could set off a tornado in Texas (Fig. 12.8).

> **Box 12.2: Chaos Theory**
>
> Lorenz (1963) presented chaos theory by using a three-variable system of nonlinear ordinary differential equations now known as the Lorenz equations. The model derives from a truncated Fourier expansion of the partial differential equations describing a thin, horizontal layer of fluid that is heated from below and cooled from above. The equations can be written as: $dx/dt = a(-x+y)$, $dy/dt = cx - y - xz$, and $dz/dt = -bz + xy$, where x represents intensity of convective motion, y is proportional to the temperature difference between the ascending and descending convective currents, and z indicates the deviation of the vertical temperature profile from linearity. The parameters b and c represent particulars of the flow geometry and rheology. Lorenz set them equal to $b = 8/3$ and $c = 10$, respectively; leaving only the Rayleigh number a to vary. For small c, the system has a stable fixed point at $x = y = z = 0$, corresponding to no convection. If $24.74 > c \geq 1$, the system has two symmetrical fixed points, representing two steady convective states. At $c = 24.74$, these two convective states lose stability; at $c = 28$, the system shows nonperiodic trajectories. Such trajectories orbit along a bounded region of 3-D space known as a chaotic attractor, never intersecting themselves (Fig. 12.9). For larger values of c, the Lorenz equations exhibit different behaviors that have been catalogued by Sparrow (1982).

It is beyond the scope of this book to discuss the theory of chaos in more detail. However, it is good to point out that chaos is closely connected to fractals and multifractals. The attractor's geometry can be related to its fractal properties. For example, although Lorenz could not resolve this from the numeric he applied, his attractor has fractal dimension equal to approximately 2.06 (Motter and Campbell 2013, p. 30). An excellent introduction to chaos theory and its relation to fractals can be found in Turcotte (1997). The simplest nonlinear single differential equation

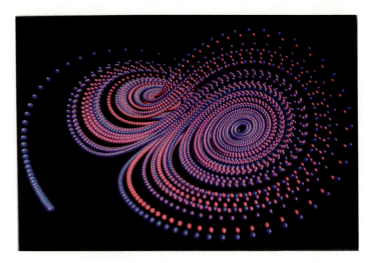

Fig. 12.9 The Lorenz attractor as revealed by the never-repeating trajectory of a single chaotic orbit (from Motter and Campbell 2013, Fig. 2). The spheres represent iterations of the Lorenz equations, calculated using the parameters originally used by Lorenz (1963). Spheres are colored according to iteration count. The two lobes of the attractor resemble a butterfly, a coincidence that helped earn sensitive dependence on initial conditions. Hence, its nickname: "the butterfly effect"

illustrating some aspects of chaotic behavior is the logistic equation that can be written as $dx/dt = x(1-x)$ where x and t are non-dimensional variables for population size and time, respectively. There are no parameters in this equation because characteristic time and representative population size were defined and used. The solution of this logistic equation has fixed points at $x = 0$ and 1, respectively. The fixed point at $x = 0$ is unstable in that solutions in its immediate vicinity diverge away from it. On the other hand, the fixed point at $x = 1$ has solutions in its immediate vicinity that are stable. Introducing, the new variable, $x_1 = x - 1$, and neglecting the quadratic term, the logistic equation has the solution $x_1 = x_{10} e^{-t}$ where x_{10} is assumed to be small but constant. All such solutions "flow" toward $x = 1$ in time. They are not chaotic.

Chaotic solutions evolve in time with exponential sensitivity to the initial conditions. The so-called logistic map arises from the recursive relation $x_{k+1} = a \cdot x_k (1 - x_k)$ with iterations for $k = 0, 1, \ldots$. May (1976) found that the resulting iterations have a remarkable range of behavior depending on the value of the constant a that is chosen. Turcotte (1997, Sect. 10.1) discusses in detail that there now are two fixed points at $x = 0$ and $1 - a^{-1}$, respectively. The fixed point at $x = 0$ is stable for $0 < a < 1$ and unstable for $a > 1$. The other fixed point is unstable for $0 < a < 1$, stable for $1 < a < 3$, and unstable for $a > 3$. At $a = 3$ a so-called flip bifurcation occurs. Both singular points are unstable and the iteration converges on an oscillating limit cycle. At $a = 3.449479$, another flip bifurcation occurs and there suddenly are four limit cycles. Writing $a_1 = 3$ and $a_2 = 3.449459$, it turns out that the constants a_i ($i = 1, 2, \ldots, \infty$) define intervals with 2^{a_i} limit cycles that satisfy the iterative relation $a_{i+1} - a_i = F^{-1} \cdot (a_i - a_{i-1})$

where $F = 4.669202$ is the so-called Feigenbaum constant. In the limit, $a_\infty = (F-1)^{-1} \cdot (Fa_i - a_{i-1}) = 3.569946$. In the region $3.569946 < a < 4$, windows of chaos and multiple cycles occur. Logistic maps for different values of a including some that show chaotic behavior are shown by Turcotte (1997, Figs. 10.1, 10.2, 10.3, 10.4, 10.5, and 10.6). Sornette et al. (1991) and Dubois and Cheminée (1991) have treated the return periods for eruptions of the Piton de la Fournaise on Réunion Island and Mauna Lao and Kilauea in Hawaii as return maps that resemble chaotic maps resulting from the logistic model.

As pointed out in the chapters on fractals and multifractals, nonlinear process modeling is providing new clues to answers of where the randomness in nature comes from. From chaos theory it is known that otherwise deterministic Earth process models can contain terms that generate purely random responses. The solutions of such equations may contain unstable fixed points or bifurcations. In the previous chapter it was shown that multifractals provide a novel way of approach to problem-solving in situations where the attributes display strongly positively skewed frequency distributions. The preceding examples of deterministic processes result in chaotic results. However, one can ask the question of whether there exist deterministic processes that can fully explain fractals and multifractals? The power-law models related to fractals and multifractals can be partially explained on the basis of the concept of self-similarity (Box 11.1). Various chaotic patterns observed for element concentration values in rocks and orebodies could be partially explained as the results of multiplicative cascade models such as the model of de Wijs. These models invoke random elements such as increases of element concentration that are not deterministic.

As already discussed in Sect. 10.1, successful applications of non-linear modeling in geoscience include the following: Rundle et al. (2003) showed that the Gutenberg-Richter frequency-magnitude relation is a combined effect of the geometrical (fractal) structure of fault networks and the non-linear dynamics of seismicity. Most weather-related processes taking place in the atmosphere including cloud formation and rainfall are multifractal (Lovejoy and Schertzer 2007; Sharma 1995). Other space-related non-linear processes include "current disruption" and "magnetic reconnection" scenarios (Uritsky et al. 2008). Within the solid Earth's crust, processes involving the release of large amounts of energy over very short intervals of time including earthquakes (Turcotte 1997), landslides (Park and Chi 2008), flooding (Gupta et al. 2007) and forest fires (Malamud et al. 1998) are non-linear and result in fractals or multifractals.

12.4 Three-Parameter Model of de Wijs

This section is concerned with the lognormal, and its logbinomial approximation, in connection with a three-parameter version of the model of de Wijs. The three parameters are: overall average element concentration value (ξ), dispersion index (d), and apparent number of subdivisions of the environment (N). Multifractal theory produces new methods for estimating the parameters of this model. In practical applications, the frequency distribution of element concentration values for small rock samples is related to self-similar spatial variability patterns of the element in large regions or segments of the Earth's crust. The approach will be illustrated by application to spatial variability of gold and arsenic in glacial till samples from southern Saskatchewan. It will be shown that for these two elements the model of de Wijs is satisfied on a regional scale but degree of dispersion decreases rapidly toward the local, sample-size scale. Thus the apparent number of subdivisions (N) is considerably less than would be expected if degree of dispersion were to extend from regional to local scale as generally assumed in the past (cf. Agterberg 2007a).

12.4.1 Effective Number of Iterations

Multifractal modeling offers an independent method to verify validity of the model of de Wijs and to estimate d. In principle, the value of n could be made infinitely large. However, the logarithmic variance (σ^2 in the variance equation of de Wijs) then also becomes infinitely large and the frequency distribution of the element concentration values would cease to exist. Application of the method of moments in multifractal analysis results in a multifractal spectrum that is a limiting form for infinitely large n. The frequency distribution corresponding to this limiting form cannot exist in reality because it has infinitely large variance. In practice, any set of element concentration values for very small blocks of rock collected from a very large environment has a frequency distribution with finite logarithmic variance.

Suppose that the generating process of subdividing blocks under the same dispersion index (d) ceases to be operative for blocks that are larger than the very small blocks used for chemical analysis. In examples of application to be discussed in this section, d at the regional scale does not apply at the local scale (for small blocks used for chemical analysis). Locally, d is either zero or much smaller than d at the regional scale. Under these conditions, an apparent maximum number of subdivisions N can be estimated. Self-similarity at scales exceeding a critical lower limit results in a model of de Wijs with three parameters: ξ, d, and N.

The multifractal method used for estimating d is similar to the method for separation of geochemical anomalies from background introduced by Cheng (1994) and Cheng et al. (1994). As previously discussed in Agterberg (2001a, b, 2007a), validity of the model of de Wijs for the larger concentration values results in:

$$\lim_{q \to \infty} \frac{d\tau(q)}{dq} = \alpha_{\min} = \log_2 \left(\frac{\eta+1}{\eta} \right)^2$$

In 2-D applications, chemical element measures are formed by multiplying average cell concentration values by cell areas. Raising these measures to relatively high powers q filters out the influence of smaller concentration values. Thus our estimate of d is based on parameters estimated from the relatively large concentration values of an element. The lower-value tails of the observed frequency distributions could

12.4 Three-Parameter Model of de Wijs

Table 12.2 Gold and arsenic in till samples, Southern Saskatchewan: results of graphical estimation of parameters for the model of de Wijs

	Gold	Arsenic
Straightline on Q-Q plot: intercept	0.8006	−5.3035
Ditto, slope	0.4222	2.7611
Mean	2.48	7.29
Logarithmic variance	5.61	0.131
Cell data Q-Q plot: intercept	−0.2121	12.399
Ditto, slope	0.8784	6.0591
Logarithmic variance of cell averages	1.296	0.0272
Straightline on Tau vs q plot: intercept	−5.3035	−1.4232
Ditto, Slope	2.7611	1.8075
Dispersion index (Eq. 3)	0.433	0.069
Apparent number of subdivisions – till samples	26.12	27.5
Number of subdivisions, cell data	6.03	5.71

Gold concentration values in ppb; As concentration values in ppm (Source: Agterberg 2007a, Table 1)

be mixtures of separate populations, and the exact nature of these mixtures can be left undetermined.

If the model of de Wijs is satisfied, the mass exponents $\tau(q)$ become linearly related to q when q is large. The parameter d then can be estimated from the slope of the resulting straight line. Together with the estimate of logarithmic variance provided by the original variance equation of de Wijs, this yields an estimate of apparent maximum number of subdivisions N.

12.4.2 Au and As in South Saskatchewan till Example

The approach is illustrated by application to gold and arsenic content of till samples obtained during systematic ultra-low density (1 site per 800 km^2) geochemical reconnaissance, Southern Saskatchewan. There are 389 observed gold concentration values ranging from below detection limit to 77 ppb. All Au values below detection limit (=2 ppb) were set equal to 1 ppb. The arithmetic mean gold concentration value is 2.5 ppb. In total, only about one third of 389 observed values exceed this average. Overall average gold value of 2.5 ppb slightly overestimates true mean gold concentration value because most gold values below 2 ppb are probably less than 1 ppb.

For arsenic, all observed values are above the detection limit with overall mean of 7.9 ppm and ranging from about 2–44 ppm. The main difference between frequency distributions of Au and As is that gold has positively skewed distribution whereas arsenic shows approximately symmetric, normal distribution. Most diagrams in this section are for Au only. Similar results (not shown as diagrams) were obtained for As. Estimates for both Au and As are listed in Table 12.2.

The method used here resembles the one used by Brinck (1974). Lognormal Q-Q plots for Au and As are shown in Fig. 12.10. The first step in Brinck's approach

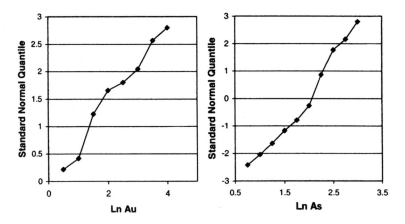

Fig. 12.10 Cumulative frequency distributions (lognormal Q-Q plots) of gold and arsenic in 379 till samples, Southern Saskatchewan. Units of Au and As concentration values are ppb and ppm, respectively (Source: Agterberg 2007a, Fig. 7)

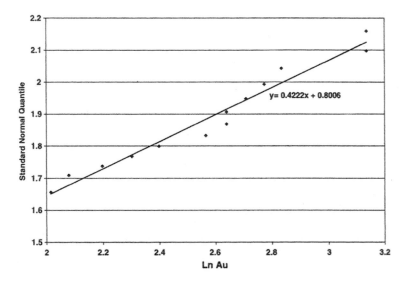

Fig. 12.11 Straight line fitted to part of lognormal Q-Q plot of gold values shown in Fig. 12.12. Inverse of slope provides estimate of logarithmic (base e) standard deviation (Source: Agterberg 2007a, Fig. 8)

consists of fitting a straight line on the Q-Q plot. Figure 12.11 shows such a line of best fit for Au. Because natural logarithms are used, the standard deviation of logarithmically transformed concentration values can be set equal to the inverse of the slope of the best-fitting straight line. It amounts to 2.37 ppb for Au. Difference in shape of frequency distribution is reflected in the fact that the logarithmic variance of Au is more than 40 times as large as the logarithmic variance of As

12.4 Three-Parameter Model of de Wijs

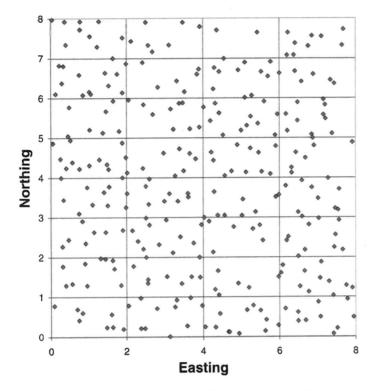

Fig. 12.12 Study area of 64 cells with locations of till samples, Southern Saskatchewan. Unit of distance for east–west and north–south directions is 78.125 km (Source: Agterberg 2007a, Fig. 9)

(see Table 11.1). Original sampling design for the geochemical reconnaissance study in Southern Saskatchewan is described by Garrett and Thorleifson (1995). An 80 × 80 km grid was selected as the basic sampling cell. It was successively subdivided into 40 × 40 km, 20 × 20 km and finally 10 × 10 km grid cells for sampling with randomly selected 1 × 1 km target cells. Compositional data for many other elements and minerals were determined as well. Gold and arsenic were selected for this study because their frequency distributions seem to be unimodal, and these elements had been used extensively in other multifractal studies (Cheng 1994; Agterberg 2001a, b).

For the case study described here, a modified grid was used for Southern Saskatchewan resulting in a square-shaped study area consisting of 64 square cells measuring slightly less than 80 km on a side (see Fig. 12.12). The purpose of this re-definition of boundaries of study area was to minimize edge effects in multifractal analysis. Data from outside the square study area were not used. In total, 290 of the original 389 (on average, about 4.5/cell) were retained. None of the cells inside the study area of Fig. 12.12 are empty. Figure 12.13 is a Q-Q plot of gold cell average concentration values with best-fitting straight line.

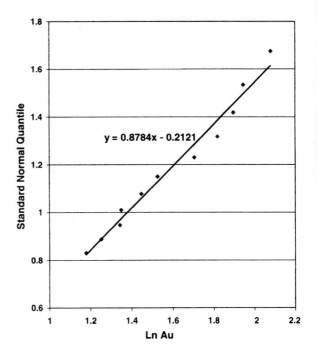

Fig. 12.13 Straight line fitted to part of lognormal Q-Q plot for 64 average gold concentration values for cells (Source: Agterberg 2007a, Fig. 10)

The square arrays of Au and As cell average concentration values were subjected to multifractal analysis using the method of moments. Figure 12.14 shows mass-partition function results for gold. For relatively large values of q there is approximate linear relationship between the mass exponent $\tau(q)$ and q for Au (Fig. 12.15). The slopes of the best-fitting straight lines are 0.962 for Au and 1.808 for As, respectively (Table 11.1). These slopes were converted into dispersion index estimates of $d = 0.433$ for Au, and $d = 0.069$ for As. Straight line fitting to lognormal Q-Q plots for the 64 cell average concentration values gave logarithmic variances of 1.292 for Au, and 0.029 for As, respectively. Using the original variance equation of de Wijs then yields estimates of n equal to 6.0 for Au, and 5.7 for As, respectively. Both estimates are nearly equal to 6 representing their expected value for the model of de Wijs because $2^6 = 64$. This agreement between results indicates that the model of de Wijs is approximately satisfied on a regional scale for both gold and arsenic.

Logarithmic variances estimated from the 379 original till samples were 2.369 for Au and 0.362 for As, respectively. Using the variance equation of de Wijs, this yields estimates of N equal to 26.1 for Au, and 27.5 for As, respectively. These apparent maximum numbers of subdivisions of the environment would correspond to square cells measuring approximately 75 m, and 40 m on a side, respectively. Clearly, these cells are much larger than the very small areas where the till was actually sampled for the purpose of chemical analysis. It shows that, for both Au and As, the regional model of de Wijs does not apply at the local sampling scale.

12.4 Three-Parameter Model of de Wijs

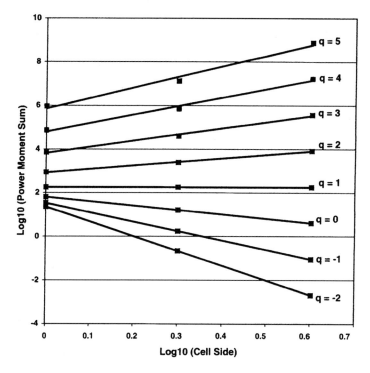

Fig. 12.14 Method of moments applied to 64 cell average gold concentration values. A property of self-similarity is that power moment sums of cell masses raised to powers of q are related to cell side according to power cells with mass exponents $\tau(q)$ given by the slopes of the straight lines. Because of uncertainties related to gold detection limit, results for $q \geq 0$ can be used only in this application (Source: Agterberg 2007a, Fig. 11)

Logbinomial distributions for gold and arsenic are shown in the lognormal Q-Q plots of Fig. 12.16. Because the negative binomial is discrete, the preceding estimates of N were rounded off to the nearest integer. The parameters of the distributions shown are $\xi = 1.82$ ppb, $d = 0.433$, and $N = 26$ for gold (Fig. 12.16a); and $\xi = 7.05$ ppm, $d = 0.069$, and $N = 28$ for arsenic (Fig. 12.16b). Mean values ξ were determined by forcing the log-binomials and their lognormal approximations through the centers of clusters of points on lognormal Q-Q plots for original data. These results remain approximate but are interesting because they permit shape comparison of logbinomial and lognormal tails.

The dispersion index of gold (=0.433) is much larger than that of arsenic (=0.069), but the patterns on Q-Q plots in Fig. 12.16 for these two elements are similar. Both are approximately lognormal over the ranges of observed values. Outside these ranges the logbinomial upper (and lower) tails become increasingly thinner than their lognormal approximations. However, the resulting differences have relatively little practical significance. For example, the logbinomial would predict that 1.45 % of all possible till samples have gold concentrations greater than 32 ppb, whereas the corresponding lognormal prediction (=1.56 %) is hardly different from this.

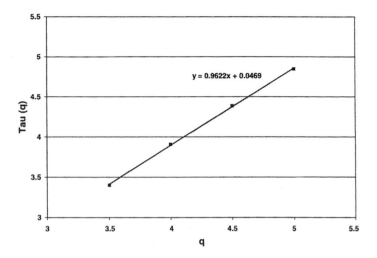

Fig. 12.15 Mass exponent τ(q) plotted against relatively large values of q. Slope of best-fitting straight line is used to estimate the dispersion index $d = 0.433$ for gold (Source: Agterberg 2007a, Fig. 12)

12.5 Other Modifications of the Model of de Wijs

The following one-dimensional computer simulation experiment for the model of de Wijs was previously described in Agterberg (1994). Suppose $\mu(S_A)$ represents the measure of a set S in a segment of $\mathfrak{R}^1$. A line segment of length L can be partitioned into $N(\varepsilon)$ cells (intervals) of equal size ε; let $\mu_i(\varepsilon)$ denote the measure on S for the i-th cell of size ε in $(0, L)$ with $i = 1, 2, \ldots, N(\varepsilon)$. A simple stochastic version of the multiplicative cascade model in $\mathfrak{R}^1$ then is as follows. At the first stage ($k = 1$) in a process of n stages, the interval $(0, L)$ with measure $\xi \cdot L$ is subdivided into two equal intervals: (a) $(0, L/2)$ with measure $(1+B) \cdot \xi \cdot L$, and (b) $(L/2, L)$ with $(1-B) \cdot \xi \cdot L$, where B is a random variable with probabilities $P(B = d) = P(B = -d) = 1/2$ ($d > 0$). At stage 2 these two intervals are halved again with new measures for the halves defined in the same way as at stage 1. The process is repeated at stages $k = 3, 4, \ldots$ At stage k the i-th subinterval with concentration value $X_i(\varepsilon)$ has size $\varepsilon = L/2^k$, and $E\{X_i(\varepsilon)\} = \xi$. The frequency distribution of $X_i(\varepsilon)$ is logbinomial, tending to become lognormal in the center as $\varepsilon \to 0$ and, depending on the direction of ordering, slightly weaker than lognormal in both tails. Figure 12.17 shows a realization of this process for $n = 8$, $\xi = 1$, and $d = 0.4$. An obvious drawback of the original model of de Wijs is that, if the dispersion index d applies at one stage, it is unlikely to apply at later steps because d generally must be a random variable itself. In this section d will be replaced by the random variable D.

12.5 Other Modifications of the Model of de Wijs

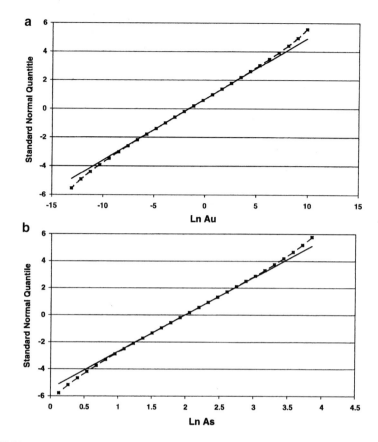

Fig. 12.16 Lognormal Q-Q plots for discrete binomial frequency distributions of (**a**) gold and (**b**) arsenic concentration values resulting from the model of the Wijs. (**a**) For Au: overall average element concentration value $\xi = 1.82$ ppb, dispersion index $d = 0.433$, and apparent number of subdivisions of the environment $N = 26$; (**b**) For As: $\xi = 7.05$ ppm, $d = 0.069$, and $N = 28$ (Source: Agterberg 2007a, Fig. 13)

12.5.1 Random Cut Model

In the original model of de Wijs illustrated in Fig. 12.16, the dispersion index d is a constant. The lower and higher values generated at successive iterations were assigned random spatial locations with respect to one another by introducing the random variable B with $P(B=d) = P(B=-d) = 1/2$ $(d>0)$. A different way of formulating this type of randomness is to introduce a new random variable $D^* = 1+D$ that assumes either the value d or the value $-d$ with equal probabilities. Consequently, $E(D^*) = 1$ and $E(D) = 0$. In the variant of the model of de Wijs resulting in Figs. 12.17 and 12.18, D^* is bimodal. Its two separate peaks, which satisfy normal distributions with standard deviations equal to 0.1, are centered about 0.4 and 1.4, respectively. It is more realistic to replace this bimodal model by a unimodal frequency density model centered about $E(D^*) = 1$ (see later).

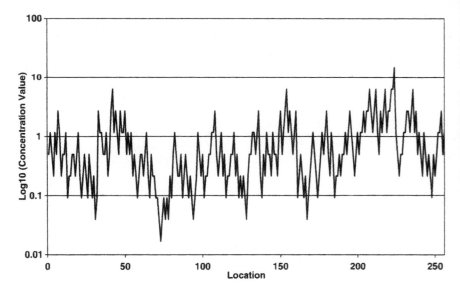

Fig. 12.17 Artificial sequence of 256 values generated by model of de Wijs; dispersion index $d = 0.4$; overall mean $\mu = 1$ (Source: Agterberg 2007a, Fig. 1)

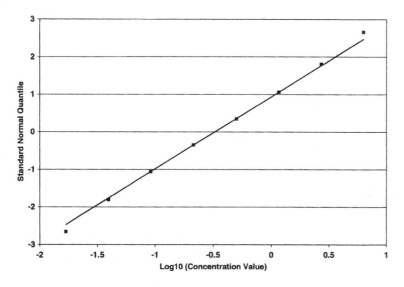

Fig. 12.18 Lognormal Q-Q plot of 256 values of Fig. 12.17; Straight line representing lognormal distribution was fitted to central part of logbinomial. Note that logbinomial tails are thinner than lognormal tails (Source: Agterberg 2007a, Fig. 2)

12.5 Other Modifications of the Model of de Wijs

A good question to ask is: would it be possible to estimate the frequency distribution of D^*. The following idea was originally proposed by Matheron (1962, pp. 310–311). It is based on the concept of relative concentration value $Y_{12} = X_1/X_2$ where X_1 represents concentration value of a small block with volume V_1 that is randomly located within a larger block with concentration value X_2 and volume V_2. In the computer simulation experiments of Fig. 12.17 V_1 is exactly half of V_2 but V_1 can be much smaller than half of V_2. If there is self-similarity, Y_{12}, on average, only depends on the ratio V_1/V_2 of the block volumes. It is independent of the volumes themselves. Suppose that smaller blocks are contained within larger blocks so that $V_1/V_2 = V_2/V_3 = \ldots = V_n/V_{n-1} = \lambda$. Consequently, $V_n/V_1 = \lambda^{n-1}$. For the examples in $\Re^1$, $V_n/V_1 = 2^n$. All relative concentration values $Y_{i,i+1}$ ($i = 1, \ldots, n-1$) are realizations of the same random variable, and $Y_{1n} = X_1/X_n = Y_{12} \cdot Y_{23} \cdot \ldots \cdot Y_{n-2,n-1}$. Suppose that the logarithmic variance of $Y_{i,i+1}$ is written as h^2, then σ^2 representing the logarithmic variance of Y_{1n} satisfies: $\sigma^2 = (n-1)h^2 = \alpha^* \ln\{V_n/V_1\}$, where $\alpha^* = h^2/\ln \lambda$. This expression is equivalent σ^2 in the original variance equation of de Wijs. However, the underlying rationale is different. It now is assumed that $D^* = 1 + D$ is a unimodal random variable with expected value equal to 1 and variance $\sigma^2(D^*)$. If D^* has normal frequency distribution, its mean deviation is equal to $\sigma\sqrt{2/\pi}$. This statistic would be equivalent to the constant dispersion index d in the original model of de Wijs.

Suppose that the mean deviation is set equal to the dispersion index $d = 0.4$ as used for constructing Fig. 12.16. Then $\sigma = 0.5013$. However, D^* cannot be negative nor greater than 2. In the following computer simulation experiment, occurrence of very small or large values of D^* is prevented by setting random normal numbers less than 0.1 equal to 0.1, and values greater than 1.9 equal to 1.9. The resulting approximately normal distribution has standard deviation slightly less than 0.5013. Replacement of this value by 0.51 results in an approximately normal distribution with mean deviation equal to 0.40. Figures 12.17 and 12.18 show results for this random-cut model with $n = 8$, $E(D^*) = \xi = 1$, and $\sigma(D^*) = 0.51$. Arithmetic instead of logarithmic scale for concentration values is used in Fig. 12.18.

The preceding computer simulation experiments illustrate that the original model of de Wijs (Fig. 12.16) with a single value for the dispersion index d produces results that are similar to more realistic models in which is D is a random variable resulting in different values of d for every division of blocks into halves. Most practical applications continue to use the original model of de Wijs because it provides valid approximations for large values of n. Moreover, multifractal theory is well established for the original model of de Wijs and can be used without difficulty.

However, the alternative approach of allowing $D^* = 1 + D$ to be a random variable allows empirical estimation of the frequency distribution of relative concentration values $Y_{12} = X_1/X_2$ where X_1 is the concentration value of a small block with volume V_1 that is randomly located within a larger block with concentration value X_2 and volume V_2. If $V_1/V_2 = 0.5$, measurements of Y_{12} would provide direct estimates of values that are realizations of the random variable D^*. This approach, with V_1 much smaller than V_2, now will be applied to relative Au concentration values of till samples in the next example.

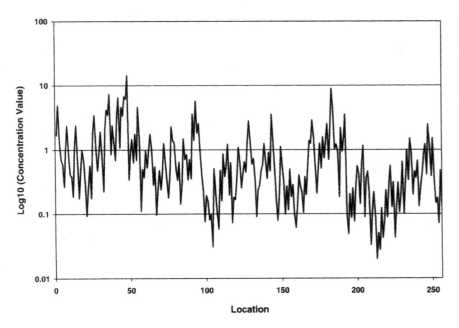

Fig. 12.19 Computer simulation experiment of Fig. 12.17 repeated after replacing dispersion index $d=0.4$ by normal random variable D with $E(D)=0.4$ and $\sigma(D)=0.1$. Overall pattern resembles pattern of Fig. 12.19 (Source: Agterberg 2007a, Fig. 3)

Relative element concentration values $Y_{1n} = X_1/Xn_2$ were calculated for gold, where X_1 is the concentration value of a till sample with volume V_1 that is randomly located within a cell belonging to the grid of cells shown in Fig. 12.11 with concentration value X_n cell size V_n. Figure 12.19 is a lognormal Q-Q plot of these 290 relative Au concentration values. The straight line provides a good fit except on the left side of Fig. 12.19 where there is some bias related to the Au detection limit and at the ends where frequencies of relative gold concentration values are small.

The logarithmic variance derived from the straight line of Fig. 12.19 is 0.5717. This illustrates that positive skewness of the relative gold concentration values is significant. Setting V_n/V_1 equal to 2^{20} yields an estimated value of $\alpha^* = 0.04124$. The logarithmic variance of Y_{12} would amount to $h^2 = 0.5717/19 = 0.03009$. Because it can be assumed that Y_{12}, like Y_{1n}, has lognormal distribution, its variance is estimated to be $\sigma^2(Y_{12}) = 0.2326$. Because this value is rather small, Y_{12} is approximately normally distributed with standard deviation $\sigma(Y_{12}) = 0.482$. It follows that mean deviation from $E(Y_{12}) = 1$ amounts to m.d. $= 0.39$. This would be a crude estimate of average dispersion index in the random-cut model. It is close to $d = 0.43$ estimated for the original logbinomial model of de Wijs (Table 12.2).

Although m.d. and d are different parameters one would expect them to be approximately equal to one another because of convergence of logbinomial and lognormal distributions for increasing number of subdivisions (n). It was attempted to repeat the preceding analysis for As but the 290 values $Y_{1n} = X_1/Xn_2$ for As do not show a simple straight line pattern, and m.d. could not be estimated with sufficient precision (Figs. 12.20, 12.21, 12.22, and 12.23).

12.5 Other Modifications of the Model of de Wijs

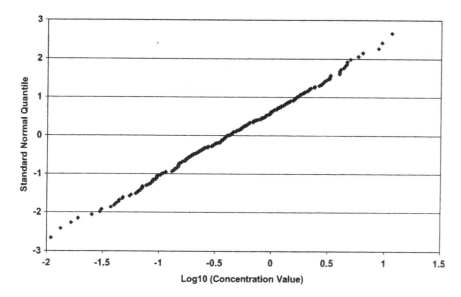

Fig. 12.20 Lognormal Q-Q plot of 256 values of Fig. 12.19; frequency distribution is approximately lognormal (Source: Agterberg 2007a, Fig. 4)

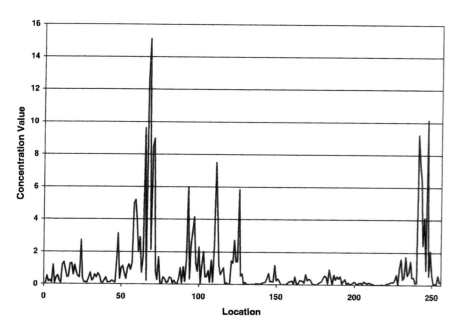

Fig. 12.21 Computer simulation experiment of Fig. 12.17 repeated after replacing constant dispersion index $d = 0.4$ by approximately normal random variable D^* with $E(D^*) = 1$ and $E\{|D^* - E(D^*)|\} = 0.4$, which is equivalent to $\sigma(D^*) = 0.51$ (Source: Agterberg 2007a, Fig. 5)

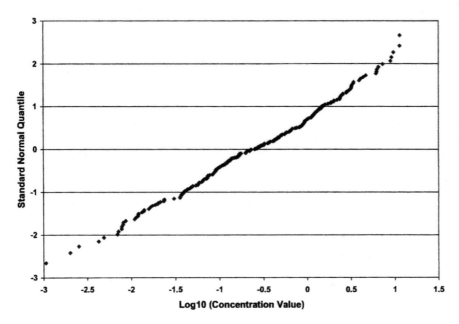

Fig. 12.22 Lognormal Q-Q plot of 256 values of Fig. 12.21 (Source: Agterberg 2007a, Fig. 6)

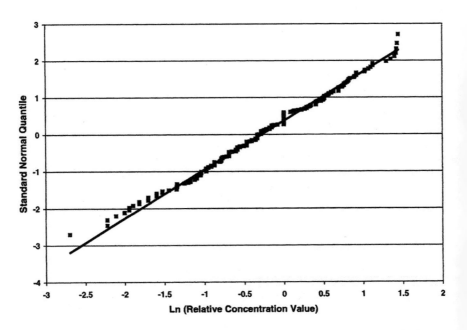

Fig. 12.23 Lognormal Q-Q plot for 290 relative gold concentration values of Fig. 12.12. Best fitting line indicates approximate lognormality and is used to estimate mean deviation of variable dispersion index D^* in random-cut model (Source: Agterberg 2007a, Fig. 14)

12.5 Other Modifications of the Model of de Wijs

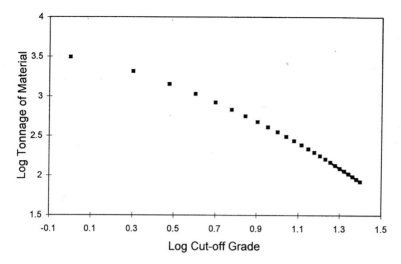

Fig. 12.24 United States uranium resources (unit = 10^6 t) versus cut-off grade (in percentage U_3O8) from Harris (1984) on log-log paper (logs base 10). Last eight values satisfy Pareto distribution model. The decrease in slope toward the origin probably is caused by the fact that lower grade deposits are underreported because it is not economic to mine them (Source: Agterberg 1995, Fig. 5)

12.5.2 Accelerated Dispersion Model

A common problem in mineral resource studies is that frequently the high-value tail of the observed size-frequency distribution is thicker than lognormal which, in turn, is thicker than logbinomial (*cf.* Sect. 10.4). Another example of this is shown in Fig. 12.24 for U.S. uranium resources. The high-value tail in this log-log diagram is approximately according to a straight line indicating a Pareto distribution. The following relatively simple modification of the model of de Wijs results in thicker than lognormal frequency distribution tails (*cf.* Agterberg 2007b). It could be assumed that the coefficient of dispersion d increases as a function of element concentration value ξ. For example, suppose that the first derivative of a chemical element's dispersion coefficient is a linear function of ξ so that $d = d_0 \exp(p \cdot \xi)$ where p is a constant. Setting $p = 0.01$ and re-running the 2-D experiment previously resulting in Fig. 10.22 ($d_0 = 0.4$; $N = 14$) yielded the pattern for ($\xi + 4$) shown in Fig. 12.25 where the vertical scale is logarithmic. The logarithmically transformed concentration values of Fig. 12.25 are normally distributed except for the largest values. Figure 12.26 is a comparison of the largest log-concentration values with those arising when there would be no acceleration of dispersion. Values with $\log_{10}(\xi) \approx 2$ are larger than expected in comparison with patterns such as Fig. 10.22 resulting from the model of de Wijs with $p = 0$. Only relatively few very large values emerge from the approximately lognormal background if p is small.

Fig. 12.25 Realization of accelerated dispersion model for $d_0 = 0.4$, $N = 14$, and $p = 0.01$. Vertical scale is for $4 + \log_{10}$ (Value) (Source: Agterberg 2007b, Fig. 5)

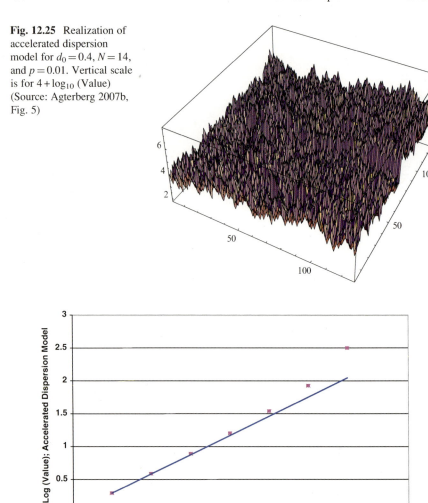

Fig. 12.26 Comparison of largest logarithmically transformed concentration values (base 10) in experiment of Fig. 12.25 (Source: Agterberg 2007b, Fig. 6)

Because p is small, most element concentration values after 14 iterations are nearly equal to values that satisfy the original model of de Wijs. As illustrated in Fig. 12.26, exceptions are that the largest values, which include $(1+d)^{14} = 316$, significantly exceed values that would be generated by the original model, which include $(1+d_0)^{14} = 111$. A similar effect would occur in the low-value tail. The frequencies of very small values that are generated exceed those generated by the ordinary model of de Wijs. Because the lowest concentration values cannot become negative, they accumulate within an interval that is close to zero. When a logarithmic scale is used for value, the pattern that is expected would be similar to that previously shown in Fig. 3.11 for gold values in Merriespruit Mine, Witwatersrand goldfield, South Africa. It is noted, however, that values in the lower value tail cannot be less than a minimum value that is greater than zero.

12.6 Trends, Multifractals and White Noise

Regional geology generally can be explained as the result of deterministic processes that took place millions of years ago. How does the well-documented and explained mosaic constituting the upper part of the Earth's crust fit in with the concept of fractals and multifractals?

Brinck's (1974) model constituted an early application of the model of de Wijs (*cf.* Sect. 11.1.2). Estimation of parameters in this model including d could be improved by adopting the multifractal modeling approach explained in Sect. 12.4. At first glance, the Brinck approach seems to run counter to the fact that mineral deposits are of many different types and result from different genetic processes. However, Mandelbrot (1983) has shown that, for example, mountainous landscapes can be modeled as fractals (*cf.* Sect. 10.1). Smoothed contours of elevation on such maps continue to exhibit similar shapes when the scale is enlarged, as in Krige's (1966) example for Klerksdorp gold contours (Sect. 11.1.1). Lovejoy and Schertzer (2007) argued convincingly that the Earth's topography can be modeled as a multifractal, both on continents and ocean floors in accordance with power-law relations originally established by Vening Meinesz (1964) as explained in Sect. 10.1.1. These broad-scale approaches to the nature of topography also seem to run counter to the fact that landscapes are of many different types and result from different genetic processes. Nevertheless, it can be assumed as Brinck did that chemical elements within the Earth's crust or within smaller, better defined environments like the Witwatersrand goldfields can be modeled as multifractals.

In early applications of mathematical statistics to geoscience, it often was assumed that regional features can be modeled by using deterministic functions (e.g., in trend surface analysis, Chap. 4). Residuals from the trends were assumed to be white noise with the properties of uncorrelated (iid) random variables. It gradually became clear that residuals often are better modeled as stationary random functions with a spatial covariance function or semi-variogram. Universal kriging (cf. Sect. 7.2) is an approach that embodies the three components consisting of deterministic trends (or drifts), stationary random functions and white noise (nugget effect). A frequently used geostatistical model is that the semivariogram shows nugget effect at the origin, a range that can be modeled for use is kriging and a sill related to regional mean. Nested designs of superimposed models of this type also are frequently used. Multifractal modeling can help to refine the nugget effect. Based on the concept of self-similarity (or scale independence), the spatial autocorrelation function can be extrapolated to very short distances. Use of the Chen algorithm (Sect. 11.6) resulted in identification of white noise components that are much smaller than previously suggested by semivariogram or correlogram. It was emphasized in Chap. 11 that this approach is close to the original Matheron (1962) approach to ore reserve calculations.

The applications of Chap. 11 make use of multifractal spectra that are symmetrical. It was assumed that the central part of these spectra with shapes estimated by the method of moments (Sect. 11.2.1) is according to the model of de Wijs (binomial p/model) or one of its generalizations, although commonly the minimum and maximum singularities, with fractal dimensions close to zero, are outside the range predicted by the model of de Wijs. They represent extreme events in the tails of the frequency distributions of the phenomena that are being modeled. In practice, multifractal spectra can be asymmetrical, display negative fractal dimensions in the tails or suggest the existence of infinitely large or small singularities (cf. Mandelbrot 1999). In Agterberg (2001a) it was shown by means of computer simulations experiments that 2-D element concentration patterns generated by the model of de Wijs with superimposed regional trends can result in asymmetrical multifractal spectra and negative fractal dimensions that are artefacts.

In this section, white noise will be added to a 1-D pattern generated by the model of de Wijs that seems to show some broad systematic variations resembling regional trends. It will be shown that the white noise component can be extracted from the data by means of the Chen algorithm but after a limited number of iterations.

12.6.1 Computer Simulation Experiment

Figure 12.27 shows an artificial series of 250 values generated by using a one-dimensional version of the model of de Wijs with dispersion index $d = 0.4$ (cf. Agterberg 2012b). The pattern of Fig. 12.27 was generated in the same way as that of Fig. 12.17 using different random numbers. These hypothetical element concentration values were logarithmically (base 10) transformed and a Gaussian white noise component with zero mean and standard deviation equal to 0.25 was added. The result is shown in Fig. 12.28. The antilogs (base 10) of the values shown in Fig. 12.28 were subjected to local singularity analysis. Patterns of the smoothing coefficients c_k are shown in Fig. 12.29 for $k = 1, 5, 10, 120,$ and $1,000$, respectively. Obviously, degree of smoothing increases when k is increased. The purpose of the iterative algorithm is to optimize local singularity α_k rather than c_k which in the limit would become a straight-line pattern with values close to average element concentration value (cf. Agterberg 2012a, b). The result for $k = 1$ represents ordinary local singularity analysis. The other patterns are for larger values of k obtained by means of the Chen algorithm. Figures 12.30, 12.31 and 12.32 show logarithmically transformed input values plotted against singularities for $k = 1, 120,$ and $1,000$, respectively. In these three diagrams, the best-fitting straight line obtained by least squares is also shown. It can be assumed that the logarithmically transformed input value is the sum of a "true" value that in this application satisfies the model of de Wijs for $d = 0.4$, plus a random error with zero mean and standard deviation of 0.25 corresponding to the Gaussian white noise component.

12.6 Trends, Multifractals and White Noise

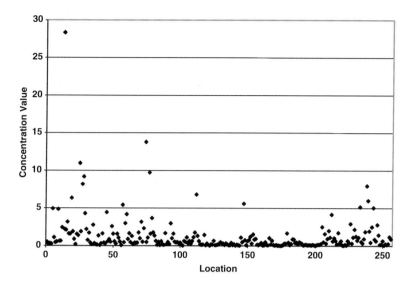

Fig. 12.27 Artificial series of 256 values satisfying model of de Wijs with overall mean set equal to 1 and index of dispersion $d = 0.4$ (Source: Agterberg 2012b, Fig. 4)

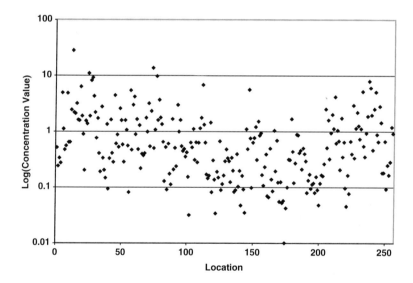

Fig. 12.28 Logarithmically transformed values of Fig. 12.27 after addition of random normal numbers with zero mean and standard deviation equal to 0.25. The purpose of this experiment is to create an artificial series that has two components of which the only first one (Fig. 12.27) has autocorrelated values (Source: Agterberg 2012b, Fig. 5)

The estimated singularities should provide approximately error-free estimates of the logarithmically transformed input values. Standard deviations of differences between logarithmically transformed simulated element concentration values and estimated singularities on the regression lines shown in Figs. 12.30, 12.31, and

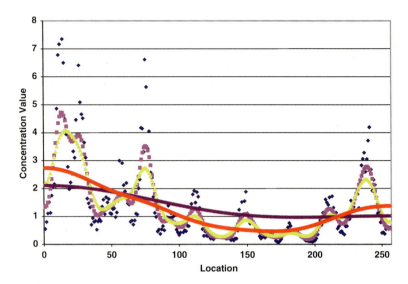

Fig. 12.29 Successive patterns of c_k obtained after 1 (*diamonds*), 5 (*squares*), 10 (*triangles*) 120 (*solid line*) and 1,000 (*smoothest curve*) iterations (Source: Agterberg 2012b, Fig. 6)

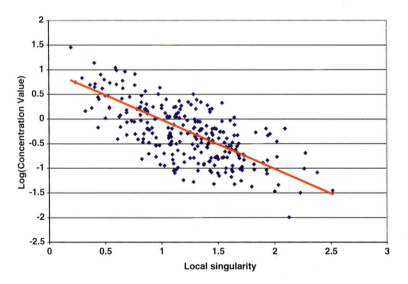

Fig. 12.30 Logarithmically (base 10) transformed simulated element concentration versus local singularity after single iteration ($k = 1$) with regression line (singularity assumed to be error-free). The standard deviation of residuals from the best-fitting straight line amounts to 0.42, which is greater than 0.25 (Source: Agterberg 2012b, Fig. 7)

12.32 for $k = 1$, 120 and 1,000 were 0.42, 0.25 and 0.16, respectively. Three other residual standard deviations estimated after $k = 10$, 100 and 200 iterations were 0.33, 0.26 and 0.22, respectively. It can be concluded that the original white noise

12.6 Trends, Multifractals and White Noise

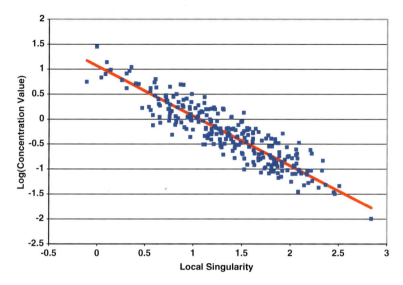

Fig. 12.31 Logarithmically (base 10) transformed simulated element concentration versus local singularity after single iteration ($k = 120$) with regression line (singularity assumed to be error-free). The standard deviation of residuals from the best-fitting straight line amounts to 0.25, which is equal to standard deviation of Gaussian white noise component added in Fig. 12.28 (Source: Agterberg 2012b, Fig. 8)

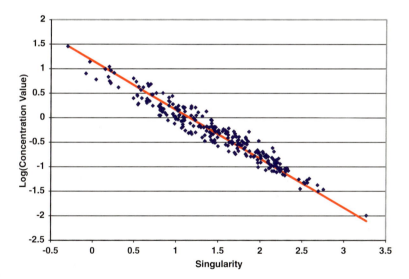

Fig. 12.32 Logarithmically (base 10) transformed simulated element concentration versus local singularity after single iteration ($k = 1{,}000$) with regression line (singularity assumed to be error-free). The standard deviation of residuals from the best-fitting straight line amounts to 0.16, which is less than 0.25 (Source: Agterberg 2012b, Fig. 9)

component with standard deviation of 0.25 was retrieved after about $k = 120$ iterations. It corresponds to the line for $k = 120$ in Fig. 12.29. Other values of k reproduced standard deviations for the white noise component that are either too large or too small.

The solid line in Fig. 12.29 can be regarded as a smoothed version of the original concentration values and c_k patterns for $k = 1, 5$ and 10 shown in the same diagram. The experiment indicates that $k = 120$ represents an approximate optimum upper limit for the number of iterations to be used in the Chen algorithm. Further increases of k yield patterns of c_k that are smoother, do not seem to provide a good fit with variances that are smaller than 0.25. Ultimately, with k approaching infinity, the pattern of c_k would become a straight line as in the previous application to the Pulacayo orebody (Agterberg 2012a, Fig. 19). The computer experiment described in this section indicates that full convergence is not desirable in all applications.

12.7 Universal Multifractals

Already in the 1980s, Schertzer and Lovejoy (1985) pointed out that the binomial/p model (model of de Wijs) can be regarded as a "micro-canonical" version of their α-model in which the strict condition of local preservation of mass is replaced by the more general condition of preservation of mass within larger neighborhoods (preservation of ensemble averages). Cascades of this type can result in lognormals with Pareto tails. In the three-parameter Lovejoy-Schertzer $\boldsymbol{\alpha}$-model, $\boldsymbol{\alpha}$ (bold alpha) does not represent the singularity α but represents the Lévy index that, together with the "codimension" C_1 and the "deviation from conservation" H, characterizes a universal multifractal field. Examples of applicability of universal multifractals to geological processes that took place in the Earth's crust have been given by Lovejoy and Schertzer (2007).

The model originally developed by Schertzer and Lovejoy (1991a, b) is based on the concept of a multifractal field ρ_λ with codimension $C_1(\gamma)$ where $\lambda = L/\epsilon$ is the so-called scale ratio with L representing the largest scale that can be set equal to 1 without loss of generality (Cheng and Agterberg 1996). The field ρ_λ can be a flux or density in physics. In geochemical applications it is the element concentration value of a chemical element. It is characterized by its probability distribution $P(\rho_\lambda \lambda^\gamma) \propto \lambda^{-C_1(\gamma)}$ with statistical moments $E(\rho_\lambda^q) \propto \lambda^{K(q)}$. The relations between $K(q), C_1(\gamma)$ and the field order γ are:

$$K(q) = \max_\gamma \{q \cdot \gamma - C_1(\gamma)\}; C_1(\gamma) = \max_q \{q \cdot \gamma - K(q)\}$$

Cheng and Agterberg (1996) have made a systematic comparison of the binomial/p model with the multifractal spectrum discussed in Sect. 11.2.1 and the universal multifractal model of Schertzer and Lovejoy (1991a, b).

12.7 Universal Multifractals

In Sect. 11.6.1 it was pointed out that estimates of α_{min} and α_{max} derived for the Pulacayo orebody by means of singularity analysis differ greatly from previous estimates based on the binomial/p model. From application of the method of moments it would follow that $d = 0.121$, $\alpha_{min} = 0.835$, and $\alpha_{max} = 1.186$. The latter two estimates differ not only from $\alpha_{min} = 0.835$ and $\alpha_{max} = 1.719$ derived in Sect. 11.6.1; $\alpha_{max} = 1.186$ also is less than the estimate $\alpha_{max} = 1.402$ on the right side of the multifractal spectrum in Fig. 11.7. Obviously, the estimate $d = 0.121$ is much too small. Using absolute values of differences between successive values, de Wijs (1951) had already derived the larger value $d = 0.205$. Use of the estimates of α_{min} or α_{max} based on the full convergence local singularities derived by the Chen algorithm yielded $d = 0.369$ and $d = 0.392$, respectively (Sect. 11.6.1). Clearly, the binomial/p model has limited range of applicability and a more flexible model with additional parameters should be used. The accelerated dispersion model of Sect. 12.5.1 offers one possible explanation. Another approach consists of using the Lovejoy-Schertzer α-model. These authors have successfully applied this model to the 118 Pulacayo zinc values as will be discussed in the next section.

The three-parameter α-model is a generalization of the conservative two-parameter $(0, C_1, \alpha)$ universal canonical multifractal (Schertzer and Lovejoy 1991b, p. 59) model. So-called extremal Lévy variables play an important role in a two-parameter approach. In their large-value tail these random variables have probability distributions of the Pareto form $P(X \ x) \propto |x|^{-\alpha}$ ($\alpha < 2$). They can be used to generate multiplicative cascades starting from uniform random variables as outlined in Box 12.3. Frequency distributions of values belonging to multifractal fields of this type already can have high-value tails that are thicker than logbinomial and lognormal tails.

Box 12.3: Extremal Lévy Variables

Suppose W is a uniform random variable within the interval [0, 1] and $V = W^{-1/\alpha}$; so that the probability $P(V = v) = \alpha \cdot v^{-1-\alpha}$ if $v \geq 1$, and $P(V = v) = 0$ if $v < 1$ (cf. Wilson et al. 1991, APPENDIX B). Then, $\sum_{i=1}^{n} V_i$ representing the sum of n such random variables V_i ($i = 1, 2, \ldots, n$) has a Lévy limit distribution with high-value tails. The Lévy index α defines another index α' by means of the relation $1/\alpha + 1/\alpha' = 1$. Suppose that μ and c are two constants and a new random variable is defined as $Y = c \cdot (\mu - V)$. Then, for large n: $Y = \frac{1}{n^{1/\alpha}} \sum_{i=1}^{n} c \cdot \left(\frac{\alpha}{\alpha - 1} \cdot w_i^{-1/\alpha} \right)$ with Laplacian characteristic function $E(e^{Y \cdot q}) = \exp\{\alpha \cdot \Gamma(-\alpha) \cdot (cq)^{\alpha}\}$. It follows, after some manipulation, that $c = \left(\frac{C_1}{\Gamma(2-\alpha)} \right)^{1/\alpha}$ resulting in the two-parameter universal form $K(q) = \frac{C_1 \alpha'^{C}}{\alpha} (q^{\alpha} - q); 0 \ \alpha \ 2; 0 < C_1 < E$.

The Lévy index α characterizes the degree of multifractality and the codimension C_1 characterizes the sparseness of the mean field (Lovejoy and Schertzer 2010). In the three-parameter Lovejoy-Schertzer α-model a third parameter H is added that

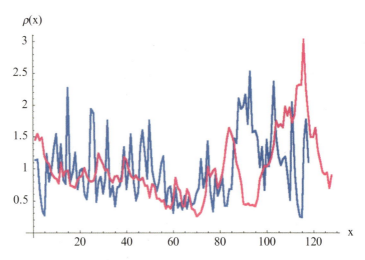

Fig. 12.33 Blue: original Pulacayo zinc values, with x representing horizontal distance in units of 2 m. Pink: universal multifractal simulation results for Levy index $\alpha = 1.8$; co-dimension $C_1 = 0.03$ characterizing the "sparseness" of the mean field; and deviation from conservation $H = 0.090$. Both patterns were normalized to unity (mean $= 15.6$ % Zn) (Source: Lovejoy and Schertzer 2007, Fig. 3a)

characterizes the deviation from conservation. A space-time universal multifractal field for element concentration can be subject to change of its mean value in the course of time. Even in static 3-D situations, incorporation of H into the model provides increased flexibility. Lovejoy and Schertzer (2007) provide examples of geophysical data for which the universal multifractal approach is applicable. This includes applications to magnetic susceptibility data studied by Pilkington and Todoeschuck (1995).

12.7.1 Pulacayo Mine Example

Figure 12.33 taken from Lovejoy and Schertzer (2007) shows a realistic universal multifractal simulation for the Pulacayo orebody using the following three parameters: Lévy index $\alpha = 1.8$, codimension $C_1 = 0.03$ and deviation from conservation $H = 0.090$. This approach is explained in detail and illustrated by means of other applications in a large number of publications including Lovejoy et al.(2008), Lovejoy et al. (2001), Schertzer and Lovejoy (1991a), Schertzer and Lovejoy (1997), Schertzer et al. (1997) and Lovejoy and Schertzer (2013). The codimension C_1, which characterizes sparseness of mean field, and deviation from conservation H can be derived as follows. First a log-log plot of the so-called "first order structure function" (*cf.* Monin and Yaglom 1975) is constructed (Fig. 12.34). Successive moments are obtained for absolute values of differences between concentration values for points that are distance h apart by raising them to the powers

12.7 Universal Multifractals

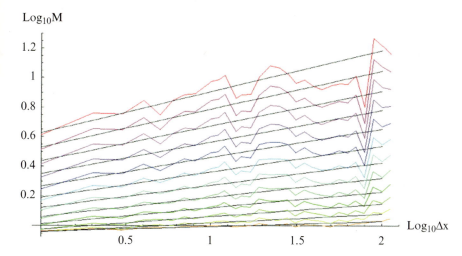

Fig. 12.34 Log-Log plot of first order structure function for Pulacayo zinc values for $q = 0.25$, $0.5, \ldots, 3$. Fourth line from top ($q = 2$) is variogram; first point on fourth line from bottom ($q = 1$) represents de Wijs dispersion index d (Source: Lovejoy and Schertzer 2007, Fig. 26a)

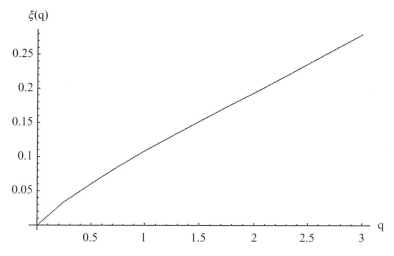

Fig. 12.35 Slopes $\xi(q)$. Slope and value near $q = 1$ yield $H = 0.090$, $C_1 = 0.018$, because $H = \xi$ (1) and $C_1 = H - \xi'(1)$; H represents deviation from conservation of pure multiplicative process for which $H = 0$ (Source: Lovejoy and Schertzer 2007, Fig. 26b)

q (=0.25, 0.5, ..., 3 for the 118 zinc values). The resulting pattern for $q = 2$ represents the variogram and the first point on the pattern for $q = 0$ is the de Wijs index of dispersion d. Straight lines are fitted to all patterns and a new diagram is constructed with the slopes of the lines (ξ_q) plotted against q. Slope and value of this new line near $q = 1$ yielded $H = 0.090$ and $C_1 = 0.02$ for the Pulacayo orebody because $H = \xi_I$ and $C_1 = H - \xi'_I$ where ξ'_I is the first derivative of ξ_q with respect to q (Fig. 12.35).

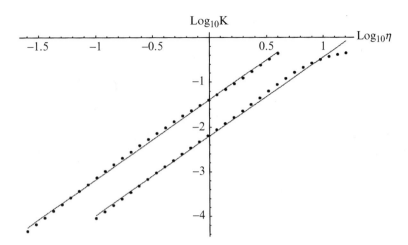

Fig. 12.36 Double Trace Moment analysis of the Pulacayo zinc values with $q=2$ (left), $q=0.5$ (right). Slopes yield Levy index $\alpha = 1.76$, 1.78; and co-dimension $C_1 = 0.023$, 0.022, respectively (Source: Lovejoy and Schertzer 2007, Fig. 27)

Use of the so-called "double trace moment" method (cf. Lavallée et al. 1992) yielded estimates of the Lévy index equal to $\alpha = 1.76$ and $\alpha = 1.78$, and codimension $C_1 = 0.023$, 0.022, respectively (Fig. 12.36). In general, a relatively small value of C_1 with respect to H indicates that the multifractality is so weak that deviation from conservation (H) will be dominant except for quite high moments (Lovejoy and Schertzer 2007, p. 491). In the Pulacayo application the estimated value of H is small so that the two-parameter model of Box 12.3 would be approximately satisfied. It was shown before that the binomial/p model produced inconsistencies between results for very small and very large moments. Universal multifractal modeling is more flexible and produces realistic zinc concentration variability. On the other hand, the estimate for the second order moment obtained by the method of moments $\tau(2) = 0.979 \pm 0.038$ produced a realistic autocorrelation function including the nugget effect, which affects the power spectrum for high frequencies as will be discussed next.

Another important tool in universal multifractal modeling is spectral analysis. Theoretically, this approach results in a spectrum consisting of a straight line with slope $-\beta$. This parameter can either be estimated directly or indirectly using $\beta = 1 - K_2 + 2H$ where K_2 representing the "second characteristic function". Lovejoy and Schertzer (2007) estimated $K_2 = 0.05$ by double trace moment analysis (Fig. 12.37). With the previously mentioned estimate $H = 0.090$ this yielded $\beta \approx 1.12$ in good agreement with the experimental spectrum for the 118 zinc values.

Spectral analysis of the 118 logarithmically zinc values was discussed previously in Sect. 6.2.7. In Agterberg (1974, Fig. 67) the discrete Fourier transform was taken of autocorrelation correlation coefficients with lag distances <32 m after applying a cosine transformation in order to largely eliminate distortions according

12.8 Cell Composition Modeling

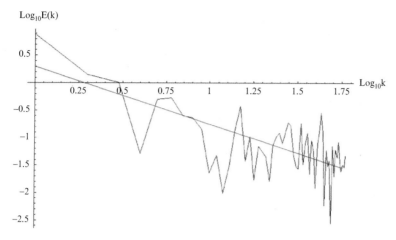

Fig. 12.37 Pulacayo zinc spectrum according to Lovejoy and Schertzer (2007, Fig. 3b). Red line is theoretical with slope $\beta = 1 - K(2) + 2H = 1.12$; $K(q)$ is "second characteristic function"; $K(2) = 0.05$ was determined by double trace moment analysis; deviation from conservation $H = 0.090$ was derived from first order structure functions (Source: Lovejoy and Schertzer 2007, Fig. 3b)

to the "hanning" method (Blackman and Tukey 1958). In a discussion of this result, Tukey (1970) pointed out that the resulting spectrum "drooped" although it was within the 90 % confidence interval around the theoretical spectrum for the signal-plus-noise model with negative exponential autocorrelation function. Re-plotting the earlier results on a log-log plot shows a linear pattern with straight line of best fit yielding $\beta = 0.79$. Although the straight-line model provided a good fit in this plication, this estimate of β is somewhat less than that obtained by Lovejoy and Schertzer (2007).

12.8 Cell Composition Modeling

In various geomathematical applications to map data, the study area is subdivided into square cells belonging to a regular grid. If random variables are defined for such cells it is of interest to know how the parameters of these variables depend on cell size. In 3-D applications the theory applies to blocks instead of grid cells. The relationship between block values and block size has been considered by Matheron (1976). One of his models (the discrete Gaussian model) is taken as a starting point for most applications in this section. Some frequency distributions have the property of "permanence". This concept resembles that of limit distributions for sums of independent random variables. Six types of permanent random variables were considered in Agterberg (1984) where further discussions can be found. Only three types of permanent frequency distributions (lognormal, probnormal and asymmetrical bivariate binomial distributions) will be exemplified in this section. Other types of orthogonal polynomials are required for other types of permanent frequency distributions. Books covering theory of orthogonal polynomials include Beckmann (1973) and Szegö (1975).

Theory will be summarized and applied in case history studies dealing with the areal distribution of acidic volcanics in the vicinity of Bathurst, New Brunswick, and in the Abitibi area on the Canadian Shield. Both examples have been used extensively in previous chapters in order to exemplify a variety of statistical techniques and methods of mathematical morphology. These rocks constitute favorable environments for the occurrence of volcanogenic massive sulphide deposits. It should be kept in mind that the information used as input is 2-D only. In reality these rock formations have 3-D structures that are only known to a limited extent although during the past 10 years significant progress has been made in 3-D geological map construction (cf. Chap. 1).

12.8.1 Permanent Frequency Distributions

Suppose that a rock or geological environment is sampled by randomly superimposing on it a large block with average value X_2, and that a small block with value X_1 is sampled at random from within the large block. Then the expected value of X_1 is X_2, or $E(X_1 \mid X_2) = X_2$. Let $f(x_1)$ and $f(x_2)$ represent the frequency density functions of the random variables X_1 and X_2, respectively. Suppose that X_1 can be transformed into Z_1 by $X_1 = \psi_1(Z_1)$, and X_2 into Z_2 by $X_2 = \psi_2(Z_2)$ so that the random variables Z_1 with marginal density function $f(z_1)$, and Z_2 with $f(z_2)$, satisfy a bivariate density function of the type

$$f(z_1, z_2) = f_1(z_1) \cdot f_2(z_2) \left[1 + \sum_{j=1}^{\infty} \rho^j Q_j(z_1) S_j(z_2) / h_{1j} h_{2j} \right]$$

In this equation, ρ represents the product–moment correlation coefficient of Z_1 and Z_2. $Q_j(z_1)$ and $S_j(z_2)$ are orthogonal polynomials with $f_1(z_1)$ and $f_2(z_2)$ as weighting functions, and with norms h_{1j} and h_{2j}, respectively. It is implied that

$$E[Q_j(Z_1)|Z_2] = \rho^j S_j(z_2) h_{1j} / h_{2j}$$

In most applications, $f(z_1, z_2)$ is symmetric with $f_1(z_i) = f_2(z_i)$ for $i = 1, 2$. For example, if $f_1(z_1)$ and $f_2(z_2)$ are standard normal, $Q_j(z_1) = H_j(z_1)$ and $S_j(z_2) = H_j(z_2)$ are Hermite polynomials with squared norms $j!$ for both $Q_j(z_1)$ and $S_j(z_2)$. Then the preceding expression for the bivariate density function becomes the well-known Mehler identity.

12.8 Cell Composition Modeling

> **Box 12.4: Hermite Polynomials**
>
> Hermite polynomials $H_j(u)$ satisfy $\sqrt{j!} \cdot H_j(u) = u^j - \frac{j^{(2)}}{2} u^{j-2} + \frac{j^{(4)}}{2^2 \cdot 2!} u^{j-4} - \frac{j^{(6)}}{2^3 \cdot 3!} u^{j-6} + \ldots$, where $j^{(t)} = j(j-1)(j-2)\ldots(j-t+1)$. For example: $H_0(u) = 1; H_1(u) = u; H_2(u) = u^2 - 1; \quad H_3(u) = u^3 - 3u; H_4(u) = u^4 - 6u^2 + 3; H_5(u) = u^5 - 10u^3 + 15u$. If Z_1 and Z_2 are two standard normal random variables with zero means, unit variances and correlation coefficient ρ, their bivariate density function is: $\varphi(z_1, z_2) = \frac{\exp[-(z_1^2 - 2\rho z_1 z_2 + z_2^2)/2(1-\rho^2)]}{2\pi\sqrt{1-\rho^2}} = \frac{\exp[-\frac{1}{2}(z_1^2 + z_2^2)]}{2\pi} \left[1 + \sum_{j=1}^{\infty} \rho^j H_j(z_1) H_j(z_1)\right]$ representing the Mehler identity. Other properties include $E\{H_j(X)\} = \mu^j$ where X is a normal random variable with mean μ.

When $f(x_1)$ is known, $f(x_2)$ can be derived in combination with $E(X_1 \mid X_2) = X_2$. If $f(x_1)$ and $f(x_2)$ are of the same type, their frequency distribution is permanent. Examples include the lognormal and logbinomial distributions. Permanence of these two distributions was first established by Matheron (1974, 1980). It means that blocks of different sizes have the same type of frequency distribution; for example, all can be lognormals with the same mean but different logarithmic variances.

The regression of X_1 on X_2, or $E(X_1 \mid X_2) = X_2$, results in the following two relations between block variances:

$$\sigma^2(X_1) = \sigma^2(X_1 - X_2) + \sigma^2(X_2); \sigma^2(X_2) = \rho_x^2 \sigma^2(X_1)$$

with $\rho_x > 0$ representing the product–moment correlation coefficient of X_1 and X_2. The first part of this equation can be rewritten as $\sigma^2(v, V) = \sigma^2(v) + \sigma^2(V)$. Suppose that v is contained within a larger volume v' that, in turn, is contained within V. Then, $\sigma^2(v, V) = \sigma^2(v, v') + \sigma^2(v', V)$ These results are general in that they do not only apply to original element concentration values but also to transformed element concentration values; e.g. $E(\ln X_1 \mid \ln X_2) = \ln X_2$. The resulting logarithmic variance relationship for average values in blocks of three different sizes was originally discovered by Krige (1951) and was called "Krige's formula for the propagation of variances" by Matheron (Sect. 2.1.2).

This result also can be derived as follows. Using Hermite polynomials the lognormal random variable X_1 can be written as $X_1 = \mu \cdot \exp(\sigma_1 Z_1 - \frac{1}{2}\sigma_1^2) = \mu \cdot \sum_{j=0}^{\infty} \sigma_1^j H_j(Z_1)/j!$. From results for correlation between X_1 and X_2 described in Box 12.5 it follows that X_2 satisfies the same expression when σ_1 is replaced by σ_2, and $\sigma_2^2 = \sigma_1^2 \cdot \rho$.

It is possible to use the relation $\sigma^2(\ln X_1) = \rho_x^2 \cdot \sigma^2(\ln X_2)$ to estimate the correlation coefficient ρ_x. The following example of application was described in Agterberg (1977). Probability indices for occurrence of large copper deposits in

$(20 \times 20$ km) and $(40 \times 40$ km) cells in the Abitibi area on the Canadian Shield, similar to those discussed in Sect. 4.3.2, are approximately lognormally distributed with logarithmic variances equal to 2.150 and 0.920, respectively. Consequently, the correlation coefficient between values for $(20 \times 20$ km) and $(40 \times 40$ km) cells is estimated to be equal to 0.654.

Box 12.5: Correlation between Original and Standardized Cell Composition Data

Suppose the mean value of the two random variables X_1 and X_2 (for smaller and larger cells) is written as $\mu = E(X_1) = E(X_2)$. In general, the transformations ψ_1 and ψ_2 can be expanded by using the orthogonal polynomials Q_j and S_j with $x_1 = \psi_1(z_1) = \sum_{j=1}^{\infty} c_j Q_j(z_1); x_2 = \psi_2(z_2) = \sum_{j=1}^{\infty} c_j^* S_j(z_2)$ where x_1 and x_2 are ordinary continuous variables. The coefficients differ from one another except for $c_0 = c_0^* = \mu$. In general, $\int_{-\infty}^{\infty} f_1(z_1) Q_i(z_1) Q_j(z_1) dz_1 = \delta_{ij} \cdot h_{1j}^2$ where $\delta_{ij} = 0$ if $i \neq j$, and $\delta_{ij} = 1$ if $i = j$. Also, $x_2 = \int_{-\infty}^{\infty} \psi_1(z_1)[f(z_1, z_2)/f_2(z_2)] dz_1$

$= \int_{-\infty}^{\infty} \psi_1(z_1) f_1(z_1) \left[\frac{\sum_{j=0}^{\infty} \rho^j Q_j(z_2) S_j(z_2)}{h_{1j} h_{2j}} \right] dz_1 = \psi_2(z_2) = \sum_{j=0}^{\infty} \rho^j c_j S_j(z_2) h_{1j}/h_{2j}$ or c_j^*

$= \rho^j c_j h_{1j}/h_{2j}$. For the larger cells, $E(X_2^2) = \int_{-\infty}^{\infty} x_2^2 f_{X_2}(x_2) dx_2 = \int_{-\infty}^{\infty} \psi_2^2(z_2) f_2(z_2) dz_2$

$= \int_{-\infty}^{\infty} f_2(z_2) \left[\sum_{j=0}^{\infty} \rho^j c_j S_j(z_2) h_{1j}/h_{2j} \right] dz_2 = \sum_{j=0}^{\infty} \rho^{2j} c_j^2 h_{1j}^2$

so that their variance satisfies $\sigma^2(X_2) = E(X_2^2) - \mu^2 = \sum_{j=1}^{\infty} \rho^{2j} c_j^2 h_{1j}^2$ or $\rho_x^2 \sigma^2(X_2) = \sum_{j=1}^{\infty} \rho^{2j} c_j^2 h_{1j}^2$.

12.8.2 The Probnormal Distribution

In an application of the discrete Gaussian model approach, Matheron (1974, pp. 57–61) considered the following cell-value distribution model for binary patterns: $X = \Phi \left[\frac{\rho Z - \beta}{\sqrt{1-\rho^2}} \right]$. In this equation, X is a random variable for cell values generated from a binary variable X_0, Z is a standard normal random variable, while β and ρ are constants and Φ represents the quantile of a normal distribution function.

Cell values are compositional data in that they represent proportions of cell area underlain by the binary pattern (e.g., for a rock type on a geological map) that is

12.8 Cell Composition Modeling

being studied. The cell can be square, rectangular or have another shape. One property of this model is that the complement $(1 - X)$ also is the quantile of a normal random variable with different mean but the same variance. The fact that percentage values for a rock type and its complement can both have a frequency distribution which satisfies the same equation may be of interest in the study of closed number systems. If the values for a set of random variables sum to one at all observation points, these variables cannot have the same type of distribution if one or more of them has a normal, lognormal, or gamma distribution. On the other hand, each variable in a set of random variables summing to one can have a frequency distribution function given by $X = \Phi\left[\frac{\rho Z - \beta}{\sqrt{1-\rho^2}}\right]$. An example of a complement $(1 - X)$ will be provided in Fig. 12.53. The close resemblance of probits to logits was discussed in Sect. 12.2.1 (Fig. 12.8). The logit of $(1 - X)$ is $-$ logit (X).

If this model is valid, observed cell values should have a frequency distribution that plots as a straight line on "prob-prob" paper with standard normal quantile scales along both X- and Y-axis. Then the parameters β and ρ can be estimated from the intercept and slope of this straight line. In this respect the approach seems to form a natural extension of two Q-Q straight line fitting techniques that are widely applied: use of log-log paper in fractal-multifractal analysis, and log-prob paper in fitting a lognormal frequency distribution to data.

In the next section, two previously used data sets (felsic metavolcanics in the Bathurst and Abitibi areas) will be re-analyzed. The multifractal method of moments also will be applied to the larger of these two data sets (Abitibi felsic metavolcanics).

Matheron's (1974, 1976) discrete Gaussian model can be applied to cell values (*cf.* Agterberg 1981, 1984) as follows: Let X_1 with density function $f(x_1)$ represent a random variable for average concentration values of small cells with size S_1 and X_1, with $f(x_2)$, that of large cells with size S_2. Suppose that X_1 can be transformed into Z_1 by $X_1 = \psi_1(Z_1)$ and X_2 into Z_2 by $X_2 = \psi_2(Z_2)$ so that the random variable Z_1, with marginal density function $\varphi(z_1)$, and Z_2 with $\varphi(z_2)$, together satisfy the bivariate standard normal density function

$$\varphi(z_1, z_2) = \frac{1}{2\pi\sqrt{1-\rho^2}} \exp\{-(z_1^2 - 2\rho z_1 z_2 + z_2^2)/2(1-\rho^2)\}$$

where ρ represents the product–moment correlation coefficient of Z_1 and Z_2. In general, if the regression of X_1 on X_2 satisfies $E(X_1|X_2) = X_2$, $f(x_2)$ can be derived from $f(x_1)$. The interpretation for a binary pattern is as follows. Suppose that X_2 represents the average value for amount of pattern in a large cell superimposed on the map, while X_1 is this value for a small cell sampled at random from within the large cell. If the small cell is made infinitely small, X_1 becomes a binary random variable that can be written as X_0 for presence or absence of the pattern at a point. Consequently,

$$\varphi(z_2|z_1) = \frac{\varphi(z_1,z_2)}{\varphi(z_1)} = \varphi\left[\frac{(z_2-\rho z_2)}{2\pi\sqrt{1-\rho^2}}\right]$$

As discussed in more detail before, the Mehler identity is:

$$\varphi(z_1,z_2) = \varphi(z_1)\cdot\varphi(z_2)\left[1+\sum_{j=1}^{\infty}\rho^j H_j(z_1)H_j(z_2)/j!\right]$$

where $H_j(z_1)$ and $H_j(z_2)$ are Hermite polynomials with the property:

$$\int_{-\infty}^{z_2} H_j(z_2)\varphi(z_2)dz_2 = H_{j-1}(z_2)\varphi(z_2)$$

Use of this property during integration with respect to z_2, replacement of z_1 by the standard normal random variable Z, and replacing z_2 by the constant β yields the random variable

$$X = \Phi\left[\frac{\rho Z - \beta}{\sqrt{1-\rho^2}}\right] = 1 - \Phi(\beta) - \varphi(\beta)\sum_{j=1}^{\infty}\rho^j H_j(Z)H_{j-1}(\beta)/j!$$

This is equivalent to assuming that the "probit" of X has normal distribution (Fig. 12.38). The binary random variable X_1 for points randomly located within cells with values assumed by X can be defined by letting ρ tend to one. Then ρ can be interpreted as the correlation coefficient between X_1 and X. The mean of X satisfies $\mu = 1 - \Phi(\beta)$ and its variance is $\sigma^2 = \varphi^2(\beta)\sum_{j=1}^{\infty}\rho^{2j}[H_{j-1}(\beta)]^2(j!)^{-1}$. By means of these two equations it is possible to obtain estimates of β and ρ from estimates of μ and σ^2 as is shown graphically in Fig. 12.39. Of course, it also is possible to estimate these parameters directly by fitting a straight line to a Q-Q plot of cell value percentage values and their observed frequencies using prob-prob paper.

12.8.3 Bathurst Area Acidic Volcanics Example

Occurrences of volcanogenic massive sulphides and their relationship with acidic volcanics in the Bathurst area, New Brunswick were taken for example in Sect. 1.5.2 to illustrate various Minkowski operations. In this section the same example is used to model the pattern of acidic volcanics (Fig. 1.14d) by means of the probnormal model.

12.8 Cell Composition Modeling

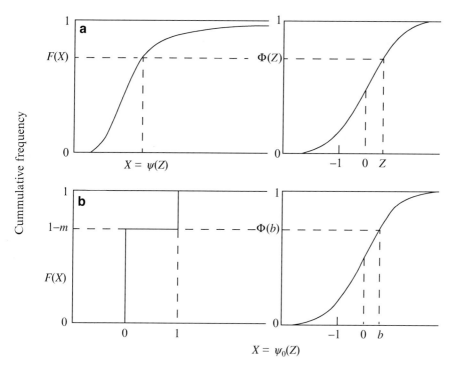

Fig. 12.38 Graphical representation of the transformation ψ in $X = \psi(Z)$. Distribution functions $\Phi(Z)$ on the right-hand side are for standard normal random deviates. A value $F(X)$ on the left-hand side of (**a**) is equal to $\Phi(Z)$ if $X = \psi(Z)$; (**b**) Shows the distribution function of a binary random variable which can also be related to a standard normal random variable (Source: Agterberg 1981, Fig. 6)

The geometrical covariance $K_\alpha(h)$ satisfies $K_\alpha(h) = \text{mes } A\Theta B'$ where A is the original pattern and B' an operator set consisting of two points. One point is the origin of B and the other point occurs at a distance in the direction α. The accent on B denotes reflection of B with respect to its origin. $K_\alpha(h)$ is shown in Fig. 12.40 for the east–west direction. These measurements were obtained on the Quantimet 720 with linear correlator module at the Ecole Polytechnique in Montreal (Agterberg and Fabbri 1978). In order to obtain the corresponding statistical covariance, the values of Fig. 12.40 were first increased by the factor mes T_0/mes $T_0 \Theta B'$ where T_0 represents a square study area around A which measures exactly 80 km on a side. The statistical covariances were obtained by subtracting m^2 from the corrected geometrical covariances where $m = \text{mes } A/\text{mes } T_0$ is the proportion of the study area underlain by acidic volcanics. The statistical covariances were divided by the variance $C_0 = m - m^2$ and this gave the autocorrelation coefficients plotted along the vertical axis with logarithmic scale in Fig. 12.41. The signal-plus-noise model with $r_h = c \cdot \exp(-p \cdot |h|)$ with $c = 0.87$ and $p = 0.194$ provides a reasonably good fit.

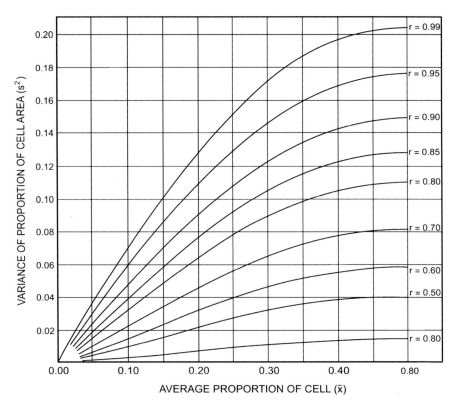

Fig. 12.39 Relationship between estimates of mean (x), and variance (s^2) of proportion of cell area underlain by the rock type considered, for various values of r (Source: Agterberg 1981, Fig. 3)

The preceding measurements on the Quantimet 720 were amplified by analysis of binary images obtained on the Flying Spot Scanner with results shown in Fig. 12.42. This two-dimensional geometrical covariance function is isotropic by good approximation. Geometrical covariances for the east–west direction were converted into autocorrelation coefficients as before and are close to the estimates based on Quantimet 720 readings and the fitted signal-plus- noise model of Fig. 12.41. Therefore, the semi-exponential autocorrelation function with $c = 0.87$ and $p = 0.194$ can be used for further computations. However, there was a discrepancy in the estimate of mes A (1,319 km^2 on the Quantimet 720 vs 1,264 km^2 on the Flying Spot Scanner). Because the Quantimet reading is less satisfactory in this instance (cf. Agterberg 1978a), the Flying Spot Scanner reading was adopted. It is close to an estimate of mes $A = 1,262.5$ km^2 originally obtained by point-counting the original geological maps at points on a grid with 500 m spacing in the north–south and east–west directions. The study area T_0 of 80×80 km consists of 64 (10×10 km) cells belonging to Zone 19 of the Universal Transverse Mercator grid shown on 1:250,000 topographic maps. These 64 cells were point-counted separately (400 counts per cell) and a histogram of the 64 values for acidic

12.8 Cell Composition Modeling

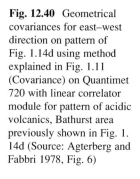

Fig. 12.40 Geometrical covariances for east–west direction on pattern of Fig. 1.14d using method explained in Fig. 1.11 (Covariance) on Quantimet 720 with linear correlator module for pattern of acidic volcanics, Bathurst area previously shown in Fig. 1.14d (Source: Agterberg and Fabbri 1978, Fig. 6)

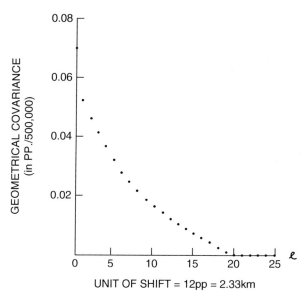

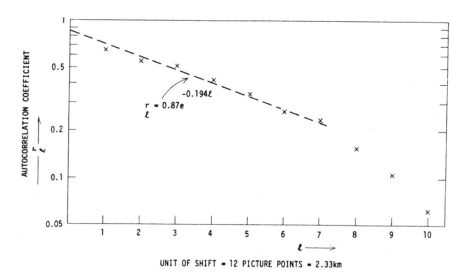

Fig. 12.41 Autocorrelation coefficients computed from first ten geometrical covariances of Fig. 12.39 with best-fitting semi-exponential model (Source: Agterberg and Fabbri 1978, Fig. 7)

volcanics per (10 × 10 km) cell is shown in Fig. 12.43. It is noted that 28 cells in the study area contain no acidic volcanics. The frequency of empty cells depends strongly on the choice of the boundaries of the study area T_0 which is rather arbitrary. It will be seen, however, that this choice is not of critical importance in the statistical model to be used. The mean m (=mes A/mes T_0) and variance s^2

518   12 Selected Topics for Further Research

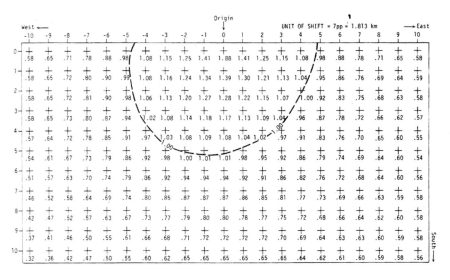

Fig. 12.42 Two-dimensional geometrical covariance function. Values are for pp/10,000. Approximate isotropy is indicated (Source: Agterberg and Fabbri 1978, Fig. 8)

Fig. 12.43 Histogram of 64 percentage of acidic volcanics/cell values for Bathurst example (Source: Agterberg and Fabbri 1978, Fig. 10)

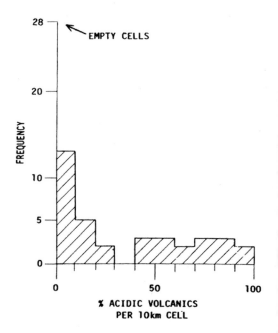

12.8 Cell Composition Modeling

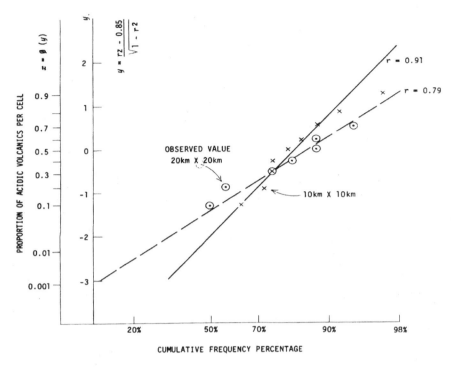

Fig. 12.44 Theoretical frequency distributions for amount of acidic volcanics per cell compared with 64 10-km cell vales in Fig. 12.43 and 16 20-km cell values (Source: Agterberg and Fabbri 1978, Fig. 11)

estimated from the 64 values are $m = 0.1973$ and $s^2 = 0.0919$, respectively. The variance of a square cell with side of length a satisfies

$$C_0(a) = \frac{C_0}{a^4} \cdot \int dx \int c \cdot e^{-p|x-y|} dy$$

where x and y denote vectors in the plane and the integrations are carried out over the cell with side a (cf. Agterberg 1978a). This 2-D equation is in accordance with basic 3-D geostatistical theory (Box 6.7).

Substitution of the coefficients $c = 0.87$ and $p = 0.194$ obtained previously and setting $C_0 = m - m^2 = 0.1584$, yields $C_0(a) = 0.0910$ which is close to $s^2 = 0.0919$ estimated from the 64 values for (10 × 10 km) cells. A further test of the model consisted of dividing the (80 × 80 km) test area into 16 (20 × 20 km) cells. This gave 16 percentage values for acidic volcanics in (20 × 20 km) cells with mean $m = 0.1973$ and $s^2 = 0.0622$. The corresponding theoretical variance is $C_0(a) = 0.0627$ (Fig. 12.44).

The preceding results indicate that it is possible to model the variance of the random variable X which represents the relative amount of acidic volcanics per

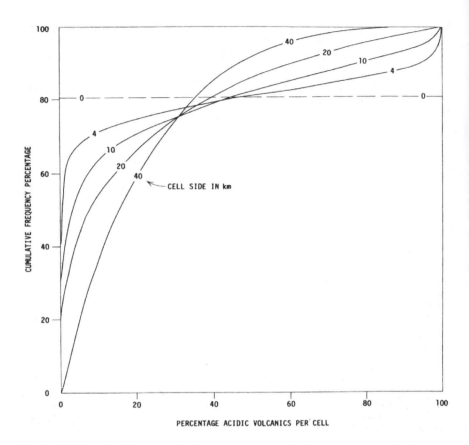

Fig. 12.45 Theoretical frequency distributions of amount of acidic volcanics per cell of variable size. Includes two frequency distributions of Fig. 12.44 (Source: Agterberg and Fabbri 1978, Fig. 12)

random cell of variable size. In order to model the frequency distribution of X, the probnormal model can be used. Suppose that the random cell i with value $x_i = h_2(z_i)$ contains a much smaller cell with value $h_1(u_i)$ which occurs at random within the larger cell i; $h_1(u_i)$ is either one or zero. The pair of values (z_i, u_i) is assumed to be a realization of a standard bivariate normal distribution with correlation coefficient r. Thus z_i and u_i are realizations of random variables Z and U with standard normal distributions. The transformed random variables $h_1(U)$ and $X = h_2(Z)$ can be expanded in terms of Hermite polynomials.

Figure 12.45 shows a sequence of frequency distributions arising from the random cell with variable side a. For any value of a, the frequency distribution of X is determined by the parameters m, c, and p because $C_0(a)$, b, and r are derived from these three basic parameters. The five examples in Fig. 12.45 are for $a = 0$, 4, 10, 20, and 40 km, respectively. When a tends to zero ($a \to 0$), $r \to 1$ and in the limit ($a = 0$) the density function of X consists of separate spikes at $x = 0$ (with

12.8 Cell Composition Modeling

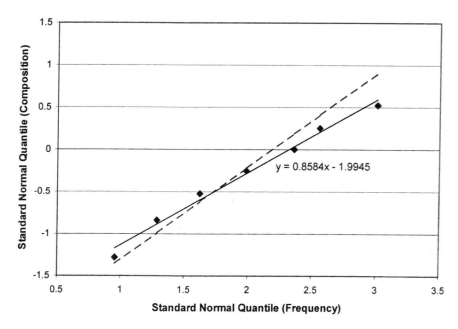

Fig. 12.46 Content of felsic metavolcanics in 64 square cells (Bathurst area, New Brunswick) measuring 10 km on a side; Q-Q plot of cell values in Fig. 12.44 compared with line of best fit (Source: Agterberg 2005, Fig. 1)

frequency $1 - m$) and at $x = 1$ (with frequency m). For increasing cell size, for example when $a = 4$ km, the density function of X is U-shaped with densities tending to infinity as $x \to 0$ and as $x \to 1$. For larger cells, for example when $a = 20$ km, the distribution of X resembles a gamma distribution with density tending to infinity as $x \to 0$. Finally, for very large cells (not shown in Fig. 12.45), the corresponding density function would become unimodal.

It is noted that this example was re-analyzed by Agterberg (1981, Fig. 5) with slightly different results as shown in the prob-prob plot of Fig. 12.46. The broken line in Fig. 12.46 is as in Fig. 12.44 obtained by estimating β and ρ from estimated values of the mean and variance of the 64 cell values using Fig. 12.39 that gave the estimates $b = 0.85$ and $r = 0.91$, respectively. Figure 12.46 also shows a best-fitting straight line obtained by linear regression of the quantile for frequency (Y-axis) on the quantile for composition (X-axis). The intercept and slope of this best-fitting line yielded estimates of 0.82 and 0.86 for β and ρ, respectively. The difference between the two straight lines in Fig. 12.46 is probably due to the fact that on the Q-Q plot more weight is given to points with larger relative frequencies. The best-fitting straight line gives estimates that are probably slightly better than the straight line derived from estimates of the mean and variance. This would be because the smallest percentage values have greater relative errors than the larger percentage values.

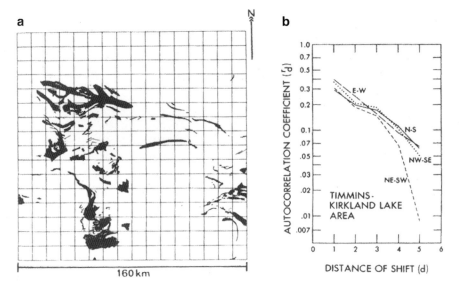

Fig. 12.47 (a) Areal distribution of acidic volcanics in a 160 km by 160 km area surrounding Timmins and Kirkland Lake, east–central Ontario. UTM grid with 10-km cells is superimposed. The geometrical covariance of the black and white pattern was measured on a Quantimet 720 in four directions. (b) Autocorrelation coefficients (r_d) obtained from centered covariances with unit of distance equal to 4.694 km (Source: Agterberg 1984, Fig. 7)

12.8.4 Abitibi Acidic Volcanics Example

The second example is for a larger area measuring 480 km in the east–west direction and 160 km in the north–south direction located within the Abitibi Subprovince on the Canadian Shield. This study area was subdivided into 768 square (10 × 10 km) cells and amount of acidic volcanics was measured for each cell. The central third with 256 (10 × 10 km) cells is shown in Fig. 12.47 along with autocorrelation coefficients for four different directions estimated by the same method used in the previous section for acidic volcanics in the Bathurst area, New Brunswick. Percentage values for the larger Abitibi array of (40 × 40 km) cells are shown in Fig. 12.48. Best estimates of parameters for the approximately semi-exponential autocorrelation function $r_h = c \cdot \exp(-p \cdot |h|)$ were $c = 0.36$ and $p = 0.45$ (Agterberg 1978a, 1984).

The (10 × 10 km) cell value frequency distribution is shown in Fig. 12.49. Only one-third of the 10-km cells contain acidic volcanics. Gamma density functions were fitted successively, including and excluding the 512 empty 10-km cells with zeros (Agterberg 1977, 1978a). If the zeros are combined with values of less than 10 %, and a single class is formed of all values greater than 50 %, then the gamma distribution fitted to all values is more satisfactory than the gamma distribution fitted to nonzero values only. Application of a goodness-of-fit

12.8 Cell Composition Modeling

8	2			4	2		4	5	2	1	3
1	3		9	18	25	6	3	3	5	6	20
11			4	7	14	2	7	14	18	2	5
		5	5		20		5	5			1

GSC

Fig. 12.48 Array of (12 × 4) percentage values for amount of acidic volcanics per 40-km cell. Central 160 km by 160 km area is same as area shown in Fig. 12.47. Histogram values of Fig. 12.49 are for (48 × 16) array of 10-km cells from within this (12 × 4) array (Source: Agterberg 1984, Fig. 6)

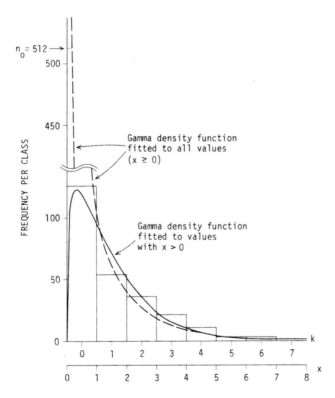

Fig. 12.49 Histogram values (k) contained in (48 × 16) array for amount of acidic volcanics per 10-km cell coinciding with array of (12 × 4) percentage values shown in Fig. 12.50. Gamma distributions were fitted to $x = k + 0.5$ with and without $n_0 = 512$ zeros (Source: Agterberg 1984, Fig. 5)

test gave $\hat{\chi}^2(3) = 14.7$, to be compared with $\chi^2_{0.05}(3) = 7.8$ (95 % fractile of χ^2 distribution with three degrees of freedom).This suggests that the fitted curve has frequencies that may be too low in the 20–50 % range. The shape of the histogram obviously depends on the boundaries of the area selected for statistical analysis. It depends even more strongly on cell size as illustrated in Fig. 12.48 in which the relative frequency of zeros is reduced from 67 to 25 %. Later in this section the probnormal frequency distribution will be fitted to these data.

A simplified fractal/multifractal analysis was performed on the larger Abitibi data set using the method of moments. Multifractal analysis is appropriate if the pattern being studied is self-similar or scale-independent. A (mono-) fractal can arise as a special case of a self-similar multifractal. Suppose that a grid of cells with length of cell side e is superimposed on a pattern, and that amount of pattern in a cell is called the cell's "measure" μ_e. A self-similar multifractal can be characterized by its multifractal spectrum in which the fractal dimension $f(\alpha)$ is plotted against the singularity (or Hölder exponent) α. The measure satisfies $\mu_e \sim e^\alpha$ where $\sim$ denotes proportionality, and the fractal dimension $f(\alpha)$ satisfies $N_e \sim e^{f(\alpha)}$ where N_e represents number of cells with singularity approximately equal to α.

The method of moments to determine the multifractal spectrum consists of three consecutive steps. Initially, power moment sums are calculated for different cell sizes. In our application, square cells with sizes of 10×10 km, 20×20 km and 40×40 km were used. As unit for the measure (amount of acidic volcanics per cell), decimal fraction per (10×10 km) cell was used. The power moment sums are plotted against length of cell side using log-log paper. Figure 12.50 shows results for power moments q between -1 and 5. Logarithms (base 10) were used to plot the power moment sums, and the three cell sizes are labelled 1, 2, and 3 in Fig. 12.51. The underlying pattern would be multifractal if the power moment sums exhibit straight line patterns on log-log paper. This condition seems to be satisfied in Fig. 12.50. The second step in the method of moments consists of assuming that the straight line slopes represent multifractal mass exponents (τ or "tau") of the pattern for the power moments q. The first derivative of $\tau(q)$ with respect to q then yields an estimate of $\alpha(q)$ representing the singularity.

Figure 12.51 is the plot of $\tau(q)$ versus q using the slope estimates of Fig. 12.50. The pattern of Fig. 12.51 is closely approximated by the straight line $\tau(q) = 0.4149 \cdot q - 0.4136$. This would imply $\alpha(q) = 0.41$ representing a (mono-) fractal with constant singularity of 0.41 rather than a multifractal with different values of α and $f(\alpha)$. The final step in the method of moments consists of constructing the multifractal spectrum that is a plot of $f(\alpha)$ versus α using the relation: $f(\alpha) = q \cdot \alpha(q) - \tau(q)$. In the application to Abitibi felsic metavolcanics, the multifractal spectrum is reduced to a single spike representing a fractal. The intercept of the straight line, which also is 0.41, can be regarded as the (constant) fractal dimension $f(\alpha)$ of the pattern.

As in the example of Fig. 12.46, the broken line in Figure 12.52 was derived previously (Fig. 12.39) by estimating β and ρ from the mean and variance. In this earlier application, the variance was estimated from the sample of 768 cell values.

12.8 Cell Composition Modeling

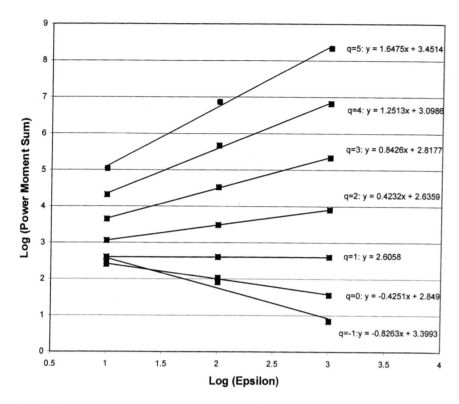

Fig. 12.50 First step of multifractal analysis (power-moment sums) of acidic volcanics in 768 square cells (Abitibi area, Canadian Shield) measuring 10 km on a side of example of Figs. 12.47 and 12.48. Logarithms base 10 (Source: Agterberg 2005, Fig. 3)

The resulting estimates of β and ρ were 1.62 and 0.74, respectively; versus 1.51 and 0.65 as derived from the lest-fitting line. The difference between the two straight lines is probably due to differences in weighting of the observed frequencies.

In the two case history studies on 2-D occurrences of acidic volcanics in the Bathurst and Abitibi areas it was shown that the probnormal model is applicable in modeling the frequency distribution of amounts of acidic volcanics contained in cells of variable size. In Fig. 12.53 these probnormals are shown together with results of a number of other applications. In each case, slope and intercept of the straight line were computed from mean and variance using Fig. 12.39. It has been pointed out that the probnormal generally also can be obtained simply by fitting a straight line to the points in a prob-prob plot. In the last section another example of cell compositional data will be discussed (asymmetric bivariate binomial distribution). It also leads to good results for acidic volcanics in the Abitibi area. This model has potential for modeling the frequencies of empty small empty cells of variable cells because it has one more parameter than the probnormal.

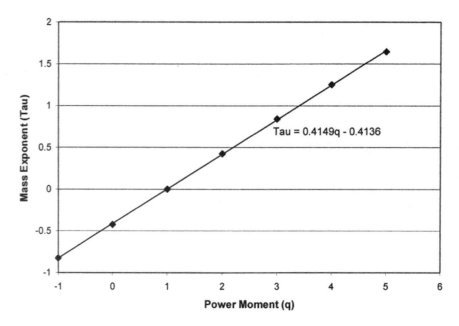

Fig. 12.51 Second step of multifractal analysis showing slopes of straight lines in Fig. 12.50 plotted against moment q. Result is a fractal with multifractal spectrum (not shown) reduced to single spike for $f(\alpha)$ at $\alpha = 0.41$ (Source: Agterberg 2005, Fig. 4)

12.8.5 Asymmetrical Bivariate Binomial Distribution

Matheron (1980) has suggested the use of discrete orthogonal polynomials of the binomial distribution for empirical determination of the coefficients c_j and c_j^* in $x_1 = \psi_1(z_1) = \sum_{j=0}^{N} c_j Q_j(z_1)$ and $x_2 = \psi_2(z_2) = \sum_{j=0}^{N} c_{1j}^* S_j(z_2)$. Krawtchouck polynomials can be used with, for example, N set equal to 10 or 20. An advantage of this approach is that an arbitrary number of zeros can be accommodated by choosing $q = 1 - p$ for the binomial distribution do that $n_0 = nq^N$ (n = total number of cells). Relative frequency of zeros decreases when cell size is increased. With N remaining constant, q must decrease. An asymmetrical binomial distribution can be used in which z_2 is assigned a parameter q_2 that differs from that of z_2. Matheron (1980) has shown that setting $\rho = \sqrt{\frac{q_2 p_1}{p_2 q_1}}$ then results in a suitable model. Using Krawtchouck polynomials k_j with squared norm

$$h_j^2 = \left\{ \binom{N}{j} p^j q^j \right\}^{-1}$$

it follows that $x_1 = \psi_1(z_1) = \sum_{j=0}^{N} c_j k_j(z_1)$ and $x_2 = \psi_2(z_2)$

$$= \sum_{j=0}^{N} c_j \left\{ \frac{q_2}{q_1} \right\}^j k_j(z_2).$$

12.8 Cell Composition Modeling

Box 12.6: Algorithm for Computation of Coefficients

In order to perform the required calculations it is convenient to define an auxiliary function $W_j(i,q)$ with as properties $W_{-1}(i,q) = 0; W_0(i,q) = W_i(q) = \binom{N}{j} p^i q^i W_i(q) k_j(i,q)$ $(j = 1, 2, \ldots N)$. From the recurrence formula for Krawtchouck polynomials, it can be derived that $(j+1)W_{j+1}(i,q) + [i - Np + j(p-q)]W_j(i,q) + (N-j+1)pqW_{j-1}(i, of\ q) = 0$. Consequently, the coefficients of $\psi_{1i} = \sum_{j=1}^{N} c_j k_j(i, q_1)$ satisfy $c_j = \sum_{j=1}^{N} \psi_{1i} W_j(i, q_1)$. The correlation coefficient ρ then can be obtained from the variance $\sigma_2^2 = \sum_{j=1}^{N} c_j^2 \rho^{2j} W(q_1)$ and, finally, the values ψ_{2i} corresponding to ψ_{1i} become

$$\psi_{2i} = \sum_{j=0}^{N} \binom{i}{j} \left\{\frac{p_1}{p_2}\right\}^j \left\{1 - \frac{p_1}{p_2}\right\}^{i-j} \psi_{1j}.$$

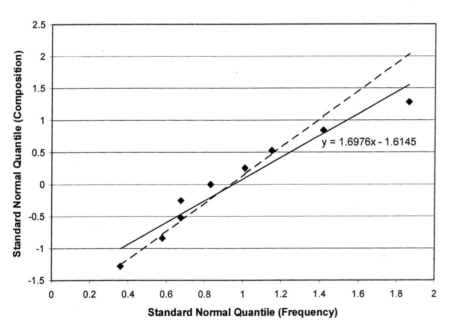

Fig. 12.52 Content of acidic volcanics in 768 square cells (Abitibi area, Canadian Shield) measuring 10 km on a side; Q-Q plot of cell values similar to Fig. 12.46 (Source: Agterberg 2005, Fig. 2)

For example, it will be attempted to determine the Abitibi acidic volcanics frequency distribution of the 48 values for 40-km cells shown in Fig. 12.48 from the frequency distribution of the 768 values for 10-km cells shown in Fig. 12.49 plus an estimate of variance $s_2^2 = 0.00386$ for the 40-km cells. It can be assumed

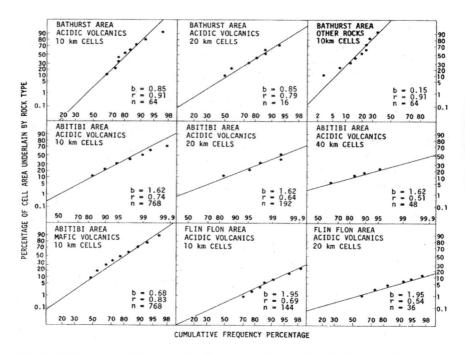

Fig. 12.53 Experimental frequency distributions for nine samples of cell values plotted along axes with two normal probability scales (from Agterberg 1981). The values of b and r were obtained from the means and variances of the cell values using Fig. 12.53. They determine the positions and slopes of the lines for probnormal distributions (Source: Agterberg 1984, Fig. 9)

Table 12.3 Estimation of the values ψ_{1i} (Source: Agterberg 1984, Table 2)

i	0	1	2	3
$_NC_i q_1^{N-i} p_1^i$	q_1^{15}	$15\,q_1^{14} p_1$	$105\,q_1^{13} p_1^2$	$455\,q_1^{12} p_1^3$
$f_i(z_1)$	0.66667	0.27400	0.05255	0.00624
$F_i(z_1)$	0.66667	0.94066	0.99322	0.99946
$x_1(F_i)$	0	0.0282	0.0540	0.0765
ψ_{1i}	0	0.0141	0.0411	0.06525

that, in general, the relationship between variance and cell size can be determined independently by geostatistical methods of 2-D integration such as the one explained in Sect. 12.8.3 for acidic volcanics in the Bathurst area. There are 512 empty cells in Fig. 12.49. Therefore, the frequency of zero values satisfies $q_1^N = 512/786 = 0.66667$. Several values of $N \geq 5$ were tried yielding nearly the same results. Setting $N = 15$ gives $q_1 = 0.97333$ and $p_1 = 1 - q_1 = 0.02667$. A sequence of values $f_i(z_1) = \binom{N}{i} q_1^{N-1} p_1^i$ the corresponding values of ψ_{1i} is shown in Table 12.3.

The cumulative frequency $F_i(z_1)$ in Table 12.3 is close to 1.0 for $i = 3$. The cumulative frequency distribution for 10-km cells was plotted and the corresponding values of $x_1(F_2)$ were read from the resulting graph. The ψ_{1i} values

Table 12.4 Relationship between ψ_{2i} and $f_1(z_2)$ (Source: Agterberg 1984, Table 3)

i	0	1	2	3
ψ_{2i}	0	0.04757	0.10983	0.18072
$f_i(z_2)$	0.29078	0.37437	0.22493	0.08366
$F_i(z_2)$	0.29078	0.66515	0.89008	0.97374

Table 12.5 Asymmetrical bivariate binomial model: comparison of observed and theoretical frequencies for 40-km cells (Source: Agterberg 1984, Table 4)

x_2	Observed frequency	Theoretical frequency
0	12	13.96
0–0.4757	17	17.97
0.4757–0.10983	12	10.80
0.10983–1.0	7	5.28

($i = 1, 2, 3$) satisfy $2\psi_{1i} = x_1(F_i) + x_1(F_{i-1})$. These values are slightly too large because curvature of the frequency density function was neglected. The three values of ψ_{1i} in Table 12.3, together with $N = 15$ and $q_1 = 097333$, were used as input to a computer program in which $\sigma_2^2 = \sum_{j=1}^{N} c_j^2 \rho^{2j} W(q_1)$ was tabulated as a function of ρ. By means of this special-purpose table, the variance $s_2^2 = 0.00386$ was converted into $r = 0.565$. From $\rho = \sqrt{\frac{q_2 p_1}{p_2 q_1}}$ it follows that $p_2 = 0.07905$ and $q_2 = 0.92095$. Now the equation for ψ_{2i} derived in Box 12.6 can be used resulting in the values shown in the first row of Table 12.4 in relation to $f_1(z_2) = \binom{N}{i} q_2^{N-1} p_2^i$.

The resulting theoretical frequencies are compared to the observed frequencies for Fig. 12.48 in Table 12.5. The degree of correspondence is fairly good. In a goodness-of-fit test, $\widehat{\chi}2 = 1.02$ is less than $\chi_{0.05}^2(1) = 3.84$. More general applicability of the asymmetrical bivariate binomial model remains to be investigated. This method can be useful for modeling arbitrary frequencies of empty cells as a function of different cell size. The approach taken in Sect. 12.8 relies strongly on application of the method of diagonal expansion applied to bivariate distributions. Kotz (1975, p. 259) considered this method to provide the most powerful unified attack on the structure of bivariate distributions (cf. Hutchinson and Lai, 1991, p. 222). In Sect. 12.8 this concept was adopted to quantify the spatial distribution of rock types represented on geological maps.

References

Agterberg FP (1970) Autocorrelation functions in geology. In: Merriam DF (ed) Geostatistics. Plenum, New York, pp 113–142
Agterberg FP (1973) Probabilistic models to evaluate regional mineral potential. In: Proc Symp Mathematical Methods in the Geosciences, Příbram, Czechoslovakia, pp 3–38
Agterberg FP (1974) Geomathematics. Elsevier, Amsterdam
Agterberg FP (1977) Frequency distributions and spatial variability of geological variables. In: Ramani RV (ed) Application of computing methods in the mineral industry. American Institute of Mining, Metallurgical, and Petroleum Industry, New York, pp 287–298

Agterberg FP (1978a) Use of spatial analysis in mineral resource evaluation. Math Geol 16(6):565–589

Agterberg FP (1978b) Quantification and statistical analysis of geological variables for mineral resource evaluation. Mém Bur Rech Géol Min 91:399–406

Agterberg FP (1981) Cell value distribution models in spatial pattern analysis. In: Craig RG, Labovitz ML (eds) Future trends of geomathematics. Pion, London, pp 5–28

Agterberg FP (1984) Use of spatial analysis in mineral resource evaluation. Math Geol 16:565–589

Agterberg FP (1994) Fractals, multifractals, and change of support. In: Dimitrakopoulos R (ed) Geostatistics for the next century. Kluwer, Dordrecht, pp 223–234

Agterberg FP (1995) Multifractal modeling of the sizes and grades of giant and supergiant deposits. Int Geol Rev 37:1–8

Agterberg FP (2001a) Multifractal simulation of geochemical map patterns. In: Merriam D, Davis JC (eds) Geologic modeling and simulation: sedimentary systems. Kluwer, New York, pp 327–346

Agterberg FP (2001b) Aspects of multifractal modeling. In: Proceedings of the 7th conference of the International Association for Mathematical Geology (CD-ROM), Cancun, Mexico

Agterberg FP (2005) Spatial analysis of cell composition data. In: Proceedings of CoDaWork'05, University of Girona, Spain (CD-ROM)

Agterberg FP (2007a) New applications of the model of de Wijs in regional geochemistry. Math Geol 39(1):1–26

Agterberg FP (2007b) Mixtures of multiplicative cascade models in geochemistry. Nonlinear Process Geophys 14:201–209

Agterberg FP (2012a) Sampling and analysis of element concentration distribution in rock units and orebodies. Nonlinear Process Geophys 19:23–44

Agterberg FP (2012b) Multifractals and geostatistics. J Geochem Explor 122:113–122

Agterberg FP, Fabbri AG (1978) Spatial correlation of stratigraphic units quantified from geological maps. Comput Geosci 4:284–294

Aitchison J (1986) The statistical analysis of compositional data. Chapman and Hall, London

Beckmann P (1973) Orthogonal polynomials for engineers and physicists. Golem Press, Boulder

Beirlant J, Goegebeur Y, Segers J, Teugels J (2005) Statistics of extremes. Wiley, New York

Blackman RB, Tukey JW (1958) The measurement of power spectra. Dover, New York

Brinck JW (1974) The geochemical distribution of uranium as a primary criterion for the formation of ore deposits. In: Chemical and physical mechanisms in the formation of uranium mineralization, geochronology, isotope geology and mineralization. In: International Atomic Energy Agency, Vienna, Proceedings Series STI/PUB/374, pp 21–32

Chayes F (1971) Ratio correlation. University of Chicago Press, Chicago

Cheng Q (1994) Multifractal modeling and spatial analysis with GIS: gold mineral potential estimation in the Mitchell-Sulphurets area, northwestern British Columbia. Unpublished doctoral dissertation. University of Ottawa

Cheng Q, Agterberg FP (1996) Comparison between two types of multifractal modeling. Math Geol 28:1001–1015

Cheng Q, Agterberg FP (2009) Singularity analysis of ore-mineral and toxic trace elements in stream sediments. Comput Geosci 35:234–244

Cheng Q, Agterberg FP, Ballantyne SB (1994) The separation of geochemical anomalies from background by fractal methods. J Geochem Explor 51(2):109–130

Chung CF, Agterberg FP (1980) Regression models for estimating mineral resources from geological map data. Math Geol 12:473–488

Davis JC (2002) Statistics and data analysis in geology, 3rd edn. Wiley, New York

De Wijs HJ (1951) Statistics of ore distribution I. Geol Mijnbouw 13:365–375

Dubois J, Cheminée JL (1991) Fractal analysis of eruptive activity of some basaltic volcanoes. J Volcan Geotherm Res 45:197–208

Efron B (1982) The jackknife, the bootstrap and other resampling plans. SIAM, Philadelphia

Egozcue JJ, Pawlowsky-Glahn V, Mateu-Figueras G, Barceló-Vidal C (2003) Isometric logratio transformations for compositional data analysis. Math Geol 35(3):279–300

References

Fabbri AG, Chung CF (2008) On blind tests and spatial prediction models. In: Bonham-Carter G, Cheng Q (eds) Progress in geomathematics. Springer, Heidelberg, pp 315–332

Filzmoser P, Hron K (2011) Robust statistical analysis. In: Pawlowsky-Glahn V, Buccianti A (eds) Compositional data analysis – theory and applications. Wiley, New York, pp 59–71

Filzmoser P, Hron K, Reimann C (2009) Univariate statistical analysis of environmental (compositional) data: problems and possibilities. Sci Total Environ 407(6):6100–6108

Garrett RG, Thorleifson LH (1995) Kimberlite indicator mineral and till geochemical reconnaissance, southern Saskatchewan. Geological Survey of Canada Open File 3119, pp 227–253

Greenacre M (2009) Distributional equivalence and subcompositional coherence in gthe analysis of compositional data, contingency tables and ratio-scale measurements. J Classification 26(1):29–54

Grunsky EC, Kjarsgaard BA (2008) Classification of distinct eruptive phases of the diamondiferous Star kimberlite, Saskatchewan, Canada based on statistical treatment of whole rock geochemical analyses. Appl Geochem 23:3321–3336

Gupta VK, Troutman B, Dawdy D (2007) Towards a nonlinear geophysical theory of floods in river networks: an overview of 20 years of progress. In: Tsonis AA, Elsher JB (eds) Nonlinear dynamics in geosciences. Springer, New York, pp 121–150

Harris DP (1984) Mineral resources appraisal. Clarendon Press, Oxford

Hutchinson TP, Lai CD (1991) The engineering statistician's guide to continuous bivariate distributions. Rumsby Scientific Publishing, Adelaide

Jöreskog KE, Klovan JE, Reyment RA (1976) Geological factor analysis. Elsevier, Amsterdam

Kotz S (1975) Multivariate distributions at a cross road. In: Patil GP, Kotz S (eds) A modern course on distributions in scientific work I – models and structures. Reidel, Dordrecht

Krige DG (1951) A statistical approach to some basic valuation problems on the Witwatersrand. J S Afric Inst Mining Metallurgy 52:119–139

Krige DG (1966) A study of gold and uranium distribution pattern in the Klerksdorp goldfield. Geoexploration 4:43–53

Lavallée D, Lovejoy S, Schertzer D, Schmitt F (1992) On the determination of universal multifractal parameters in turbulence. In: Moffatt HK, Zaslavsky GM, Conte P, Tabor M (eds) Topological aspects of the dynamics of fluids and plasmas. Kluwer, Dordrecht, pp 463–478

Lorenz EN (1963) Deterministic nonperiodic flow. J Atmos Sci 20:130–141

Lovejoy S, Schertzer D (2007) Scaling and multifractal fields in the solid earth and topography. Nonlinear Process Geophys 14:465–502

Lovejoy S, Schertzer D (2010) On the simulation of continuous in scale universal multifractals, part I: spatially continuous processes. Comput Geosci 36:1393–1403

Lovejoy S, Schertzer D (2013) The weather and climate. Cambridge University Press, Cambridge

Lovejoy S, Currie WJS, Tessier Y, Claereboudt MR, Bourget E, Roff JC, Schertzer D (2001) Universal multifractals and ocean patchiness: phytoplankton, physical fields and coastal heterogeneity. J Plankton Res 23:117–141

Lovejoy S, Gaonac'h H, Schertzer D (2008) Anisotropic scaling models of rock density and the earth's surface gravity field. In: Bonham-Carter G, Cheng Q (eds) Progress in geomathematics. Springer, Heidelberg, pp 151–194

Malamud BD, Morein G, Turcotte DL (1998) Forest fires: an example of self-organized critical behavior. Science 281:1840–1842

Mandelbrot BB (1983) The fractal geometry of nature. Freeman, San Francisco

Mandelbrot BB (1999) Multifractals and $1/f$ noise. Springer, New York

Matheron G (1962) Traité de géostatistique appliquée. Mém BRGM 14, Paris

Matheron G (1974) Les fonctions de transfert des petits panneaux, Note Géostatistique 127. Centre de Morphologie Mathématique, Fontainebleau

Matheron G (1976) Forecasting block grade distributions: the transfer functions. In: Guarascio M, David M, Huijbregts C (eds) Advanced geostatistics in the mining industry. Reidel, Dordrecht

Matheron G (1980) Models isofactoriels pout l'effet zero, Note Géostatistique 659. Centre de Morphologie Mathématique, Fontainebleau

May RM (1976) Simple mathematical models with very complicated dynamics. Nature 261:459–467

Monin AS, Yaglom AM (1975) Statistical fluid mechanics 2. MIT Press, Cambridge, MA
Mosteller F, Tukey JW (1968) Data analysis including statistics. In: Lindzey G, Aronson E (eds) Handbook of social psychology, 2nd edn, vol 2. Addison-Wesley, Reading, pp 80–123
Motter E, Campbell DK (2013) Chaos at fifty. Phys Today 66:27–33
Park N-W, Chi K-H (2008) Quantitative assessment of landslide susceptibility using high-resolution remote sensing data and a generalized additive model. Int J Remote Sens 29(1):247–264
Pawlowsky-Glahn V, Buccianti A (eds) (2011) Compositional data analysis – theory and applications. Wiley, New York
Pawlowsky-Glahn V, Olea RA (2004) Geostatistical analysis of compositional data. Oxford University Press, New York
Pearson K (1897) On a form of spurious correlation which may arise when indices are used in the measurements of organs. Proc R Soc Lond 60:489–498
Pilkington M, Todoeschuck J (1995) Scaling nature of crustal susceptibilities. Geophys Res Lett 22:779–782
Poincaré H (1899) Les methods nouvelles de la mécanique celeste. Gauthier-Villars, Paris
Quenouille M (1949) Approximate tests of correlation in time series. J R Stat Soc Ser B 27:395–449
Reyment RA, Mosoyana I, Oka M, Tanaka Y (2008) A note on seasonal variation in radiolarian abundance. In: Bonham-Carter G, Cheng Q (eds) Progress in geomathematics. Springer, Heidelberg, pp 413–430
Rundle JB, Turcotte DL, Shcherbakov R, Klein W, Sammis C (2003) Statistical physics approach to understanding the multiscale dynamics of earthquake fault systems. Rev Geophys 41:1019
Schertzer D, Lovejoy S (1985) The dimension and intermittency of atmospheric dynamics multifractal cascade dynamics and turbulent intermittency. In: Launder B (ed) Turbulent shear flow 4. Springer, New York, pp 7–33
Schertzer D, Lovejoy S (eds) (1991a) Non-linear variability in geophysics. Kluwer, Dordrecht
Schertzer D, Lovejoy S (1991b) Non-linear geodynamical variability: multiple singularities, universality and observables. In: Schertzer D, Lovejoy S (eds) Non-linear variability in geophysics. Kluwer, Dordrecht, pp 41–82
Schertzer D, Lovejoy S (1997) Universal multifractals do exist! Comments on "A statistical analysis of mesoscale rainfall as a random cascade". J Appl Meteorol 36:1296–1303
Schertzer D, Lovejoy S, Schmitt F, Chigiranskaya Y, Marsan D (1997) Multifractal cascade dynamics and turbulent intermittency. Fractals 5:427–471
Schucany WR, Gray HL, Owen DB (1971) On bias reduction in estimation. J Am Stat Assoc 66:524–533
Sharma AS (1995) Assessing the magnetosphere's nonlinear behavior: its dimension is low, its predictability high. Rev Geophys 33:645
Sornette A, Dubois J, Cheminée JL, Sornette D (1991) Are sequences of volcanic eruptions deterministically chaotic. J Geophys Res 96(11):931–945
Sparrow C (1982) The Lorentz equation: bifurcations, chaos, and strange attractors. Springer, New York
Szegö G (1975) Orthogonal polynomials, 9th edn, American Mathematical Society Colloquium Publications 23. American Mathematical Society, Providence
Thompson RN, Esson J, Duncan AC (1972) Major element chemical variation in the Eocene lavas of the Isle of Skye, Scotland. J Petrol 13:219–253
Tukey JW (1970) Some further inputs. In: Merriam DF (ed) Geostatistics. Plenum, New York, pp 173–174
Turcotte DL (1997) Fractals and chaos in geology and geophysics, 2nd edn. Cambridge University Press, Cambridge
Uritsky VM, Donovan E, Klimas AJ (2008) Scale-free and scale-dependent modes of energy release dynamics in the night time magnetosphere. Geophys Res Lett 35(21):L21101, 1–5
Vening Meinesz FA (1964) The earth's crust and mantle. Elsevier, Amsterdam
Wilson J, Schertzer D, Lovejoy S (1991) Continuous multiplicative cascade models of rain and clouds. In: Schertzer D, Lovejoy S (eds) Non-linear variability in geophysics. Kluwer, Dordrecht, pp 185–207

Author Index

A
Aczélm, J., 415
Adams, K.C., 24, 373
Afzal, P., 396
Ahlfield, F., 67
Ahmed, S.M., 261
Ahrens, L.H., 90, 399
Ailleres, L., 142
Aitchison, J., 3, 88, 212, 399, 474, 475
Aki, K., 229
Albanese, S., 396
Alghalandis, Y.F., 396
Allais, M., 100, 113
Alldrick, D.J., 396
Allègre, C.J.U., 399
Alloway, B.V., 357
Alroy, J., 309
Altural, T., 142
Alvarez, W., 88
Amos, D.E., 191
Andersen, E.B., 164
Anderson, R.G., 395, 396
Anderson, R.Y., 191, 200
Ansorge, J., 297, 298
Antevs, E., 196
Arias, M., 419
Armstrong, R.I., 74, 355, 361, 362
Arroyo, D., 47
Assad, R., 140

B
Baddeley, A., 383
Bailey, E.H., 382
Bain, G.W., 257
Bak, P., 371
Ballantyne, S.B., 376–380, 393, 394, 423, 484
Ballard, R.D., 181
Banerjee, I., 191–195, 197–203
Bao, Z., 444
Bárány, I., 383
Barceló-Vidal, C., 3, 474
Bardossy, G., 112
Barlclay, J.E., 36
Barnett, P.J., 150
Barnsley, M., 370
Barrio-Parra, F., 142
Barton, C.C., 371
Basset, M.G., 356
Bateman, A.M., 112
Batschelet, E., 279
Bawiee, W.J., 382
Bean, C., 230
Beaver, R.J., 347
Becker, R.T., 361
Beckmann, P., 510
Behnia, P., 181
Beirlant, J., 480
Bell, L.I., 36
Bellefleur, G., 113
Benciolini, L., 294
Benson, W.T., 212, 225
Benton, M.J., 307
Berkson, J., 165
Bernknopf, R., 116
Berry, P.A.M., 197
Berti, M., 142
Betts, P.G., 142
Beu, A.G., 309, 357
Bickel, P.J., 56, 109, 358, 415
Bilali, H.E., 202
Billa, M., 142
Billingsley, P., 57
Bishop, Y.M.M., 64, 143, 152, 159

Blackman, R.B., 191, 509
Blenkinsop, T.G., 370, 371, 385
Bliss, C.I., 57
Bloomfield, P., 46, 190
Bonham-Carter, G.F., 3, 81, 112, 114, 135, 141–153, 161, 163, 177, 181, 395–397, 439
Borgatti, L., 142
Borrini, D., 299
Boucher, A., 180
Bourget, E., 506
Bradley, P.G., 54, 100
Bramlette, M.N., 317
Brennand, T.A., 150
Brier, G.W., 196
Brillinger, D.R., 190
Brinck, J.W., 399, 417–419, 485, 499
Brower, J.C., 306, 308, 345
Brown, A., 433, 434, 436–439
Brown, J.A.C., 88, 212, 399
Buccianti, A., 474
Buhmann, J.H., 3
Bul, J.M., 385
Burnett, A.I., 24, 373

C
Cabilio, P., 136
Caers, J., 42
Campbell, D.K., 481, 482
Campbell, H.J., 357
Campbell, R.I., 36
Cantelli, L., 296, 298–301
Cargill, S.M., 382
Carlson, A., 384–386
Carranza, E.J.M., 112, 116, 142, 370, 396
Carsteanu, A., 370
Cassard, D., 142
Castellarin, A., 291, 296–301
Caumon, G., 22, 180
Cedillo-Pardo, E., 88
Cervi, F., 142
Chamberlain, V.E., 362, 363
Chamberlin, T.C., 2, 10
Channell, J.E.T., 298, 299
Chayes, F., 3, 50, 51, 473
Cheminée, J.L., 483
Chemingui, N., 229
Chen, J., 3, 66, 215, 444, 448, 451–456
Chen, Y., 3
Chen, Z., 66, 215, 217
Cheng, Q.M., 66, 112, 143, 144, 149, 150, 154, 156, 163, 181, 214, 215, 217, 306, 310, 313, 314, 324, 333, 335, 338, 340, 352, 353, 370–372, 375–380, 384, 386, 388,
 389, 393–398, 403, 405, 406, 419, 423, 425, 427–434, 436–442, 444, 445, 448–456, 458, 480, 484, 487, 504
Chenm, Z., 444, 452–456
Chhabra, A., 420
Chi, K.-H., 483
Chigiranskaya, Y., 506
Chigirinskaya, Y., 397, 398
Cho, S.H., 142
Christensen, R., 164, 181
Chung, C.F., 105, 140, 162, 178, 260–264, 266, 389, 470
Cicchella, D., 396
Claereboudt, M.R., 506
Claeys, W., 88
Clark, I., 44
Cloetingh, S., 297
Cody, R.D., 309
Coolbaugh, M.F., 142
Cooley, J.W., 224, 267
Coons, R.L., 236
Cooper, R.A., 309, 357, 360, 362
Corrigan, D., 19, 162
Corsini, A., 142
Cowie, J.W., 356
Cowie, P.A., 433
Cox, A.V., 74–76, 79, 80, 355, 357, 361, 362
Cox, D.R., 162
Craig, L.A., 74, 355, 361, 362
Cramér, H., 63, 64
Crampton, J.S., 309, 357
Cressie, N.A.C., 3, 45, 46, 190, 204, 245, 249, 253, 383, 384, 428
Croll, J., 4
Crouch, E.M., 357
Crovelli, R.A., 382
Crundwell, M.P., 357
Cunningham, D., 297, 298
Currie, W.J.S., 506
Curtis, G.H., 88

D
Dacey, M.F., 54
Daemen, J.J.K., 433
Dalrymple, G.B., 74
Darnley, A.G., 445
Davaud, E., 317, 318
David, M., 126, 212
Davidsen, J., 370
Davis, J.C., 59, 108, 237, 238, 474
Dawdy, D., 371
De Boor, C., 327, 328
De Finetti, B., 11, 12
de Kemp, E.A., 16, 20, 112, 113

Author Index 535

De Vivo, B., 396
de Wijs, H.J., 45, 66, 67, 207, 397, 399, 414, 451, 480, 505
DeGeoffroy, J.G., 101
Deino, A.L., 88, 358
Deming, W.E., 106, 195
Deng, J., 396
Deng, M., 181
Denham, C.R., 289
Dennison, J.M., 52, 53
DePaolo, D.J., 88, 358
Derry, D.R., 258
Deutsch, C.V., 47
Dhombres, J., 415
Diggle, P.J., 32, 33, 383, 386, 387, 442
D'Iorio, M.A., 345
Divi, S.R., 26, 214, 382
Doglioni, C., 298, 299
Doksum, K.A., 56, 109, 415
Dong, Y., 214
Donovan, E., 371, 483
Doveton, J.H., 55
Dowd, P.A., 47
Dowds, J.P., 141
Draper, N.R., 117
Drew, L.J., 61, 382
du Toit, S.R., 212
Dubois, D., 42
Dubois, J., 483
Dubrule, O., 323
Duncan, A.C., 475

E
Edwards, L.E., 317, 347
Efron, B., 468–470
Egozcue, J.J., 3, 474
Eisenhart, M.A., 43
Embleton, B.J.J., 278, 283
Emery, X., 47
Endt, P.M., 358
Engel, A.L., 212
Esson, J., 475
Eubank, R.I., 323, 327
Eubank, R.L., 46
Everitt, R.A., 433
Evertsz, C.J.G., 370, 403, 419, 420, 425, 449, 451, 458

F
Fabbri, A.G., 26, 105, 140, 274, 389, 470, 515, 517–520
Falck, H., 112

Falconer, K., 390, 449
Fantoni, R., 296, 298–301
Favini, G., 140
Feder, J., 372, 383, 384, 419, 421, 423, 449
Feller, W., 49, 50, 100
Ferris, D.R., 142
Fienberg, S.E., 64, 152
Filzmoser, P., 474, 479
Findlay, D.C., 13
Finley, D., 150
Finney, P.J., 90
Fisher Box, J., 43, 44
Fisher, N.I., 278, 283
Fisher, R.A., 3, 10, 43, 73, 108, 110, 283, 286
Fodor, J., 112
Foote, M., 307, 309
Ford, A., 370
Fowler, A.D., 204, 371, 435
Francheteau, J., 181
Freeden, W., 374
Freiling, E.C., 382
Freund, Y., 3, 181
Friedman, J.H., 3, 181
Frisch, W., 299
Fűgenschuh, B., 297, 300
Furcolo, P., 397, 415

G
Galbrun, B., 88
Gandin, L.S., 46
Gaonac'h, H., 370, 506
Gardin, S., 88
Garrett, R.G., 399, 418, 445, 487
Gebrande, H., 291, 297, 299
Ghosh, A., 433
Giardino, J.R., 142
Gillespie, P.A., 435
Girard, E., 113
Gnedenko, B.V., 60
Goegebeur, Y., 480
Goehring, L., 371
Goff, J.A., 208, 229, 230
Gohrband, T.M., 356
Gonçalves, M.A., 419
González-Álvarez, I., 142
Good, D., 433, 434, 436–439
Good, G.J., 253
Good, H., 141
Goodacre, A., 161
Goodman, 191
Goovaerts, P., 47, 204
Gorney, R.M., 142
Gorrel, G., 150

Goutier, J.E., 113
Gradstein, F.M., 6, 87, 306, 308–310, 313, 314, 318, 324, 327, 332, 333, 335, 338–340, 342–345, 349–362
Graham, I.J., 357
Grajales-Nishimura, J.M., 88
Granger, C.W.J., 208
Gray, H.L., 469
Graybill, F.A., 3, 43
Grayson, C.I. Jr., 11, 12
Greenacre, M., 474
Grenander, U., 208
Griffiths, J.C., 3, 10, 54, 100, 101, 113
Grollimund, B., 300
Grunsky, E.C., 113, 162, 393, 475–479
Guex, J., 308, 317, 318, 321, 347
Gumiel, P., 419
Gupta, V.K., 370, 371, 397, 483
Gy, P., 66

H

Hacking, I., 42, 43
Hacquebard, P.A., 265
Haining, R.P., 253
Hald, A., 65, 94
Hale, M., 445
Hall, G.E.M., 445
Hammer, O., 307–309, 358–361
Hamon, B.V., 191
Hannan, E.J., 191
Harbaugh, J.W., 81
Hardenbol, J., 87, 361, 362
Harff, J., 22
Harland, W.B., 74, 75, 77–79, 81, 355, 357, 361, 362
Harper, C.W. Jr., 317
Harper, D.A.T., 307–309
Harris, 497
Harris, D.P., 113, 399, 417, 418
Harris, J.R., 112
Harrison, J.M., 6, 7
Harrison, T.M., 358
Harwood, D.M., 309
Hastie, T., 181
Hatanaka, M., 208
Hay, W.W., 52, 317
Heien, D.M., 198
Heinen, H.J., 212
Heiskanen, W.A., 15, 374
Hershery, G., 236
Herzfeld, U.C., 419
Heuhuber, S., 197

Hilgen, F.J., 6, 88, 358
Hillier, M., 16, 18–20
Hills, L.V., 279, 281
Hinnov, I.A., 6
Hinnov, L.A., 88
Hiromu, D., 142
Hochard, C., 298
Hogarth, D.D., 214
Hohenegger, J., 197
Hohn, M.E., 308
Holland, P.W., 64, 153
Hollinger, K., 208, 229
Hollis, C.J., 357
Holmes, A., 354, 355
Hood, K.C., 308
Hosmer, D.W., 162
Howard, C.B., 435
Hron, K., 474, 479
Huang, J., 3
Huang, Z., 87, 361, 362
Hubbard, B.B., 202
Hudson, C.B., 317, 347
Hughes, O.L., 196
Huijbregts, C.J., 204, 212
Hurst, V.J., 236
Husson, D., 88
Hutchinson, T.P., 529

I

Isaaks, E.H., 44, 204

J

Jackson, P.W., 355
Jacobi, J., 42
Jäger, E., 358
Janovskaya, T.B., 12
Jefferson, C., 112
Jenkins, G.M., 202, 224, 228
Jennings, T.W., 229
Jensen, R.V., 420
Jizba, Z.V., 96
Johansson, S., 8
Johnson, N.L., 96
Jones, C.M., 357
Jordan, T.H., 229
Jöreskog, K.E., 474
Jorgens, M.L., 323
Journel, A.G., 204, 212
Journel, A.J., 180
Jowett, G.H., 205
Juteau, T., 181

K

Kamineni, D.C., 433
Kaminski, M.A., 332, 339, 349–351
Kamp, P.J.J., 357
Kapp, H.E., 91
Kapteyn, J.C., 90
Karner, D.B., 88, 358
Kaufman, G.M., 36, 37, 90
Kaye, B.H., 379
Kazutoki, A., 142
Kelly, A.M., 105, 135, 140, 389
Kemple, W.G., 308, 347, 360
Kendall, M.G., 43, 76, 78, 96, 191, 202
Kendall, W.S., 33, 383, 390
Kerswill, J., 181
Khakzad, A., 396
Kim, Il., 419
Kissling, E., 297, 298
Kjarsgaard, B.A., 475–479
Klein, W., 371, 483
Klimas, A.J., 371, 483
Klimeš, L., 229
Klovan, J.E., 474
Knill-Jones, R.P., 141
Koch, G.S. Jr., 65, 279
Kolmogorov, A.N., 11, 44, 48, 90
Koopmans, L.H., 191, 200
Korvin, G., 383, 385, 415, 433
Köster, E., 279
Kotz, S., 529
Krige, D.G., 44, 45, 70, 90, 95, 96, 98, 106, 107, 212, 213, 250, 267, 268, 416, 417, 499, 511
Krijgsman, W., 358
Krishnan, S., 180
Krumbein, W.C., 3, 43, 54, 55, 90, 235, 278
Kuenen, P.H., 193
Kuiper, F., 88
Kuiper, K., 358
Kwon, J., 358

L

La Pointe, P.R., 371
Lai, C.D., 529
Lambert, A., 142
Lambert, B.S.J., 362, 363
Lammerer, B., 299
Lane, H.R., 308, 356
Langousis, A., 397, 398, 415
Lapin, L.I., 59
Laskar, J., 88
Laubscher, H., 299, 301
Lauziére, K., 113

Lavallée, D., 508
Lee, J., 36
Lee, P.J., 36
Lee, Y.W., 46
Lemeshow, S., 162
Lemke, W., 22
Lemkov, D., 112
Lerche, I., 435
Levander, A., 229
Levy, R.H., 309
Lew, S.N., 306, 318, 327, 343
Lewin, E., 399
Lewis, T., 278, 283
L'Heureux, I., 370, 371
Li, N., 396
Li, Q., 393
Li, X.-P., 229
Lima, A., 396
Lindley, D.V., 48
Lindsay, M.D., 142
Link, R.F., 65, 279
Linzer, G., 299
Lippitsch, R., 297, 298
Liu, G., 22, 306, 310, 313, 314, 324, 333, 335, 338–340, 352, 353
Liu, J., 144
Liu, W., 370
Llevellyn, P.G., 74
Llewellyn, P.G., 355, 357, 361, 362
Lliboutry, L.A., 204
Locklair, R.E., 88
Loéve, M., 60
Looney, C.G., 142
Lorenz, E.N., 481, 482
Loudon, T.V., 17, 286
Lovejoy, S., 66, 132, 217, 225, 370, 371, 374, 397, 398, 415, 419, 458, 483, 499, 504, 506, 508, 509
Ludwig, K.R., 358
Lűschen, E., 291, 297, 299
Luth, S., 297
Lybanon, M., 363
Lydon, J.W., 127
Lyell, C., 10, 356

M

Ma, X.P., 22
MacGregor, I.D., 239
Mahalanobis, P.C., 92
Malamud, B.D., 371, 483
Mallet, J.L., 18, 20–22
Mamuse, A., 142
Mancktelow, N.S., 300

Mandelbrot, B.B., 131, 229, 370, 372, 373, 379, 381, 384, 397, 401, 403, 419–421, 425, 449, 451, 458, 499, 500
Manenti, F., 142
Mann, K.O., 308
Mao, X., 22
Margolis, S.V., 88
Mark, D.F., 88
Markwitz, V., 142
Marlow, A.R., 12
Marsan, D., 230, 397, 398, 506
Martin-Izard, A., 419
Martinsson, A., 356
Matalas, N.C., 245
Matérn, B., 46, 210, 221
Mateu-Figueras, G., 3, 474
Mateus, A., 419
Matheron, G., 3, 22, 45, 46, 66, 70, 190, 205–210, 212, 214–216, 218, 219, 250, 415, 430, 493, 500, 509, 511–513, 526
Mattinson, J.M., 358
Maurrasse, R., 88
May, R.M., 482
McCallum, A., 181
McCrossan, R.G., 90
McCuaig, T.C., 142
McDougall, I., 358
McKerrow, W.S., 362, 363
McWilliams, M.O., 88
Meakin, P., 370, 419
Mecke, J., 33, 383, 390
Meneveau, C., 419
Menzie, W.D., 112, 382, 391, 392
Merle, O., 299
Merriam, D.F., 2
Michelsen, O., 356
Milankovitch, M., 4
Mildenhall, D.C., 357
Millahn, K., 291, 297, 299
Miller, A.J., 307
Millington, J.D., 142
Min, K., 358
Mitchell, W.S. III., 88
Moarefvand, P., 396
Møller, J., 383
Monecke, T., 113
Monin, A.S., 506
Montanari, A., 88
Moratti, G., 297, 298
Morein, G., 371, 483
Morgan, L.E., 88
Morgans, H.E.G., 357
Morris, S.W., 370, 371
Mortensen, J.K., 361
Mosoyana, I., 474

Mosteller, F., 468
Motter, E., 481, 482
Mouterde, R., 330
Mu, X., 445
Mueller, U.A., 162
Müller, W., 300
Mundil, R., 88, 358

N
Naish, T.R., 357
Nazli, K., 331
Nel, L.D., 307
Neubauer, F., 291, 297
Neubauer, K., 299
Neuhäuser, B., 142
Nicolich, R., 291, 296–301
Nieuwenkamp, W., 8, 9
Norwark, W., 181
Novillo, C.J., 142

O
Oberholzer, WJ., 212
Odin, G.S., 362
Ogg, G., 6, 309, 324, 354–357, 359, 360
Ogg, J.G., 6, 87, 309, 324, 354–357, 359–362
Oka, M., 474
Olea, R.A., 47, 474
Oliveira, V., 419
Oliver, J., 306, 318, 327, 342
Omran, N.R., 396
Ondrick, C.W., 54
Orcutt, J.A., 419
Ortiz, J.M., 180
Ortiz, M., 47
Osadetz, K.G., 36
Ostwald W., 10
Over, T.M., 397
Overbeck, C., 419
Owen, D.B., 469
Owens, T.L., 88, 358
Ozdemir, A., 142

P
Pahani, A., 163, 419
Pálfy, J., 361
Palmer, A.R., 307, 339, 347, 349
Pardo-Igúzquiza, E., 47
Park, N.-W., 483
Parker, R.L., 289
Parzen, E., 202
Pawlowsky-Glahn, V., 3, 474
Pearson, K., 43, 473

Author Index

Peláez, M., 47
Peredo, O., 47
Perrin, J., 374
Perron, G., 113
Peters, S.A., 309
Pettijohn, F.J., 278
Picardi, L., 297, 298
Pickering, G., 385
Pickton, C.A.G., 74, 355, 357, 361, 362
Pilkington, M., 506
Pinto-Vásquez, J., 67
Plackett, R.I., 43
Plant, J.A., 396, 445
Plotnick, R.E., 309
Poincaré, H., 480
Poles, A.G., 383
Poli, M.E., 294
Popper, K., 11
Porwal, A., 142
Potter, P.E., 278
Poudyal, N.C., 142
Prade, H., 42
Pregibon, D., 162, 181
Prey, A., 374
Price, P.R., 36
Prokoph, A., 202

Q

Quandt, R.E., 392
Quenouille, M., 469

R

Rabeau, O., 180
Raine, J.I., 357
Raines, G.L., 142, 181, 370, 386, 388, 389
Ramsay, J.G., 289, 290
Rangan, C., 181
Rao, C.R., 81, 83
Ratschbacher, L., 299
Reddy, R.K.T., 135
Regmi, N.R., 142
Reiche, P., 283
Reimann, C., 479
Reinsch, C.H., 323
Reinson, G.E., 36, 37
Remane, J., 356
Ren, T., 445
Renne, P.R., 88, 358
Reyment, R.A., 474
Richter, C., 294, 295
Ring, U., 294, 295

Ripley, B.D., 32, 33, 108, 109, 363, 371, 383, 386, 387, 440
Roach, DE., 435
Roberts, R.K., 142
Robertson, J.A., 256
Robinson, S.C., 6, 164
Roff, J.C., 506
Romero-Calcerrada, R., 142
Roncaglia, I., 357
Ronchetti, F., 142
Root, D.H., 382
Rosenblatt, M., 208
Rowlingson, B.S., 383
Rowlinson, B.S., 32, 33
Ruget, C., 330
Rundle, J.B., 371, 483
Russell, H.A.J., 150
Ruzicka, V., 418

S

Sadler, P.M., 308, 309, 347, 357, 360
Sagar, B.S.D., 22, 26
Sammis, C., 371, 483
Sander, B., 283
Sanderson, D.J., 385, 419
Sani, F., 297, 298
Sawatzky, D.I., 142, 163
Schaeben, H., 181
Schapire R.E., 3, 181
Scheicher, R., 383
Scheidegger, A.E., 14, 17, 286
Scherer, W., 54
Schertzer, D., 66, 132, 217, 371, 374, 397, 398, 415, 419, 458, 483, 499, 504, 506, 508, 509
Schetselaar, E.M., 16, 113
Schiøler, P., 357
Schmid, S.M., 297
Schmidegg, O., 302
Schmitt, F., 397, 398, 506, 508
Schmitz, M.D., 6, 309, 324, 354–360
Schoenberg, I.J., 323
Schove, D.J., 196, 197
Schreiner, M., 374
Schucany, W.R., 469
Schuenemeyer, J.H., 61, 382
Schulz, B., 302
Schwarzacher, W.S., 4, 196, 306, 345
Scot, G.H., 357
Scott, D.J., 19
Segers, J., 480
Sella, M., 296, 298–301

Selli, M., 296, 298–301
Serra, J., 22, 26, 27, 205, 221
Seward, D., 300
Shannon, C.E., 94
Shapire, R.E., 3
Sharma, A.S., 371, 483
Sharma, S., 370
Sharpe, D.R., 150, 154
Shaw, A.G., 307, 308, 339, 347, 349
Shcherbakov, R., 370, 371, 483
Sheshpari, M., 113
Sichel, H.S., 90, 96, 212
Sigismund, S., 302
Sikorsky, R.I., 434, 439
Simpson, T., 43
Sinclair, A.J., 375, 399
Singer, D.A., 112, 136, 382, 391, 392
Skinner, R., 26, 27
Slichter, L.B., 100
Smit, J., 88
Smith, A.G., 74, 309, 355–357, 361, 362
Smith, C.H., 13, 91, 239
Smith, D.G., 74, 355, 361, 362
Smith, H., 117
Smith, P.I., 361
Sokoutis, D., 297
Sommerfeld, A., 202
Song, H., 142
Soonawala, N.M., 433
Sornette, A., 483
Sornette, D., 433, 483
Sparrow, C., 481
Spiegelhalter, D.J., 141, 145, 159, 179
Springer, J.S., 105, 140, 389
Sreenivasan, K.R., 419
Srivastava, R.M., 44, 204
Stam, B., 330, 331
Stampfli, G.M., 298
Stanley, H., 370, 419
Stanton, R.L., 257
Stattegger, K., 22
Staub, R., 4
Steenken, A., 302
Steiger, R.H., 358
Steinhaus, H., 374
Stigler, S.M., 43
Stöckli, G., 300
St-Onge, M.R., 19, 20, 116
Stoyan, D., 33, 383, 390
Strauss, D.G., 308, 347, 360
Strong, C.P., 357
Stuart, A., 76, 191, 202
Sullivan, F.R., 317, 318, 320, 321

Sumio, M., 142
Sundius, N., 8
Sutton, C., 181
Swisher, C.C. III., 88, 358
Szegö, G., 510

T
Tanaka, Y., 474
Tarquis, A., 370
Tarvainen, T., 396
Terhorst, B., 142
Tessier, Y., 506
Teugels, J., 480
Thiart, C., 163, 168, 177
Thibault, N., 88
Thierry, J., 361, 362
Thierry, T., 87
Thompson, M., 108, 363
Thompson, R.N., 475
Thorleifson, L.H., 487
Tian, Y., 22
Tibshirani, R., 181
Tintant, H., 330
Todoeschuck, J., 506
Trettin, H.P., 279, 281
Tribus, M., 95
Trollope, F., 36
Trooster, G., 16
Troutman, B., 371
Tukey, J.W., 3, 65, 164, 167, 191, 224, 253, 267, 468, 509
Turcotte, D.L., 370–372, 382, 398, 401, 433, 480–483
Turneaure, F.S., 67
Turner, R., 383

U
Ueckermann, H.J., 70, 267
Ueno, H., 68
Uhler, R.S., 54, 100
Uritsky, V.M., 370, 371, 483
Usio, K., 142

V
Vai, G.B., 298
Van Bemmelen, R.W., 4, 7, 9–11
van den Boogaart, K.G., 181
van der Leun, C., 358
van Veen, P., 87, 361, 362
Vanneste, C., 433

Veneziano, D., 397, 398, 415
Vening Meinesz, F.A., 15, 374, 375, 499
Villa, I.M., 300
Villaipando, B., 68
Viola, G., 300
Visman, J., 261
Visonà, D., 294
Vistelius, A.B., 3, 12, 65, 90, 91, 214, 236, 399
Vitek, J.D., 142
Vittori, E., 297, 298
Von Buch, L., 8, 9
Von Kármán, T., 210, 229

W
Wagreich, M., 197
Wahba, G., 327
Walsh, J.H., 261, 385
Walsh, J.J., 435
Walters, R., 74, 355, 357, 361, 362
Wan, L., 396
Wang, C., 444, 448, 451, 452
Wang, N., 356
Wang, Q., 396
Wang, W., 144
Wang, Z., 214
Ward, A.D., 142
Warters, W., 36
Watson, G.S., 3, 26, 27, 236, 245, 253, 287, 289, 300, 323
Watson, M.I., 212
Watterson, J., 385, 435
Watts, D.G., 202
Webster, R., 44
Weertman, J., 204
Wegener, A., 4
Wegmann, E., 10
Wein, A., 116
Wellmer, F.W., 126
Westergård, A.H., 8, 9
Whalley, B.J.P., 261
Whittaker, E.T., 323
Whitten, E.H.T., 236–268, 295
Whittle, P., 209, 210
Wiener, N., 191
Wijbrans, J.R., 358
Wilks, S.S., 60
Williams, L.R., 142

Williamson, M.A., 306, 318, 327, 342
Williamson, M.J., 339
Willingshofer, E., 297
Wilson, G.J., 357
Wilson, G.S., 357
Wilson, J., 505
Winchester, S., 3
Wodika, N., 19
Wong, A.S., 26
Woolard, G.P., 236
Worsley, T.R., 323
Wright, C.W., 10
Wright, D.F., 112, 141, 143, 144, 161, 181, 386, 388, 389, 423, 439
Wu, C., 22
Wu, G., 444, 448, 451, 452
Wu, S.M., 101
Wu, S.-S., 229

X
Xia, Q.L., 112, 396, 444
Xiao, F., 3, 444, 448, 451, 452
Xie, S., 66, 215, 444, 452–456
Xie, X., 445
Xu, D., 214
Xu, Y., 393
Xu, Z., 229

Y
Yaglom, A.M., 194, 202, 506
Yielding, G., 385
Yu, C., 445, 447
Yule, G.U., 143

Z
Zanferrari, A., 294
Zehner, R.E., 142
Zhang, D., 179, 185
Zhang, T., 309
Zhang, X.L., 22
Zhang, Z., 444, 448, 451, 452
Zhao, B., 3
Zhao, J., 144, 396
Zhao, P.O., 3, 112
Zhou, D., 307
Zuo, R.G., 179, 396, 444

Subject Index

A
Abitibi acidic volcanics, 527
Abitibi acidic volcanics example, 522–526
Abitibi area, 113, 118, 140, 165, 236, 470, 510, 511
Abitibi copper deposits example, 118–123
Abitibi copper potential map, 472–473
Abitibi lode gold deposits example, 381–383
Abitibi volcanic belt, 383
Abitibi volcanogenic massive sulphides, 471
Abitibi volcanogenic massive sulphides example, 470–473
Acadian anticlines, 144, 147
Accelerated dispersion model, 497–499
Accuracy and precision, 358
ActiveX (OCX) Development Environment, 306
Additive log-ratio *(alr)*, 474
Adria microplate, 4, 5, 284, 297, 298, 301
Adriatic indenter, 299
AECL. *See* Atomic Energy of Canada Limited (AECL)
Aeromagnetics, 170
Affine transformation, 387
African plate, 4
Age determinations, 62, 305
Agordo, 301
Agordo area, 298, 301
Alpine orogeny, 295
Analysis of variance, 43, 61, 117, 249, 280, 343
Analysis of variance table, 110, 117
Anisotropy, 386–388
Apparent number of subdivisions, 484, 491
Appearance event ordination, 309
$^{40}Ar/^{39}Ar$ method, 355
Arbuckle formation, 235, 468
Archeological seriation, 308

Arnisdale, 289, 290
Arnisdale Gneiss example, 289–290
Arsenic C-A plot, 450
Artesian aquifers, 152
Astrochronology, 6, 357
Astronomical calibrations, 6
Asymmetrical bivariate binomial distribution, 510, 526–529
Atomic Energy of Canada Limited (AECL), 433
Austroalpine units, 296
Autocorrelation, 13, 46, 70, 189–230
 coefficient, 249
 data, 249
 function, 202, 267
 of residuals, 235, 245
Autocovariance, 191, 194
Autoregressive process, 202
Avalanche, 371
Axes of pebbles in glacial drift, 277, 279
Axioms of mathematical statistics, 48

B
Background, 484
Backward elimination, 117
Basement of the Italian Dolomites, 297–300
Basin and Range Province, 384
Bathurst area, 510, 514
 and Abitibi area, 525
 area acidic volcanics example, 514–522
B-axes, 283, 285, 290, 294
Bayesian approach, 180
Bayes' rule, 142
Bayes' theorem, 48–49
Bernoulli distribution, 50
Bernoulli trial, 50, 53

Subject Index

Besicovitch process, 421
Bessel's correction, 47, 58
Best-fitting unit vector field, 288
Best Linear Unbiased Estimator (BLUE), 261
Bias, 45, 468–473
   correction, 436
   reduction, 470
Bifractal model, 385
Bifurcations, 482, 483
Binary random variable, 514
Binary variables, 143
Binomial distribution, 51
Binomial measure, 421
Binomial/p model, 451, 504
Binomial test, 321
Biostratigraphy, 308
   events, 306
   uncertainty, 310
Bivariate density function, 510
Bivariate frequency distribution, 106, 107
Bivariate normal, 106
Bjorne Formation paleodelta example, 277, 279–282
Blind subset, 468
B-lineations, 278, 286, 289, 291
BLUE. *See* Best Linear Unbiased Estimator (BLUE)
Bohemian Massif, 459
Boosting, 3, 181
Bootstrap, 3, 360, 468, 469
Borehole samples, 51
Borel sets, 48
Box-counting, 371, 384, 388
British Newfoundland Exploration Company, 245
Brixen area, 294, 299
Brixen granodiorite, 292
Brownian landscape, 372
Bruneck (Brunico), 291, 292, 300
Buffer zones, 150, 156
Butterfly effect, 481, 482

C

Cabrito Formation, 331
Caerfai-St David's boundary, 78, 79
Caerfai-St David's boundary example, 77
Californian Eocene nannofossils example, 317–323
Cambrian-Ordovician boundary, 359
Cambrian Riley Formation, 339
C-A method. *See* Concentration-area (C-A) method

Campanian-Maastrichtian boundary, 350
Canadian Appalachian region, 382
Canadian Shield, 7, 61–62
Cape Breton Development Corporation, 259
C-A plot, 446
Carnian Alps, 298, 301
CASC. *See* Correlation and scaling (CASC)
Cascade process, 467
Cell composition modeling, 509–529
Cenozoic, 75, 88, 356
   isochrons, 342
   microfossils, 332
   microfossils example, 334–339
   stage boundaries, 357
Central difference technique, 424
Central-limit theorem, 43, 51, 56, 59
Central Texas Cambrian Riley Formation Example, 346–349
Centred log-ratio (clr), 474
Channel sampling, 66, 68, 106, 417
Chaos, 372
Chaos theory, 481
Characteristic function, 56
Chen algorithm, 449–463, 499, 500, 505
Chi-square, 361
   distribution, 274
   test for goodness of fit, 249, 333
   tests, 43, 55, 61, 63, 92
Chrome spinel, 239
Chronogram, 74, 75, 77, 79, 87–88, 361
Chronostratigraphic boundaries, 74, 80, 357
Chronostratigraphic divisions, 355
CI. *See* Conditional independence (CI)
Cima d'Asta, 294, 298, 301
CI test, 177, 182
Classicists, 10
Clasts, 281
Closed number system, 480
Cloud formation, 371
CLUST, 440
Cluster density determination, 384
Cluster fractal dimension, 384
Coastline of Britain, 372
Coast of Brittany, 374
Coast of Norway, 372
Cochrane readvance, 196, 204
Codimension, 420, 504, 505
Coefficient of variation (CV), 70, 94
Coherence, 200, 201
Columnar basalt joints, 371
Complementary error function, 80
Complete spatial randomness (CSR) of points, 440

Subject Index 545

Composite standard, 308
Compositional data analysis, 3, 473–480, 474
Compound distributions, 73–102
Computer simulation, 81
Computer simulation experiment, 81–83, 490, 500–504
Concentration-area (C-A) method, 369, 394–399
Concentration-area relation, 377
Conceptual models, 3, 8
Conditional independence (CI), 143, 161, 163, 173, 180
Conditional probability, 48–49
Confidence belt, 65, 110
Confidence intervals, 46, 47, 58
CONOP, 308, 347, 360
Consistent dates, 74
Continental drift, 4, 11
Contingency table, 44
Continuous frequency distributions, 55–61
Continuous random variable, 56
Contrast C, 115, 147
"Control" area, 161, 176–179, 473
Control cells, 470–471
Copper deposits, Abitibi area, 382
Correlation, 105–136
   coefficient, 43, 107, 108, 200
   matrix, 108, 474
Correlation and scaling (CASC)
   correlation, 310, 338, 350
   correlation lines, 347
   multiwell comparison, 342, 344
Correlogram, 46, 195, 199, 245
Correspondence analysis, 308, 474
Co-spectrum, 200
Covariance, 107
Cratonic mantle peridotite, 479
Crest-lines, 274
Cretaceous, 6
   Greenland-Norway Seaway Microfossils Example, 349–352
   zonal model, 350
Cretaceous seaway between Norway and Greenland, 305, 339, 349
Crocodile model, 299
Cross-correlation, 191
Cross-correlation function, 200
Cross-correlogram, 202
Cross-over inconsistencies, 350
Cross-spectral analysis, 191, 198–202, 200
Cross-spectrum, 200
Cross validation, 359, 360, 468
Crustal shortening, 4

Crystalline basement of the Dolomites, 283
Cubic hypersurface, 251
Cubic polynomial curve, 293
Cubic smoothing splines, 289, 328
Cumulative RASC distance, 345
Cut-off grades, 114
Cuttings, 309
CV. *See* Coefficient of variation (CV)
Cycling events, 322, 323
Cyclostratigraphy, 4

D

Data acquisition, 2
Data analysis, 2
2-D autocorrelation functions, 235, 252, 258
Decay constant, 358
Deductions, 10–11
Deductive method, 11
Deep seismic lines, 297
Defereggen Schlinge example, 302
Degree of trend surface, 236
Degrees of freedom, 43–45, 58, 64, 235, 248, 249
de Moivre-Laplace theorem, 56, 415
Dendrogram, 324
Dependent variable, 108
Design of experiments, 43–44
Deviation from conservation, 504, 506, 508
Devonian granite, 144, 146, 148, 153
3-D geological mapping, 113
2-D harmonic analysis, 236
Diachronism, 350
Differentiability, 374
Diffusion processes, 12–13
Dinarides, 297, 298
Diopsidic clinopyroxene, 239
Directed lines, 277, 278
Directional features, 277–302
Discrete data, 265–266
Discrete Fourier transform, 508
Discrete frequency distributions, 48–56
Discrete Gaussian model, 509, 512
Discrete multivariate analysis, 164
Discretization, 177, 178
Discriminant analysis, 162
Dispersion index, 451, 492, 493
3-D map making, 1, 18
Dolomites in northern Italy, 277, 286, 296, 301, 302
   Synclinorium, 299
Double Fourier series, 235
Double trace moment, 508

Drill-core samples, 66
Drilling, 6
Ductile extrusion model, 299

E

Earthquakes, 161
Earth's topography, 372–375, 499
East-Central Ontario Copper and Gold Occurrence Example, 273–274
Eastern Alps, 4, 5, 284, 297, 299
East Pacific Rise, 181
East Pacific Rise seafloor example, 180–185
Economic cut-off effect, 383
Edge effect correction, 440
Edge effects, 435, 487
Eggerberg, 295
Elliot Lake area, 235, 257
Ensemble average, 190
Enstatite, 239
Eocene nannofossils, 319
Equal duration of stages hypothesis, 357
Equiprobability, 49
Equivalent number of independent observations, 46, 249
Ergodicity, 190
Error bar, 342, 344
Eskers, 196
ESS, 239
Eurasian plate, 4, 5
Explanatory variables, 108, 162
Exploratory data analysis, 3
Exploratory drilling, 51
Exploratory wells, 307
Exponential functions, 13
Extended RASC range, 346
Extreme events, 480

F

Factor analysis, 108, 474
Fast Fourier Transform, 267
Filtering, 46, 68
First appearance datum (FAD), 307
First common occurrence (FCO), 350
First order structure function, 506
Fisher distribution, 277, 283
Fixed points, 482, 483
Flattening, 350
Flooding, 371
Flowing wells in the Greater Toronto Area, 150–152
Flow lines, 281
Fluence monitor, 358
Flying Spot Scanner, 516
Foraminifera, 331, 332
Forest fires, 371
Forsterite (FO), 111, 239, 307
Fort à la Corne kimberlite field, 476
Forward selection, 117
Fourier series, 13
Fourier transformation, 194, 202
Fractals
  dimension, 370–383, 420, 481
  geometry, 374
  modeling of fractures, 433
  perimeter-area relation, 379
  point patterns, 385
Fractile, 63, 83
F-ratio, 248
Frequency bands, 200, 202
F-test, 61
Full convergence, 454
Fuzzy logic, 142

G

Gaderbach, 291, 300
Gamma distribution, 57, 337, 513, 521, 522
Gaussian autocorrelation function, 259
Gaussian distribution, 77
Gaussian weighting function, 80
Gaussian white noise, 500
Gejiu area, 413, 445
Gejiu Batholith, 445
Gejiu mineral district example, 445–448
General linear model, 43, 105–136, 164–168, 236
Genetic algorithms, 309
Geochemistry, 170
  anomalies, 484
  anomaly separation, 439
  reconnaissance, 487
  signature, 146, 160
Geographic Information Systems (GIS's), 142, 168, 172, 179, 376
Geological cross-sections, 8–9
Geological facts, 9
Geological framework, 113
Geological map, 3, 6
Geological Survey of Canada, 118
Geologic timescale, 74, 109, 305–363, 323, 352
Geologic TimeScale Foundation, 355
Geomathematics, 3

Subject Index

Geometrical covariance, 515, 516
Geometric distribution, 54, 55
Geomodeling, 10
Geophysical exploration, 6
Geostatistical model, 499
Geostatistics, 46, 189–230, 190
GIS's. *See* Geographic Information Systems (GIS's)
Giudicaria Fault, 301, 302
Giudicaria Line, 299
Glacial drift, 150–151
Glacial Lake Barlow-Ojibway example, 191–198
Glacial varve-thickness data, 189
GLADYS expert system, 145, 179
Global Boundary Stratotype Section and Point (GSSP), 88, 356
Gold assays, 45
Gold deposits in Meguma Terrane, 145
Gold deposits in Iskut map area, 413
Golden spike, 356
Goldenville formation, 148
Goldenville/Halifax contact, 153
Gold in balsam fir, 147, 151
Gold "reef", 70
Goodness of fit, 43, 55, 63
Gosaldo area, 294
Gowganda area, 177
Gowganda area gold occurrences, 168–173
GPS data, 297
Gradstein-Thomas database, 342
Grand Banks, 330, 339, 341
   Basin, 331
   Foraminifera, 339–346
GraphCor, 308
Graphic correlation, 307
Gravitational concept, 4
Greenschist facies metamorphism, 295
Grenville Province, 61–62
Grid spacing, 179
Grouped Jackknife, 468–473
Gutenberg-Richter earthquake frequency-magnitude relation, 371

H

Halifax Formation, 146
Hanning, 191, 203, 509
Harbour seam, 235, 259
Harmonics, 196
   analysis, 236, 267, 272
   trend analysis, 273
   trend surface analysis, 266–274

Hay example, 317, 319, 324
Heinrich event, 204
Hercynian minor folds, 277, 286
Hercynian schistosity planes, 291
Hermite polynomials, 514, 520
Hindsight study, 178
Histogram methods, 425–427
Hölder exponent, 420, 524
Hydrocarbon deposits, 51
Hydrothermal precious-metal mines, 384
Hypothesis testing, 2

I

Ice ages, 11
Image analysis, 1, 22
Inch-dwt values, 106
Inch-pennyweight value, 70
Inconsistent dates, 74, 80
Incorporation of missing data, 152–161
Incorporation of uncertainty, 159
Independent and identically distributed (iid) observations, 43, 46, 70, 245, 468
Index of dispersion, 507
Indicator map patterns, 139, 144
Indirect method, 328
Inductions, 10, 11
Infinitely large variance, 191
Integrability, 374
Intensity function, 386
Intensity of exploration, 114
Interevent distance, 314
International Commission on Stratigraphy, 355
Interpolation spline, 324
Intrinsic stationarity, 191
Intrinsic variance, 114
Inverse distance weighted moving average method, 68, 445
Inverse Fourier transform, 194, 267
Iridium anomaly, 356
Iskut River map gold occurrences, 439–442
Iskut River map sheet, 384
Island of Newfoundland, 265
Isometric log-ratio (ilr), 474
Iterative process, 453

J

Jackknife, 3, 178, 468–470, 473
Jurassic, 6
Jurassic Foraminifera, 330

K

K-Ar age determinations, 74
K-function, 371
Kilauea, 483
Kimberlite, 479
Kinnehulle area, 8–9
Kirkland Lake area, 386–389
Klerksdorp goldfield, 45, 416, 499
"Knowledge-driven" methods, 112
Kolmogorov-Smirnov statistic, 152, 164
Kolmogorov-Smirnov test, 64, 66, 147, 163, 178, 181
Krawtchouck polynomials, 526
Krige's formula, 45
Krige's formula for the propagation of variances, 511
Kriging, 45–46, 68, 107, 235, 250–265
   interpolation problem, 252
   mean, 262
   of residuals, 252
   variance, 45, 161
KTB borehole, 443, 459
KTB copper example, 459–463
k-th order spline, 328
Kummell's equation, 106

L

Labrador and northern Grand Banks wells, 332
Labrador Shelf, 339
Lac du Bonnet Batholith, 434
Lac du Bonnet Batholith fractures example, 433–439
Lacunarity, 386, 387
LAD. *See* Last appearance datum (LAD)
Lagrangian multipliers, 95, 251, 261
Lake Barlow-Ojibway, 196
Land-ice retreat, 204
Landslides, 371
Laplacian characteristic function, 505
Large copper deposits, 470, 511
Last appearance datum (LAD), 307
Last common occurrence (LCO), 349–350
Last occurrence (LO), 307
Late Alpine Tectonics, 300–301
Late Miocene, 5
Late Wisconsin ice-sheet, 191
LCO. *See* Last common occurrence (LCO)
Least-squares, 109
Leave-off-one cross-validation, 359, 468
Leduc reef pools, 90
Level of significance, 63

Lévy index, 504
Linear regression, 105–136, 106, 108, 109
Line of correlation, 307
Lingan Mine, 235, 259
Lingan Mine Example, 259–265
Lithospheric peridotite, 479
LO. *See* Last occurrence (LO)
Local residuals, 237–238
Local singularity analysis, 372, 413–463, 500
Logarithmic series distribution, 56, 333
Logarithmic transformation, 193
Logbinomial, 414
Logbinomial frequencies, 427
LOGDIA, 162
Logic of Scientific Discovery, 11
LOGIST, 162
Logistic equation, 480–482
Logistic map, 482
Logistic model, 164–168, 265
Logistic regression, 115, 142, 161, 162
Logistic regression diagnostics, 162, 181
Logistic trend surface analysis, 265–266
Logits, 143, 165, 480
Log-likelihood function, 75, 76, 80, 83
Loglinear model, 163
Log-log plots, 377, 383, 435
Lognormality, 45, 61, 73–102, 420, 510
   distributions, 90, 383
   frequency distribution, 88
   Q-Q plot, 494
LOGPOL, 162, 163, 180
Longitudinal section, 270
Lorentz attractor, 481–483
Lorenz equations, 482
Lovejoy-Schertzer *a*-model, 504
Lower Tertiary nannofossils, 317
Lusitanian Basin, 331

M

Maastrichtian-Paleocene boundary, 88, 356
Magmatic nickel-copper deposits, 167
Major axis, 108, 278
Map layers, 161
Marked point processes, 384
Marker events, 334
Marker horizons, 310, 336
Markov chain, 55
Markov process, 199
Markov property, 46, 249
Markov random field, 181
Mass exponents, 421, 488, 490, 524

Subject Index 549

Mass fractal dimension, 435
Mass-partition function, 420, 428
Mathematical models, 3
Mathematical statisticians, 3
Mathematical statistics, 43–44, 49, 499
Matinenda formation, 235, 256
Matinenda formation example, 256–259
Mauna Lao, 483
Maximal horizons, 318
Maximum entropy, 94
Maximum likelihood for functional
 relationship (MLFR), 105, 108, 109
Maximum likelihood method, 47, 73–102, 74,
 80, 87, 90, 162, 163, 361, 469
Mean, 46
Measure of association, 143
Media Agua Creek section, 317, 320, 321
Mega-surge, 204
Meguma Terrane gold deposits, 144, 163
Mehler identity, 510, 514
Melville Island, 277, 279
Melville Peninsula, 162
Merriespruit Mine, 498–499
Mesozoic stage boundaries, 87
Method of doubling the angle, 278
Method of least squares, 105–136, 115
Method of moments, 422
Method of scoring, 83
Microfossils, 51
Microfossil taxa, 51
Milankovitch theory, 4
MIMIC, 418
Mineral potential, 112
 estimation, 113
 map, 113, 178
Mineral prospectivity, 142
Mineral resource estimation, 112–115, 114
Minimum number of sections (MNS), 334, 347
Minimum variance property, 73, 469
Mining applications, 45
Minkowski operations, 514
Miocene sinistral strike-slip movement, 299
Missing patterns, 159
Mississippi Valley-type lead-zinc deposits, 371
Mitchell-Sulphurets area, 377, 422
Mitchell-Sulphurets example, 375–381
Mitchell-Sulphurets mineral district, 376
MLFR. *See* Maximum likelihood for functional
 relationship (MLFR)
MNS. *See* Minimum number of sections
 (MNS)
Model of de Wijs, 414, 417, 449, 467, 483,
 492, 498, 499

Modified RASC, 345
Modified weights-of-evidence, 179–185
Moine Series, 289
Moment, 421
 generating function, 56
 of a random variable, 47
Monofractal, 384
Montejunto area, 330
Morien Group, 259
Mosaic model, 115–116
Mount Albert, 243
 peridotite example, 239–243
 peridotite intrusion, 235, 243
Mount Etna, 9
Moving average, 68, 273
Moving-average data, 268
Multifractals, 4, 370, 384, 413–463, 487,
 499–505, 524
 analysis, 442, 524
 autocorrelation model, 432
 cluster, 385
 correlogram, 429
 modeling, 499
 semivariogram, 428, 439, 480
 spatial correlation, 427–432
 spectrum, 419–427, 451, 504, 524
Multiple-point geostatistical simulation, 47
Multiple regression, 144, 236
Multiple squared correlation coefficient, 110
Multiple working hypotheses, 10
Muskox intrusion, 93, 95
Muskox layered ultramafic gabbroic
 intrusion, 13, 91

N

Negative binomial distribution, 54
Neo-Alpine anticline, 285
Neo-Alpine compression, 277, 297
Neo-Alpine orogeny, 297
Neogene, 6, 298
Neptunism, 8–10
Neural networks, 142
Newton-Raphson iteration, 162, 163
New Zealand Geologic Time Scale, 357
Neyman-Scott models, 384
Noise, 257
Non-linear process modeling, 112, 370,
 480, 481
Non-parametric regression, 327
Norian-Rhaetian boundary, 79
Normal distribution, 43, 51, 57
Normality assumption, 62

North Sea wells, 332
Northwestern Atlantic margin, 305, 333
Nugget effect, 459
Null-hypothesis, 58
NW Atlantic Margin, 339–346

O

Oak Ridge Moraine (ORM), 150, 154
Odds, 49, 143
Oficina formation, 54, 55
Oil and gas pools, 371
Oligocene, 4, 5
Oligocene-Miocene dextral shear movements, 299
Olivine, 239, 473, 479
Omnibus test, 163
Ontology, 10
Oppel zone, 318
Optimum smoothing factor, 359
Ordovician time scale, 362
ORM. *See* Oak Ridge Moraine (ORM)
Orthopyroxene, 239
Oscillating limit cycle, 482
Outlier, 360–361
Overthrust sheets, 4, 5

P

Paleocene-Eocene boundary, 321
Paleocurrents, 279, 280
Paleogene, 6
Paleomagnetic measurements, 3
Pan-sedimentary theory, 8
3-Parameter lognormal, 416
Pareto, 381
 distribution, 383, 420
 tails, 415
Partial differential equations, 12–13
PAST, 308
Pebbles in glacial drift, 279
Periadriatic Lineament, 284, 291, 296, 299–301, 302
Permanence, 45, 509
 frequency distributions, 510–512
 random variables, 509
Permotriassic, 283, 296, 300
Petrographic modal analysis, 50, 51
Phase, 200
 angle, 203
 difference, 201
 lag, 202
Philosophical concepts, 8

Piton de la Fournaise, 483
Plate tectonics, 5, 11
Pleistocene, 357
Pleistocene ice ages, 4
Plotting percentages, 64–65
Plutonism, 10
Point patterns, 383, 388
Point-process models, 383
Poisson distribution, 51, 52
Poisson models, 113, 384, 440
Polarity chronozones, 75
Polynomial, 13
 trend analysis, 236–250
 trend surface, 235
 unit vector field fitting, 284–285
Posterior probabilities, 114, 160, 163, 173, 181
Posterior probability map, 178
Potassium/argon age determinations, 62
Potassium-Argon method, 356
Power density, 203
Power-law, 375, 377, 415
Power spectrum, 191, 200
PPLs. *See* Probable position lines (PPLs)
Prediction, 11–12, 480
 of occurrence of discrete events, 139–185
 potency, 179
Predictor map, 148
Preferred orientation, 283
Primeval basalt, 8–9
Principal component (PC) analysis, 108, 144, 170, 278, 475
Principle of conservation of total mass, 423
Principles of Geology, 10
Prior probability, 159
Probabilities, 143
 calculus, 44, 48–56
 generating functions, 49–50
 index, 113
Probable position lines (PPLs), 312, 313, 334, 335, 350
Probits, 57, 165, 315, 480, 514
Probit transformation, 315
Probnormal distribution, 510, 512–514
"Prob-prob" paper, 513, 514
Prognosis-diagnosis method, 11
Prognostic contour map, 140
Projection pursuit, 3
Pseudo-values, 470–471
Pulacayo Mine, 66–70, 423
Pulacayo Mine example, 429–432, 453, 506–509
Pulacayo ore, 473
Pulacayo orebody, 504

Pulacayo zinc concentration values, 425, 506
Pulacayo zinc spectrum, 509
Pustertal (Pusteria), 291, 298, 300
   Lineament, 292
   quartzphyllite belt, 283
   quartzphyllites, 299
   shortening, 300
   tectonites, 295

Q

Q-Q plots, 64, 337, 485–487, 521
Quadratic exponential trend surface, 263
Quadrature spectrum, 200
Quantimet 720, 515
Quantitative prediction, 2
Quantitative stratigraphy, 306, 307
Quartzphyllites, 283, 294
Quirke Syncline, 256

R

Radial basis function, 3
Radiometric decay constants, 358
Radiometric methods, 357
Radiometric re-calibration, 358
Rainfall, 371
Ramping, 361
Random cut model, 491–497
Random field model, 253
Random numbers, 500
Random sampling, 43, 179
Random variables, 12, 46, 49, 70, 76
Ranked optimum sequence, 306, 337
Ranking scattergram, 338
RASC, 308, 321, 333
   biozonation, 305, 339
   & CASC Version 20, 306
   clusters, 334
   computer program, 306
   dictionary, 318
   distance, 326
   distance scale, 325
   variance analysis, 349
RASC sequence (SEQ) files, 319
Reconnaissance geochemical surveys, 445
Rectangular frequency distribution, 360
Recursive relation, 482
Reduced major axis, 108
Redundancy, 248
Reefs, 416
Regionalized random variables, 3, 45, 46, 70
Regional trend component, 237–238

Regression analysis, 44–45
Regression line, 110
Re-proportioning the Relative Geologic
   Time Scale, 362–363
Residuals, 236, 238
Residual variance, 110, 117
Rhenium-Osmium, 355
Rieserferner tonalite intrusion, 299
Riley Composite Standard (RST), 339, 347
Ripley's $K(r)$ function, 385–387
Roll-off effects, 371, 374, 385
Romanticists, 10
Rose diagram, 279
Rothamsted Experimental Station, 44
RST. *See* Riley Composite Standard (RST)
Rubidium-Strontium method, 356

S

Sampling, 2, 66
   error, 417
   size, 52, 334
   variance, 58
San Stefano area, 284, 287, 288, 298
San Stefano quartzphyllites, 300
San Stefano quartzphyllites example, 285–289
Scaled residuals, 359
Scale-independence, 370
Scale-models, 3
Scaling, 306, 314
Schistosity planes, 285, 286
Schlinge, 302
Schmidt net, 283, 287
Scoring method, 85
Scotian Shelf, 339
Screeplot, 477
SDM. *See* Spatial data modeller (SDM)
Seafloor example, 181
Second characteristic function, 508
Second-order intensity function, 386
Second-order mass exponent, 423, 428
Second-order stationary process, 289
Sedimentary basin, 51
Self-organized criticality, 371
Self-similarity, 370, 414–416
Semi-exponential filter, 195
Semivariogram, 428, 499
Sequence stratigraphy, 350
Serpentinization, 235, 239
SF. *See* Smoothing factor (SF)
Shortening in the Alps, 297
Signal-plus noise model, 194, 509, 515
Signals, 68, 194, 253, 257, 259

Significance test, 43–44, 58
Silt-clay couplets, 193
Sine wave, 271
Singularities, 420, 456, 524
Singularitiy analysis, 445
Skye lavas, 475
Smoothing factor (SF), 327, 343, 359
Smoothing splines, 323, 331
Solar radiation, 4
Soli-lunar cycles, 196
South Saskatchewan till example, 485–490
Space series, 204
SPANS. *See* Spatial Analysis System (SPANS)
Sparky sandstone, 373
Spatial Analysis System (SPANS), 376, 433
Spatial autocorrelation, 70
Spatial correlation, 146
Spatial data, 45
Spatial data modeller (SDM), 142
Spatial periodicities, 267
Spatial point process, 386
Spatial statistics, 44–46, 190
Specific gravity trend, 242
Spectral analysis, 198, 508
Sphalerite, 67, 473
Spherical harmonics, 374
Spline-fitting, 359
Splining, 46, 87, 305–363, 323, 358–360
    curve, 323, 343
    smoothing, 327
Spurious correlation, 473
Squared multiple correlation coefficient, 117
$^{87}Sr/^{86}Sr$ stable isotope scale, 357
Stage boundary, 75
Stage boundary age, 74
Star kimberlite, 475
Stationarity, 190–191
Stationary random process model, 235
Statistical analysis, 10, 61–66
Statistical inference, 55–61
Statistical moments, 504
Statistical point-process modeling, 386
Stepwise regression, 117, 118, 470
Stereographic projection, 283
Stochastic differential equation, 199
Stochastic model, 200, 202–204
STRATCOR, 308
Stream sediments, 445
Strigno area, 4, 5, 298
Strontium, 13
Structured Field Interpolation (SFI), 7
Structural geology, 283
Student's t-test, 43, 61, 159, 181

Student's t-value, 446, 450
Subduction, 297
Sullivan database, 318, 320
Sulphur in coal, 259–265
Sum of two random variables, 60
Sunspot cycles, 197, 199, 201, 204
Supergiants, 114
Superpositional relationships, 311, 322, 339
Surface fractures, 413, 433
Sydney coalfield, 260

T
Tajo vein, 67, 68, 456
Target area, 179
Tauern Window, 300
Taylor series approximation, 93
Tectonites, 278
Tele-seismic tomographic data, 297
Temperature fluctuations on Earth, 196
Ternary pattern, 144, 160
Tertiary, 10
Tetrahedrite, 68
Textural fractal dimension, 379
Theory of probability, 11
Theory of proportionate effect, 90
Three-body problem, 480
Three-parameter lognormal distribution, 95–96
Three-parameter model of de Wijs, 483–490
Timescale history, 354–357
Time series analysis, 46, 190
Tin deposits, 445
Tojeira sections correlation example, 329–332
Top of Arbuckle Formation, 237–238, 252
Total dollar value, 113
Training cells, 176–179
TRANSALP, 291, 292, 296, 297, 299, 300, 302
    profile, 283, 284
    profile example, 290–295
    transect, 290
Traveling salesman (TSM) problem, 309
"Treasure Hunt" database, 170
Trend-line, 12–14
Trends, 2, 12, 236, 257, 499–504
    analysis, 111
    drift, 12
    elimination, 198–202
    surface analysis, 252, 264, 280, 284, 499
Trial age, 82
Triassic, 6
Trilobites, 339
Truncated Fourier expansion, 481
Truncated lognormal distribution, 94

Subject Index

TSM problem. *See* Traveling salesman (TSM) problem
t-test, 63
Turbidity currents, 193
t-value map, 114
Two-dimensional autocorrelation function, 263, 271
Two-dimensional lognormal model, 382
Two-dimensional power spectrum, 269, 271

U

Unbiased estimator, 469
Underground drill-hole data, 270
Underground Research Laboratory (URL), 433
Undirected lines, 277, 278
Uniform distribution, 60, 77
Uniform random variable, 505
Unique conditions, 162, 163, 172
Unique event option, 336
Unitary associations, 308
Unit cell, 163
Unit regional value, 113
Unit vector fields, 284, 297–300
Unit vectors, 277, 284
Universal canonical multifractal, 505
Universal kriging, 253, 261, 499
Universal multifractal simulation, 506
Unweighted distance analysis, 326
Upper Canada de Santa Anita, 317
Uranium-Lead, 355
Uranium resources, 418
URL. *See* Underground Research Laboratory (URL)
U.S. uranium resources, 497

V

Valsugana Lineament, 298, 299
van der Pol equation, 480
Variance, 46
    analysis, 14
    of the contrast C, 146
    equation of de Wijs, 484, 488
    formula of de Wijs, 414, 416
Variance-covariance matrix, 108, 475
Variogram, 46
Varves in Lake Barlow-Ojibway, 203
Varve time series, 196
Vector mean, 283

Venn diagrams, 142
Verrucano conglomerate, 286
Vibroseis, 290, 296, 297, 299
Villnösz Fault, 300
Virginia Gold Mine Example, 267–268
Virginia Mine, 70
Vistelius model, 90, 91, 383
Vistula, 374
Volcanogenic massive sulphide deposits, 167

W

Wavelet analysis, 202
Weakly stationary random variable, 236
Weighted distance analysis, 326
Weighted logistic regression (WLR), 161–179, 162, 163, 168, 173, 180
Weighting function, 76, 79, 80
Weight of evidence, 49
Weights-of-evidence (WofE), 114, 141, 142, 146, 163, 168, 173, 179, 180
Welsberg, 283, 295
Werner's theory, 9
Western Drauzug, 296
Whalesback copper deposit exploration, 268–273
Whalesback copper mine example, 243–250
Whalesback deposit, 235, 249
White noise, 68, 194, 202, 253, 459, 499–504
White noise component, 455
Wildcats, 48
Witwatersrand goldfields, 47
Witwatersrand gold mine, 106
WLR. *See* Weighted logistic regression (WLR)
WLR likelihood function (WLR), 180
WofE. *See* Weights-of-evidence (WofE)
Working hypotheses, 10
Worldwide uranium resources, 417
Wulff net, 283

X

X-ray fluorescence spectroscopy, 459

Z

Zhejiang Province Pb-Zn example, 448–449
z-test, 147

Printed by Printforce, the Netherlands